Keshav P. Dahal, Kay Chen Tan, Peter I. Cowling (Eds.)

Evolutionary Scheduling

Studies in Computational Intelligence, Volume 49

Editor-in-chief
Prof. Janusz Kacprzyk
Systems Research Institute
Polish Academy of Sciences
ul. Newelska 6
01-447 Warsaw
Poland
E-mail: kacprzyk@ibspan.waw.pl

Further volumes of this series can be found on our homepage: springer.com

Vol. 31. Ajith Abraham, Crina Grosan, Vitorino Ramos (Eds.)
Stigmergic Optimization, 2006
ISBN 978-3-540-34689-0

Vol. 32. Akira Hirose
Complex-Valued Neural Networks, 2006
ISBN 978-3-540-33456-9

Vol. 33. Martin Pelikan, Kumara Sastry, Erick Cantú-Paz (Eds.)
Scalable Optimization via Probabilistic Modeling, 2006
ISBN 978-3-540-34953-2

Vol. 34. Ajith Abraham, Crina Grosan, Vitorino Ramos (Eds.)
Swarm Intelligence in Data Mining, 2006
ISBN 978-3-540-34955-6

Vol. 35. Ke Chen, Lipo Wang (Eds.)
Trends in Neural Computation, 2007
ISBN 978-3-540-36121-3

Vol. 36. Ildar Batyrshin, Janusz Kacprzyk, Leonid Sheremetor, Lotfi A. Zadeh (Eds.)
Preception-based Data Mining and Decision Making in Economics and Finance, 2006
ISBN 978-3-540-36244-9

Vol. 37. Jie Lu, Da Ruan, Guangquan Zhang (Eds.)
E-Service Intelligence, 2007
ISBN 978-3-540-37015-4

Vol. 38. Art Lew, Holger Mauch
Dynamic Programming, 2007
ISBN 978-3-540-37013-0

Vol. 39. Gregory Levitin (Ed.)
Computational Intelligence in Reliability Engineering, 2007
ISBN 978-3-540-37367-4

Vol. 40. Gregory Levitin (Ed.)
Computational Intelligence in Reliability Engineering, 2007
ISBN 978-3-540-37371-1

Vol. 41. Mukesh Khare, S.M. Shiva Nagendra (Eds.)
Artificial Neural Networks in Vehicular Pollution Modelling, 2007
ISBN 978-3-540-37417-6

Vol. 42. Bernd J. Krämer, Wolfgang A. Halang (Eds.)
Contributions to Ubiquitous Computing, 2007
ISBN 978-3-540-44909-6

Vol. 43. Fabrice Guillet, Howard J. Hamilton (Eds.)
Quality Measures in Data Mining, 2007
ISBN 978-3-540-44911-9

Vol. 44. Nadia Nedjah, Luiza de Macedo Mourelle, Mario Neto Borges, Nival Nunes de Almeida (Eds.)
Intelligent Educational Machines, 2007
ISBN 978-3-540-44920-1

Vol. 45. Vladimir G. Ivancevic, Tijana T. Ivancevic
Neuro-Fuzzy Associative Machinery for Comprehensive Brain and Cognition Modeling, 2007
ISBN 978-3-540-47463-0

Vol. 46. Valentina Zharkova, Lakhmi C. Jain
Artificial Intelligence in Recognition and Classification of Astrophysical and Medical Images, 2007
ISBN 978-3-540-47511-8

Vol. 47. S. Sumathi, S. Esakkirajan
Fundamentals of Relational Database Management Systems, 2007
ISBN 978-3-540-48397-7

Vol. 48. H. Yoshida (Ed.)
Advanced Computational Intelligence Paradigms in Healthcare, 2007
ISBN 978-3-540-47523-1

Vol. 49. Keshav P. Dahal, Kay Chen Tan, Peter I. Cowling (Eds.)
Evolutionary Scheduling, 2007
ISBN 978-3-540-48582-7

Keshav P. Dahal
Kay Chen Tan
Peter I. Cowling
(Eds.)

Evolutionary Scheduling

With 198 Figures and 120 Tables

Springer

Dr. Keshav P. Dahal
Modeling Optimisation Scheduling
And Intelligent Control (MOSAIC)
Research Centre
Department of Computing
University of Bradford
Bradford, West Yorkshire, BD7 1DP
The United Kingdom
E-mail: K.P.Dahal@Bradford.ac.uk

Professor Peter I. Cowling
Modeling Optimisation Scheduling
And Intelligent Control (MOSAIC)
Research Centre
Department of Computing
University of Bradford
Bradford, West Yorkshire, BD7 1DP
The United Kingdom
E-mail: P.I.Cowling@Bradford.ac.uk

Dr. Kay Chen Tan
Department of Electrical
and Computer Engineering
National University of Singapore
4 Engineering Drive 3
Singapore 117576
Republic of Singapore
E-mail: eletankc@nus.edu.sg

Library of Congress Control Number: 2006936973

ISSN print edition: 1860-949X
ISSN electronic edition: 1860-9503
ISBN-10 3-540-48582-1 Springer Berlin Heidelberg New York
ISBN-13 978-3-540-48582-7 Springer Berlin Heidelberg New York

Springer is a part of Springer Science+Business Media
springer.com

Cover design: deblik, Berlin
Typesetting by the editors
Printed on acid-free paper SPIN: 11508199 89/SPi 5 4 3 2 1 0

Preface

Evolutionary scheduling

Evolutionary scheduling is a vital research domain at the interface of two important sciences - artificial intelligence and operational research. Scheduling in its wide variety of forms is a critical problem in today's productivity-oriented world having significant economic and social consequences. Scheduling problems encompass a wide range of combinatorial optimization problems where the primary objective is to temporally or spatially accommodate a set of entities such as events, activities, people, and vehicles so that the available, and usually scarce, resources are most efficiently utilized while satisfying a set of constraints that define the feasibility of the schedule. The benefits of proper scheduling may be tangible in the form of monetary profits or reduced environmental impact or intangible in the form of higher satisfaction for individuals such as customers and employees. It is for these reasons that much effort has been expended over the years to develop algorithms for automated scheduling.

Real-world scheduling problems are generally complex, large scale, constrained, and multi-objective in nature, and classical operational research techniques are often inadequate at solving them effectively. With the advent of computation intelligence, there is renewed interest in solving scheduling problems using evolutionary computational techniques. These techniques, which include genetic algorithms, genetic programming, evolutionary strategies, memetic algorithms, particle swarm optimization, ant colony systems, etc, are derived from biologically inspired concepts and are well-suited to solve scheduling problems since they are highly scalable and flexible in terms of handling constraints and multiple objectives. They are particularly effective compared to other search algorithms on the large search spaces typical of real-world scheduling.

The purpose of this edited book is to demonstrate the applicability of evolutionary computational techniques to solve scheduling problems, not only to small-scale test problems, but also fully-fledged real-world optimization problems. The intended readers of this book are engineers, researchers, senior undergraduates, and graduate students who are interested in the field of evolutionary scheduling. The assumed background for the book is some basic knowledge of evolutionary computation. As well as its obvious value to researchers, the book should be particularly useful to practitioners and those with an operational research background looking to broaden their range of available techniques.

This book is divided into seven parts. Part I provides readers with an insight into the methodological issues of evolutionary scheduling. The opening chapter, “Memetic algorithms in planning, scheduling, and timetabling” by Cotta and Fernandez, examines the application of evolutionary computation techniques, memetic algorithms in particular, to scheduling problems and presents some guidelines to the design of a successful memetic algorithm. The other chapter, “Landscapes, embedded paths and evolutionary scheduling” by Reeves, discusses the form of the search space when heuristic search methods such as evolutionary algorithms are applied to scheduling problems.

The remaining chapters are grouped into six parts based on the type of scheduling problems they tackle. Part II focuses on classical and non-classical models of production scheduling. The first chapter in this part, “Scheduling of flow-shop, job-shop, and combined scheduling problems using MOEAs with fixed and variable length chromosomes” by Kleeman and Lamont, combines the generic flow shop and job shop problems into a novel multi-component scheduling problem and solves it using a multi-objective evolutionary algorithm. The next chapter, “Designing dispatching rules to minimize total tardiness” by Tay and Ho, presents a genetic programming approach to discover composite dispatching rules for approximating optimal solutions to the flexible job shop problem. “A robust meta-hyper-heuristic approach to hybrid flow-shop scheduling” by Rodríguez and Salhi, presents the use of a hyperheuristic to generate information to assist a metaheuristic to solve the hybrid flow shop problem. The chapter, “Hybrid particle swarm optimizers in the single machine scheduling problem: an experimental study” by Cagnina et al, discusses and compares three particle swarm optimizers that have been designed to solve the single machine total weighted tardiness problem. The final chapter in this part, “An evolutionary approach for solving the multi-objective job-shop scheduling problem” by Ripon et al, presents a jumping gene genetic algorithm to solve the multi-objective job shop problem. The algorithm is shown to be capable of finding a set of diverse solutions, providing the decision maker with a wide range of choices.

Part III is concerned with timetabling problems. Two real university class timetabling problems are solved with great success using a multi-objective evolutionary algorithm in "Multi-objective evolutionary algorithm for university class timetabling problem" by Datta et al. The same problem is considered in "Metaheuristics for university course timetabling" by Lewis et al, where an example algorithm is shown to be able to perform well across a range of benchmark problems.

Energy-related applications are considered in Part IV. In "Optimum oil production planning using an evolutionary approach" by Ray and Sarker, a practical oil production planning problem, where the objective is to identify the optimal amount of gas that needs to be injected into each oil well to maximize the amount of oil extracted, is discussed and solved using an evolutionary algorithm. The following chapter, "A hybrid evolutionary algorithm for service restoration in power distribution systems" by Watanabe et al, describes a hybrid evolutionary algorithm for addressing service restoration problems in power distribution systems. The chapter, "Particle swarm optimization for operational planning: unit commitment and economic dispatch" by Sriyanyong et al, proposes the application of a particle swarm optimization algorithm to unit commitment and economic dispatch problems in the operational planning of a power system. The last chapter of this part, "Evolutionary generator maintenance scheduling in power systems" by Dahal and Galloway, considers the development of metaheuristic and evolutionary-based methodologies to solve the generator maintenance scheduling problem of a centralized electrical power system.

Part V covers network-related applications. "Evolvable fuzzy scheduling scheme for a multiple-channel packet switching network" by Li et al, considers the application of evolutionary fuzzy systems for cell scheduling in ATM networks, while the channel routing problem, which is derived from the detailed routing model in VLSI design, is considered in "A multi-objective evolutionary algorithm for channel routing problems" by Goh et al.

Transportation problems are the focus of part VI. The chapter, "Simultaneous planning and scheduling for multi-autonomous vehicles" by Liu and Kulatunga, addresses concurrently the task allocation, path planning, and collision avoidance problems in an automated container terminal. The scheduling of production and distribution activities of a network of plants supplying rapidly perishable materials using genetic algorithms and a few fast schedule construction heuristics is considered in "Scheduling production and distribution of rapidly perishable materials with hybrid GA's" by Naso et al. The discussions on transportation issues in this part are completed by the chapter, "A scenario-based evolutionary scheduling approach for assessing future supply chain fleet capabilities" by Baker et al, which introduces the problem of optimizing vehicle fleet mixes in a military deployment.

Part VII tackles scheduling problems encountered in business management. The first chapter, "Evolutionary optimization of business process designs" by Tiwari et al, demonstrates how a business process design problem can be modeled as a multi-objective optimization problem and solved using existing evolutionary techniques. The next chapter, "Using a large set of low level heuristics in a hyperheuristic approach to personnel scheduling" by Cowling and Chakhlevitch, addresses the low level heuristic selection issues for hyperheuristics using a trainer scheduling problem. The third chapter in this part, "A genetic-algorithm-based reconfigurable scheduler" by Montana et al, presents a reconfigurable scheduler that is able to handle a wide variety of scheduling problems with different types of constraints and criteria. Finally, the chapter, "Evolutionary algorithm for an inventory location problem" by Chew et al, completes the book. It deals with minimizing the cost in a joint location-inventory model with a single supplier supplying to multiple capacitated distribution centers.

At this point, we would like to express our appreciation to everyone who has made the publication of this edited book possible. We are very grateful to all contributors, who are authorities in the field of evolutionary computation, for their expert contributions. We would also like to thank the many reviewers for their time and cooperation. We have aimed to produce a book which gives an overview of many of the current developments in the large and growing field of evolutionary scheduling. We hope the book will be of use to the research and practitioner communities for many years to come. We leave it to you, the reader, to judge whether we have succeeded in this ambitious aim.

September 2006

Keshav P. Dahal
Kay Chen Tan
Peter I. Cowling

Contents

III. Timetabling

IV. Energy Applications

V. Networks

Memetic Algorithms in Planning, Scheduling, and Timetabling

Carlos Cotta and Antonio J. Fernández

Dept. Lenguajes y Ciencias de la Computación, ETSI Informática,
University of Málaga, Campus de Teatinos, 29071 - Málaga, Spain.

{ccottap,afdez}@lcc.uma.es

Abstract. Memetic algorithms (MAs) constitute a metaheuristic optimization paradigm based on the systematic exploitation of knowledge about the problem being solved, and the synergistic combination of ideas taken from other population-based and trajectory-based metaheuristics. They have been successfully deployed on a plethora of hard combinatorial optimization problems, amongst which scheduling, planning and timetabling are distinguished examples due to their practical interest. This work examines the application of MAs to problems in these domains. We describe the basic architecture of a MA, and present some guidelines to the design of a successful MA for these applications. An overview of the existing literature on the topic is also provided. We conclude with some reflections on the lessons learned, and the future directions that research could take in this area.

1 Introduction

Back in the late 1960s and early 1970s, it began to be evident that there existed many practical problems for which neither the exact resolution nor approximate approaches with realistic performance guarantees were acceptable in practice. Motivated by this fact, several researchers laid the foundations of what we now know as *evolutionary algorithms* [1–4] (EAs). Despite some voices claiming that such approaches constituted "an admission of defeat", these techniques have steadily grown in usage and understanding to become what they nowadays represent: the cutting-edge approach to real-world optimization. Certainly, this has also been the case for other related techniques, such as *simulated annealing* [5] (SA), *tabu search* [6] (TS), etc. The term *metaheuristics* has been coined to denote them.

The development of metaheuristics in general, and EAs in particular, reached a critical point in the mid 1990s, when the need of exploiting problem knowledge was clearly exposed. The formulation of the *No Free Lunch Theorem* (NFL) by Wolpert and Macready [7] made it definitely clear that a search algorithm performs in strict accordance with the amount and quality of the problem knowledge they incorporate. Quite interestingly, this line of thinking had already been advocated by several researchers in the late 1980s and early 1990s, e.g., Hart and

C. Cotta and A.J. Fernández: *Memetic Algorithms in Planning, Scheduling, and Timetabling,* Studies in Computational Intelligence (SCI) **49**, 1–30 (2007)
www.springerlink.com

Belew [8], Davis [9], and Moscato [10]. It was precisely in the work of Moscato where the paradigm of *memetic algorithms* [11–13] (MAs) started.

MAs are a family of metaheuristics that try to blend several concepts from tightly separated –in their origins– families such as EAs and SA. The adjective 'memetic' comes from the term 'meme', coined by R. Dawkins [14] to denote an entity that is analogous to the *gene* in the context of cultural evolution. The purpose of the analogy (sometimes over-stressed in the literature) is to emphasize the departure from biologically-inspired mechanisms of evolution, to more general processes where actual information is manipulated, learned, and transmitted. Due to the way in which this can be implemented, it is often the case that MAs are used under a different name (e.g., 'hybrid EAs', 'Lamarckian EAs', etc.), and sometimes with a very restrictive meaning. At any rate, we can say that a MA is a search strategy in which a population of optimizing agents synergistically cooperate and compete [10]. These agents are explicitly concerned with using knowledge from the problem being solved, as suggested by both theory and practice [15]. A more detailed description of the algorithmic pattern of MAs is given in Sect. 2.

As mentioned above, the *raison d'être* of metaheuristics is the fact that many problems are very difficult to solve using classical approaches. This is precisely the case for many problems in the area of industrial planning, scheduling, timetabling, etc. All these problems have something in common: a set of entities have to be arranged in a time-like fashion, subject to some particular constraints (based on, e.g., precedence, resource consumption, etc.), and usually with some cost figure associated (to be optimized). Not surprisingly, MAs have been extensively used to solve this kind of problems. In this work, we shall analyze the deployment of MAs on this domain. To this end, we shall provide some general design guidelines in Sect. 3, and an overview of the relevant applications of MAs in Sect. 4, trying to highlight the successful strategies. This chapter will end with a summary of lessons learned, and some current and emerging research trends in MAs for scheduling, planning, and timetabling.

2 Memetic Algorithms

MAs are population-based metaheuristics. This means that the algorithm maintain a *population* of candidate solutions for the problem at hand, i.e., a pool comprising several tentative solutions. In the EA terminology, each of these candidate solutions is called *individual*. However, the nature of MAs suggests *agent* as a more appropriate term here. In essence, this is due to the fact that 'individual' denotes a *passive* entity, subject to some evolutionary rules, whereas 'agent' implies the existence of an *active* behavior, purposefully directed to solving the optimization problem at hand. This active behavior is reflected in several typical components of the algorithm such as (but not exclusively as) local search add-ons. We shall return to this point later in this section.

A general sketch of a MA is shown in Fig. 1. As in EAs, the population of agents is subject to processes of competition and mutual cooperation. Competi-

Memetic Algorithm

```
      INPUT:  An instance I of problem P.
      OUTPUT: A solution sol.

      // Generate Initial Population
 1:   for j ← 1:popsize do
 2:     let ind ← GenerateHeuristicSolution(I)
 3:     let pop[j] ← LocalImprover (ind, I)
 4:   endfor
 5:   repeat // Basic Generational Loop
        // Selection
 6:     let breeders ← SelectFromPopulation(pop)
        // Pipelined reproduction
 7:     let auxpop[0] ← pop
 8:     for j ← 1:#op do
 9:       let auxpop[j] ← ApplyOperator (op[j], auxpop[j − 1], I)
10:     endfor
11:     let newpop ← auxpop[#op]
        // Replacement
12:     let pop ← UpdatePopulation (pop, newpop)
        // Check Population Convergence
13:     if Converged(pop) then
14:       let pop ← RestartPopulation(pop, I)
15:     endif
16:   until MA-TerminationCriterion(pop, I)
17:   return Best (pop, I)
```

Fig. 1. The general template of a memetic algorithm

tion is achieved via the standard procedures of selection (line 6) and replacement (line 12): using the information provided by an *ad hoc* guiding function (*fitness* function in the EA terminology), the goodness of agents in *pop* is evaluated; subsequently, a sample of them is selected for reproduction according to this goodness measure. This information is later used to determine which agents will be removed from the population in order to make room for the newly created ones. In both cases – selection and replacement – any of the well-known strategies used in EAs can be utilized, e.g., tournament, ranking, or fitness-proportionate selection, plus or comma replacement, etc. In addition to these, there are other strategies that could be used here, and that have been proved successful in several scheduling domains (see Sect. 4).

As to cooperation, it is accomplished through reproduction. At this stage, new agents are created by using the existing ones. This is done by utilizing a number of reproductive *operators*. Many different such operators can be used in a MA, as illustrated in the general pseudocode shown in Fig. 1, lines 7–11. As it can be seen, an array *op* of operators can be in general assumed. These operators (whose number is denoted by $\#op$) are sequentially applied to the population in a pipeline fashion, thus resulting in several intermediate populations $auxpop[i]$, $0 \leqslant i \leqslant \#op$, where $auxpop[0]$ is initialized to pop, and $auxpop[\#op]$ is the final offspring. In practice, the most typical situation involves utilizing just three operators: recombination, mutation, and local improvement. Notice that in line 9 of the pseudocode, these operators receive not only the solutions they must act on, but also the problem instance I. This illustrates the fact that in MAs these operators are problem-aware, and base their functioning on their knowledge about the problem being solved. This is an important difference with respect to classical EAs.

Recombination is the process that best encapsulates mutual cooperation among several agents (typically two of them, but a higher number is possible [16]). This is done by constructing new solutions using the *relevant* information contained in a number of selected parents. By "relevant", it is implied that the information pieces manipulated bear some important information in order to determine the goodness (or badness) of the solutions. This is an interesting notion that departs from the classical manipulation of symbols for a certain syntax of candidate solutions, typical of plain EAs. We shall return to this in next section.

The other classical operator – mutation – plays the role of *keeping the pot boiling*, continuously injecting new material in the population, but at a low rate (otherwise the search would degrade to a random walk in the solution space). This interpretation of mutation reflects the classical view, dominant in the genetic algorithm arena [17]. Certainly, evolutionary programming practitioners [1] would disagree with this characterization, claiming the crucial role for mutation. According to this latter view, recombination is generally regarded as a *macro-mutation* process. While this is something that could be accepted to some extent in many EA applications in which recombination is a mere random-shuffler of information, the situation is quite different for MAs. Indeed, recombination is usually performed in a *smart* way as anticipated before, and hence it provides a central contribution to the search.

Lastly, one of the most distinctive components of MAs is the use of local improvers. To understand their philosophy, let us consider the following abstract formulation: first of all, assume a graph whose vertices are solutions, and whose edges connect pairs of vertices such that the corresponding solutions differ in some (typically small) amount of relevant information. Now, a local improver is a process that starts at a certain vertex, and moves to an adjacent vertex, provided that the neighboring solution is better than the current solution. Thus, the local improver tries to find an "uphill" (in terms of improving the value provided by the guiding function) path in the graph whose definition was sketched before (formally termed *fitness landscape* [18]). Notice that this description is very simplistic, and that many variations may exist in the way in which the neighbor is selected, the precise criterion under which it is accepted or rejected, and the termination condition for the local improvement.

As anticipated before, the utilization of local improvers (notice that several different ones could be used in different points of the algorithm) is one of the most characteristic features of MAs. It is mainly because of the use of this mechanism for improving solutions on a local (and even autonomous) basis that the term 'agent' is deserved. Thus, the MA can be viewed as a collection of agents performing an autonomous exploration of the search space, cooperating some times via recombination, and competing for computational resources due to the use of selection/replacement mechanisms.

There is another interesting element in the pseudocode shown in Fig. 1, namely the RestartPopulation process (lines 13–15). This process is very important in order to make an appropriate use of the computational resources: should the population reach a state in which all agents were very similar to each other, the generation of new improved solutions would be very unlikely. This phenomenon is known as *convergence*, and can be identified using measures such as Shannon's entropy [19]. If this measure falls below a predefined threshold, the population is considered at a degenerate state. This threshold depends upon the representation of the problem being used, and must therefore be determined in an *ad hoc* fashion.

It must be noted that while most MA applications can be dissected using the general template presented here, it is obviously possible to conceive algorithms somehow departing from it, that could nevertheless be catalogued as MAs. This fact notwithstanding, the general principles depicted in this section should still be applicable to these MAs.

3 Design Principles for Effective MAs

In order to solve a particular optimization problem in the area of planning and scheduling, the general template of MAs depicted in Sect. 2 must be instantiated with precise problem-aware components. No general approach for the design of effective MAs exists in a well-defined sense, and hence this design phase must be addressed from a heuristic point of view as well. Let us consider in the following the main decisions that have to be taken.

3.1 Representation

The first element that one has to decide is the *representation* of solutions. This notion must be understood not as the way solutions are encoded (something that is subject to considerations of memory consumption, manipulation complexity, etc.), but as the abstract formulation of solutions, as regarded by the reproductive operators [20]. In this sense, recall the notion of "relevant" information introduced in Sect. 2: given a certain representation of solutions, these are expressed via some *information units*; if the operators used for manipulating solutions are problem-aware, these information units they identify must be important to determine whether a solution is good or not. The evolutionary dynamics of the system would then drive the population to retain those positive information units, making negative units disappear. This is better illustrated with an example: consider a problem whose solution space is composed of all permutations of n elements; there are several types of information units in such solutions [21], e.g.,

- *positional*, i.e., element e appears in position j.
- *precedence*, i.e., element e appears before/after element e'.
- *adjacency*, i.e., element e appears next to element e'.

The relevance of each type of information unit will obviously depend on the problem being solved. For example, adjacency information is important for the Travelling Salesman Problem (TSP), but positional information is less so. On the other hand, it seems that positional information is relevant when minimizing makespan in a permutation flowshop problem [22], and adjacency information is more irrelevant in this case. Therefore, an edge-manipulation operator such as edge-recombination [23] (ER) will perform better than position-based operators such as partially-mapped crossover [24] (PMX) or uniform cycle crossover [22] (UCX) on the TSP, but the latter will be better on permutation flowshop problems.

There have been several attempts for quantifying how good a certain set of information units is for representing solutions for a specific problems. Among these we can cite *epistasis* (non-additive influence on the guiding function of combining several information units) [25, 26], *fitness variance of formae* (variance of the values returned by the guiding function, measured across a representative subset of solutions carrying this information unit) [27], and *fitness correlation* (correlation in the values of the guiding function for parents and offspring) [28, 29]. Notice that in addition to using a quality metric of representations to predict the performance of a certain pre-existing operator (i.e., *inverse analysis*), new *ad hoc* operators can be defined to manipulate the best representation (*direct analysis*) [13]. This is for example done in [22] for permutation flowshop scheduling, once the positional representation is revealed as the most promising.

It is also important to note that whatever the metric used to quantify the goodness of a particular representation is, there are other considerations that can play a central role, namely, the presence of constraints. Typically, these are handled in three ways: (1) by using penalty functions that guide the search to the feasible region, (2) by using repairing mechanisms that take infeasible solutions back to the feasible region, and (3) by defining reproductive operators that always remain in the feasible region. In the first two cases, the complexity of the representation and the operators can be kept at a lower level[1]. In the latter case, responsibility has to be taken either by representation or by operators to ensure feasibility, and this comes at the cost of an increased complexity. Focusing on representations, the use of *decoders* is a common option to ensure feasibility. The basic idea is to use a complex genotype-to-phenotype mapping that not only produces feasible solutions, but can also provide additional problem knowledge and hence solutions of better quality. For example, a greedy permutational decoder is used in [30] for a nurse scheduling problem. A related approach is used in [31–33] in the context of job shop scheduling.

---

[1] At least from the functional point of view; from the practical point of view, more sophisticated strategies can obviously result in improved performance.

3.2 Reproductive Operators

The generation of new solutions during the reproductive stage is done by manipulating the relevant pieces of information identified. To do so, the user could resort to any of the generic templates defined for that purpose, e.g., *random respectful recombination*, *random assorting recombination*, and *random transmitting recombination* among others [34]. These are generic templates, in the sense that they blindly process abstract information units. However, in order to ensure top performance, reproductive operators must not only manipulate the relevant information, but must do so in a sensible way, that is, using problem knowledge.

There are many ways to achieve this inclusion of problem knowledge. From a rather general stand-point, there are two major aspects into which problem knowledge can be injected: the selection of the parental features that will be transmitted to the descendant, and the selection of non-parental features that will be added to it. Regarding the former issue, there exists evidence that transmission of common features is beneficial for some problems (e.g., [23, 35]). After this initial transmission, the offspring can be completed in several ways. For example, Radcliffe and Surry [27] have proposed the use of local improvers or implicit enumeration schemes. These implicit enumeration schemes can also be used to find the best combination of the information units present in the parents [36] (in this case, the resulting solution would not necessarily respect common features, unless forced to do so). This operation is monotonic in the sense that any child generated is at least as good as the best parent. Ibaraki [37] uses dynamic programming for this purpose, in a single-machine scheduling problem.

To some extent, the above discussion is also applicable to mutation operators, although these exhibit a clearly different role: they must introduce new information as indicated in Sect. 2. The typical procedure is removing some information units from a single solution, and either complete it at random, or use any of the completion procedures described before. Several considerations must be made here though. The first one is the lower relevance that mutation may have in some memetic contexts. Indeed, mutation is sometimes not used in MAs, and instead it is embedded into the local search component, e.g., see [38, 39] in job shop scheduling, and [40] in single machine scheduling, among others. The reason is the widespread usage in MAs of re-starting procedures for refreshing the population when a stagnation point is reached (see Sect. 3.4). In some applications, it may be better to achieve faster convergence and then re-start, than diversifying continuously the search using mutation in pursuit of steady (yet slower) progress.

In other applications, re-start strategies are not used and mutation is thus more important. In these cases, it is not unusual to use several mutation operators, either by considering different basic neighborhoods (e.g., [41] in open shop scheduling, and [42] in single machine scheduling), or by defining *light* and *heavy* mutations that introduce different amounts of new information (e.g., [43, 44] in timetabling, and [45] in flowshop scheduling). Note that according to the operator-based view of representations presented in Sect. 3.1, the use of multiple operators may imply the consideration of different solution representations at different stages of the reproductive phase. This feature of MAs is also exhibited by other metaheuristics such as *variable neighborhood search* [46] (VNS).

Problem-knowledge can also be attained by using of constructive heuristics. These can be used for instance to create the initial population, as depicted in Fig. 1, line 2. For example, Yeh [47] uses a greedy heuristic for this purpose in a flowshop scheduling problem. Other examples of heuristic initialization can be found in [48, 31, 49] for job shop scheduling, and in [43, 50, 51] for timetabling.

3.3 Local Search

The presence of a local search (LS) component is usually regarded as the distinctive feature of MAs with respect to plain evolutionary algorithms. To some extent, it is true that most MAs incorporate LS; this said, the naïve equation MA = EA + LS is an oversimplification that should be avoided [11–13]. Indeed, there are metaheuristic approaches whose philosophy is strongly connected to that of MAs, but that cannot be called "evolutionary" unless a very broad meaning of the term (i.e., practically encompassing every population-based metaheuristic) were assumed. The scatter search metaheuristic [52] is a good example of this situation. On the other hand, there are MA that rely heavily on the use of knowledge-augmented recombination operators, rather than on LS operators, e.g., [36, 53]. Be that as it may, this is not an obstacle to state that LS is commonly used in MAs (i.e., EA + LS $\subset$ MA), and usually has a determining influence on the final performance.

As sketched in Sect. 2, LS can be typically modelled as a trajectory in the search space, that is, an ordered sequence of solutions such that neighboring solutions in this sequence differ in some *small* amount of information. Of course, some implementations can depart from this idealized description. As an example, we can consider MA applications in which the LS component is implemented via TS (tabu search, Sect. 1), such as [54, 55] in flowshop scheduling, [41] in openshop scheduling [56–60] in timetabling, or [61, 62] in maintenance scheduling, among others. Some implementations of TS are endowed with intensification strategies that resume the search from previous elite solutions (hence, rather than a linear sequence, the path traversed by TS can be regarded as a branching trajectory). Additionally, a feature of the utilization of TS – which is shared with MAs that use other LS components as SA (simulated annealing, Sect. 1), e.g., [39, 63] in flowshop scheduling, and also [44, 61] in maintenance scheduling– is the fact that the quality of solutions in the trajectory is not monotonically increasing, but can eventually decrease in order to escape from a local optimum. Obviously, at the end of the process the best solution found (rather than the last one in the trajectory) is kept.

Several issues must be considered when implementing the LS component. One of them is the termination criterion for the search. In classical hill climbing (HC) techniques, it makes sense to stop the search whenever a local optimum is found. However, this is not directly applicable to LS techniques with global optimization capabilities such as TS or SA. In these cases (and indeed in the case of plain HC) it is customary to define a maximum computational effort to be devoted to each LS invocation. On one hand, this means that the final solution need not be a local optimum, as some incorrect characterizations of MAs as EAs in the space of local optima would suggest. On the other hand, it is necessary to define an appropriate balance between the effort of LS and that of the underlying population-based search. This issue has been also acknowledged in other domains, e.g., [64], and the use of *partial Lamarckism* has been suggested, i.e., not using LS on every new solution computed, but only on some of them, selected at random with some probability or on the basis of quality, or only in every $k-$th generation; see [65] for an analysis of these strategies in a multi-objective permutation flowshop problem.

Similarly to the remaining reproductive operators, the selection of a particular LS scheme can be done in light of quality metrics computed on the resulting fitness landscape. *Fitness distance correlation* [66, 67] (FDC) has been proposed as a hardness measure. In essence, FDC is the correlation between the quality of local optima and their closeness to the global optimum. If this FDC coefficient is high (that is, quality tends to be larger for increasing closeness to the optimum), the natural dynamics of the MA (i.e., get closer to local optima by virtue of the LS component) would also lead it close to the global optimum. This *a priori* analysis can be helpful to estimate the potential effectiveness of a particular LS scheme. Its usefulness as a tool for dynamically acquiring information on the fitness landscape is much more questionable, since some general theoretical results indicate that information is conserved during optimization (i.e., information that is apparently gained during the run is in fact a result of *a priori* knowledge) [68].

Another important issue that must be considered during landscape analysis is its global topology, and more precisely, whether it is regular or not, and in the latter case whether the irregularity is in some sense connected to quality or not. This issue has been analyzed by Bierwirth *et al.* [69] for a job-shop scheduling problem. They found that high quality solutions are also highly connected, and hence there is a beneficial drift force that makes that random walks tend to land closer to high quality solutions than to low quality solutions. Of course, the contrary might be true for a certain problem, and this could hinder good performance of the LS algorithm. At any rate, the same consideration regarding the use of multiple mutation operators done in Sect. 3.2 is applicable here, and multiple LS schemes can be used within a MA, see e.g., [65, 51].

3.4 Managing Diversity

As mentioned in Sect. 3.2, there are different views on how to manage the diversity of information in the population. In essence, we can make a distinction between methods for *preserving* diversity, and methods for *restoring* diversity. Clearly, the use of mutation operators falls within the first class. This does not exhaust the possibilities though. For example, the strategy of injecting in the population "random immigrants" [70] – i.e., completely new solutions – could be used. This has been done in [71] for task allocation on multiprocessors. A much more widespread option is the utilization of structured populations [72]: rather than maintaining a panmictic pool of agents in which any two of them can mate, or in which a new solution can potentially replace any existing solution, the population is endowed with a precise topology; both mating and replacement are confined to *neighboring* agents. This causes a slowdown in the propagation of information across the population, and hence hinders the apparition of *superagents* that might quickly take the population over and destroy diversity.

Different population topologies are reported in the literature, i.e., unidirectional/bidirectional rings, grids, hypercubes, etc. In the context of MAs and scheduling, the ternary tree topology has been particularly successful, see e.g., [73–80, 45]. MAs endowed with this topology usually combine it with the use of a quality-based mechanism for placing solutions. More precisely, each internal node of the tree is forced to have a better solution than that of its immediate descendants in the tree. If at a certain point of the run an agent bears a better solution than that of its parent's, they exchange their solutions. This way, there is a continuous upwards flow of better solutions that also guarantees that when recombination is attempted, the intervening solutions are of similar quality.

The alternative (or better, the complement) to these diversity preservation mechanisms is the use of diversity restoration procedures. These are triggered whenever it is detected that the population has stagnated, or is close to a such a dead-end state. This situation can be detected by monitoring the population composition as mentioned in Sect. 2, or by analyzing the population dynamics [81]. In either case, a re-starting procedure is activated. These procedures can be implemented in different ways. A possibility is the utilization of triggered hypermutation [82] (cf. heavy mutation, see Sect. 3.2), as it is done in [45] for a flowshop scheduling problem with sequence dependent family setups. Alternatively, the population can be refreshed by using the previously mentioned random-immigrant strategy, i.e., keeping a fraction of the existing agents (typically some percentage of the best ones), and renewing the rest of the population with random (or heuristically constructed solutions).

We would like to close this section by emphasizing again the heuristic nature of the design principles described in this and previous sections. There is still much room for research in methodological aspects of MAs (e.g., see [83]), and the open-minded philosophy of MAs make them suitable for incorporating mechanisms from other optimization techniques.

4 Applications in Planning, Scheduling, and Timetabling

Scheduling problems can take many forms, and adopt many variants, so we have opted for considering four major subclasses, namely machine scheduling, timetabling, manpower scheduling, and industrial planning. Note that this classification aims to provide a general view of MA applications in this field, rather than a conclusive taxonomy.

4.1 Machine Scheduling

In a broad sense, machine scheduling amounts to organizing in a time-like fashion a set of jobs that have to be processed in a collection of machines. This general definition admits numerous variants in terms of (1) the number of machines onto which the schedule must be arranged (e.g., one or many), (2) the precise constraints involved in the arrangement (e.g., precedence constraints, setup times, etc.), and (3) the quality measure being optimized (e.g., makespan, total tardiness, number of tardy jobs, etc.). The reader may be convinced of the competence of MAs by noting that almost every conceivable instantiation of this generic problem family has been tackled with MAs in the literature.

One of the most well-studied problems in this area is *single machine scheduling* (SMS), i.e., the scheduling of n jobs on a single processor, subject to different constraints and/or cost functions. Many different SMS problems have been solved with MAs. For example, França *et al.* [74, 78] and Mendes *et al.* [79] tackle the SMS problem with sequence-dependent setup times and due dates, aiming to minimizing the total tardiness of the schedule (i.e., the sum of the tardiness of each job). This is done with a structured MA (with ternary tree topology, as described in Sect. 3.4) using two different schemes for both local search (insertion and swaps) and mutation (light and heavy, cf. Sect. 3.2). Sevaux and Dauzère-Pérès [42] also tackle the SMS problem with MAs, considering the weighted number of late jobs as quality measure. They use a low-cost local search scheme whose complexity is $O(n^2)$, n being the number of jobs, and compare different decoding functions for computing a feasible schedule from a plain permutation of the jobs. The same SMS problem with the added requirement of *robustness* (that is, high quality solutions remaining good when some changes take place in the input data) is tackled by Sevaux and Sörensen in [84]. Maheswaran *et al.* [40] consider a related objective function, namely the minimization of the total weighted tardiness. They use a simple local search scheme based on swaps, which is terminated as soon as the first fitness improvement is achieved.

Parallel machine scheduling (PMS) is the natural generalization of the SMS to multiple processors. Cheng and Gen's work [85] is one of the first memetic approaches to PMS. They consider a MA with a sophisticated decoding mechanism based on heuristics for the SMS. França *et al.* [73], Mendes *et al.* [75, 77], and Moscato *et al.* [80] tackle successfully the PMS using the same structured MA described before for the SMS. This approach is also used by these authors in flowshop scheduling [76, 45], see below. Also, Bonfim and Yamakami [86] present an interesting proposal in which a simple MA (using a plain hill-climber for local search) is endowed with a neural network in order to evaluate solutions.

Flowshop scheduling (FSS) is another conspicuous problem variant, in which the jobs have to be processed on m machines in the same order. Yamada and Reeves [87, 88, 54] consider a MA that uses with *path relinking* [89] (PR) for recombination. PR is a method that explores a sequence of solutions defined by two endpoints: given initial solution s and a final solution s', relevant attributes of the former are successively dropped, and substituted by attributes from the latter. Along this path, a local search procedure can be eventually triggered. Yeh

[47] tackles the problem with a MA that incorporates a greedy method to inject a single high-quality solution in the population, and a local search scheme based in two neighborhoods (swaps and insertions).

Several authors have also considered *hybrid flowshop scheduling* (HFSS) problems, in which jobs have to be sequenced through k stages, being a number of identical machines available for each stage (this number being in general different for each stage). Sevaux *et al.* [90, 91] have tackled this problem combining MAs with *constraint programming* (CP) (see Sect. 5.1). In this case, the CP method (actually a branch-and-bound algorithm) is used as local search mechanism. Ždánský and Poživil [55] also deal with the HFSS. The main feature of their MA is the use of TS as an embedded method for performing local search.

FSS problems can be generalized to *job-shop scheduling* (JSS) problems, in which each job follows its own technological processing sequence on the set of machines, and further to *open-shop scheduling* (OSS) problems, in which no processing sequence is imposed for jobs (i.e., a job requires being processed in some subset of machines, and this can be done in any order). The JSS problem has been dealt by Yamada and Nakano [38, 92, 93] using the PR approach mentioned before for recombination. Also for the JSS, Wang and Zheng [94, 39] consider a MA that incorporates simulated annealing as local search mechanism. Quite interestingly, they name their approach GASA or "modified GA" rather than MA. As to the OSS problem, MAs have been used by Liaw [41]. In this case, the MA features tabu search as local search mechanism, and uses two *ad hoc* heuristics for initializing the population.

4.2 Timetabling

Timetabling consists basically of allocating a number of events to a finite number of time periods (also called *slots*) in such a way that a certain set of constraints is satisfied. Two types of constraints are usually considered, the so called *hard constrains*, that is, those constraints that have to be fulfilled under all circumstances, and *soft constraints*, that is, those constraints that should be fulfilled if possible. In some cases, it is not possible to fully satisfy all the constraints, and the aim turns to be finding good solutions subject to certain quality criteria (e.g., minimizing the number of violated constraints, or alternatively maximizing the number of satisfied hard constraints, whilst the number of violated soft constraints is minimized).

Timetabling arises in many different forms that differ mainly in the kind of event (e.g., exams, lectures, courses, etc.) to be scheduled. By the mid 1990s, it was already suggested that incorporating some amount of local search within evolutionary algorithms might enhance the quality of final solutions [95, 96], and experiments with directed and targeted mutation were addressed [97]. From then on, MAs have been shown to be specially useful to tackle timetabling in each of its modalities as shown in the following.

The *university exam timetabling* (i.e., scheduling a number of exams in a given set of sessions avoiding clashes, e.g., no student has two exams at the same time) is one of the instances that have attracted more interest for evolutionary techniques. For example, Burke *et al.* [43] propose a MA that uses a combination of mutation and local search. The local search component consists of a hill climber applied after the mutation process (two operators are proposed). In general, this algorithm does not perform well in highly constrained problems, and the reason seems to be that the local search operator is less effective. A similar idea taking into account recombination operators instead of mutation operators is investigated by Burke and Newall [98] with unproductive results. They also describe in [99] a multi-stage memetic algorithm that decomposes larger problems into smaller components, and then applies to each of the subproblems a similar MA to that in [43]. The basic idea is to decompose the set of events in k subsets phases ($k = 3$ in the paper) and then schedule the original set of events in k phases. To avoid non-schedulable subsets, an idea from heuristic sequencing methods is applied, choosing subsets according to a smart ordering. This proposals follows the maxima of *divide and conquer* with the aim of reducing the complexity of the problem. Indeed, this idea improves both the time to find a solution, and the quality of solutions with respect to the original MA applied over the whole original problem. Batenburg and Palenstijn [60] describe an alternative multi-stage algorithm constructed from the replacement of the MA used in [99] by TS.

Several authors have also suggested memetic solutions to tackle the problem of producing a periodical (usually weekly) timetable to allocate a set of lecturers and/or students in a certain number of time slots and rooms, with varying facilities and student capacities (i.e., the *university course timetabling problem*). For instance, Alkan and Özcan [100] use a set of violation directed mutation operators, and a violation direct hill climbing operator in their MAs. Rossi-Doria and Paechter [51] describe a different proposal where local search consists of a stochastic process that transforms infeasible timetables into feasible ones. Local search is applied initially on each solution in the initial population, and from then on, on each child generated between generations. Wilke *et al.* [101] consider a variant of the problem in the context of high schools. Their MA incorporates a mechanism for self-adapting the parameters that control the application of local search. Additional information on memetic approaches to course and exam timetabling problems can be found in [102, 50, 103, 104].

Public transportation scheduling is another timetabling problem that is attracting increasing interest, specially in the railway area. For instance, Greistorfer [57, 58] is concerned with the problem of finding a schedule of train arrivals in a railway station with a number of different lines passing through it. To obtain such a schedule, a MA incorporating tabu search is used. Semet and Schoenauer [105] deal with the problem of minimizing the resulting delays in train timetables in case of small perturbations of the traffic. To do so, they consider an indirect approach based on permutations representing train ordering, and combine it with ILOG CPLEX, a mathematical programming tool. The cooperation is performed in an autonomous way: initially the EA computes a good solution that is then provided as input to CPLEX.

The *automatic timetabling of sport leagues* is another variant that has also been solved successfully by MAs. Costa [56] describes an evolutionary tabu search (ETS) algorithm that combines the mechanisms of GAs and tabu search. The basic idea is replacing the mutation step in the GA by a search in the space of feasible solutions commanded by the tabu search. This choice of TS for performing LS is dictated by the reported superiority of this approach over other LS techniques such as SA in the domain of graph coloring [106, 107]. The ETS is applied to construct schedules on the National Hockey League of North America. Schönberger *et al.* [108] propose a MA to schedule a regional table-tennis league in Germany. Here, constraint programming (see Sect. 5.1) is used to experiment with the order in which the decision variables are instantiated in the heuristic.

4.3 Manpower Scheduling

Manpower scheduling (also called *rostering* or *human scheduling*) is concerned with the arrangement of employee timetables in an institution. To solve the problem, a set of constraints or preferences (involving the employees, the employers, and even the customers) must be considered. The goal is to find the best shifts and resource assignments that satisfy these constraints.

One of the most popular instances is *nurse scheduling*, i.e., allocating the shifts (day and night shifts, holidays, etc.) for nurses under various constraints. Traditionally, this problem has been tackled via an integer programming formulation, although it has also attracted the attention of the evolutionary community and many proposals of MAs have been done. For instance, Aickelin [109] analyzes the effect of a family of GAs applied to nurse rostering. He concludes that basic GAs cannot solve the problem, and that the results can be improved via specialized operators and local searches. To do so, De Causmaecker and van den Berghe [110] propose the use of tabu search as a local heuristic in a MA.

Burke *et al.* [59] present a number of MAs which use a steepest descent improvement heuristic. These MAs only consider locally optimal solutions. The MAs are compared to previously published TS results and, later, hybridized with this TS algorithm (either as local improvement or as an improved method to be applied over the best solution found by the MA; in both cases this combination produces the best overall results in terms of quality of the solutions). Gröbner and Wilke [111] describe a MA that incorporates reparing operators, applied at an adaptive rate (cf. [101] for timetabling). Burke *et al.* [112] discuss a set of MAs that apply local search on every solution in the population. The range of these MAs varies from those already presented in [59] to new ones that consider random selection in different stages of the algorithm (e.g., in the local search step, in the parent selection, etc.). Özcan [113] has also tackled the nurse rostering problem via a memetic approach based on the same setting proposed in [100] for timetabling. Basically, Özcan proposes a self-adaptive violation-directed hierarchical hill climbing (VDHC) method as a part of the MA; VDHC provides a framework for the cooperation of a set of hill climbers targeted to specific constraint types. The idea is very similar to the VNS approach.

A different problem – *driver scheduling* – is tackled by Li and Kwan [114]. They present a GA with fuzzy evaluation (GAFE) that can be catalogued as a MA. The basic idea is similar to that of the GRASP metaheuristic [115] in the sense that GAFE also applies a greedy heuristic to obtain feasible solutions and performs searches based on multiple solutions to improve the local optimum. In GAFE, fuzzy set theory is applied in the evaluation of potential shifts based on fuzzified criteria represented by fuzzy membership functions. These functions are weighted and the GA is precisely employed to tune these weights. This same approach is extended in [116] by a self-adjusting approach that can be viewed as a particular hybrid of population-based search and local search.

4.4 Industrial Planning

Industrial planning comprises those activities directed to the development, optimization, and maintenance of industrial processes. Roughly speaking, this amounts to producing a list of activities (a plan) to achieve some pre-defined goal. In some contexts, planning can be considered a prior stage to scheduling, the latter being responsible for arranging in time those planned activities. However, this distinction is not always clear. An example of this can be found in *maintenance scheduling*, that is, organizing the activities required to keep a certain set of facilities at a functioning level. Typically, this involves fulfilling several constraints related to the external demands the system has to serve (e.g., an electricity transmission system must keep supplying the demanded energy even if some station is down due to maintenance reasons). Additionally, maintenance costs must be kept low, thus introducing an optimality criterion.

Several maintenance scheduling problems have been attacked with MAs. Burke and Smith [117, 118] consider the maintenance of thermal generators. A rolling maintenance plan is sought, such that capacity and output constraints are not violated, and such that the total combined cost of production and maintenance is minimized. They compared MAs incorporating either HC, SA, or TS. It was shown that the MA with TS performed slightly better than the HC-based and SA-based variants. The influence of local search was determinant in the performance, to the point that heuristic initialization of the populations seems to exert a negligible effect. Digalakis and Margaritis [61] further studied this problem from the point of view of parallel multi-population MAs. In their experiments, a MA with multiple populations using different local search techniques produces better results than homogeneous multi-population MAs, using just one kind of local improver.

Burke and Smith [44, 62] also address the maintenance problem of a British regional electricity grid. As before, MAs with HC, SA, and TS are compared. In order to alleviate the cost of local search, it is limited to a small number of iterations after the first local optimum is found. In this case, the HC-based and the TS-based MAs perform similarly, providing the best results and outperforming the SA-based MA. This result was consistent with previous work by the authors [119] indicating that TS was better than SA in terms of solution quality. TS was also slightly better than HC, but at the cost of a higher running time.

There have been other attempts to deploy MAs on industrial planning problems. Not related with maintenance scheduling, but sharing several important features, Evans and Fletcher [120] have considered the *boiler scheduling* problem. The goal is scheduling the operation of boilers in a power plant, so as to optimize the production of pressurized steam. The problem exhibits some production constraints that are considered by an *ad hoc* heuristic in order to produce an initial population of feasible solutions. Local search is implemented as a simple hill-climbing step, i.e., a solution is modified, and the change is kept only if it results in a quality improvement. Notice that this problem is also strongly related to the area of power scheduling, one of whose most conspicuous members is the *unit commitment* problem. Although MAs have been applied here as well, e.g., [121], an overview of these applications is beyond the scope of this work.

5 Directions for Future Developments

Unlike other optimization techniques, MAs were explicitly conceived as a eclectic paradigm, open to the integration of other techniques (metaheuristic or not). Ultimately, this ability to synergistically combine with diverse methods is one of the major reasons of their success. The availability of numerous alternative (and certainly complementary) optimization trends, tools, and methods thus offers a huge potential for future developments. We shall provide an overview of these possibilities in this section.

5.1 Hybridization with Constraint Programming

Most scheduling problems can be naturally formulated as constraint satisfaction problems (CSPs) involving a high number of constraints. In evolutionary approaches, constraint handling represents a difficult task that can be managed in different ways as mentioned in Sect. 3.1, e.g., using a suitable encoding in the space of feasible solutions, or integrating constraints in the evaluation process in form of penalty functions, among other approaches. In any case, dealing with constraints is essential for solving scheduling problems.

A natural way to manage constraints and CSPs is *constraint programming* [122–126] (CP). CP is a sound programming paradigm based on strong theoretical foundations [127] that represents a heterogeneous field of research ranging from theoretical topics in mathematical logic to practical applications in industry. As a consequence, CP is attracting also widespread commercial interest since it is suitable for modelling a wide variety of optimizations problems, particularly, problems involving heterogeneous constraints and combinatorial search.

Optimization in CP is usually based on a form of branch and bound (although other alternative models are also proposed, e.g. [128]), that is, as soon as a solution is found, a further constraint is added, so that from that point on, the value of the optimizing criterion must be better than the value just found. This causes the system to backtrack until a better solution is found. When no further solutions can be found the optimum value is known. CP techniques are complete methods, and thus always guarantee in optimization problems that (1) if there exist at least a solution, the solution found is optimal, and (2) if a solution is not found, it is guaranteed that no solution exist. CP techniques have already been applied to a wide range of scheduling and resource allocation problems (e.g., [129–132]), and there exist many successful applications (e.g., [133–135]).

Compared to evolutionary algorithms, CP systems present some advantages. For instance, one could argue that these systems do not require excessive tuning to fit the problem, and thus are usually easier to modify and maintain; they can also handle certain classes of constraints better than MAs, e.g., preferences [136, 137] since, these can be directly modelled as soft constraints, and one has the possibility of controlling which of them are relaxed (whereas, in general, in evolutionary techniques constraints are simply part of the evaluation function, or are present in the representation of solutions). However, the nature of complete-search techniques of CP is also its main drawback since the time needed to find the optimal solution can be prohibitive for large problems. Hence, stochastic techniques (e.g., MAs) may be better when the search space is huge.

In fact, we can say that CP and MAs are two complementary worlds that clearly can profit one from the other. The hybridization of both approaches opens very interesting lines of research. In this sense, some appealing hybrid proposals to scheduling problems have recently appeared. We have already mentioned some of these, i.e., [108, 90, 91]. Further in this line, Backer *et al.* [138] describe a method for using local search techniques within a CP framework, and apply this technique to vehicle routing problems. To avoid the search getting trapped in local minima, they investigate the use of several meta-heuristics (from a simple TS method to guided local search). Also, Yun and Gen [139] use CP techniques for dealing with the preemptive and non-preemptive case in a single machine job-shop scheduling problem. They consider constraints of different types (e.g., temporal constraints, resource constraints, etc.) and use them for generating the initial population. Merlot *et al.* [140] propose a three-phase hybrid algorithm to deal with timetabling problems. In the first phase, CP is applied with the aim of obtaining a feasible solution (if any): a specialized constraint propagation algorithm is firstly applied to reduce the domain of the constrained variables and then, when no further reduction is possible, enumeration strategies are applied to reactivate the propagation. Moreover, with the aim of improving quality, the solution obtained by the CP method is used as starting point of a simulated annealing-based phase and, in a third phase a hill climber is also used.

In general, we argue that MAs can help CP to tackle larger problems and CP can help MAs to reduce drastically the search space by removing infeasible regions what would allow to focus the evolution in the promising regions. Some initial steps have already been done in this exciting line of researching [141].

5.2 Emergent Technologies

It is clear that our world is getting increasingly complex at an accelerated rate, at least from a technological point of view (famous Moore's Law being just an example of this trend). In order to cope with the optimization problems to come, in particular those from the areas of planning and scheduling, optimization tools have to adapt to this complexity. This means that traditional, one-dimensional, sequential approaches must move aside to make room for the next generation of optimization techniques. Focusing in MAs, some of the topics that will become increasingly important in the next years are multi-objective optimization, self-adaptation, and autonomous functioning.

Starting with *multi-objective optimization* (MOO), it is clear that the existence of many different cost functions for a single problem (e.g., machine scheduling, cf. Sect. 4.1) is an indication of (1) the richness of these problems, and (2) the inappropriateness of single-objective optimization to grasp many of their practical implications. Although MOO is hardly an emerging paradigm (in the sense of having been extensively studied in the last decades), the development of multi-objective MAs for scheduling and planning is still a developing field. Several proposals have been made, e.g., [63, 65, 142], but clearly, there is still a long way to go in exploring new strategies for adapting MAs to MOO.

Another crucial feature of MAs that deserves further exploration is *self-adaptation*. As anticipated in [11], future MAs will work in at least two levels and two time scales: in the short-time scale, a set of agents would explore the search space associated to the problem; in the long-time scale the MA would adapt the heuristics associated with the agents. This idea is at the core of the *memeplexes* suggested by Krasnogor and Smith [143]. Some work has already been done in this area [116]. Very related ideas are also currently being developed in *hyperheuristics*, see e.g. [144, 145]. A hyperheuristic is a high-level heuristic which adaptively controls the combination of several low-level heuristics. Hyperheuristics have been successfully applied to scheduling problems [146–150], and offer interesting prospects for their combination with MAs.

Finally, *autonomous functioning* is another feature that has to be boosted in near-future MAs. Recall the use of the term "agent" in the description of the functional pattern of MAs (see Sect. 3). Indeed, the original conception of MAs envisioned the search as a rather decoupled process, that is, with inter-agent communication being less frequent than individual improvement. This fits very well with the behavior of multi-agent systems, which have been also applied to planning and scheduling with satisfactory results [151–153]. Enhancing the autonomous component of MA agents would redound in new possibilities for their efficient parallelization in distributed systems, as well as open a plethora of research lines such as, e.g., the use of epistemic logic systems for modelling the distributed belief of the agents on the optimal solution [154].

5.3 Other Interesting Scheduling Problems

There exist several scheduling problems that have not been treated – to the best of our knowledge – by MAs. These do not just provide challenging optimization tasks, but can also open new scenarios for further research on MAs for scheduling.

The *scheduling of social tournaments* (SST) has attracted significant attention in recent years since they arise in many practical applications, and induce highly combinatorial problems. SST problems may be considered either as instances of timetabling problems (e.g., timetabling of sport leagues) or rostering problems (e.g., judge assignments). One of the most popular SST instances is that known as the *social golfer problem* (problem #10 in the CSPLib[2]): it consists of trying to schedule $g \times s$ golfers into g groups of s players over w weeks, such that no golfer plays in the same group with any other golfer more than once. The problem can be regarded as an optimization problem if for two given numbers g and s we ask for the maximum number of weeks the golfers can play together. An instance to the problem is characterized by a triple $w - g - s$. The initial question consisted of scheduling 32 golfers in a local golf club (i.e., $g = 8$ and $s = 4$). The optimal solution for this instance is not yet known, and the current best known solution is a 9 week schedule (i.e., $w = 9$). There also exist interesting instances and variants for this problem as the *Kirkman's schoolgirl problem* [155], the *debating tournament problem* and the *judge assignment* [156].

Pattern sequencing problems have also important applications, especially in the field of production planning (for instance, in *talent scheduling* [157, 158]). Those problems generally consist of finding a permutation of predetermined production patterns (groupings of some elementary order types) with respect to different objectives. These objectives may represent, e.g., handling costs or stock capacity restrictions, which usually leads to NP-hard problems. In these problems, the use of heuristics to construct near-optimal pattern sequences is generally assumed to be appropriate [159].

6 Concluding Remarks

One of the main conclusions that can be drawn from the extensive literature on MAs for planning, scheduling, and timetabling is they constitute a versatile and effective optimization paradigm. Indeed, MAs are one of the primary weapons in our arsenal for dealing with problems in this area. They provide an appropriate framework to seamlessly integrate successful heuristics into a single search engine. In this sense, MAs should not be regarded as competitors, but as integrators. Whenever non-hybrid metaheuristics start to reach their limits, MAs are the next natural step.

There is an important empirical component in the design of MAs. However, this does not imply that MAs are just a *plug-and-play* approach. The user can benefit from the methodological corpus available for both population-based and trajectory-based search techniques. Design by analogy is another powerful strategy in this area: although very diverse at first sight, scheduling problems have strong underlying connections; hence, knowledge transfer from one subarea to another one is not just feasible, but also likely to be useful. The selection of reproductive operators and/or local-search strategies is at any rate an open problem in methodological terms. Some guidelines for LS design in specific applications are available as shown in Sect. 4, but these are very specific, and hard to

[2] `http://www.csplib.org`

generalize. For this reason, computational considerations, such as the affordable running time, remain one of the governing factors in taking decisions with this regard. For example, more sophisticated LS techniques can provide better results regarding solution quality than plain HC, but the improvement is likely to take place after longer run times. This must be taken into account when dealing with complex scheduling problems in which evaluating local moves is computationally expensive, or in which the size of neighborhoods is huge.

New computational challenges will rise in the years to come. Scheduling problems will not just become a matter of large-scale optimization, but will also become richer and more complex. Consider for example the situation in machine scheduling, where technological developments in manufacturing processes and production strategies will result in new (multiple) objectives to optimize, additional constraints to be considered, etc. New methods will start to play an essential role, e.g., safe kernelization techniques, commonly used in the realm of parameterized complexity [160]. Metaheuristics will have to adapt to this new scenario, and eclecticism appears to be essential for this. The future looks promising for MAs.

Acknowledgments

This work was partially supported by Spanish MCyT under contracts TIN2004-7943-C04-01 and TIN2005-08818-C04-01. Thanks are also due to the reviewers for their useful comments.

References

1. Fogel, L.J., Owens, A.J., Walsh, M.J.: Artificial Intelligence through Simulated Evolution. John Wiley & Sons, New York (1966)
2. Holland, J.: Adaptation in Natural and Artificial Systems. University of Michigan Press (1975)
3. Rechenberg, I.: Evolutionsstrategie: Optimierung technischer Systeme nach Prinzipien der biologischen Evolution. Frommann-Holzboog, Stuttgart (1973)
4. Schwefel, H.P.: Kybernetische Evolution als Strategie der experimentellen Forschung in der Strömungstechnik. Diplomarbeit, Technische Universität Berlin, Hermann Föttinger–Institut für Strömungstechnik (1965)
5. Kirkpatrick, S., Gelatt Jr., C.D., Vecchi, M.P.: Optimization by simulated annealing. Science **220** (1983) 671–680
6. Glover, F., Laguna, M.: Tabu Search. Kluwer Academic Publishers, Boston, MA (1997)
7. Wolpert, D.H., Macready, W.G.: No free lunch theorems for optimization. IEEE Transactions on Evolutionary Computation **1(1)** (1997) 67–82
8. Hart, W.E., Belew, R.K.: Optimizing an arbitrary function is hard for the genetic algorithm. In Belew, R.K., Booker, L.B., eds.: Proceedings of the 4th International Conference on Genetic Algorithms, San Mateo CA, Morgan Kaufmann (1991) 190–195
9. Davis, L.D.: Handbook of Genetic Algorithms. Van Nostrand Reinhold Computer Library, New York (1991)

10. Moscato, P.: On Evolution, Search, Optimization, Genetic Algorithms and Martial Arts: Towards Memetic Algorithms. Technical Report Caltech Concurrent Computation Program, Report. 826, California Institute of Technology, Pasadena, California, USA (1989)
11. Moscato, P.: Memetic algorithms: A short introduction. In Corne, D., Dorigo, M., Glover, F., eds.: New Ideas in Optimization. McGraw-Hill, Maidenhead, Berkshire, England, UK (1999) 219–234
12. Moscato, P., Cotta, C.: A gentle introduction to memetic algorithms. In Glover, F., Kochenberger, G., eds.: Handbook of Metaheuristics. Kluwer Academic Publishers, Boston MA (2003) 105–144
13. Moscato, P., Cotta, C., Mendes, A.S.: Memetic algorithms. In Onwubolu, G.C., Babu, B.V., eds.: New Optimization Techniques in Engineering. Springer-Verlag, Berlin Heidelberg (2004) 53–85
14. Dawkins, R.: The Selfish Gene. Clarendon Press, Oxford (1976)
15. Culberson, J.: On the futility of blind search: An algorithmic view of "No Free Lunch". Evolutionary Computation **6** (1998) 109–127
16. Eiben, A.E., Raue, P.E., Ruttkay, Z.: Genetic algorithms with multi-parent recombination. In Davidor, Y., Schwefel, H.P., Männer, R., eds.: Parallel Problem Solving From Nature III. Volume 866 of Lecture Notes in Computer Science. Springer-Verlag (1994) 78–87
17. Goldberg, D.E.: Genetic Algorithms in Search, Optimization and Machine Learning. Addison-Wesley, Reading, MA (1989)
18. Jones, T.C.: Evolutionary Algorithms, Fitness Landscapes and Search. PhD thesis, University of New Mexico (1995)
19. Davidor, Y., Ben-Kiki, O.: The interplay among the genetic algorithm operators: Information theory tools used in a holistic way. In Männer, R., Manderick, B., eds.: Parallel Problem Solving From Nature II, Amsterdam, Elsevier Science Publishers B.V. (1992) 75–84
20. Radcliffe, N.J.: Non-linear genetic representations. In Männer, R., Manderick, B., eds.: Parallel Problem Solving From Nature II, Amsterdam, Elsevier Science Publishers B.V. (1992) 259–268
21. Fox, B.R., McMahon, M.B.: Genetic operators for sequencing problems. In Rawlins, G.J.E., ed.: Foundations of Genetic Algorithms I, San Mateo, CA, Morgan Kaufmann (1991) 284–300
22. Cotta, C., Troya, J.M.: Genetic forma recombination in permutation flowshop problems. Evolutionary Computation **6** (1998) 25–44
23. Mathias, K., Whitley, L.D.: Genetic operators, the fitness landscape and the traveling salesman problem. In Männer, R., Manderick, B., eds.: Parallel Problem Solving From Nature II, Amsterdam, Elsevier Science Publishers B.V. (1992) 221–230
24. Goldberg, D.E., Lingle Jr., R.: Alleles, loci and the traveling salesman problem. In Grefenstette, J.J., ed.: Proceedings of the 1st International Conference on Genetic Algorithms, Hillsdale NJ, Lawrence Erlbaum Associates (1985) 154–159
25. Davidor, Y.: Epistasis Variance: Suitability of a Representation to Genetic Algorithms. Complex Systems **4** (1990) 369–383
26. Davidor, Y.: Epistasis variance: A viewpoint on GA-hardness. In Rawlins, G.J.E., ed.: Foundations of Genetic Algorithms I, San Mateo, CA, Morgan Kaufmann (1991) 23–35
27. Radcliffe, N.J., Surry, P.D.: Fitness Variance of Formae and Performance Prediction. In Whitley, L.D., Vose, M.D., eds.: Foundations of Genetic Algorithms III, San Francisco, CA, Morgan Kaufmann (1994) 51–72

28. Manderick, B., de Weger, M., Spiessens, P.: The Genetic Algorithm and the Structure of the Fitness Landscape. In Belew, R.K., Booker, L.B., eds.: Proceedings of the 4th International Conference on Genetic Algorithms, San Mateo, CA, Morgan Kaufmann (1991) 143–150
29. Dzubera, J., Whitley, L.D.: Advanced Correlation Analysis of Operators for the Traveling Salesman Problem. In Schwefel, H.P., Männer, R., eds.: Parallel Problem Solving from Nature III. Volume 866 of Lecture Notes in Computer Science., Dortmund, Germany, Springer-Verlag, Berlin, Germany (1994) 68–77
30. Aickelin, U., Dowsland, K.: Exploiting problem structure in a genetic algorithm approach to a nurse rostering problem. Journal of Scheduling **3** (2000) 139–153
31. Puente, J., Vela, C.R., Prieto, C., Varela, R.: Hybridizing a genetic algorithm with local search and heuristic seeding. In Mira, J., Álvarez, J.R., eds.: Artificial Neural Nets Problem Solving Methods. Volume 2687 of Lecture Notes in Computer Science., Berlin Heidelberg, Springer-Verlag (2003) 329–336
32. Varela, R., Serrano, D., Sierra, M.: New codification schemas for scheduling with genetic algorithms. In Mira, J., Álvarez, J.R., eds.: Artificial Intelligence and Knowledge Engineering Applications: a Bioinspired Approach. Volume 3562 of Lecture Notes in Computer Science., Berlin Heidelberg, Springer-Verlag (2005) 11–20
33. Varela, R., Puente, J., Vela, C.R.: Some issues in chromosome codification for scheduling with genetic algorithms. In Castillo, L., Borrajo, D., Salido, M.A., Oddi, A., eds.: Planning, Scheduling and Constraint Satisfaction: From Theory to Practice. Volume 117 of Frontiers in Artificial Intelligence and Applications. IOS Press (2005) 1–10
34. Radcliffe, N.J.: The algebra of genetic algorithms. Annals of Mathematics and Artificial Intelligence **10** (1994) 339–384
35. Oğuz, C., Ercan, M.F.: A genetic algorithm for hybrid flow-shop scheduling with multiprocessor tasks. Journal of Scheduling **8** (2005) 323–351
36. Cotta, C., Troya, J.M.: Embedding branch and bound within evolutionary algorithms. Applied Intelligence **18** (2003) 137–153
37. Ibaraki, T.: Combination with dynamic programming. In Bäck, T., Fogel, D., Michalewicz, Z., eds.: Handbook of Evolutionary Computation. Oxford University Press, New York NY (1997) D3.4:1–2
38. Yamada, T., Nakano, R.: A genetic algorithm with multi-step crossover for job-shop scheduling problems. In: Proceedings of the 1st International Conference on Genetic Algorithms in Engineering Systems: Innovations and Applications, Sheffield, UK, Institution of Electrical Engineers (1995) 146–151
39. Wang, L., Zheng, D.Z.: A modified genetic algorithm for job-shop scheduling. International Journal of Advanced Manufacturing Technology **20** (2002) 72–76
40. Maheswaran, R., Ponnambalam, S.G., Aranvidan, C.: A meta-heuristic approach to single machine scheduling problems. International Journal of Advanced Manufacturing Technology **25** (2005) 772–776
41. Liaw, C.F.: A hybrid genetic algorithm for the open shop scheduling problem. European Journal of Operational Research **124** (2000) 28–42
42. Sevaux, M., Dauzère-Pérès, S.: Genetic algorithms to minimize the weighted number of late jobs on a single machine. European Journal of Operational Research **151** (2003) 296–306
43. Burke, E.K., Newall, J., Weare, R.: A memetic algorithm for university exam timetabling. In Burke, E.K., Ross, P., eds.: Practice and Theory of Automated Timetabling. Volume 1153 of Lecture Notes in Computer Science., Berlin Heidelberg, Springer-Verlag (1996)

44. Burke, E.K., Smith, A.J.: A memetic algorithm to schedule planned grid maintenance. In Mohammadian, M., ed.: Computational Intelligence for Modelling, Control and Automation, IOS Press (1999) 12–127
45. França, P.M., Gupta, J.N.D., Mendes, A.S., Moscato, P., Veltnik, K.J.: Evolutionary algorithms for scheduling a flowshop manufacturing cell with sequence dependent family setups. Computers and Industrial Engineering **48** (2005) 491–506
46. Hansen, P., Mladenović, N.: Variable neighborhood search: Principles and applications. European Journal of Operational Research **130** (2001) 449–467
47. Yeh, W.C.: A memetic algorithm fo the $n/2/\text{Flowshop}/\alpha\text{F}+\ \beta\text{C}_{max}$ scheduling problem. International Journal of Advanced Manufacturing Technology **20** (2002) 464–473
48. Varela, R., Gómez, A., Vela, C.R., Puente, J., Alonso, C.: Heuristic generation of the initial population in solving job shop problems by evolutionary strategies. In Mira, J., Sánchez-Andrés, J.V., eds.: Foundations and Tools for Neural Modeling. Volume 1606 of Lecture Notes in Computer Science., Berlin Heidelberg, Springer-Verlag (1999) 690–699
49. Varela, R., Puente, J., Vela, C.R., Gómez, A.: A knowledge-based evolutionary strategy for scheduling problems with bottlenecks. European Journal of Operational Research **145** (2003) 57–71
50. Burke, E.K., Petrovic, S.: Recent research directions in automated timetabling. European Journal of Operational Research **140** (2002) 266–280
51. Rossi-Doria, O., Paechter, B.: A memetic algorithm for university course timetabling. In: Combinatorial Optimisation 2004 Book of Abstracts, Lancaster, UK, Lancaster University (2004) 56
52. Laguna, M., Martí, R.: Scatter Search. Methodology and Implementations in C. Kluwer Academic Publishers, Boston MA (2003)
53. Nagata, Y., Kobayashi, S.: Edge assembly crossover: A high-power genetic algorithm for the traveling salesman problem. In Bäck, T., ed.: Proceedings of the Seventh International Conference on Genetic Algorithms, East Lansing, EE.UU., San Mateo, CA, Morgan Kaufmann (1997) 450–457
54. Yamada, T., Reeves, C.R.: Solving the C_{sum} permutation flowshop scheduling problem by genetic local search. In: 1998 IEEE International Conference on Evolutionary Computation, Piscataway, NJ, IEEE Press (1998) 230–234
55. Žďánský, M., Poživil, J.: Combination genetic/tabu search algorithm for hybrid flowshops optimization. In: Proceedings of ALGORITMY 2002 – Conference on Scientific Computing, Vysoke Tatry, Podbanske, Slovakia (2002) 230–236
56. Costa, D.: An evolutionary tabu search algorithm and the NHL scheduling problem. INFOR **33** (1995) 161–178
57. Greistorfer, P.: Hybrid genetic tabu search for a cyclic scheduling problem. In Voß, S., Martello, S., Osman, I.H., Roucairol, C., eds.: Meta-Heuristics: Advances and Trends in Local Search Paradigms for Optimization. Kluwer Academic Publishers, Boston, MA (1998) 213–229
58. Greistorfer, P.: Genetic tabu search extensions for the solving of the cyclic regular max-min scheduling problem. In: International Conference on Operations Research (OR98), Zürich, Schwitzerland (1998)
59. Burke, E.K., Cowling, P.I., De Causmaecker, P., van den Berghe, G.: A memetic approach to the nurse rostering problem. Applied Intelligence **15** (2001) 199–214
60. Batenburg, K.J., Palenstijn, W.J.: A new exam timetabling algorithm. In Heskes, T., Lucas, P., Vuurpijl, L., Wiegerinck, W., eds.: Proceedings of the 15th

Belgian-Dutch Conference on Artificial Intelligence BNAIC'03, Nijmegen, The Netherlands (2003) 19–26
61. Digalakis, J., Margaritis, K.: A multipopulation memetic model for the maintenance scheduling problem. In: Proceedings of the 5th Hellenic European Conference on Computer Mathematics and its Applications, Athens, Greece, LEA Press (2001) 318–323
62. Burke, E.K., Smith, A.J.: A memetic algorithm to schedule planned maintenance for the national grid. Journal of Experimental Algorithmics **4** (1999) 1–13
63. Ponnambalam, S.G., Mohan Reddy, M.: A GA-SA multiobjective hybrid search algorithm for integrating lot sizing and sequencing in flow-line scheduling. International Journal of Advanced Manufacturing Technology **21** (2003) 126–137
64. Houck, C., Joines, J.A., Kay, M.G., Wilson, J.R.: Empirical investigation of the benefits of partial lamarckianism. Evolutionary Computation **5** (1997) 31–60
65. Ishibuchi, H., Yoshida, T., Murata, T.: Balance between genetic search and local search in memetic algorithms for multiobjective permutation flowshop scheduling. IEEE Transactions on Evolutionary Computation **7** (2003) 204–223
66. Jones, T.C., Forrest, S.: Fitness distance correlation as a measure of problem difficulty for genetic algorithms. In Eshelman, L.J., ed.: Proceedings of the 6th International Conference on Genetic Algorithms, Morgan Kaufmann (1995) 184–192
67. Merz, P., Freisleben, B.: Fitness landscapes and memetic algorithm design. In Corne, D., Dorigo, M., Glover, F., eds.: New Ideas in Optimization. McGraw-Hill, Maidenhead, Berkshire, England, UK (1999) 245–260
68. English, T.M.: Evaluation of evolutionary and genetic optimizers: No free lunch. In Fogel, L.J., Angeline, P.J., Bäck, T., eds.: Evolutionary Programming V, Cambridge, MA, MIT Press (1996) 163–169
69. Bierwirth, C., Mattfeld, D.C., Watson, J.P.: Landscape regularity and random walks for the job shop scheduling problem. In Gottlieb, J., Raidl, G.R., eds.: Evolutionary Computation in Combinatorial Optimization. Volume 3004 of Lecture Notes in Computer Science., Berlin, Springer-Verlag (2004) 21–30
70. Grefenstette, J.J.: Genetic algorithms for changing environments. In Männer, R., Manderick, B., eds.: Parallel Problem Solving from Nature II, Amsterdam, North-Holland Elsevier (1992) 137–144
71. Hadj-Alouane, A.B., Bean, J.C., Murty, K.G.: A hybrid genetic/optimization algorithm for a task allocation problem. Journal of Scheduling **2** (1999) 181–201
72. Tomassini, M.: Spatially Structured Evolutionary Algorithms: Artificial Evolution in Space and Time. Springer-Verlag (2005)
73. França, P.M., Mendes, A.S., Müller, F., Moscato, P.: Memetic algorithms applied to the single machine and parallel machine scheduling problems. In: Anais da Primeira Oficina de Planejamento e Controle da Produção em Sistemas de Manufatura, Campinas, SP, Brazil (1999)
74. França, P.M., Mendes, A.S., Moscato, P.: Memetic algorithms to minimize tardiness on a single machine with sequence-dependent setup times. In Despotis, D.K., Zopounidis, C., eds.: Proceedings of the 5th International Conference of the Decision Sciences Institute, Athens, Greece (1999) 1708–1710
75. Mendes, A.S., Müller, F., França, P.M., Moscato, P.: Comparing meta-heuristic approaches for parallel machine scheduling problems with sequence-dependent setup times. In: Procedings of the 15th International Conference on CAD/CAM Robotics and Factories of the Future, Águas de Lindóia, SP, Brazil (1999) 1–6

76. França, P.M., Gupta, J.N.D., Mendes, A.S., Moscato, P., Veltnik, K.J.: Metaheuristic approaches for the pure flowshop manufacturing cell problem. In: Proceedings of the 7th International Workshop on Project Management and Scheduling, Osnabrück, Germany (2000) 128–130
77. Mendes, A.S., França, P.M., Moscato, P.: Fuzzy-evolutionary algorithms applied to scheduling problems. In: Proceedings of the 1st World Conference on Production and Operations Management, Seville, Spain (2000) 1–10
78. França, P.M., Mendes, A.S., Moscato, P.: A memetic algorithm for the total tardiness single machine scheduling problem. European Journal of Operational Research **132** (2001) 224–242
79. Mendes, A.S., França, P.M., Moscato, P.: Fitness landscapes for the total tardiness single machine scheduling problem. Neural Network World **2** (2002) 165–180
80. Moscato, P., Mendes, A., Cotta, C.: Scheduling and production & control. In Onwubolu, G.C., Babu, B.V., eds.: New Optimization Techniques in Engineering. Springer-Verlag, Berlin Heidelberg (2004) 655–680
81. Cotta, C., Alba, E., Troya, J.M.: Stochastic reverse hillclimbing and iterated local search. In: Proceedings of the 1999 Congress on Evolutionary Computation, Washington D.C., IEEE Neural Network Council - Evolutionary Programming Society - Institution of Electrical Engineers (1999) 1558–1565
82. Cobb, H.G.: An investigation into the use of hypermutation as an adaptive operator in genetic algorithms having continuous, time-dependent nonstationary environments. Technical Report AIC-90-001, Naval Research Laboratory, Washington DC (1990)
83. Krasnogor, N.: Studies on the Theory and Design Space of Memetic Algorithms. PhD thesis, Faculty of Engineering, Computer Science and Mathematics. University of the West of England. Bristol, United Kingdom (2002)
84. Sevaux, M., Sörensen, K.: A genetic algorithm for robust schedules. In: Proceedings of the 8th International Workshop on Project Management and Scheduling. (2002) 330–333
85. Cheng, R., Gen, M.: Parallel machine scheduling problems using memetic algorithms. Computers and Industrial Engineering **33** (1997) 761–764
86. Bonfim, T.R., Yamakami, A.: Neural network applied to the coevolution of the memetic algorithm for solving the makespan minimization problem in parallel machine scheduling. In Ludermir, T.B., de Souto, M.C.P., eds.: Proceedings of the 7th Brazilian Symposium on Neural Networks (SBRN 2002), Recife, Brazil, IEEE Computer Society (2002) 197–199
87. Yamada, T., Reeves, C.R.: Permutation flowshop scheduling by genetic local search. In: Proceedings of the 2nd International Conference on Genetic Algorithms in Engineering Systems: Innovations and Applications, London, UK, Institution of Electrical Engineers (1997) 232–238
88. Reeves, C.R., Yamada, T.: Genetic algorithms, path relinking and the flowshop sequencing problem. Evolutionary Computation **6** (1998) 230–234
89. Glover, F.: Scatter search and path relinking. In Corne, D., Dorigo, M., Glover, F., eds.: New Methods in Optimization. McGraw-Hill, London (1999) 291–316
90. Sevaux, M., Jouglet, A., Oğuz, C.: MLS+CP for the hybrid flowshop scheduling problem. In: Workshop on the Combination of metaheuristic and local search with Constraint Programming techniques, Nantes, France (2005)
91. Sevaux, M., Jouglet, A., Oğuz, C.: Combining constraint programming and memetic algorithm for the hybrid flowshop scheduling problem. In: ORBEL 19th annual conference of the SOGESCI-BVWB, Louvain-la-Neuve, Belgium (2005)

92. Yamada, T., Nakano, R.: A fusion of crossover and local search. In: IEEE International Conference on Industrial Technology ICIT'96, Shangai, China, IEEE Press (1996) 426–430
93. Yamada, T., Nakano, R.: Scheduling by genetic local search with multi-step crossover. In Voigt, H.M., Ebeling, W., Rechenberg, I., Schwefel, H.P., eds.: Parallel Problem Solving From Nature IV. Volume 1141 of Lecture Notes in Computer Science., Berlin Heidelberg, Springer-Verlag (1996) 960–969
94. Wang, L., Zheng, D.Z.: An effective hybrid optimization strategy for job-shop scheduling problems. Computers & Operations Research **28** (2001) 585–596
95. Paechter, B., Cumming, A., Luchian, H.: The use of local search suggestion lists for improving the solution of timetable problems with evolutionary algorithms. In Fogarty, T.C., ed.: AISB Workshop on evolutionary computing. Volume 993 of Lecture Notes in Computer Science., Berlin Heidelberg, Springer-Verlag (1995) 86–93
96. Rankin, B.: Memetic timetabling in practice. In Burke, E.K., Ross, P., eds.: Proceedings of the 1st International Conference on the Practice and Theory of Automated Timetabling. Volume 1153 of Lecture Notes in Computer Science., Berlin Heidelberg, Springer-Verlag (1996) 266–279
97. Paechter, B., Cumming, A., Norman, M.G., Luchian, H.: Extensions to a memetic timetabling system. In Burke, E.K., Ross, P., eds.: Proceedings of the 1st International Conference on the Practice and Theory of Automated Timetabling. Volume 1153 of Lecture Notes in Computer Science., Berlin Heidelberg, Springer-Verlag (1996) 251–265
98. Burke, E.K., Newall, J.P.: Investigating the benefits of utilising problem specific heuristics within a memetic timetabling algorithm. Workin Paper NOTTCS-TR-97-6, dept. of Computer Science, University of Nottingham, UK (1997)
99. Burke, E.K., Newall, J.: A multi-stage evolutionary algorithm for the timetable problem. IEEE Transactions on Evolutionary Computation **3** (1999) 63–74
100. Alkan, A., Özcan, E.: Memetic algorithms for timetabling. In: Proceedings of the 2003 IEEE Congress on Evolutionary Computation, Canberra, Australia, IEEE Press (2003) 1796–1802
101. Wilke, P., Gröbner, M., Oster, N.: A hybrid genetic algorithm for school timetabling. In McKay, B., Slaney, J., eds.: AI 2002: Advances in Artificial Intelligence, 15th Australian Joint Conference on Artificial Intelligence. Volume 2557 of Lecture Notes in Computer Science., Canberra, Australia, Springer (2002) 455–464
102. Burke, E.K., Jackson, K., Kingston, J.H., Weare, R.F.: Automated university timetabling: The state of the art. The Computer Journal **40** (1997) 565–571
103. Burke, E.K., Landa Silva, J.D.: The design of memetic algorithms for scheduling and timetabling problems. In Krasnogor, N., Hart, W., Smith, J., eds.: Recent Advances in Memetic Algorithms. Volume 166 of Studies in Fuzziness and Soft Computing. Springer-Verlag (2004) 289–312
104. Petrovic, S., Burke, E.K.: University timetabling. In Leung, J., ed.: Handbook of Scheduling: Algorithms, Models, and Performance Analysis. Chapman Hall/CRC Press (2004)
105. Semet, Y., Schoenauer, M.: An efficient memetic, permutation-based evolutionary algorithm for real-world train timetabling. In: Proceedings of the 2005 Congress on Evolutionary Computation, Edinburgh, UK, IEEE Press (2005) 2752–2759
106. Hertz, A., de Werra, D.: Using tabu search techniques for graph coloring. Computing **39** (1987) 345–351

107. Costa, D.: On the use of some known methods for T-colorings of graphs. Annals of Operations Research **41** (1993) 343–358
108. Schönberger, J., Mattfeld, D.C., Kopfer, H.: Memetic algorithm timetabling for non-commercial sport leagues. European Journal of Operational Research **153** (2004) 102–116
109. Aickelin, U.: Nurse rostering with genetic algorithms. In: Proceedings of young operational research conference 1998, Guildford, UK (1998)
110. De Causmaecker, P., van den Berghe, G.: Using tabu search as a local heuristic in a memetic algorithm for the nurse rostering problem. In: Proceedings of the 13th Conference on Quantitative Methods for Decision Making (ORBEL XIII), Brussels (1999)
111. Gröbner, M., Wilke, P.: Optimizing employee schedules by a hybrid genetic algorithm. In Boers, E.J.W., et al., eds.: Applications of Evolutionary Computing 2001. Volume 2037 of Lecture Notes in Computer Science., Berlin Heidelberg, Springer-Verlag (2001) 463–472
112. Burke, E.K., De Causmaecker, P., van den Berghe, G.: Novel metaheuristic approaches to nurse rostering problems in belgian hospitals. In Leung, J., ed.: Handbook of Scheduling: Algorithms, Models, and Performance Analysis. Chapman Hall/CRC Press (2004) 44.1–44.18
113. Özcan, E.: Memetic algorithms for nurse rostering. In Yolum, P., Güngör, T., Gürgen, F.S., Özturan, C.C., eds.: Computer and Information Sciences - ISCIS 2005, 20th International Symposium (ISCIS). Volume 3733 of Lecture Notes in Computer Science., Berlin Heidelberg, Springer-Verlag (2005) 482–492
114. Li, J., Kwan, R.S.K.: A fuzzy genetic algorithm for driver scheduling. European Journal of Operational Research **147** (2003) 334–344
115. Feo, T.A., Resende, M.G.C.: Greedy randomized adaptive search procedures. Journal of Global Optimization **6** (1995) 109–133
116. Li, J., Kwan, R.S.K.: A self adjusting algorithm for driver scheduling. Journal of Heuristics **11** (2005) 351–367
117. Burke, E.K., Smith, A.J.: Hybrid evolutionary techniques for the maintenance scheduling problem. IEEE Transactions on Power Systems **15** (2000) 122–128
118. Burke, E.K., Smith, A.J.: A memetic algorithm for the maintenance scheduling problem. In: Proceedings of the 4th International Conference on Neural Information Processing ICONIP'97, Dunedin, New Zealand, Springer-Verlag (1997) 469–474
119. Burke, E., Clark, J., Smith, J.: Four methods for maintenance scheduling. In Smith, G., Steele, N., Albrecht, R., eds.: Artificial Neural Nets and Genetic Algorithms 3, Wien New York, Springer-Verlag (1998) 264–269
120. Evans, S., Fletcher, I.: A variation on a memetic algorithm for boiler scheduling. In Hotz, L., Krebs, T., eds.: Proceedings Workshop Planen und Konfigurieren (PuK-2003), Hamburg, Germany (2003)
121. Valenzuela, J., Smith, A.: A seeded memetic algorithm for large unit commitment problems. Journal of Heuristics **8** (2002) 173–195
122. Marriot, K., Stuckey, P.J.: Programming with constraints. The MIT Press, Cambridge, Massachusetts (1998)
123. Smith, B.M.: A tutorial on constraint programming. Research Report 95.14, University of Leeds, School of Computer Studies, England (1995)
124. Dechter, R.: Constraint processing. Morgan Kaufmann (2003)
125. Apt, K.R.: Principles of constraint programming. Cambridge University Press (2003)

126. Frühwirth, T., Abdennadher, S.: Essentials of constraint programming. Cognitive Technologies Series. Springer-Verlag (2003)
127. Tsang, E.: Foundations of constraint satisfaction. Academic Press, London and San Diego (1993)
128. Larrosa, J., Morancho, E., Niso, D.: On the practical use of variable elimination in constraint optimization problems: 'still life' as a case study. Journal of Artificial Intelligence Research **23** (2005) 421–440
129. Baptiste, P., Le Pape, C.: Constraint propagation and decomposition techniques for highly disjunctive and highly cumulative project scheduling problems. Constraints **5** (2000) 119–139
130. Barták, R.: Constraint satisfaction for planning and scheduling. In Vlahavas, I., Vrakas, D., eds.: Intelligent Techniques for Planning. Idea Group, Hershey, PA, USA (2005)
131. Le Pape, C.: Constraint-based scheduling: A tutorial. Proceedings of the 1st International Summer School on Constraint Programming (2005)
132. Kilby, P., Prosser, P., Shaw, P.: A comparison of traditional and constraint-based heuristic methods on vehicle routing problems with side constraints. Constraints **5** (2000) 389–414
133. Oliveira, E., Smith, B.M.: A combined constraint-based search method for single-track railway scheduling problem. In Brazdil, P., Jorge, A., eds.: Proceedings of the 10th Portuguese Conference on Artificial Intelligence on Progress in Artificial Intelligence, Knowledge Extraction, Multi-agent Systems, Logic Programming and Constraint Solving. Volume 2258 of Lecture Notes in Computer Science., Berlin Heidelberg, Springer-Verlag (2001) 371–378
134. Wallace, R.J., Freuder, E.C.: Supporting dispatchability in schedules with consumable resources. Journal of Scheduling **8** (2005) 7–23
135. Khemmoudj, M.O.I., Porcheron, M., Bennaceur, H.: Using constraint programming and local search for scheduling of electricité de france nuclear power plant outages. In: Proceedings of the Workshop on the Combination of Metaheuristic and Local Search with Constraint Programming techniques, Nantes, France (2005)
136. Bistarelli, S.: Semirings for Soft Constraint Solving and Programming. Volume 2962 of Lecture Notes in Computer Science. Springer-Verlag, Berlin Heidelberg (2004)
137. Rossi, F.: Preference reasoning. In van Beek, P., ed.: Proceedings of the 11th International Conference on Principles and Practice of Constraint Programming. Volume 3709 of Lecture Notes in Computer Science., Berlin Heidelberg, Springer-Verlag (2005) 9–12
138. Backer, B.D., Furnon, V., Shaw, P., Kilby, P., Prosser, P.: Solving vehicle routing problems using constraint programming and metaheuristics. Journal of Heuristics **6** (2000) 501–523
139. Yun, Y.S., Gen, M.: Advanced scheduling problem using constraint programming techniques in SCM environment. Computers & Industrial Engineering **43** (2002) 213–229
140. Merlot, L.T.G., Boland, N., Hughes, B.D., Stuckey, P.J.: A hybrid algorithm for the examination timetabling problem. In Burke, E.K., Causmaecker, P.D., eds.: Proceedings of the 4th International Conference on the Practice and Theory of Automated Timetabling. Volume 2740 of Lecture Notes in Computer Science., Berlin Heidelberg, Springer-Verlag (2003) 207–231

141. Terashima, H.: Combinations of GAs and CSP strategies for solving examination timetabling problems. PhD thesis, Instituto Tecnológico y de Estudios Superiores de Monterrey (1998)
142. Landa Silva, J.D., Burke, E.K., Petrovic, S.: An introduction to multiobjective metaheuristics for scheduling and timetabling. In X., G., M., S., Sörensen K. and, T.V., eds.: Metaheuristic for Multiobjective Optimisation. Volume 535 of Lecture Notes in Economics and Mathematical Systems. Springer (2004) 91–129
143. Krasnogor, N., Smith, J.E.: Emergence of profitable search strategies based on a simple inheritance mechanism. In Spector, L., et al., eds.: Proceedings of the 2001 Genetic and Evolutionary Computation Conference, Morgan Kaufmann (2001) 432–439
144. Kendall, G., Soubeiga, E., Cowling, P.I.: Choice function and random hyperheuristics. In: Proceedings of the 4th Asia-Pacific Conference on Simulated Evolution and Learning (SEAL'02). (2002) 667–671
145. Burke, E.K., Kendall, G., Newall, J., Hart, E., Ross, P., Schulenburg, S.: Hyperheuristics: an emerging direction in modern search technology. In Glover, F., Kochenberger, G., eds.: Handbook of Metaheuristics. Kluwer Academic Publishers, Boston MA (2003) 457–474
146. Cowling, P.I., Kendall, G., Soubeiga, E.: A hyperheuristic approach to scheduling a sales summit. In Burke, E., Erben, W., eds.: Selected Papers of the 3rd PATAT conference. Volume 2079 of Lecture Notes in Computer Science., Berlin Heidelberg, Springer-Verlag (2000) 176–190
147. Cowling, P.I., Kendall, G., Soubeiga, E.: Hyperheuristics: A tool for rapid prototyping in scheduling and optimisation. In Cagnoni, S., et al., eds.: Applications of Evolutionary Computing. Volume 2279 of Lecture Notes in Computer Science., Berlin Heidelberg, Springer-Verlag (2002) 1–10
148. Cowling, P.I., Kendall, G., Soubeiga, E.: Hyperheuristics: A robust optimisation method applied to nurse scheduling. In Merelo, J.J., et al., eds.: Parallel Problem Solving from Nature VII. Volume 2439 of Lecture Notes in Computer Science., Berlin Heidelberg, Springer-Verlag (2002) 851–860
149. Burke, E.K., Kendall, G., Soubeiga, E.: A tabu-search hyperheuristic for timetabling and rostering. Journal of Heuristics **9** (2003) 451–470
150. Burke, E.K., McCollum, B., Meisels, A., Petrovic, S., Qu, R.: A graph-based hyper-heuristic for educational timetabling problems. European Journal of Operational Research (2006) In press.
151. Cowling, P.I., Ouelhadj, D., Petrovic, S.: Dynamic scheduling of steel casting and milling using multi-agents. Production Planning and Control **15** (2002) 1–11
152. Cowling, P.I., Ouelhadj, D., Petrovic, S.: A multi-agent architecture for dynamic scheduling of steel hot rolling. Journal of Intelligent Manufacturing **14** (2002) 457–470
153. Ouelhadj, D., Petrovic, S., Cowling, P.I., Meisels, A.: Inter-agent cooperation and communication for agent-based robust dynamic scheduling in steel production. Advanced Engineering and Informatics **18** (2005) 161–172
154. Cotta, C., Moscato, P.: Evolutionary computation: Challenges and duties. In Menon, A., ed.: Frontiers of Evolutionary Computation. Kluwer Academic Publishers, Boston MA (2004) 53–72
155. Barnier, N., Brisset, P.: Solving Kirkman's schoolgirl problem in a few seconds. Constraints **10** (2005) 7–21
156. Dotú, I., del Val, A., van Hentenryck, P.: Scheduling social tournaments. In van Beek, P., ed.: Proceedings of the 11th International Conference on Principles and

Practice of Constraint Programming. Volume 3709 of Lecture Notes in Computer Science., Berlin Heidelberg, Springer-Verlag (2005) 845
157. Cheng, T.C.E., Diamond, J.: Optimal scheduling in film production to minimize talent hold cost. Journal of Optimization Theory and Applications **79** (1993) 197–206
158. Smith, B.M.: Constraint programming in practice: scheduling a rehearsal. Research Report APES-67-2003, APES group (2003)
159. Fink, A., Voß, S.: Applications of modern heuristic search methods to pattern sequencing problems. Computers & Operations Research **26** (1999) 17–34
160. Downey, R., Fellows, M.: Parameterized Complexity. Springer-Verlag (1998)

Landscapes, Embedded Paths and Evolutionary Scheduling

Colin R. Reeves

Applied Mathematics Research Centre
Department of Mathematical Sciences
Coventry University
Coventry, UK
Email: C.Reeves@coventry.ac.uk

Summary. Heuristic search methods such as evolutionary algorithms have often been applied to scheduling problems. Whenever we apply such an algorithm, we implicitly construct a *landscape* over which the search traverses. The nature of this landscape is not an invariant of the problem instance, but depends on a number of algorithmic choices—most obviously, the type of neighbourhood operator used as a means of exploring the search space.

In this chapter, after discussing the basic ideas of a landscape, we show how this variation in the landscape of a particular instance manifests itself in terms of an important property—the number of local optima—and discuss ways in which local optima can be avoided. We then review evidence for a particular conformation of local optima in a variety of scheduling and other combinatorial problems—the 'big valley' property.

We then turn to the question of how we can exploit such effects in terms of a fruitful search strategy—embedded *path tracing*—for flowshop scheduling problems. While many approaches could be taken, the one described here embeds the path tracing strategy within a genetic algorithm, and experimental evidence is presented that shows this approach to be capable of generating very high-quality solutions to different versions of the permutation flowshop scheduling problem.

Finally, some recent research is reported into the use of data from the search trace that provides some clues as to the quality of the results found. In particular, it is possible to use data on the re-occurrence of previous local optima to estimate the *total* number of optima and, indirectly, to quantify the probability that a global optimum has been reached.

1 Introduction

Evolutionary algorithms (EAs) have become increasingly popular for finding near-optimal solutions to large combinatorial optimization problems (COPs) such as scheduling and timetabling. At the same time other techniques (sometime called 'metaheuristics'—simulated annealing (SA) and tabu search (TS),

C.R. Reeves: *Landscapes, Embedded Paths and Evolutionary Scheduling*, Studies in Computational Intelligence (SCI) **49**, 31–48 (2007)
www.springerlink.com

for example) have also been found to perform very well, and some attempts have been made to combine them. For a review of some of these techniques see [1, 2, 3, 4].

Central to most heuristic search techniques is the concept of neighbourhood search (NS). Evolutionary algoritms are sometimes thought to be an exception, but in fact there are some similarities that can be put to profitable use, as we shall see later. If we assume that a solution is specified by a vector $\mathbf{x}$, that the set of all (feasible) solutions is denoted by $\mathcal{X}$, which we shall also call the *search space*, then every solution $\mathbf{x} \in \mathcal{X}$ has an associated set of *neighbours*, $N(\mathbf{x}) \subset \mathcal{X}$, called the neighbourhood of $\mathbf{x}$. Each solution $\mathbf{x}' \in N(\mathbf{x})$ can be reached directly from $\mathbf{x}$ by an operation called a *move*. Many different types of move are possible in any particular case, and we can view a move as being generated by the application of an operator ω. For example, if $\mathcal{X}$ is the binary hypercube $\mathbb{Z}_2^l$, a simple operator is the bit flip $\phi(k)$, given by

$$\phi(k) : \mathbb{Z}_2^l \rightarrow \mathbb{Z}_2^l \qquad \begin{cases} z'_k = 1 - z_k & \\ z'_i = z_i & \text{if } k \neq i \end{cases}$$

where $\mathbf{z}$ is a binary vector of length l. In other words, the neighbours of a given solution that are generated by applying this operator are all those at Hamming distance 1 from it.

A more relevant example for readers of this book is the backward shift operator for the case where $\mathcal{X}$ is $\boldsymbol{\Pi}_n$—the space of permutations $\boldsymbol{\pi}$ of length n. The operator $\mathsf{BSH}(i,j)$ (where we assume $i < j$) is given by

$$\mathsf{BSH}(i,j) : \boldsymbol{\Pi}_n \rightarrow \boldsymbol{\Pi}_n \qquad \begin{cases} \pi'_{k+1} = \pi_k \text{ if } i \leq k < j \\ \pi'_i = \pi_j \\ \pi'_k = \pi_k \quad \text{otherwise} \end{cases}$$

Thus if we applied $\mathsf{BSH}(2,5)$ to the permutation

$$\pi = (1,2,3,4,5,6)$$

the result would be

$$\pi' = (1,5,2,3,4,6)$$

In general, there are $\binom{n}{2}$ neighbours of a given permutation. An analogous forward shift operator $\mathsf{FSH}(i,j)$ can similarly be described, the composite of BSH and FSH being denoted by SH. Another alternative for such problems is an exchange operator $\mathsf{EX}(i,j)$ which simply exchanges the elements in the ith and jth positions.

A typical NS heuristic operates by generating neighbours in an iterative process where a move to a new solution is made whenever certain criteria are fulfilled. There are many ways in which candidate moves can be chosen for consideration, and in defining criteria for accepting candidate moves, but most need an evaluation of the *cost* of solution $\mathbf{x}$. Perhaps the most common case is that of *ascent*, in which the only moves accepted are to neighbours that improve the current solution, i.e., reduce its cost. *Steepest* or *best improving* ascent corresponds to the case where all neighbours are evaluated before a move is made—that move being the best improvement available. *First improving* ascent is similar, but the candidates are examined in some pre-defined sequence, and the first that improves the current solution is accepted, without necessarily examining the complete neighbourhood. Normally, the search terminates when no moves can be accepted. The trouble with NS is that at the point of termination, the solution generated is usually only a *local* optimum—a vector none of whose neighbours offer an improved solution.

2 Landscapes

As is implicit in the description of the previous section, the idea of a local optimum only has meaning with respect to a particular neighbourhood. Other associated concepts are those of 'landscapes', 'valleys', 'ridges', and 'basins of attraction' of a particular local optimum. However, these may alter in subtle ways, depending on the neighbourhood used, and the strategy employed for searching it. Formally, we can describe a landscape as follows. We have the search space $\mathcal{X}$, and a function

$$f : \mathcal{X} \mapsto I\!R.$$

where the problem is to find

$$\arg\max_{\mathbf{x}\in\mathcal{X}} f.$$

for the *objective function* f. The necessary changes for the case of minimization are always obvious. In the case of scheduling, we have a variety of different possibilities for f: makespan, mean flowtime, average tardiness etc. For a comprehensive discussion see, for example, [5].

The final piece of the jigsaw is a distance measure that enables us to relate different points in $\mathcal{X}$ topologically. Thus the landscape $\mathcal{L} = (\mathcal{X}, f, d)$ is defined by imposing on the search space a distance measure $d : \mathcal{X} \times \mathcal{X} \to I\!R^{+} \cup \{\infty\}$ for which it is required that

$$\left.\begin{array}{l} d(\mathbf{u},\mathbf{v}) \geq 0; \\ d(\mathbf{u},\mathbf{v}) = 0 \Leftrightarrow \mathbf{u} = \mathbf{v}; \\ d(\mathbf{u},\mathbf{w}) \leq d(\mathbf{u},\mathbf{v}) + d(\mathbf{v},\mathbf{w}); \end{array}\right\} \quad \forall \mathbf{u},\mathbf{v},\mathbf{w} \in \mathcal{X}.$$

Stadler [6] has shown that it is possible to decompose $\mathcal{L}$ into a superposition of 'elementary landscapes', which are related to non-linearities in the function f. For a simplified introduction to this topic, see [7, 8].

The obvious question in any concrete algorithm concerns the nature of d. In the case of methods based on neighbourhood search, the distance measure is generally induced by applying an operator ω. This may be thought of as a function $\omega : \mathcal{X} \times \mathcal{K} \mapsto \mathcal{X}$, where $\mathcal{K}$ defines a parameter set. By varying the parameters across their whole range, we generate a set of *neighbours*

$$N(\mathbf{x}) = \bigcup_{k \in \mathcal{K}} \omega(\mathbf{x}, k)$$

The induced 'canonical' distance measure d_ω can be constructed from

$$d_\omega(\mathbf{u}, \mathbf{v}) = 1 \Leftrightarrow \mathbf{v} \in N(\mathbf{u}),$$

the distance between non-neighbours being defined as the length of the shortest path between them.[1] Thus, under the BSH operator described earlier, the neighbourhood of $(1,2,3,4)$ is generated by the pairs of numbers $(i,j) = \{(1,2),(1,3),(1,4),(2,3),(2,4),(3,4)\}$, forming the set

$$\left\{\begin{matrix} (2,1,3,4) \\ (3,1,2,4) \\ (4,1,2,3) \\ (1,3,2,4) \\ (1,4,2,3) \\ (1,2,4,3) \end{matrix}\right\}$$

These are all 1 unit distant from $(1,2,3,4)$, whereas a permutation such as $(2,4,1,3)$, being one of the neighbours of $(2,1,3,4)$, is 2 units distant from $(1,2,3,4)$.

A local search procedure uses an operator ω and a selection strategy S to generate a sequence of points $\mathbf{x}_0, \mathbf{x}_1, \ldots, \mathbf{x}_n$, terminating at a local optimum. Even in the case of simple local search there are different strategies. For example, a best improving strategy finds $\mathbf{x}_{i+1}$ such that

$$\mathbf{x}_{i+1} = \arg \max_{\mathbf{y} \in N(\mathbf{x}_i)} f(\mathbf{y}) \quad \text{and} \quad f(\mathbf{x}_{i+1}) > f(\mathbf{x}_i),$$

stopping when the second condition cannot be met. In contrast, a first improving strategy searches $N(\mathbf{x}_i)$ in some order until it finds

$$f(\mathbf{y}) > f(\mathbf{x}_i),$$

upon which it sets $\mathbf{x}_{i+1} = \mathbf{y}$, or stops if no such $\mathbf{y}$ can be found after searching the whole neighbourhood. Such a search can be thought of as a function

$$\mu : \mathcal{X} \mapsto \mathcal{X}$$

where if $\mathbf{x}$ is the initial point, $\mathbf{x}^o = \mu(\mathbf{x})$ is the local optimum that it reaches. Each optimum $\mathbf{x}_1^o, \ldots, \mathbf{x}_\nu^o$ (where ν is the total number of optima) then has a basin of attraction whose normalized size is

[1] There is generally at least one path between any pair of points for standard operators—but we have allowed the possibility of an infinite distance in our formalization

$$p_i = \frac{|\{\mathbf{x} : \mu(\mathbf{x}) = \mathbf{x}_i^o\}|}{|\mathcal{X}|}$$

While these quantities are well defined for the case of best improvement, using other strategies may affect the effective basin sizes, although they do not change the number of optima.

It should be clear that, in using NS to solve COPs, the choice of N (or ω) and S may have a significant effect on the quality of solution obtained. It would be useful if it could be arranged that these decisions would lead to a landscape in which the total number of optima was small, and in which the global optimum had a large basin. Unfortunately, in the current state of knowledge concerning landscapes, it is not generally possible to make such choices. There are, however, some methods that can be employed to improve the chances of finding high-quality solutions, and some of these will be addressed in the remainder of this chapter.

3 Avoiding Local Optima

In recent years many strategies have been suggested for the avoidance of local optima. For completeness, we refer here briefly to some of the most popular techniques.

At the most basic level, we could use *iterative restarts* of NS from many different initial points in what we hope are different basins, thus generating a collection of local optima from which the best can be selected.

Simulated annealing uses a controlled randomization strategy—inferior moves are accepted probabilistically, the chance of such acceptance decreasing slowly over the course of a search. By relaxing the acceptance criterion in this way, it becomes possible to move out of the basin of attraction of one local optimum into that of another.

Tabu search adopts a deterministic approach, whereby a 'memory' is implemented by the recording of previously-seen solutions. This record could be explicit, but is often an implicit one, making use of simple but effective data structures. These can be thought of as a 'tabu list' of moves which have been made in the recent past of the search, and which are 'tabu' or forbidden for a certain number of iterations. This prevents cycling, allows escape from basins that are not too wide, and also helps to promote a diversified coverage of the search space.

Perturbation methods improve the restart strategy: instead of retreating to an unrelated and randomly chosen initial solution, the current local optimum is perturbed in some way and the heuristic restarted from the new solution. Perhaps the most widely-known of such techniques is the 'iterated Lin-Kernighan' (ILK) method introduced by Johnson [9] for the travelling salesman problem.

There are other possibilities besides a random perturbation of the population. Glover and Laguna [10] mentioned an idea called 'path relinking', which suggests an alternative means for exploring the landscape. The terminology does not describe exactly what is meant in the context of this paper: it is points that are being linked, not paths; nor are the points being *re*-linked. For this reason we have simply called it 'path tracing' [11].

Evolutionary algorithms (EAs) differ in using a population of solutions rather than moving from one point to the next. Furthermore, new solutions are generated from two (or, rarely) more solutions by applying a 'crossover' operator. Comments are often made to the effect that genetic or evolutionary algorithms can 'avoid local optima'. As an example of this, we quote none other than John Holland himself—the 'father of genetic algorithms':

> *[A genetic algorithm] is all but immune to some of the difficulties—false peaks, discontinuities, high dimensionality, etc.—that commonly attend complex problems.* [12]

Sadly, this is untrue. In practice, the points to which an evolutionary population converge are often not local optima at all, in a well-defined sense, as is shown by the many papers on EA applications where it is recommended that a local search operator is applied to the 'final' population in order to improve it. Recent theoretical work [13] has also shown that at least in the case of binary strings, the 'attractors' of a GA (i.e., the points to which GAs converge) *even in the best case* of an infinite population are merely local optima of the associated Hamming landscape. For real, finite populations, the attractors may have nothing to do with the concept of a local optimum.

EAs do not really have any inherent advantage over NS techniques, but they can be encompassed within the NS framework, especially if we take the path tracing analogy, as we shall discuss later in this chapter.

4 'Big Valleys'

Recent empirical analyses [14, 15, 16] have shown that, for many instances of (minimization) COPs, the landscapes induced by some commonly-used operators have a 'big valley' structure, where the local optima occur relatively close to each other, and to a global optimum. This obviously suggests that in developing algorithms, we should try to exploit this structure.

There is as yet no well-defined mathematical description of what it means for a landscape to possess a 'big valley'. The idea is a fairly informal one, based on the observation that in many combinatorial optimization problems local optima are indeed not distributed uniformly throughout the landscape. In the context of landscapes defined on binary strings, Kauffman has been the pioneer of such experiments, from which he suggested several descriptors of a big valley landscape [17]. (Because he was dealing with fitness maximization, he used the term 'central massif', but it is clear that it is the same phenomenon.)

In the context of scheduling problems, Reeves [15] addressed the question of the structure of the permutation flowshop scheduling problem (PFSP). This

problem can be defined as follows:
if we have processing times $p(i,j)$ for job i on machine j, and a job permutation $\boldsymbol{\pi} = \{\pi_1, \pi_2, \cdots, \pi_n\}$, where there are n jobs and m machines, then we calculate the completion times $C(\pi_i, j)$ as follows:

$$
\begin{aligned}
C(\pi_1, 1) &= p(\pi_1, 1) \\
C(\pi_i, 1) &= C(\pi_{i-1}, 1) + p(\pi_i, 1) \quad \mathit{for}\ i = 2, \ldots, n \\
C(\pi_1, j) &= C(\pi_1, j-1) + p(\pi_1, j)\ \mathit{for}\ \ j = 2, \ldots, m \\
C(\pi_i, j) &= \max\{C(\pi_{i-1}, j), C(\pi_i, j-1)\} + p(\pi_i, j) \\
&\qquad \mathit{for}\ i = 2, \ldots, n;\ j = 2, \ldots, m
\end{aligned}
$$

We define the *makespan* as

$$C_{max}(\boldsymbol{\pi}) = C(\pi_n, m).$$

The *makespan version* of the PFSP is then to find a permutation $\boldsymbol{\pi}^*$ that minimizes this value. While it is the problem with the makespan objective that has received most attention, other objectives can also be defined. For example, we could seek to minimize the mean *flow-time* (the time a job spends in process), or the mean *tardiness* (assuming some deadline for each job). If the release dates of all jobs are zero (i.e., all jobs are available at the outset), the mean flow-time objective reduces to minimizing

$$C_{sum}(\boldsymbol{\pi}) = \sum_{i=1}^{n} C(\pi_i, m).$$

We call this the *flowsum version* of the PFSP. Methods of solving this problem and many others are comprehensively described in [5].

The research reported in [15] showed that for the well-known benchmark problem instances complied by Taillard [18] this big valley structure was indeed evident[2]. As the idea of the big valley is not mathematically well-defined, a set of statistical methods is generally used for establishing its existence. By repeatedly sampling from different start points, it is possible to draw some conclusions. The methodology used assumes we can measure the distance between points—in particular, the distance of a given point from the global optimum. This raises two questions: firstly, we may not know where the global optimum is; secondly, although the landscape formulation is posed in terms of a distance measure d, the practical instantiation of a given algorithm is in terms of a neighbourhood operator ω. Of course, this implies a distance

[2] Taillard's benchmarks are widely used. They were obtained by generating a large number of instances for selected values of n and m, using independently and identically distributed processing times p_{ij}—the distribution being $U(1,99)$. The benchmarks were then the 10 instances for each (n,m) that proved hardest to solve using a state-of-the-art TS method.

measure d_ω, but we need to calculate a shortest path in terms of the number of applications of ω, whereas we would like a simple formula. This is especially the case in the context of scheduling and permutation operators.

4.1 Practical distance measures

In [15] four easily calculated distance measures were investigated between two solutions, represented by the permutations $\boldsymbol{\pi}$ and $\boldsymbol{\pi}'$. Of these, the following two measures seemed to provide sensible results:

- The **precedence-based measure** counts the number of times job j is *preceded* by job i in both $\boldsymbol{\pi}$ and $\boldsymbol{\pi}'$; to obtain a 'distance', this quantity is subtracted from $n(n-1)/2$.
- The **position-based measure** compares the actual *positions* in the sequence of job j in each of $\boldsymbol{\pi}$ and $\boldsymbol{\pi}'$. For a sequence $\boldsymbol{\pi}$ its inverse permutation $\boldsymbol{\sigma}$ gives the position of job π_i (i.e, $\sigma_{\pi_i} = i$). The position-based measure is then just
$$\sum_{k=1}^{n} |\sigma_k - \sigma'_k|.$$

While these measures are simple to calculate, they are independent of the operator ω, and the question remains as to whether they express the features of the landscape appropriately. Recently, Schiavinotto and Stützle [19] have given algorithms for the efficient calculation of operator-dependent distances. They further show that the precedence-based and position-based measures are not necessarily well-aligned with these 'true' distances measured in terms of ω, which they therefore recommend. However, whether that is actually relevant is still a moot point. In unpublished work by Yamada and Reeves, prior to the research reported in [20], an intensive examination of the local optima of one PFSP instance (Taillard's problem #11) was carried out. It was found that the distribution of the best local optima relative to different distance measures was much more highly correlated with the two operator-independent measures described above. In other words, the big valley was *less* visible if we insisted on measuring distances in terms of ω—in this case EX. Table 1 presents these data as a table in order to emphasize this point.

The distances in this table have been normalized by dividing by the maximum value of the respective distance measure. As can readily be seen, many—in fact, most—of the very good local optima are quite far away from the global optimum: 9.44% are in fact almost as far as it is possible to be. In contrast, using d^{prec} ensures that more than 99% of the local optima are within 40% of the maximum distance from the global optimum. Admittedly, this is based on a single instance, but it demonstrates that making assumptions about landscapes carries a certain risk!

distance measure	x-distance				
	$0 < x < 0.2$	$0.2 < x < 0.4$	$0.4 < x < 0.6$	$0.6 < x < 0.8$	$0.8 < x < 1$
d^{prec}	38.16	61.12	0.72		
d^{EX}	0.72	3.48	22.96	63.40	9.44

Table 1. Contingency table for the distribution of the best local optima in the landscape (defined as those within 20% of the value of the global optimum). The x-distance here is (normalized) d^{prec} and d^{EX} respectively, measured from the global optimum. The table entries are the percentages of the total number of local optima falling in each cell.

4.2 Experimental results

A series of experiments was carried out in [15] in order to establish the existence of a big valley, at least in an intuitive sense. Correlation coefficients between distance (in the sense of the position or precedence measures) and quality (measured as distance from the global optimum) were calculated from 50 independent local searches in 50 problem instances of varying size [18]. The significance of these correlations was assessed by means of a randomization test; in nearly all cases they were adjudged to be very different from zero. This supports the conjecture that local optima cluster in the search space relatively close to the global optimum (or to a global optimum—the possibility of there being several global optima was also considered).

5 Path Tracing for the PFSP

Similar results concerning the existence of big valleys have been reported for scheduling—and indeed, for other COPs—by several different research teams [14, 16, 21, 22]. The point of analyses such as these is not just their intrinsic interest, but also the fact that they stimulate ideas for algorithmic development.

In the context of evolutionary algorithms, the notion of crossover is well-positioned for the exploitation of big valley structures. It can readily be seen that for most crossover operators, the children constructed are 'between' their parents, in the sense that as we trace a path from one parent to the other one, the set of points through which we pass are the potential children from that operation. Thus genetic crossover can be embedded in a local search framework. In the initial stages, the search concentrates on finding a population of diverse but high-quality (probably locally-optimal) points. One of these is selected as a parent for 'seeding' a short-term local search, in which another is the 'goal'. The search proceeds by tracing a path that approaches the goal, using an operator-independent measure of distance. A diagrammatic representation of this approach is displayed in Figure 1. If a promising region is encountered along the path—as might be expected if a 'big valley' exists—there is the opportunity to find better solutions than those already in the population.

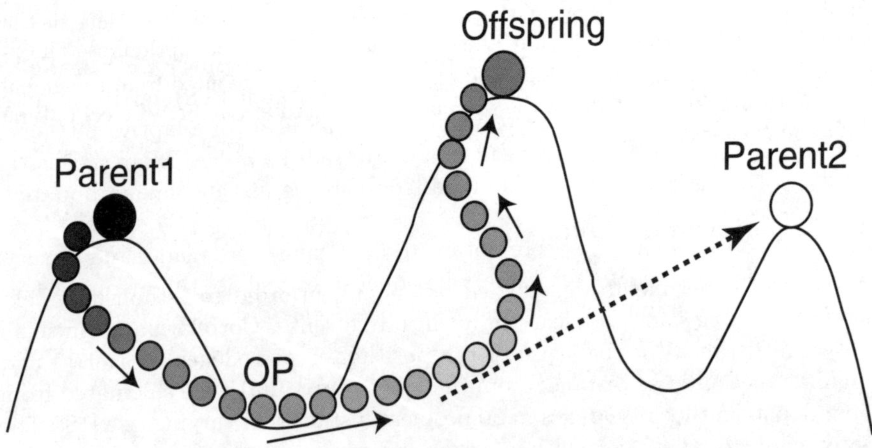

Fig. 1. Path tracing and local search: The search traces a path from the initial parent towards the target using a neighbourhood operator 'OP'. In the 'middle' of the search, good solutions may be found somewhere between the parents. A local search can then exploit this new starting point by climbing towards the top of a hill (or the bottom of a valley, if it is a minimization problem)—a new local optimum.

The algorithm described in [20] made use of this interpretation of crossover (and a similar one for mutation), and produced very high-quality results for the makespan version of the PFSP. As these are readily available in the literature, and have in any case since been superseded by the work of Grabowski and Wodecki [23][3], we will not repeat them here.

However, the results for the corresponding flowsum problem instances may be less well-known.

5.1 The flowsum case

The flowsum version of the PFSP has not been subject to the same intensity of investigation as the makespan version. Optimal solutions for the Taillard benchmarks are not known for this case. The problems are more difficult to optimize, mainly because the calculation of the objective function is more time consuming, good lower bounds are hard to generate, and problem specific knowledge such as critical blocks cannot be used. Some constructive algorithms based on heuristic rules have been suggested [24], and improved [25],

[3] This algorithm is a highly tailored tabu search method based on the idea of 'critical blocks' to focus the search more efficiently

but better quality solutions can be generated by the approach described in this paper.

Basically, the approach to the flowsum problem was the same as the one tested on the makespan version. Some differences were inevitable—for example, in the makespan case, we could exploit the idea of critical blocks to speed up the computation, but this is not relevant to the flowsum case.

Also, it was found that the search could cycle, revisiting points using a standard pattern of moves. This is classic tabu search (TS) territory, and this implementation enables us to make use of TS ideas of adaptive memory [10], which would be less convenient in a more traditional EA. More extensive details of the algorithm are available in [26]; here we just display an updated version of some of the results.

Quite consistent results were obtained, i.e. 6 runs were made, and almost all of them converged to the same job sequence in a short time (from a few seconds to a few minutes). The best results (and they are also the average results in most cases) are reported in Table 2 together with the results obtained by a more recent constructive method (LR) as reported by Liu and Reeves [25].

20×5			20×10			20×20		
prob	best	LR	prob	best	LR	prob	best	LR
1	14033	14226	11	20911	21207	21	33623	34119
2	15151	15446	12	22440	22927	22	31587	31918
3	13301	13676	13	19833	20072	23	33920	34552
4	15447	15750	14	18710	18857	24	31661	32159
5	13529	13633	15	18641	18939	25	34557	34990
6	13123	13265	16	19245	19608	26	32564	32734
7	13548	13774	17	18363	18723	27	32922	33449
8	13948	13968	18	20241	20504	28	32412	32611
9	14295	14456	19	20330	20561	29	33600	34084
10	12943	13036	20	21320	21506	30	32262	32537

Table 2. Taillard's benchmark results (20 jobs) compared with best values found by the Liu-Reeves (LR) constructive algorithm.

The next group of problems (50×5 and 50×10) are much more difficult. In each run the best results were different. Ten runs were carried out for each problem with different random seeds. A population size of 30 was used; other parameters are stated in [26]. It takes 45 minutes per run for 50×5 instances and 90 minutes for 50×10.

It is not certain how good these solutions are. Certainly, they improve on the constructive method by a useful margin, but the lower bounds are some way off (on average around 30%—but they are probably not very good bounds). Even for the easier problems in Table 2, there is no guarantee that

50 × 5				50 × 10			
prob	best	mean	LR	prob	best	mean	LR
31	64860	64934.8	65663	41	87430	87561.4	88770
32	68134	68247.2	68664	42	83157	83305.8	85600
33	63304	63523.2	64378	43	79996	80303.4	82456
34	68259	68502.7	69795	44	86725	86822.4	89356
35	69491	69619.6	70841	45	86448	86703.7	88482
36	67006	67127.6	68084	46	86651	86888.0	89602
37	66311	66450.0	67186	47	89042	89220.7	91422
38	64412	64550.1	65582	48	86924	87180.5	89549
39	63156	63223.8	63968	49	85674	85924.3	88230
40	68994	69137.4	70273	50	88215	88438.6	90787

Table 3. Taillard's benchmark results (50 jobs) compared with the best values found by the Liu-Reeves (LR) constructive algorithm.

the best solutions obtained so far are optimal, although they are closer to their lower bounds. For the problems in Table 3, the best results could probably still be improved by increasing the amount of computation. For example, a solution to problem 31 was found with $C_{sum} = 64803$ by an overnight run.

5.2 Constructive heuristics and the big valley

If available computational effort is limited, there may be advantages in adopting a constructive approach. We have already mentioned the method of Wang *et al.* [24], which produces reasonably good solutions to the flowsum problem rather cheaply. In [25] this method was further improved using several different modifications; although the results obtained were still typically 2-3% worse than those obtained by path tracing (as quoted in Tables 2 and 3 above), they were obtained extremely quickly.

However, the interesting feature of these results was that the quality of the different constructive heuristics was also related to the big valley. For each of the Taillard 20-machine problems (30 instances in all), the solutions found by the different heuristics were stored, and their average distance from a set of 50 distinct local optima (obtained by restarts) was calculated. For comparison, the distance of a random permutation from the same set of local optima was also calculated. The distance measures used were the 'precedence' and 'position' based measures already defined, and the results showed that the sequences generated by the new heuristics were much closer to the local optima than those generated by the older ones, and very much closer than a random permutation. On other words, good constructive heuristics also tend to find solutions near a big valley. Thus, an initial sequence generated by these new heuristics would provide a closer (and thus probably better) starting point for a subsequent local search than a random permutation. Not only will the search then take fewer steps, but by starting nearer the big valley, it also increases the chance of converging to a 'good' local optimum rather than to a poor one. Thus the better performance of the new heuristics could be attributed to their ability to locate the 'big valley' better than the older ones, even though no reference to notions of a landscape was made in constructing them.

5.3 Instances without big valleys

Although the big valley seems a constant feature of almost all randomly generated instances of a diverse set of problems, it should not be assumed that such a structure necessarily exists in any COP. At least in the case of scheduling problems, we have some evidence to identify the factors that influence whether or not a big valley exists.

Watson *et al.* [22] carried out extensive experimental investigations into this question for the PFSP. Rinnooy Kan [27] seems to have been the first to note that the existence of structural relationships in the processing times of jobs may have an important effect on algorithm performance. He highlighted two aspects of non-random structure: processing times with a gradient of times across machines, and a correlation of times within jobs. Later Reeves [28] also noticed that problem instances generated to have these characteristics were relatively easy to solve—especially those with time gradients, and speculated that

> *the topography of the solution space for such cases is fairly smooth, so that getting stuck in a deep local minimum is unlikely.*

Watson *et al.* set out to examine this conjecture by generating a large number of PFSP instances with differing degrees of both job-related and machine-related correlation, as well as a group of problem instances exhibiting degrees of both types of correlation. They discovered that the big valley (as measured by fitness-distance correlations amongst local optima) became less and less evident as the amount of non-random structure in the generating mechanism was increased. Partly this seemed to be associated with the emergence of 'plateaux' in the landscape—regions where neighbouring search points had identical fitness values. Some of the plateaux were very extensive, making the search for 'exits' to regions of lower value quite time-consuming. It might be conjectured that algorithms such as tabu search and path tracing should have an advantage in such cases, in that they encourage wide exploration. However, Watson *et al.* found that, as the big valley faded away, the instances became generally easier to solve, in the sense that finding a 'good' solution (one very close to a lower bound) could be achieved quite quickly by an unsophisticated local search without the extra complexity associated with GAs, tabu search or path tracing. It would appear that structure of the type they investigated encourages the development of a smooth and shallow landscape with relatively little difference between good and bad solutions. The only fitness objective considered in these experiments was makespan, so it is not clear whether landscapes associated with other objectives would behave similarly.

The makespan objective, for example, might be much more likely to give rise to the occurrence of plateaux than the flowsum.

6 Estimating the Number of Optima

A common theme in all heuristic methods for scheduling is the occurrence of local optima. Evolutionary methods, despite optimistic statements to the contrary by some very eminent people[4], do not avoid this problem. In fact, on their own, EAs may not even converge on a local optimum. It is for this reason that many authors have advocated the use of local search operators *within* an EA framework, either as a final refinement mechanism, or as an integral part of the algorithm. These ideas have become so widely advocated that there is now a sub-branch of EAs wholly devoted to such methods, which usually go under the name of 'memetic algorithms'[5].

Among the properties that make a problem instance difficult, it is reasonable to suppose that the number of local optima is one, although not the only one. It is important to emphasize again that optima are not independent properties of the problem instance, as the landscape being searched is in part determined by the actual operators and search strategy used. That some operators are more effective than others is borne out by those (e.g. Weinberger [30] and Stadler [6]) who have advocated the use of autocorrelation measures to compare the landscapes that they induce. However, this approach is indirect; *counting* the number of optima (the parameter ν of section 2) in the landscape induced by different operators would provide an objective way of comparing operators. The problem is, of course, that counting them all means visiting every point in the search space.

It is possible, however, to obtain some statistical estimates, both of the total number of optima, and of the probability of having found all the optima during a less-than-exhaustive search. A recent review of these methods can be found in [31]; this chapter will treat them briefly in the context of scheduling.

It is assumed that a heuristic search method can be restarted many times using different initial solutions. Given the landscape framework we have discussed above, by randomly generating initial solutions, we will sample many basins of attraction. Suppose that after r restarts we obtain a set of local optima $\{\mathbf{x}_1^o, \ldots, \mathbf{x}_k^o\}$, i.e. the number of *distinct* optima is k. We would like to

[4] The remark of Holland, quoted in section 3, is one example, but many others can be found in the literature without too much searching. Numerous authors, for instance, blithely describe EAs as a *global* rather than a local search method.

[5] The terminology arises from the word 'meme' invented by Richard Dawkins. The notion of a meme is controversial, being described by some as hilariously simplistic and even totally vacuous. There is a lot more to be said about memes [29], but this is not the place for a philosophical debate. The term 'memetic' is at least a useful one to identify this particular area of EAs.

use these data to estimate the actual number of optima ν. Various statistical models may be used.

6.1 Waiting-time model

We can formulate the distribution of the waiting-time to find all optima. If r exceeds k substantially, we can use this fact to estimate the probability that all optima have been found. This would also imply, *a fortiori*, that a global optimum had been found, and thus provides us with an objective confidence level concerning the quality of the best solution obtained.

6.2 Counting model

In the event, unfortunately common, that k is not much smaller than r, it is unlikely that we have seen many of the optima. However, a counting model can be used to estimate the value of ν, in a similar way to those models used by ecologists to estimate the size of an unknown animal population. This can be quite illuminating in showing the differences between landscapes generated by different neighbourhood operators.

The approach was applied to solve the *flowsum* version of the PFSP for some of the Taillard benchmarks. This showed that the estimated number of optima induced by the 5 different operators defined in [15] varied substantially. Estimates for a few representative problem instances are shown in Table 4.[6]

Problem	INV	FSH	EX	BSH	SH	composite
ta01	335535	220753	26715	13210	2033	1259
ta02	† 2000333	† 2000333	99274	399457	73264	8015
ta03	† 2000333	† 2000333	124276	79510	8755	2913
ta04	† 2000333	† 2000333	82646	28292	5486	4530
ta05	671080	10192	6026	3871	675	442
ta06	399457	986749	9077	5960	1114	622
ta07	671080	987240	29177	220753	33222	2860
ta08	986749	987746	42799	54828	4090	1834
ta09	987240	† 2000333	284360	58252	11229	5076
ta10	† 2000333	† 2000333	36392	110376	9631	2842

Table 4. Estimated numbers of optima for the 20-job, 5-machine instances devised by Taillard [18], designated ta01-ta10 respectively. The estimates were based on 2000 independent restarts. Far from finding all optima, in several cases, marked as †, all 2000 optima found were distinct, so no estimate was possible at all. The values for these cases are therefore merely estimated lower bounds for ν.

[6] While these results have been reported verbally at a conference, they have not formally been published before.

As these benchmarks have $20! \approx 2.4 \times 10^{18}$ solutions, it is impossible in any reasonable computing time to verify the accuracy of these estimates. However, for some 10-job problems it was possible to enumerate the search space and check the total number of optima: the estimates tended to be negatively biased—i.e., the estimate of ν was consistently smaller than the true value. However, the effect was fairly uniform across all operators, so the rank ordering was not affected.

There is clearly a considerable, and perhaps unexpected, difference between the landscapes induced by different operators in Taillard's first 10 benchmarks. Those associated with INV and FSH clearly have an extremely large number of optima, and while, taken overall, BSH is roughly comparable with EX, FSH is clearly inferior. Yet the size of all the neighbourhoods generated by these operators is of the same order [15], although SH, being composed of two other neighbourhoods, is roughly twice as large as all the others. However, although FSH on its own appears rather feeble, when combined with BSH (in the form of SH) the estimated number of optima reduces significantly. Finally, using a composite of all 5 neighbourhoods is very worthwhile; as observed in [15], the sets of local optima found by the different operators are fairly disjoint. The complexity of searching the composite neighbourhood is not increased (although of course the actual run time is), yet the landscape induced seems considerably less rugged.

6.3 Non-parametric estimates

Fairly restrictive assumptions are needed in order to obtain tractable statistical models of landscapes, and as indicated above, the evidence suggests this results in a negative bias. Removing the assumptions by creating more powerful parametric models leads to theoretical and numerical difficulties, but *non-parametric* approaches are possible, and have been found to provide improved estimates of ν, although the problem of negative bias remains [31]. Different sampling methods may also improve performance, as suggested (in a different context) by [32].

Further work is necessary in this area, but as well as providing a clear basis for discriminating between operators, there is the prospect of more principled techniques for assessing the quality of heuristic solutions, and for formulating termination criteria.

It is also true that the work reported here relates more strictly to local search than to evolutionary algorithms. However, as pointed out earlier, EAs have 'attractors', and they appear to be closely linked to the concept of a local optimum in a related landscape [13]. It is therefore entirely plausible to apply the same methods to the problem of estimating the number of attractors in an EA 'landscape'.

7 Conclusions

Evolutionary scheduling is a fast-moving field, with many different aspects seeing significant progress. This chapter has reviewed the application of landscape-related concepts, and some ways in which they can help in the development of new algorithms. In particular, the notion of an embedded path seems a rather fruitful way to understand and enhance the idea of crossover. Secondly, the number of local optima in the landscape can be estimated in order to improve (or possibly, to provide for the first time) quality assurance for the results of heuristic search methods.

Further theoretical work is needed: it would be helpful if we could provide a formal definition of what it means for a 'big valley' structure to exist, and how it could be related to mathematical constructs associated with neighbourhood structures. Can we more rigorously characterise classes of problems and neighbourhood structures for which it does not occur, as suggested by Watson *et al.* [22]?

More generally, it is clear that crude correlation measures are a very rough guide to the nature of a landscape instance, and we need to find better ways of characterising landscapes from empirical measurements. Extension of the idea of a 'barrier tree' [33] to scheduling problems may be of assistance here.

References

1. Reeves CR (ed) (1993) Modern Heuristic Techniques for Combinatorial Problems. Blackwell Scientific Publications, Oxford, UK (Re-issued by McGraw-Hill, London, UK (1995).)
2. Aarts E, Lenstra JK (eds) (1997) Local Search in Combinatorial Optimization. John Wiley & Sons, Chichester
3. Glover F, Kochenberger GA (eds) (2002) Handbook of Metaheuristics, Kluwer Academic Publishers, Norwell, MA
4. Burke EK, Kendall G (eds) (2005) Search Methodologies: Introductory Tutorials in Optimization and Decision Support Methodologies. Springer, New York
5. Morton TE, Pentico DW (1993) Heuristic Scheduling Systems, Wiley, NY
6. Stadler PF (1995) Towards a theory of landscapes. In: Lopéz-Peña R, Capovilla R, García-Pelayo R, Waelbroeck H, Zertuche F (eds) Complex Systems and Binary Networks. Springer, Berlin
7. Reeves CR, Rowe JE (2002) Genetic Algorithms - Principles and Perspectives. Kluwer Academic Publishers, Norwell, MA
8. Reeves CR (2005) Fitness Landscapes In: Burke EK, Kendall G (eds) (2005) Search Methodologies: Introductory Tutorials in Optimization and Decision Support Methodologies. Springer, New York
9. Johnson DS (1990) Local optimization and the traveling salesman problem. In: Goos G, Hartmanis J (eds) (1990) Automata, Languages and Programming, Lecture Notes in Computer Science 443. Springer, Berlin
10. Glover F, Laguna M (1993) Tabu Search. In: Reeves CR (ed) (1993) Modern Heuristic Techniques for Combinatorial Problems. Blackwell Scientific Publications, Oxford, UK

11. Reeves CR, Yamada T (1999) Goal-Oriented path tracing methods. In: Corne DA, Dorigo M, Glover F (eds) (1999) New Methods in Optimization, McGraw-Hill, London
12. Holland JH (1986) Escaping brittleness: The possibilities of general-purpose learning algorithms applied to parallel rule-based systems. In: Michalski RS, Carbonell JG, Mitchell TM (1986) Machine Learning II. Morgan Kaufmann, Los Altos, CA
13. Reeves CR (2003) The 'crossover landscape' and the Hamming landscape for binary search spaces. In: De Jong KA, Poli R, Rowe JE (eds.) Foundations of Genetic Algorithms 7. Morgan Kaufmann, San Francisco, CA
14. Boese KD, Kahng AB, Muddu S (1994) Operations Research Letters 16:101-113
15. Reeves CR (1999) Annals of Operational Research 86:473-490
16. Merz P, Freisleben B (1998) Memetic algorithms and the fitness landscape of the graph bi-partitioning problem. In: Eiben AE, Bäck T, Schoenauer M, Schwefel H-P (eds) (1998) Parallel Problem-Solving from Nature—PPSN. Springer, Berlin
17. Kauffman S (1993) The Origins of Order: Self-Organisation and Selection in Evolution. Oxford University Press, Oxford
18. Taillard E (1993) European J Operational Research 64:278-285
19. Schiavinotto T, Stützle T (2006) To appear in: Computers and Operations Research. (Available online at doi:10.1016/j.cor.2005.11.022)
20. Reeves CR, Yamada T (1998) Evolutionary Computation 6:45-60
21. Levenhagen J, Bortfeldt A, Gehring H (2001) Path tracing in genetic algorithms applied to the multiconstrained knapsack problem. In: Boers EJW *et al.* (eds) Applications of Evolutionary Computing. Springer, Berlin
22. Watson J-P, Barbulescu L, Whitley LD, Howe AE (2002) INFORMS J Computing 14: 98-123
23. Grabowski J, Wodecki, M (2004) Computers and Operations Research 31:1891-1909
24. Wang C, Chu C, Proth J (1997) European J Operational Research 96:636-644
25. Liu J, Reeves CR (2001) European J Operational Research 132:439-452
26. Yamada T, Reeves CR (1998) Solving the C_{sum} permutation flowshop scheduling problem by genetic local search. In: Proc. of 1998 International Conference on Evolutionary Computation, 230-234. IEEE Press
27. Rinnooy Kan AHG (1976) Machine Scheduling Problems: Classification, Complexity and Computations. Martinus Nijhoff, The Hague, NL
28. Reeves CR (1995) Computers & Operations Research 22:5-13
29. McGrath A (2005) Dawkins' God: Genes, Memes and the Meaning of Life. Blackwell, Oxford, UK.
30. Weinberger ED (1990) Biological Cybernetics 63:325-336
31. Reeves CR, Eremeev AV (2004) J Operational Research Society 55:687-693
32. Reeves CR, Aupetit-Bélaidouni MM (2004) Estimating the number of solutions for SAT problems. In Yao X *et al.* (eds.) (2004) Parallel Problem-Solving from Nature—PPSN VIII, LNCS3242. Springer, Berlin
33. Reidys CM, Stadler PF (2002) SIAM Review 44:3-54

Scheduling of Flow-Shop, Job-Shop, and Combined Scheduling Problems using MOEAs with Fixed and Variable Length Chromosomes

Mark P. Kleeman and Gary B. Lamont

Department of Electrical and Computer Engineering
Graduate School of Engineering
Air Force Institute of Technology, Wright-Patterson AFB, Dayton, OH 45433, USA,
mark.kleeman@afit.edu, gary.lamont@afit.edu *

Abstract. This chapter introduces the novel multi-component scheduling problem, which is a combination of the generic flow-shop and job-shop (or open-shop) problems. This chapter first presents an overview of five common scheduling models and examples of how they are solved. A description of some of the most common chromosome representations and genetic operators is also presented. The chapter also discusses some of the real-world problems that can be modelled using the proposed multi-component scheduling model. In particular, the multi-component engine maintenance scheduling problem is presented and solved using a multi-objective evolutionary algorithm (MOEA) called GENMOP. A variable length chromosome is used by the MOEA in order to address problem specific and generic characteristics. The experimental results compare favorably to baseline values, indicating that GENMOP can effectively solve multi-component scheduling problems. Overall, this chapter introduces a new category of scheduling problems that is quite common in real world problems and presents an example of the problem. By introducing this new category, which can have peculiarities that differ from other scheduling categories, researchers can build upon work done by others in this field.

1 Introduction

Scheduling problems are quite common in the evolutionary algorithm (EA) literature. The focus of much of the literature is on solving a particular type of scheduling problem, such as a flow-shop or job-shop problem. While many problems can be modelled as a "single" type of scheduling problem, there are real-world problems that require a combination of scheduling descriptions in order to model the problem domain. These types of problems are typically NP-Complete. Thus, stochastic methods are typically used. For scheduling problems that have

* The views expressed in this article are those of the authors and do not reflect the official policy of the United States Air Force, Department of Defense, or the United States Government.

M.P. Kleeman and G.B. Lamont: *Scheduling of Flow-Shop, Job-Sho*[illegible] *Scheduling Problems using MOEAs with Fixed and Variable Length C*[illegible]
Intelligence (SCI) **49**, 49–99 (2007)
www.springerlink.com

multiple objectives, a multi-objective EA (MOEA) is suggested. In this chapter, generic scheduling models (flow-shop, flexible flow-shop, job-shop, flexible job-shop, and open-shop) are discussed from which the multi-component scheduling problem is derived.

Section 2 elaborates various scheduling problems and introduces the multi-component scheduling model. Section 3 discusses various chromosome representations and operators that researchers have applied to scheduling problems. Section 4 provides examples of how some researchers have approached these scheduling problems. Section 5 lists several examples of where the multi-component scheduling model would be useful in the real-world. It also explains why MOEAs are useful for solving the multi-component scheduling problem, and a specific algorithm is selected. Additionally, a description of a specific real-world multi-component scheduling problem that is solved is presented. Of particular interest is a variable length chromosome structure and an associated repair function. Section 6 discusses the design of the experiments, and provides an analysis of the experimental results. The experiments are based on the scheduling of five and ten aircraft engines for maintenance. Section 7 presents an analysis of the results and future work.

2 Scheduling Problems

In this section, several of the most common scheduling problems are defined. In particular, five different scheduling models (flow-shop, flexible flow-shop, job-shop, flexible job-shop, and open-shop) are presented [1]. Each of these models are used to find the best schedule based on the particular objective functions the researcher chooses to optimize. Common objectives are finding the schedule with the minimum completion time for the last job (makespan), the total amount of time each job is processed (total flow time), and the sum of the weighted completion times of all the jobs (total weighted flow time) [2]. The evolution of the new multi-component scheduling problem, as derived from these models, is reflective of a more realistic scheduling paradigm. This is because in the real world, many problems are actually a mix of two (or more) models. In this section, a graphical depiction of each model is presented, with a reference to mathematical models. Since different authors describe the same models in a variety of ways, multiple references are provided for the detailed description of the models. Also note that a scheduler typically takes input data, such as a chromosome, or some set of dispatching rules, and determines the order in which jobs are processed on machines. State information may or may not be used in order to determine what jobs to schedule next.

2.1 Flow-Shop

The flow-shop scheduling problem consists of m machines and n jobs. The scheduler's objective is to find an optimal ordering of m machines for the n jobs. All m machines are situated in a defined series. All n jobs have to be processed on each machine. Therefore, each job has to go through m operations. All the jobs must follow the same routing along the series of machines. Once a job is completed on one machine, it is placed into the queue of the next machine in the series. Normally, jobs are removed from the queue on a *first-in, first-out* (FIFO) basis, but this can be modified to fit the needs of the problem, such as higher priority jobs could be bumped to the front of the queue.

An example of a flow-shop problem is an assembly line. A factory may want to produce 1000 identical cars. To do this, a scheduler starts the 1000 jobs at a machine and once the operation is complete, the car is sent to the next station. This process continues until the car has been to every station. Detailed descriptions of this problem can be found in [1, 3]. Figure 1 is an example of the flow-shop example where each job flows in an orderly fashion from one machine to the next. In the example, there are j number of jobs and m number of machines, where J_1 is the first job and M_1 is the first machine.

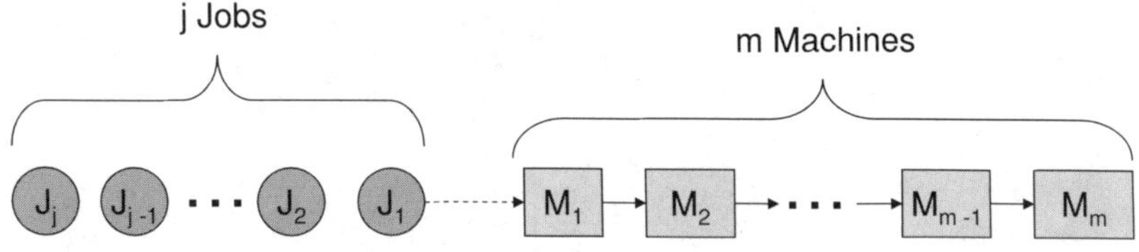

Each job is processed through a series of m machines in a set order

Fig. 1. Diagram of the generalized flow-shop problem.

The search landscape for this problem is generally very smooth, where small permutations typically lead toward better solutions. For small problem sizes, deterministic approaches, such as branch and bound algorithms [4] are used. But for larger instances, stochastic methods, such Tabu search [5], ant colony optimization [6], and EAs [7] are used. Since the landscape for most problems is relatively smooth, algorithms with local search techniques typically have the best results.

2.2 Flexible flow-shop

The flexible flow-shop problem is an extension of the flow-shop problem. This model includes the use of parallel machines in combination with the flow-shop problem. So instead of there being m machines in series, there is a series of m stages, with each stage having one or more machines. The scheduler's objective is to find an optimal ordering through m stages for the n jobs, by taking advantage of the multiple machines in one or more stages. All the jobs still have to be processed by one machine in each stage, but by having multiple machines doing the same job, bottlenecks can be alleviated. Detailed descriptions of this problem can be found in [1, 8]. Figure 2 shows an example of the flexible flow-shop problem where multiple machines can do the operation in order to limit bottlenecks in the process. The machines has a label, M_{x-y} where the x signifies the stage the machine belongs to and the y is the machine number in that stage. So M_{1-h} is the h machine in stage 1. Note that the stages may have a differing number of machines. This is why different variables are used for the y values of the last machine in each stage.

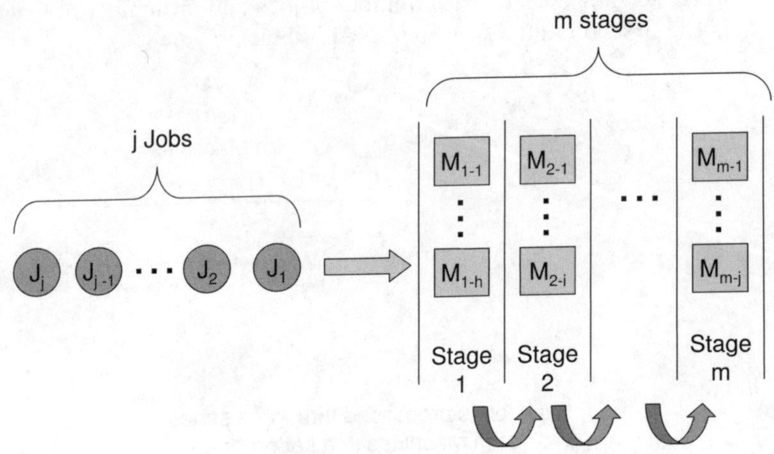

Fig. 2. Diagram of the generalized flexible flow-shop problem.

Flexible flow-shop problems are very similar to flow shop problems. This is due in part to the search landscapes being similar to one another. As such, the algorithms that are effective for solving flow-shop problems, are typically also effective on flexible flow-shop problems.

2.3 Open-shop

The open-shop scheduling problem consists of m machines and n jobs. Each of the n jobs has to be processed on each of the m machines. But this requirement is not steadfast, as the processing time for some jobs can be zero on certain machines. There is no set order for routing the jobs through the machines. The scheduler is determining the order each job is processed by the machines. This allows for different jobs to have different routes. Detailed descriptions of this problem can be found in [1, 9, 10]. Some examples of applications of the open-shop problem can be found in [9, 11–13].

The open-shop problem is decidedly different than the job-shop problem. Some jobs can skip machines and they can be scheduled in any order. Note that the open-shop problem has no constraints, so all tasks are independent of each other. See Figure 3 for an example of the open-shop problem, where each job can be processed through the machines in no particular order.

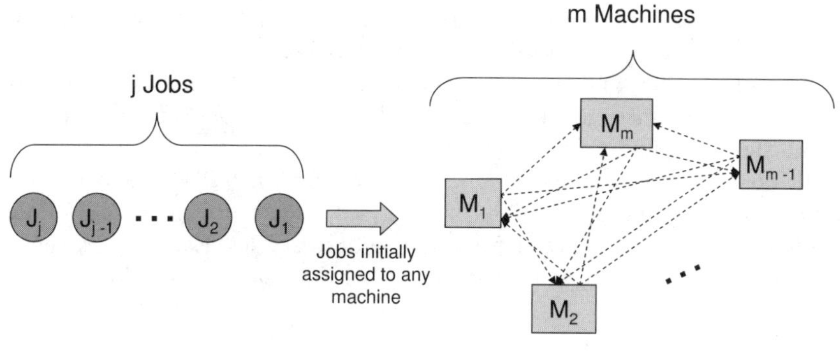

Fig. 3. Diagram of the generalized open-shop problem.

Like the flow-shop problem, exact methods, such as branch and bound algorithms [14], are typically used to solve small instances of this problem. Similarly, Tabu search [15] and hybrid EAs [10] are commonly used to find solutions to larger problems. But unlike the previous two problems, the search landscape for the open shop problem is a little more rugged, with more peaks and valleys. This is largely due to the number of combinations that the machines can be traversed. The problem domain also plays a big part in the smoothness of the search space. If the problem domain creates few conflicts between the jobs, then the search space will typically be smoother than a problem that has many conflicting possibilities.

2.4 Job-shop

For the job-shop problem model, unlike the flow-shop problem, each job has its own route to follow. The scheduler's objective is to find an optimal ordering of all the jobs with respect to their varied routing requirements through the machines.

With job-shop problems, they can be modelled by having jobs visit any machine at most one time, or they can be created to allow for multiple visits to machines [1].

The job-shop is more applicable to real-world problems than the open-shop, since each job follows its own predetermined route. This model takes into consideration any dependencies a task may have with respect to another task. An example of this would be an assembly line that processes multiple products at the same time. Not all products may need to be processed by the same machines, therefore their routes would be different based on the needs of the job. This problem has received considerable attention in the literature [1]. Detailed descriptions of this problem can be found in [1, 3]. Figure 4 shows an example of the job-shop problem, where each job follows its own path through the various machines. The machines in the example are labelled as $M_{x,y}$ where the x represents the job number and y represents the location of the machine with respect to the other machines. So machine $M_{1,3}$ be used for job 1 before $M_{1,4}$, since it is the third machine in the route and $M_{1,4}$ is the fourth machine. Note that not all jobs require the same number of machines. In order to show that each route may have a different number of machines, each route ends with a different variable for y. Also note that there are m total machines. Each row in the figure represents the ordering of a job with respect to the same m machines.

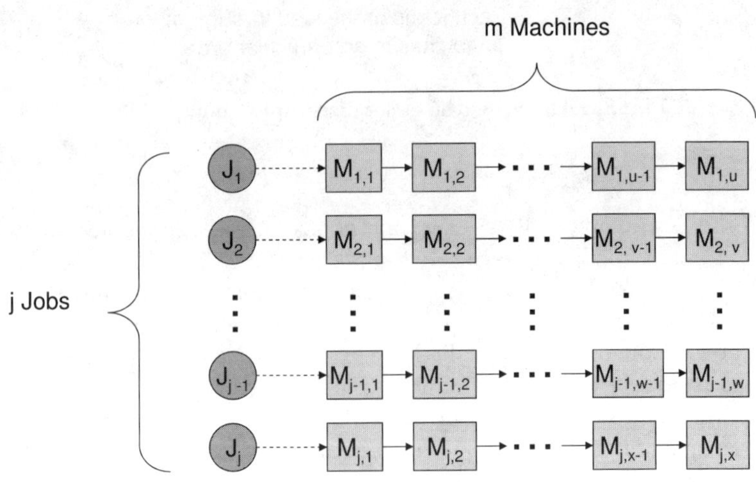

Fig. 4. Diagram of the generalized job-shop problem.

The search landscape for the job shop problem is typically more rugged than the flow shop problem due to the number of machine combinations, but it is usually not as rough as the open shop problem. Deterministic methods [16] are typically used for solving small instances of the flow-shop problem. Larger instances use stochastic methods such as EAs [17]. A good taxonomy of how job-shop problems are represented in EAs is presented in [18].

2.5 Flexible job-shop

The flexible job-shop problem is an extension of the job-shop problem. This model, like the flexible flow-shop, uses parallel machines in combination with the job-shop problem, which has a total of m possible stages (or workcenters [19]). Each stage consists of a set of $m_{i,j} \subseteq m$ machines of which one machine is chosen to perform the task (operation) [1]. The scheduler's objective is to find an optimal ordering of the machines for the jobs, given the fact that some machines are parallel with others and some jobs have different routes to follow. Kacem *et al.* [20], present a mathematical formulation of this model. This model is particularly useful when it is employed to overcome bottlenecks by adding machines in parallel where slow downs occur in the process. Figure 5 shows a diagram of the flexible job-shop where stages of parallel machines are used in an effort to overcome bottlenecks. The machines are labelled as M_{x,y_1-y_2} where x represents the job number, y_1 represents the stage number, and y_2 represents the number of machines in a particular stage. So machine $M_{4,3-5}$ represents the fifth machine in third stage of the route for job 4. Note that not only can each stage have a varying number of machines, but each route can have a varying number of machines. And since each job can follow a different route than the other jobs, stage 1 of J_1 may have a different number of machines that stage 1 of job J_4. This is why Figure 5 has so many variables.

Flexible job-shop problems have many of the same characteristics as the job-shop problem. The search landscape is similar, but may be a little smoother since parallel machines may help prevent bottlenecks. As such, the algorithms that are best for the job-shop problem usually work well for the flexible job-shop problem.

2.6 Multi-component scheduling

This new model is a hybrid of the approaches described in the previous subsections. This problem model embeds one scheduling paradigm into another. This combined approach is called the multi-component scheduling problem. Note that the literature does not reflect any research that deals with this scheduling model. Yet, the real-world has several areas where this type of problem model would be more perfect fit over the other problem models. Section 5 lists some real-world examples where the multi-component scheduling problem would be useful to employ.

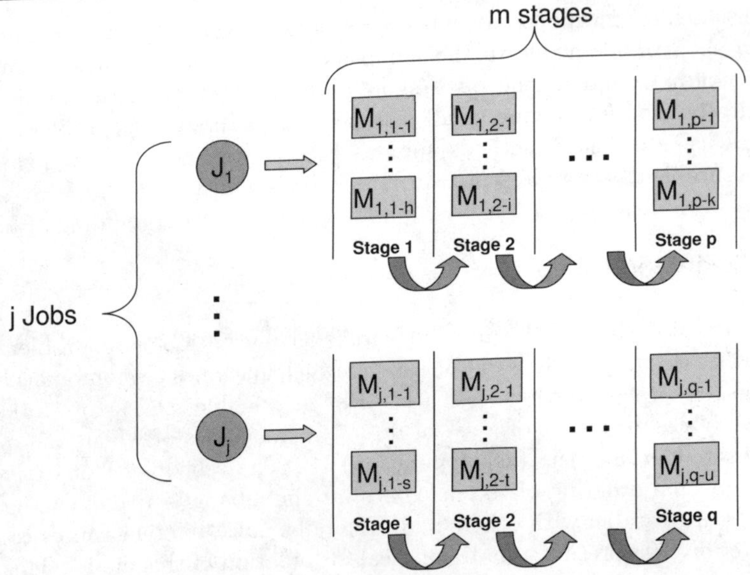

Fig. 5. Diagram of the generalized flexible job-shop problem.

Our definition of a multi-component scheduling problem is any scheduling problem that consists of jobs which are being scheduled based on the requirements of smaller subcomponents. While each job can be modelled as a flow-shop or a flexible flow-shop problem, the underlying components of each job do not follow the same order. Each subcomponent can be modelled as either a job-shop, flexible job-shop, or possibly an open-shop problem. But after the subcomponent is returned to the job for scheduling, it follows the flow-shop pattern of the assigned job.

This new model can be viewed as a type of meta scheduling problem where the main component follows a flow-shop model, while its underlying components follow a job-shop model. Figure 6 shows a diagram of a multi-component scheduling problem. In this diagram, a flow-shop problem is combined with a job-shop problem. The *machines* in the flow-shop portion of the problem are referred to as stages, where each job must flow through each stage in order. The machines in the flow-shop portion of the problem are labelled as FM_i, where FM signifies that the machine is in the flow-shop portion of the problem and the i is the machine order. So a value of FM_4 signifies the fourth machine (stage) in the flow-shop. Some stages of the flow-shop may not have a job-shop associated with them, while others may have a large job-shop that needs to be completed before the job is ready for the next stage of the flow-shop.

The job-shop machines are labelled as $JM_{x,y}$ where the JM signifies the machine is part of the job-shop portion of the problem, x represents the job number and y represents the machine number. So $JM_{3,5}$ is the fifth machine of the third job in the job-shop problem. A particular machine may be specified as $FM_i - JM_{x,y}$. So $FM_4 - JM_{2,3}$ signifies the third machine for the second job in the job-shop that is embedded in the fourth stage of the flow-shop. Reviewing the real-world examples in Section 5 may help in the understanding of how this problem flows.

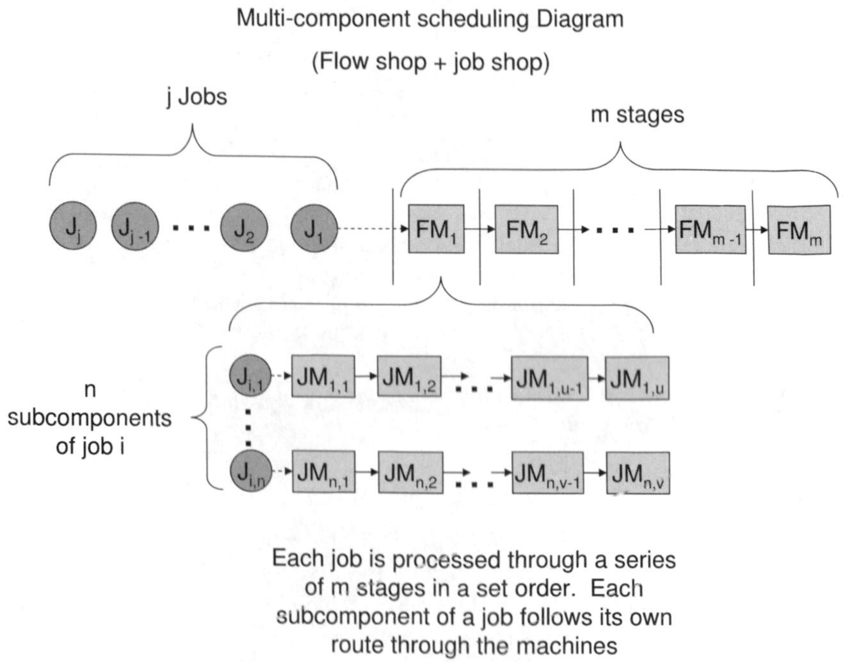

Fig. 6. Diagram of the generalized multi-component scheduling problem with a flow-shop and a job-shop combined.

Figure 7 shows another variation. In this variation, a main flow-shop model is combined with an open-shop problem. The example uses the same labelling for the flow-shop machines, and uses OM to reference machines in the open-shop portion of the problem. In both instances of the multi-component scheduling problem, the flow-shop problem is the top level scheduling problem. This is typical of real world scenarios, where a job has a definitive repair process, but the subcomponent repair routes can vary with each job.

There are many examples of this type of scheduling problem in the real world. They are mostly associated with repair scheduling. For example, a television repair shop has multiple television jobs, each job follows the same basic flow,

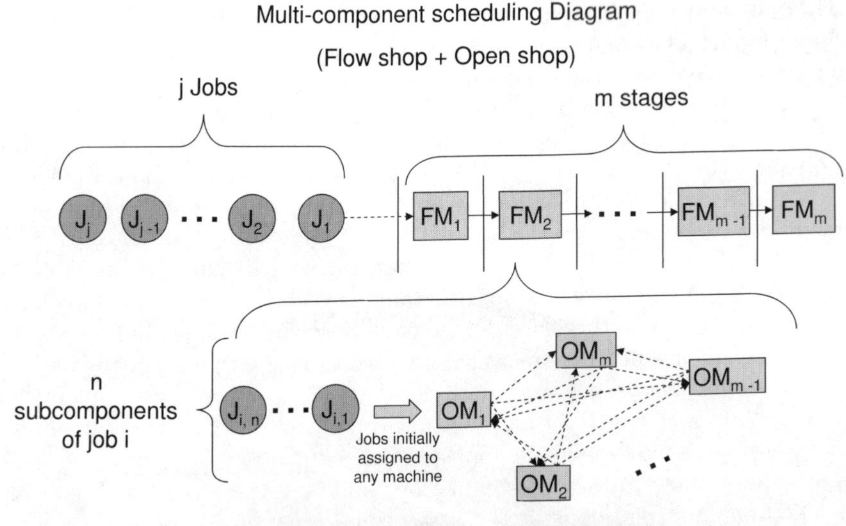

Fig. 7. Diagram of the generalized multi-component scheduling problem with a flow-shop and an open-shop combined.

troubleshoot, repair or replace, and test. But the repair and replace stage can have various subcomponents being tested/repaired on different machines and in different orders. This layer of the problem can be modelled as a job-shop or an open-shop, depending on the independence that each subcomponent has with respect to the machines. For example, a subcomponent that has to follow a certain machine order for repair uses a job-shop. While a subcomponent that has no particular machine order uses an open-shop.

The multi-component scheduling problem can be used to model numerous other real world problems, such as automobile repair, electronic system repair (compact disk player, computer, etc.), or the aircraft engine maintenance example presented in this chapter. In general, this model can represent any multi-component system that requires the processing of multiple subcomponents on different machines, but the overall system itself follows a predefined flow.

3 Scheduling Problems and Evolutionary Algorithms

With the exception of the newly introduced multi-component scheduling problem, many researchers use EAs and MOEAs to solve the afore mentioned scheduling problems. The key elements to EAs and MOEAs are the type of chromosome representation that is used, the type of crossover and mutation operators used, the selection method, and the use of elitism in the algorithm. This section briefly discusses some of the chromosome representations that researchers use and also provides an overview of the mutation and crossover operators employed.

3.1 Chromosome Representations

Chromosome representation can be extremely important when trying to find solutions to a problem. The data structures, such as the chromosome, plus the algorithm combine to make efficient programs. A bad chromosome representation can increase the size of the search space or slow down the algorithm if too many repair operators are needed to ensure the chromosome is valid. For the majority of scheduling problems, a fixed length chromosome is appropriate. But in some instances a variable length chromosome may be the best fit. For example, suppose sending a job through a machine multiple times creates a better product, but at the same time, slows down the completion time. A variable length chromosome can allow for multiple loops in the system and create a more diverse solution set.

Cheng *et al.* [18] introduced a taxonomy of how EAs represent job-shop problems. While his list focused on only job-shop problems, many of these representations have been used in other scheduling problems as well. These representations can be classified as either directly encoded approaches or indirectly coded approaches. With a direct approach, a schedule is encoded into the chromosome. The EA then operates on these schedules in an effort to find the best schedule. For direct approaches, there are five different ways the EA can be encoded [18]:

- Operation-based
- Job-based
- Job pair relation-based
- Completion time-based
- Random keys

For operation-based representations, the chromosome encoding is a schedule based on a sequence of operations. For example, Figure 8 shows how a three-job, three-operation job-shop chromosome may be encoded. The encoding is based on the job and the operations are ordered. So the second 2 in the chromosome is interpreted as the second operation of job J_2. The operations are scheduled in the order they are listed in the chromosome. The chromosome length is $n \times m$, where n is the number of jobs and m is the number of machines. All permutations of the chromosome yield a valid schedule.

Job-based representations merely encode the chromosome according to the job sequence. Figure 9 shows an encoding example for a three-job, three-operation job-shop chromosome. In this example all operations of the third job are scheduled first while the first job's operations are scheduled second, followed by the second job.

The job pair relation-based representation uses a binary matrix to encode a schedule [3, 18]. The values of the matrix are determined as follows:

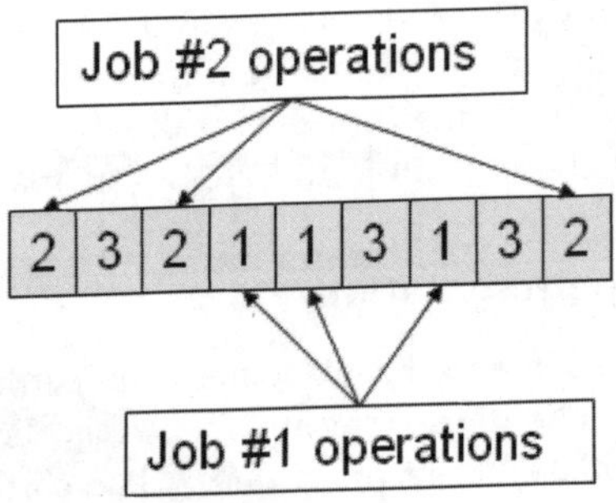

Fig. 8. Example of an operation-based chromosome representation for a three-job, three-operation job-shop.

3	1	2

Fig. 9. Example of a job-based chromosome representation for a three-job, three-operation job-shop.

$$x_{ijm} = \begin{cases} 1: & \text{if job } i \text{ is processed} \\ & \text{before job } j \text{ on machine } m \\ 0: & \text{otherwise} \end{cases}$$

A matrix for a 3-job, 3-machine problem can be designed as follows [18]:

$$\begin{pmatrix} x_{121} \; x_{122} \; x_{123} \\ x_{131} \; x_{132} \; x_{133} \\ x_{231} \; x_{233} \; x_{232} \end{pmatrix} = \begin{pmatrix} 0\,1\,0 \\ 1\,0\,1 \\ 1\,1\,0 \end{pmatrix}$$

where the sequence of the variable x_{ijm} in the matrix is kept consistent with the sequence of operations of the first job of the job pair. This is why the last line of the matrix, representing job pair (j_2, j_3) is different. It is assumed in this example that the sequence of operations in the second job is different than the first job.

This representation can be codified into a chromosome as shown in Figure 10, where each chromosome is a job pair and the sub-chromosome is the processing order of the two jobs on each machine. This representation is complex and redundant. Additionally, many chromosomes produced via population generation or through the genetic operators, are illegal [18]. As such, a repair function or a penalty function need to be implemented.

The completion time-based representation encodes the chromosome with an ordered list of completion times of the operations [21]. These times can be obtained through the generation of an active schedule by using the Giffler &

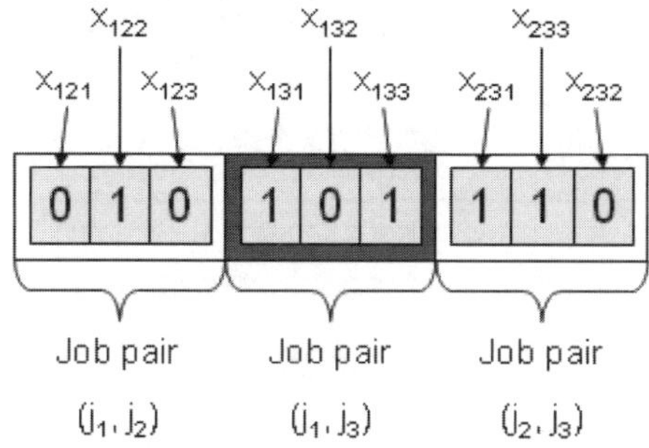

Fig. 10. Example of a job pair relation-based chromosome representation for a three-job, three-operation job-shop.

Thompson algorithm [22]. Figure 11 shows an example of a chromosome representation of a three-job, 3-machine job-shop problem [18]. Note that each chromosome is represented by a a completion time, c_{jom}, where j is the job number, o is the operation of job j, and m is the machine that accomplishes the operation. In [21], each individual represents an active schedule using the completion times, c_{jom}, as elements.

Fig. 11. Example of a completion time-based chromosome representation for a three-job, three-operation job-shop.

The random key chromosome representation [23] encodes each allele with a random number. The values are then sorted in order to determine determine the actual operation sequence. Typically, a random key is composed of two parts. The first part is an integer depicting which machine is assigned for that job. The second part is a random number between (0, 1). The smallest number for a machine is the first one scheduled. Figure 12 depicts a three-job, four machine example of the representation. The top drawing in the figure represents the encoded chromosome while the bottom part of the figure represents the order of precedence.

Indirect approaches are chromosome representations that do not directly encode the schedule into the chromosome. There are four indirect approaches used in EAs [18]:

- Preference list-based
- Priority rule-based

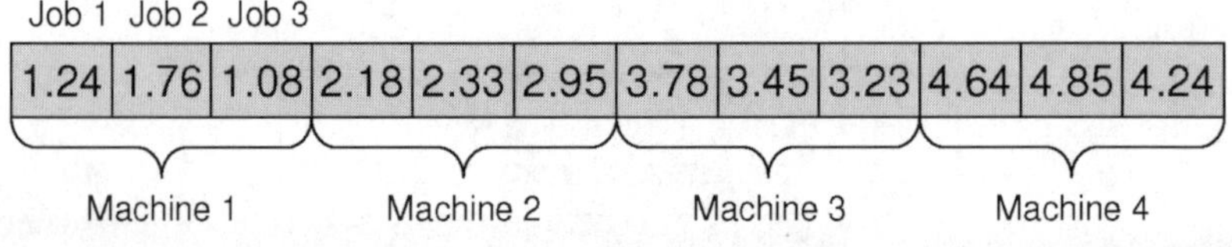

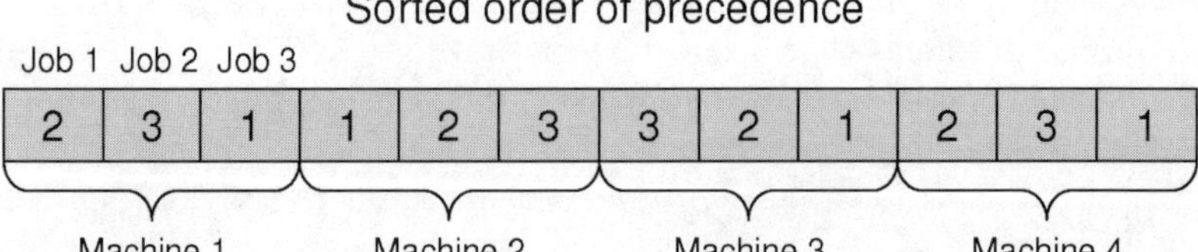

Fig. 12. Example of a random key chromosome representation for a three-job, four-operation job-shop.

- Disjunctive graph-based
- Machine-based

The preference list-based representation consists of a chromosome of size m where m is the number of machines. Each allele of the chromosome is a sub-chromosome, which lists the preference that each machine has for certain jobs. Figure 13 shows an example of a three-job, four-machine job-shop encoding. The chromosome has four allele values, one for each machine (operation). Each allele contains a sub-chromosome that shows the precedence order of the jobs in the machine. Note that this can be easily represented by a matrix.

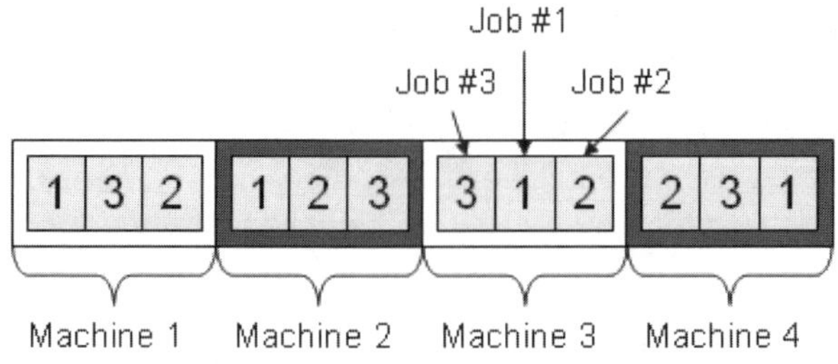

Fig. 13. Example of a preference list-based chromosome representation for a three-job, four-operation job-shop.

The priority rule-based representation encodes a chromosome as a sequence of dispatching rules. The job schedule is created using a heuristic based on the dispatching rules sequence. Priority dispatching rules, proposed by Giffler and Thompson [22], are a commonly used heuristic because of their ease of use [18]. The EA searches the best rules to deconflict scheduling problems when using the Giffler and Thompson algorithm. As mentioned earlier, the chromosome is made up of dispatching rules. The chromosome's length is $n \times m$ where n is the number of jobs and m is the number of machines. The ith position of each allele in the chromosome corresponds to the rule in the ith iteration of the Giffler and Thompson algorithm. So the rule in that position is used to deconflict any scheduling problems that occur during ith iteration. The rules are typically represented via a table. See [18, 24] for more information.

The disjunctive graph-based representation encodes the chromosome as a binary string that corresponds to an order list of disjunctive arcs connecting the different operations to be processed by the same machine [3]. The disjunctive graph can be defined as $G = (N,\ A,\ E)$, where N is the set of nodes which represent each operation, A is the set of arcs that connect the sequence of operations of the same job, and E is the set of disjunctive arcs, which connect the operations that are to be processed by the same machine. Figure 14 shows an example of a disjunctive graph for a three-job, three-machine problem. Note that the set of nodes are the circles, the set of arcs are the solid lines, and the disjunctive arcs are the dashed lines. In this example, operations 1, 5, and 9 are all processed on the same machine.

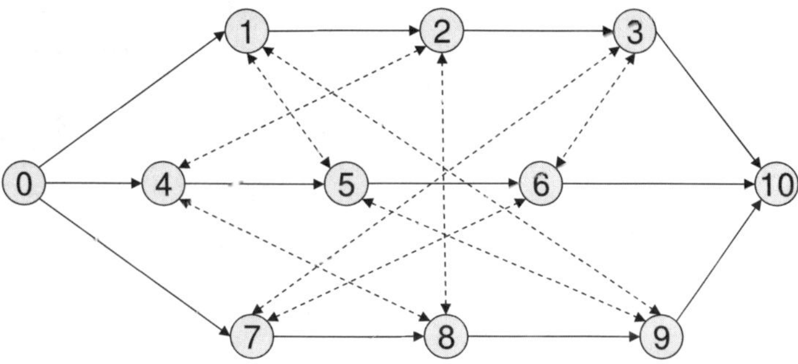

Fig. 14. Example of a disjunctive graph for a three-job, three-operation job-shop.

The chromosome is a binary string that determines which operation has precedence on a machine based on the disjunctive arcs. Each disjunctive arc is an allele in the chromosome. A disjunctive arc is represented as e_{ij} where i and j are the nodes (operations) connected with a disjunctive arc. The allele that represents e_{ij} contains a 1 if operation i goes before operation j. The allele contains a 0 if operation j goes before operation i. Figure 15 shows an example of a chromosome. Note that the chromosome does not represent a schedule, but is only used as a decision preference. A critical path based procedure is typically used to create the schedule. The chromosome is only used to deconflict machine precedence when there is a possibility of two operations occurring on the same machine [18],

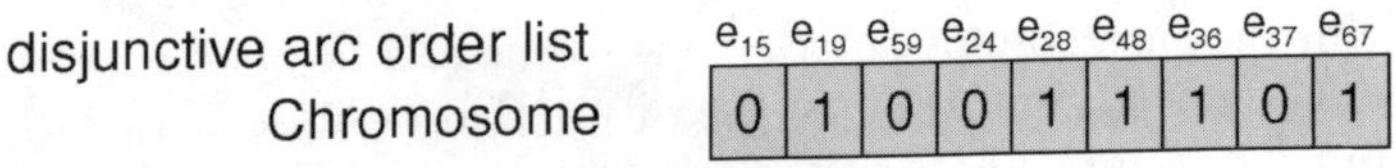

disjunctive arc order list	e_{15}	e_{19}	e_{59}	e_{24}	e_{28}	e_{48}	e_{36}	e_{37}	e_{67}
Chromosome	0	1	0	0	1	1	1	0	1

Fig. 15. Example of a disjunctive graph-based chromosome representation for a three-job, three-operation job-shop.

The machine-based representation encodes the chromosome as a sequence of machines. A shifting bottleneck heuristic [25] constructs a schedule based on the machine order. The shifting bottleneck heuristic initially takes the sequenced set of machines and identifies a bottleneck machine and sequences it optimally based on the time it takes to process all operations. The next step is to reoptimize the sequence of each critical machine in order, while keeping the rest of the sequences fixed. If all the machines have been sequenced, then the heuristic is done, if not, the heuristic goes back to the initial step and identifies the next bottleneck machine. See [18, 25] for more details on this heuristic. Figure 16 shows an example of a machine-based three-job, three-machine chromosome encoding where each allele represents a machine. Note that this encoding is similar to the job-based chromosome shown in Figure 9. The difference between the two is what each allele represents. In the job-based chromosome, the numbers represent the priority of the jobs. The numbers in the machine-based chromosome represent the sequence of the machines.

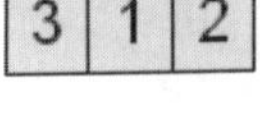

3	1	2

Fig. 16. Example of a machine-based chromosome representation for a three-job, three-operation job-shop.

While the chromosome representation is important, the operators used in the algorithm are equally important. The next section discusses some of the common crossover and mutation operators used for scheduling problems.

3.2 Common Operators

Several researchers have developed EA operators for use with scheduling problems. These operators are based upon the representations previously discussed. Cheng *et al.* [26] provide a good summary of the operators used in job-shop problems. The use of these operators is not limited to the job-shop alone, since they have been applied to other types of scheduling problems as well. A discussion of these operators is presented in this section.

Crossover operators The following is a list of crossover operators that have been used in scheduling problems:

- Partial-mapped crossover (PMX)
- Order crossover (OX)
- Position-based crossover
- Order-based crossover
- Cycle crossover (CX)
- Linear order crossover (LOX)
- Subsequence exchange crossover(SXX)
- Job-based order crossover (JOX)
- Partial schedule exchange crossover

Partially-mapped crossover was developed to tackle a blind travelling salesman problem [27]. This problem is similar to the travelling salesman problem (TSP) with the added constraint that the distance the salesman travels is not known until a tour is completed. Since both problems operate on permutations, a typical one-point crossover method can be highly destructive. The PMX method aligns two strings based on their location in the chromosome. Then two crossing sites are picked at random over the two strings. The two points are used to define a matching section [27]. In that matching section, the numbers that match represent which numbers are swapped in each parent chromosome. Figure 17 shows an example of the PMX operator. In this example, the matching section of parents A and B link the following values for exchange: $5 \leftrightarrow 3$, $4 \leftrightarrow 8$, $6 \leftrightarrow 9$, and $2 \leftrightarrow 2$. The parents swap each of these numbers in the chromosome to create the two children. Note that no repair operator is needed with this type of crossover, whereas a more common crossover operator may invalidate a permutation based chromosome and require repair.

Order crossover [28] is similar to the PMX operator. The OX operator selects two parents and then randomly picks a set of positions. The positions can be a consecutive group of alleles, as suggested in [27], or they can be picked randomly throughout the chromosome as in [28]. In fact there are multiple variations of the OX operator [29]. Figure 18 shows an example of the OX operator similar to the PMX operator. In the example, two random cutpoints are chosen. The alleles in between the cutpoints are swapped between the parents. Then, starting at the last cutpoint, the allele values are filled in a manner to preserve the relative ordering of the parent. This is done by filling in all the unused allele values from

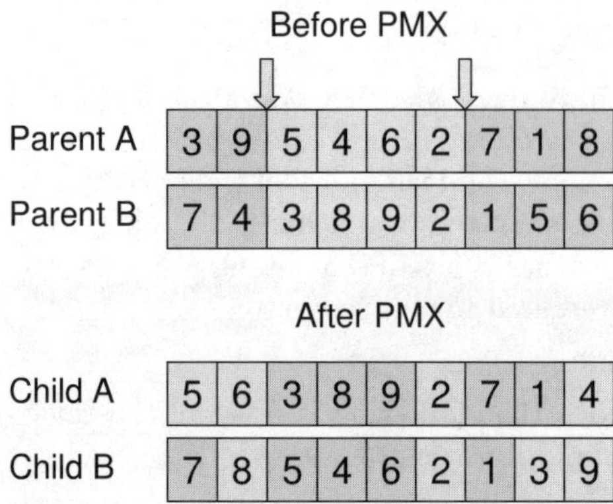

Fig. 17. Example of the partially mapped crossover operator.

the parent based on their previous order, starting at the cutpoint. This considers the sequence as a ring, and as such can cause more disruption than desired [29]. So in our example, the first child is created by starting after the swap values, 3-8-9-2. The first number in the parent after the cutpoint is a 7. Since this number is not one of the swapped values, it is inserted into the chromosome. The same goes for 1. But the next value in the parent is 8. This value is one of the values that was swapped, and since it cannot be duplicated, the algorithm cycles back to the first value, 3, which is also one of the swapped values. It continues until it finds a value in the parent that hasn't been used, in this example, that value is 5. After 5 is placed the chromosome still needs the first two positions filled, so the algorithm pulls the number after 5 in Parent A and compares it to the swapped values. Since this value, 4, hasn't been used, it is placed in the chromosome. The last value, 6, is placed in a similar manner. Child B is created the same way.

To prevent some of this disruption, Davis [28] introduced a variant. It proceeds the same as the original version, but instead of maintaining the relative ordering from the second cutpoint, the start of the chromosome is used. Figure 19 shows an example this variant of the OX operator. According to Cotta *et al.* [29] this variant is better than the one shown in Figure 18.

Position-based crossover is a variation of the uniform crossover operator for permutations [26]. For this operator, two parents are chosen and a random mask is generated to determine which alleles to swap between them. After the swap occurs, the remaining allele values fill the rest of the positions in a manner that maintains their relative order. Figure 20 shows an example of this crossover method. In the example, Child 1 starts with only the swap values in the chromosome. Starting from the first location in the chromosome, the child pulls the first values from Parent 1 and verifies it is not one of the swap values. Since 3 is not, it is placed in the first location. The child then goes to the next open location in the chromosome, which is the third spot in our example. The algorithm then pulls the 9 from Parent A and since it is not one of the swapped values,

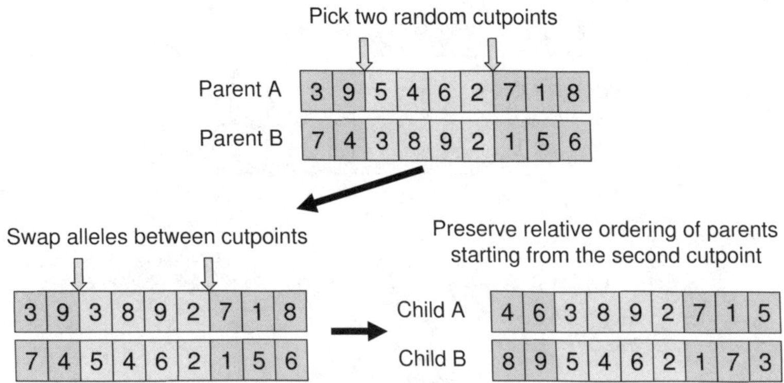

Fig. 18. Example of the order crossover operator.

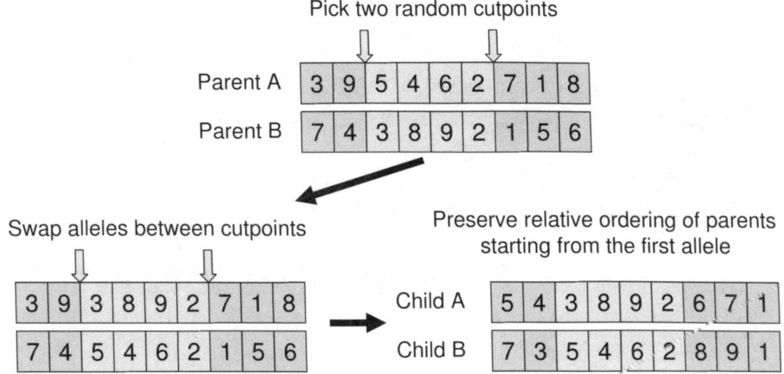

Fig. 19. Example of a variant of the order crossover operator. The relative ordering of the parent is referenced to the start of the chromosome.

it is placed in the chromosome. If the child encounters a value that has already been placed, it proceeds to the next value in the parent. The algorithm does this until all the chromosome values of the child have been filled. This operator has been found to be similar to a variant of order crossover.

Order-based crossover [28] is a variation of the position based crossover. The method uses a binary template where a zero represents swaps for one parent and a one represents swaps for the other parent. After the swaps are accomplished, the the missing values fill the empty child chromosome positions in the order they were found in the parent. Figure 21 shows an example of this operator. In this example, a 0 indicates a swap for Child B and a 1 indicates a swap for Child A. This means that after the swap, Child A contains only the values from Parent B that were indicated by a 1 from the crossover template. Likewise Child B has only values from Parent A that were swapped using the crossover

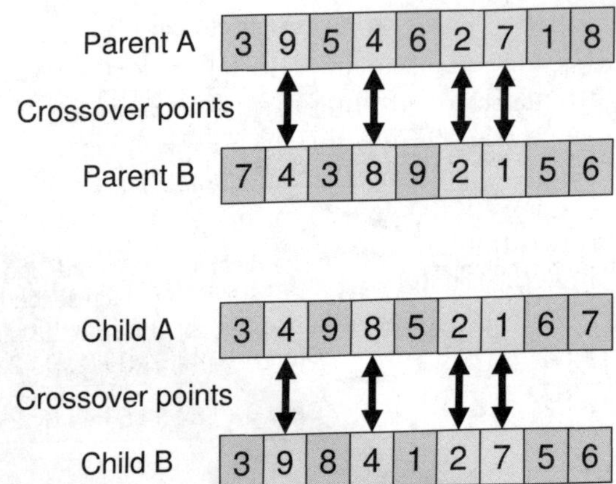

Fig. 20. Example of the position-based crossover operator.

template. The remaining chromosome locations of the children are then filled in the same manner used by the previous two crossover methods.

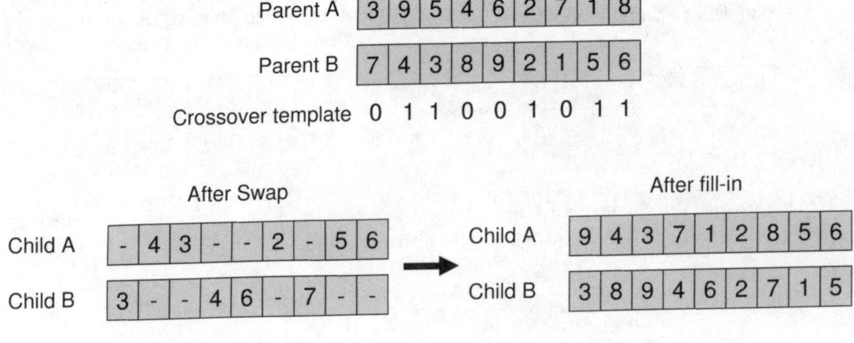

Fig. 21. Example of the order-based crossover operator.

Cycle crossover (CX) [27] creates the children in cyclic fashion. The operator starts by placing the first allele value of the first parent into the first child. The first allele value in the second chromosome determines the next value to pull from the first parent and put into the child. In the example shown in Figure 22, 7 is the value found in Parent B. This value is placed into the child in the same position as it was in the parent. In the example, the cycle continues with the values 1 and 5 being placed into the child. A cycle ends when the allele value in the second parent has a value already placed in the child. One a cycle ends, a new cycle begins with the second parent copying the first unused allele value into the corresponding position of the child. So for our example, after 5 is placed in the chromosome, the value 3 is to be placed. But since it was already placed, the cycle ends and first unused value from Parent B is placed in the chromosome. In the example, this value is 4. During this cycle, the first parent's corresponding allele values determine which value to copy next. So after 4 is placed in the example, the next value is 9, followed by 6 then 8. The operator continues until the child chromosome is fully instantiated. The second child is done in a similar manner, but this time, the second parent supplies the first allele value. Note that the numbers near the dashed lines indicate the order in which the children are filled.

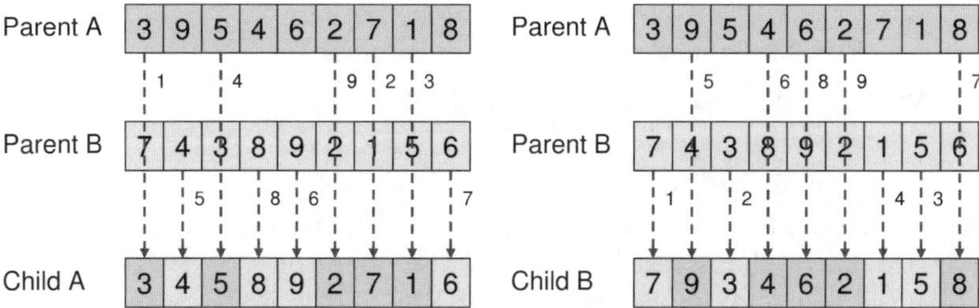

Fig. 22. Example of the cycle crossover operator. The dashed lines show which parent fills the child and the number next to the dashed line indicates the order the child is filled.

Linear order crossover (LOX) [30] is a modified version of the order crossover operator. Recall that the order crossover operator treats the chromosome as a circular string, in which it wraps around from the end of the chromosome back to the beginning. This circular assumption may not play a big role in the TSP, but for job shop problems, it can have a larger impact [26]. As such, the LOX operator treats the chromosome as a linear entity. For this operator, the swap occurs in the same fashion as it occurs in the OX operator, but when sliding the parent values around to fit in the remaining open slots of the child chromosome, they are allowed to slide to the left or right. This allows the chromosome to maintain its relative ordering and at the same time preserve the beginning and ending values. Figure 23 shows an example of this operator. In the example, after the values are swapped, there are two open spaces in the front of the chromosome and three open spaces at the end. The algorithm then goes through Parent A and finds the first two values that were not part of the swap, in this example they are 5 and 4. These values are shifted left to fill the first two chromosome locations. The final three locations are filled in a similar manner.

Note that this operator produces the same results as those produced with the second variant of OX, and shown in Figure 19. This is because the second variant of OX starts filling in the chromosome from the beginning rather than from

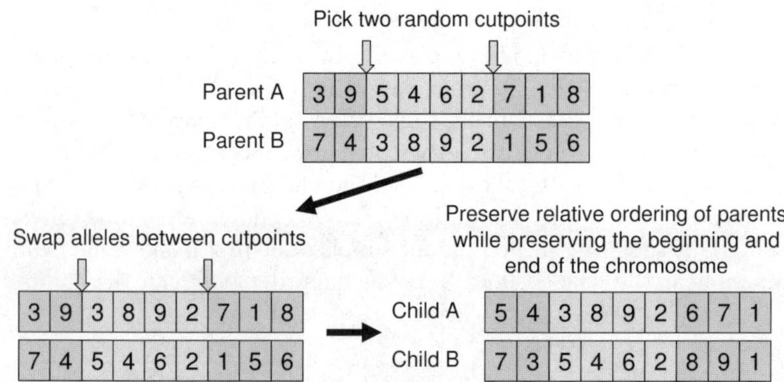

Fig. 23. Example of the linear order crossover operator.

the second cutpoint. By filling in the values from the beginning, the operator preserves the linear characteristics of the LOX operator.

The subsequence exchange crossover (SXX) [31] operator was designed for job shop problems that use the job sequence matrix as a way to represent their solutions. The job sequence matrix is an $n \times m$ matrix where n is the number of jobs and m is the number of machines. Each row of the matrix specifies the job sequence for each machine. Subsequences are defined as a set of jobs that both parents process in a consecutive manner, but can be in a different order. When subsequences are found, the children are created by swapping the subsequences of the two parents. Figure 24 shows an example of the operator with a 4-job, 4-machine matrix representation. Note how the largest possible subsequences are chosen in each row. So the first three rows can create subsequences that contain three of the four values, while the last row has no subsequences that can be exchanged. According to [31], the computational complexity of this operator is $O(mn^2)$, where m is the number of machines, and n is the number of jobs.

Job-based order crossover (JOX) [32] is a variation of the SXX operator. The operator starts by randomly picking which jobs should preserve their locus in the chromosome. These jobs are copied from the parent directly into the same position in the child. The remaining chromosome positions of the first child are filled according to their order in the second parent. Figure 25 shows an example of the JOX operator with a 4-job, 4-machine matrix representation. In this example, values from Parent A are passed on to Child A. So the first row keeps the location of values 3 and 4 from Parent A, the second row keeps values 4 and 3 and so on. The values not kept in each row are filled in based on their order in Parent B. Since 2 comes before 1 in the first row of Parent B, the 2 fills the first open location in Child A, with 1 filling the remaining slot. This continues until all the rows have been filled. Child B is filled in a similar manner.

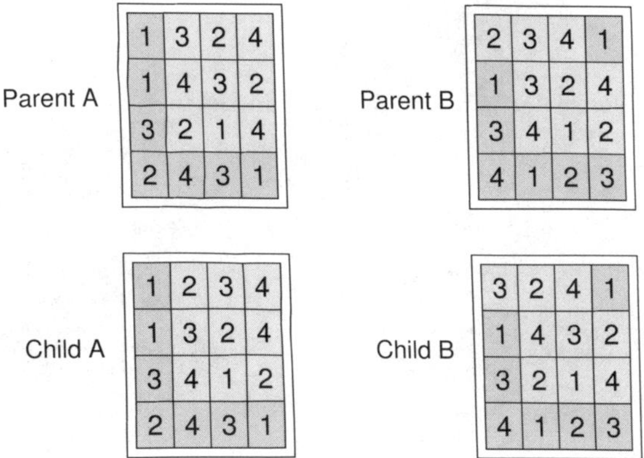

Fig. 24. Example of the subsequence exchange crossover operator for a 4-job, 4-machine example.

The partial schedule exchange crossover [33] operator has been applied to chromosomes with the operation-based encoding. Figure 8 shows an example of an operation-based encoded chromosome. Figure 26 shows an example of this operator for a 3-job, 3-machine problem. The operator starts by randomly selecting a partial schedule in the parents. In the example, the partial schedule for Parent A is 1-3-1 and the partial schedule for Parent B is 1-1. The partial schedules must have the same job in the first and last position. The partial schedules of the two parents typically differ in length. These partial schedules are then swapped to create two children. These children may have lengths that are too big or too small, and they may also have too many operations for one job and too few for another. So a repair procedure is used to fix the children and attempt to keep the operation order as similar to the parents as possible. In our example, after the swap Child A needs an operation for job 3 and Child B needs to delete an operation from job 3. The repair mechanism employed can vary, but in the example the missing value is inserted into the chromosome in a location that is nearest to where the value was located in the parent. In Child A, the 3 was placed just before the swap so it was only one location off its location in the parent. Likewise, the value that is deleted is the one that maintains most of the original ordering of the parent. In the example, the second 3 is removed from Child B. This type of repair operator is less destructive than ones that don't take into account the relative ordering of the parent.

Crossover operators are used as a means to combine good building blocks from the parents. The crossover operators discussed in this section, typically try to limit the amount of repair that is needed for the permutation. The next set of operators, the mutation operators, are used to add more variability to the search.

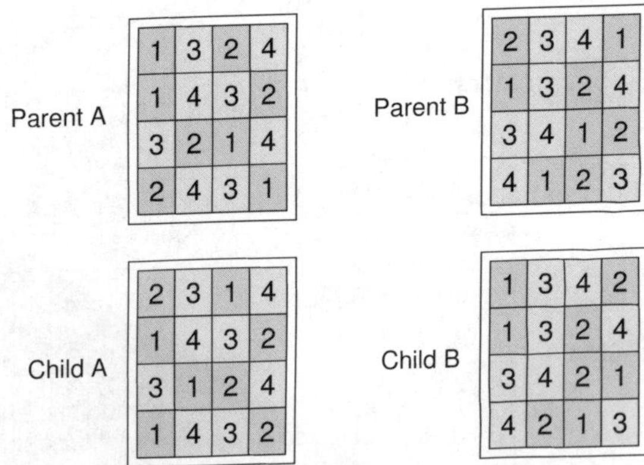

Fig. 25. Example of the job-based order crossover operator for a 4-job, 4-machine example.

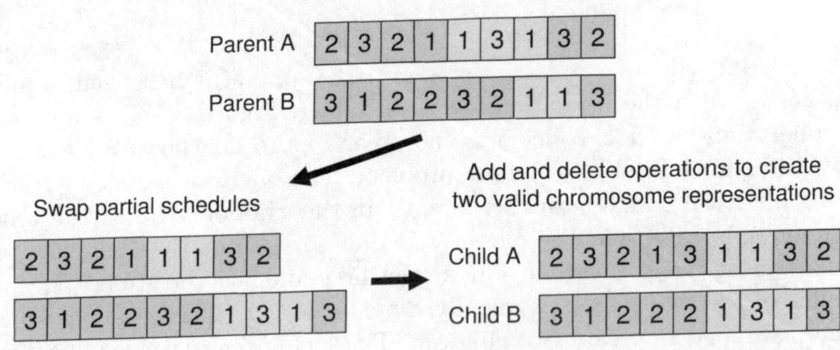

Fig. 26. Example of the partial schedule exchange crossover operator for a 3-job, 3-machine example.

Mutation Operators The following is a list of mutation operators used for permutation representations:

- Inversion mutation
- Insertion mutation
- Displacement mutation
- Reciprocal exchange mutation (swap mutation)
- Shift mutation

The inversion mutation operator [27] randomly selects two points in the chromosome. The allele values that are in between these two points are then inverted to form a new chromosome. Figure 27 shows an example of this mutation.

The insertion mutation operator selects a gene at random and then inserts it at a random position. Figure 28 shows an example of this type of mutation.

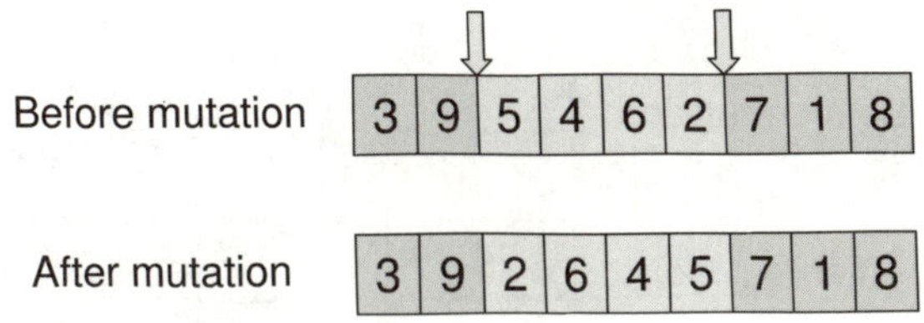

Fig. 27. Example of inversion mutation for a 9-job, job-based chromosome.

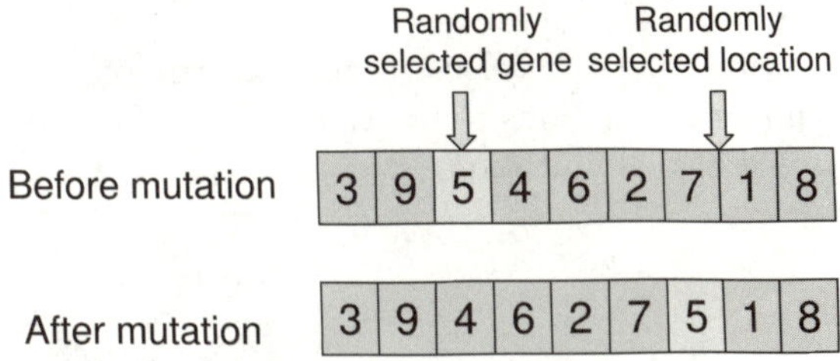

Fig. 28. Example of insertion mutation for a 9-job, job-based chromosome.

The displacement mutation operator randomly selects two points in a chromosome and then moves the substring to another randomly selected location in the chromosome. This mutation operator is similar to the insertion operator, but instead of moving only one allele, a substring is moved. Figure 29 shows an example of the displacement mutation operator.

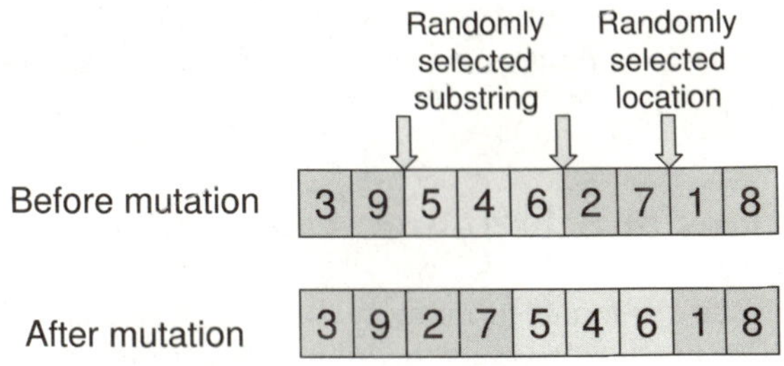

Fig. 29. Example of displacement mutation for a 9-job, job-based chromosome.

The reciprocal exchange mutation operator (also known as swap mutation) selects two alleles at random and then swaps them. Figure 30 shows an example of this type of mutation.

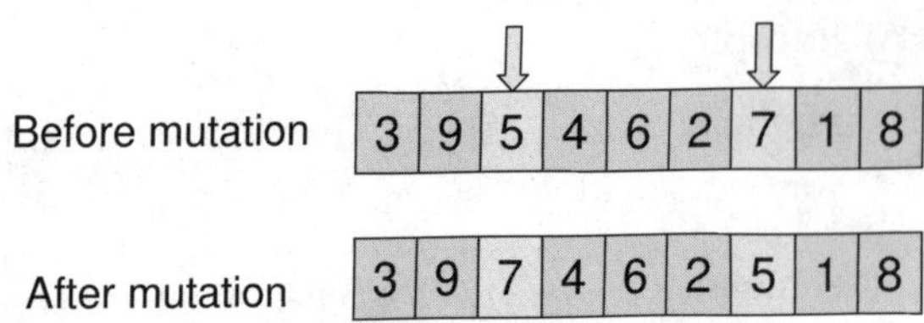

Fig. 30. Example of reciprocal mutation for a 9-job, job-based chromosome.

The shift mutation operator selects a gene at random and then shifts the gene a random number of positions to the right or left of its present spot. Figure 31 shows an example of this mutation.

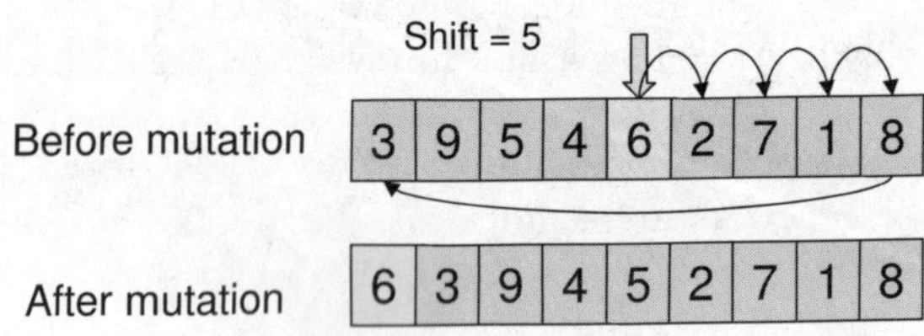

Fig. 31. Example of shift mutation for a 9-job, job-based chromosome.

Note that no repair is needed with these mutations when they are applied to a job-based encoded chromosome. But other chromosome encodings may require a repair operator when a mutation creates an invalid chromosome. The next section provides a few examples for each type of scheduling algorithm that has been discussed thus far.

4 Applications of Scheduling Algorithms

Scheduling problems are found in many real world problems. This is why many researchers work to produce better scheduling algorithms. In this section, some of the problems presented in literature are discussed. Specifically, the problems described in Section 2 are presented. While different methods of solving these problems are addressed, the primary focus is on EA and MOEA techniques. These examples give a small sample of attempted methods applied and the results of these methods.

4.1 Flow-shop Examples

Researchers use various methods to attack the flow-shop problem. Some researchers use ant colony optimization methods [6, 34]. The results from these experiments are promising.

Murata *et al.* [35] presented a performance evaluation of GAs with flow-shop scheduling problems. They looked at some of the common crossover and mutation operators used in flow-shop problems. They compared these various operators and found a version of two-point crossover worked the best for their 100 randomly generated problems. The shift change mutation was found to be the best mutation operator. They compared the GA with a local search, a tabu search, and a simulated annealing algorithm. They found that the simulated annealing and tabu search algorithms performed the best. Using these results, they developed a hybrid algorithm that combined a GA with a local search and another GA with a simulated annealing algorithm. They found their hybrid algorithms performed well with the flow-shop problem.

Ishibuchi *et al.* [36] combined local search techniques with an MOEA. They found that the hybridization algorithm improved convergence toward the Pareto front. But they found that their computation time was increased. In order to balance efficiency with effectiveness, they modified the algorithm so there was a good balance of genetic search and local search. The authors use two-point crossover and insertion (shift) mutation. They choose these operators based on the good results reported by Murata *et al.* [37] in their flow-shop research. This work appears to be an extension of previous work done by the authors [38].

Ishibuchi *et al.* [39] have also discussed how EAs and MOEAs can be applied to the flow-shop problem. They compare various permutation-based crossover and mutation operators and provide an analysis as to which ones work best for the group of test problems they selected. A version of one-point crossover performed the best when minimizing makespan was the single objective. But when minimizing maximum tardiness was the objective, there was no clear cut best operator for all the test problems. But the insertion mutation operator was found to be the best mutation operator in all instances of the single objective problem. In the MOEA realm, the Non-dominated Sorting Genetic Algorithm - II (NSGA-II) was compared to the results of single objective GAs, weighted single objective GAs, and a weighted single objective local search algorithm. The results were mixed, the highly exploitive single objective versions typically found better solutions in certain regions of the Pareto front, but the never found as many points or covered a greater region than the MOEA.

4.2 Flexible Flow-shop Example

Rodriguez *et al.* [8] tests four variants of the Single Stage Representation Genetic Algorithm (SSRGA) against the flexible flow-shop problem. In addition to the four variants of SSRGA, four dispatching rules (first-in first-out, shortest processing times, longest processing times, and shortest due dates) along with the shifting bottleneck procedure (SBP) are also applied to randomly generated instances of the flexible flow-shop problem. The SSRGA uses a random key representation, such as the one depicted in in Figure 12. The crossover operator selects two parents at random and then generates a random number between 0 and 1 for each allele position. If the value is below 0.7, then the value from the first parent is copied into the child. If the value is 0.7 or above, the value from the second parent is copied to the child. The authors do not explain why they chose a value of 0.7 as the cutoff point. While the SSRGA variants performed better than the dispatching rules and SBP, this was not the main goal of the research. The main goal was to show that a GA, which is easy to implement, is robust and simpler than SBP, and its computation time is reasonable [8].

4.3 Open-shop Examples

Dorndorf *et al.* [9] proposed a branch-and-bound method for solving the open-shop problem. Their approach focused on reducing the size of the search space through the use of constraint propagation based methods. Their "shaving" technique can be time consuming, but they found that the amount of search space reduced greatly offset this time. The results show that their new algorithm is able to solve many problem instances to optimality in a relatively short amount of time.

Liaw [10] developed a hybrid genetic algorithm to solve the open-shop problem. His algorithm combined a basic genetic algorithm with a local search procedure based on tabu search. The chromosome representation was an operation-based representation, similar to the one in Figure 8. After testing several crossover methods (PMX, OX, CX, order-based, and position-based), he found linear order crossover (LOX) worked best for the problem being considered. Figure 23 shows an example of the LOX operator. For mutation, he chose to use two mutation methods, insertion mutation and swap mutation (also known as reciprocal exchange mutation). His algorithm selected one of the two mutation operators with equal probability. The algorithm was compared to four other algorithms and it performed very well. It was able to find optimal solutions for benchmark problems that had never been solved to optimality before. The algorithm performed much better than existing methods with respect to solution quality.

4.4 Job-shop Examples

Many researchers have studied the job-shop problem. Jain *et al.* [40] provides an extensive list of methods by which the job-shop problem has been approached. Branch and bound techniques, priority dispatch rules, bottleneck based heuristics, AI techniques (constraint satisfaction and neural networks), threshold algorithms, simulated annealing, genetic algorithms, and tabu search are all discussed and compared. The conclusion is made that meta-heuristic techniques appear to work best for the job-shop problem. The EAs reviewed performed poorly, which seems to support the hypothesis that EAs are not "well suited for fine tuning structures that are very close to optimal solutions" [24, 40, 41]. But adding a local search element to the algorithm appears to improve the results.

Jensen [17] is concerned with robust scheduling with EAs . He highlights a number of EAs and special operators that have been applied to the job shop problem. He also performs experiments with coevolutionary algorithms and neighborhood based robustness. The results show that the neighborhood based robustness approach improves the robustness and flexibility for problems that factor machine breakdowns into the equation. The coevolutionary algorithm created schedules that guaranteed a set level of worst case makespan performance.

After reviewing these papers, it appears that EAs typically work best in this problem domain when they are used in conjunction with a local search algorithm. The memetic EAs provide enhanced local search techniques that enables the algorithm to more effectively search the neighborhoods surrounding good solutions.

4.5 Flexible Job-shop Examples

Researchers use many varying methods to solve the flexible job-shop problem. For example, Wu *et al.* [19] uses a multiagent scheduling method to solve the problem. The agents are job agents and machine agents. Each of the agents has its own goals, knowledge base, functional component, and control unit. The knowledge base is simply domain knowledge and/or data. The functional component has the mathematical procedures the agent needs for decision making. While the control unit contains the protocols that allow the agents to communicate with one another. The job agents are created when a job is released to the shop. They communicate with the machine agents in order to select machines for each operation. The machine agents are responsible for sequencing the jobs. Routing and sequencing are determined by the functional components of each agent. The algorithm outperformed the other baseline methods and it performed fast enough to make it a viable option for real-world problems [19].

Kacem *et al.* [20] used a hybrid method to solve the the flexible job-shop problem. They used a combination of fuzzy logic and an EA as their hybrid method. The goal is to exploit the representation capabilities of fuzzy logic and the adaptive abilities of EAs. The algorithm is composed of two stages: a fuzzy multi-objective evaluation stage and an evolutionary multi-objective optimization stage. The fuzzy stage computes the different weights for each objective function and measures the quality of each solution. The evolutionary stage has two sub-stages: approach by localization (AL) and the controlled genetic algorithm (CGA). The AL stage generates individuals that are in the most "interesting" zones of the search space. These individuals become the population of the CGA. The CGA is an EA that utilizes crossover and mutation operators. No mention is made in the experiments about how many generations are run for each problem. The algorithm appears to be effective in converging toward the lower bound of each objective function.

Mastrolilli *et al.* [42] used local search techniques (tabu search) and two neighborhood functions to solve the flexible job-shop problem. Neighbors are obtained by moving and inserting an operation in an allowed machine sequence.

The local search algorithm starts with an initial solution and travels from neighbor to neighbor in an effort to find the best solution. Tabu search is an effective method of escaping local minima, so it was picked as the local search algorithm. For this problem, one neighborhood function was found to perform faster and as effectively as the other neighborhood function, so it's results were presented in the article. The results of the experiments were compared to three other Tabu search algorithms. The new algorithm with the addition of the neighborhood function, performed better than the other versions of Tabu search on four different instances.

4.6 Automated Planning Systems

For completeness, automated planning systems need to be addressed. This is because scheduling problems are closely related to planning problems. While scheduling problems are typically interested in metrics such as makespan, planning problems are generally more broad in scope. Planning systems are designed to generate a plan or policy to achieve some set of user defined goals or objectives. Automated planning systems can be classified as follows [43]:

- **Domain-independent planners** – Only input into the planner is the description of the planning problem to solve.
- **Tunable domain-independent planners** – Similar to domain-independent tuners, but can be tuned for better results.
- **Domain-configurable planners** – Input into the planner includes domain specific control knowledge to aid the planner in developing solutions in the particular domain.
- **Domain-specific planners** – Planner is designed for a given domain and would require major modifications to work in other domains.

Of particular note are hierarchical task network (HTN) or AI planners, which are considered domain-configurable planners. Examples of these types of planners include the Open Planning Architecture (O-Plan2) [44], the System for Interactive Planning and Execution (SIPE-2) [45], and the Simple Hierarchical Ordered Planner (SHOP2) [46]. In HTN planners, the system formulates a plan by decomposing tasks into smaller subtasks until it reaches a level that the plan executer can perform. Researchers use this type of system to solve many different types of planning problems. Some examples include [43]: evacuation planning, evaluating terrorist threats, fighting forest fires, controlling multiple UAVs, and distributed planning. This type of system could also be applied to many types of scheduling problems. Unfortunately, some of the planning systems, such as the SHOP2, do not explicitly represent time and concurrency, which are essential items for determining makespan and implementing concurrent actions. These items must be aggregated into their code [43].

5 Real-world Multi-component Scheduling Problems

This work introduces the multi-component scheduling problem. While the existing scheduling models work well for some applications, they may provide only a limited view of the overall problem for other applications. By introducing this model, researchers can compare their work with others in the field and build upon the advances of one another. This is important because the multi-component scheduling problem is a common one in society, particularly in the field of electronics repair. For example, a computer repair shop would be best served using the multi-component scheduling problem versus a standard flow-shop or job-shop problem. Because the computers themselves would follow a flow-shop paradigm, where they would first be put through a series of diagnostic tests, then any problems that were found would be fixed, and then the computers would be tested again to ensure the repair actually fixed the problem. But, the actual repair step in the flow shop can be decomposed into a job-shop, based on the particular repairs that need to be done for each computer. Some computers may require new processors, others new memory, while some may need new hard drives. These multiple repair paths cannot be modelled in the flow-shop problem. So this step of the problem needs to be modelled with an open-shop or job-shop problem.

But electronic repair isn't the only real world example. Auto repair is another example of where this scheduling paradigm comes into play. Like the computers, the problem with the car must first be diagnosed, then the problem must be fixed, and then the car needs to be checked out to make sure it is working properly. And like the computer example, there are multiple paths that the repair operation can take. The car may need new brakes, a new battery, or possibly some engine work. So the multi-component scheduling problem would work well for this type of problem.

Another example is a distribution model that is similar to the TSP. Suppose a shipping agency, such as Federal Express has to get many packages to various destinations. The packages start in many drop-off locations throughout all major cities in the country. The packages are then gathered up by trucks and brought to a processing center. The packages are then sent to other processing centers throughout the country via truck and plane. Once the packages arrive at the new processing center, they are then sent out to the delivery address.

A flow-shop problem could be used to model an aircraft starting with the packages at one processing center and then going across the United States to deliver the packages. But this does not provide the necessary detail, since the packages haven't reached their destination yet. A multi-component model can take into account the embedded scheduling (or TSP) that is necessary to get the packages from their drop off points to their final destinations.

One final example could be an employee work schedule. For example, an employee may have the flow of their day planned, to include meetings, lunch, projects, etc. But in actuality, some of the events in the flow of the day are better scheduled using a job-shop or open-shop model. By using the embedded model, a work schedule could be designed in a more efficient manner.

5.1 Why MOEAs are Appropriate for Multi-Component Scheduling Problems

EAs and MOEAs have been applied to scheduling problems with much success. An advantage the MOEAs have over EAs is that they can generate solutions that allow the decision maker to decide which trade-offs he is willing to make. For example, suppose two important issues at hand are getting everything out as fast as possible and ensuring high priority items are done within certain time constraints. The decision maker must decide either *a priori* or after optimization what balance is best for the situation. He may decide that his number one concern is getting the high priority items out as quickly as possible. But more probably, he may want to find a balance between where the high priority items are completed by their allotted time and all the items are finished as fast as possible. Using an EA won't give a decision maker the necessary detail that an MOEA can.

5.2 MOEA Algorithm Selection

The general multi-component scheduling problem is an NP complete problem since it embeds other NP complete scheduling problems. These scheduling problems can sometimes have rough search terrains that are hard to traverse using only a local search algorithm. Thus a multi-objective evolutionary algorithm (MOEA) is selected to solve this problem due to the generic combinatorics [47] and search landscape. While many MOEAs, such as the NSGA-II [48], the Strength Pareto Evolutionary Algorithm - 2 (SPEA2) [49], or the Multi-Objective Genetic Algorithm (MOGA) [50] would be good to use, the General Multi-objective Parallel Genetic Algorithm (GENMOP)[51–54] algorithm is applied to this problem. GENMOP was originally designed as a real-valued, parallel MOEA. A parallel algorithm can improve efficiency for problems with large search landscapes, so GENMOP was chosen over the rest.

GENMOP is an implicit building block MOEA that attempts to find good solutions with a balance of exploration and exploitation. It is a Pareto-based algorithm that utilizes real values for crossover and mutation operators. The MOEA is an extension of the single objective Genetic Algorithm for Numerical Optimization Problems (GENOCOP) [55, 56]. Constraint processing is added to enable the software to handle the multi-component scheduling problem.

GENMOP has been used to solve a variety of problems. It was initially used to optimize flow rate, well spacing, concentration of injected electron donors, and injection schedule for bioremediation research [51, 57]. Specifically, it was used to maximize perchlorate destruction in contaminated groundwater and minimize the cost of treatment. The algorithm is also used to optimize quantum cascade laser parameters in a successful attempt to find viable solutions for a laser that operates in the terahertz frequency range [54, 58].

GENMOP is now able to handle permutations, which are common in most scheduling problems. Since chromosomes are generated randomly, a repair mechanism or penalty function must be used in order to elicit a valid permutation.

A repair operator immediately repairs invalid permutations, so only valid permutations are compared. While a penalty function allows for invalid permutations to be compared, but a penalty (cost) is added to the fitness of the invalid permutation. For this implementation, a repair operator was chosen.

The algorithm flow is similar to most MOEAs. First, the input is read from a file. Next, the initial population is created and each chromosome is evaluated. The population is then ranked based on the Pareto-ranking of the individuals. Then a mating pool is created and only the most fit individuals are chosen to be in the mating pool. Crossover and mutation are performed on the members of the mating pool. The children created are then evaluated and saved. These children are then combined with the rest of the population. This is known as a $(\mu + \lambda)$ type of genetic algorithm, where the child population λ is combined with the parent population μ. There is another type of algorithm that replaces the parents with their children. This type of algorithm is referred to as a (μ, λ) algorithm.

After the children are combined with the parents, the population is then put into Pareto-rank order. The program then checks to see if the program has run through its allotted amount of generations. If it has, the program exits. If it has not, the program creates another mating pool and goes through the process again. Figure 32 shows the flow of the GENMOP algorithm.

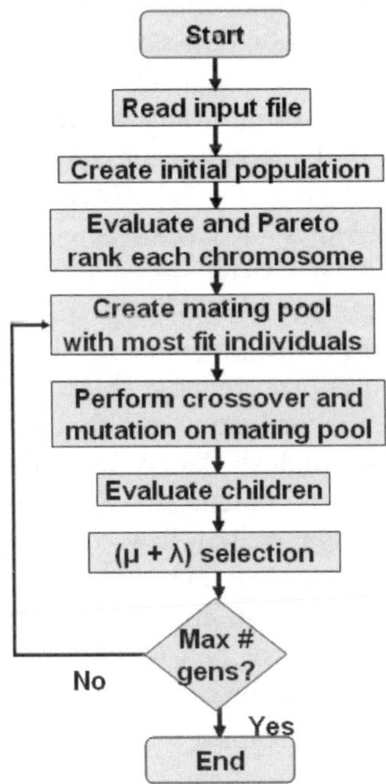

Fig. 32. Flow diagram for the GENMOP algorithm.

There are other considerations that also must be addressed, but most of these are problem specific. In the following section, a specific multi-component scheduling problem is described and specific GENMOP parameters are presented along with problem specific elements, including a repair operator.

5.3 Aircraft Engine Maintenance Scheduling Problem

An example of the multi-component scheduling problem is the aircraft engine maintenance scheduling problem. This problem deals with aircraft engine maintenance at a depot level. The engines are received based on contractual schedules from various organizations. They follow the multi-component scheduling model in that the engines follow a higher level flow-shop through what is called the assembly shops: where the engine is broken down for repair and pieced back together and tested. The components are then repaired in the backshops, where they follow more of a job-shop model, until they are reattached to the engine [52]. For our problem, a list of engines is read in, with each engine listing its arrival time, due time, priority (weight), and mean time before failure[1] (MTBF) for each of its components.

Maintenance Flow of the Engines When an engine comes into a logistics workcenter for repair, it is first logged into the system. Aircraft engines are commonly divided into smaller subcomponents which can be worked on individually and then recombined. For this problem, the engine is divided into five logical subcomponents: fan, core, turbine, augmenter, and nozzle. It is assumed that the maintenance shop has one specific work area, or backshop, for each of the components. In reality, an engine component may visit several backshops. For simplicity, it is assumed that they only visit one but varying times for repair are added to take the other backshop visits into account. This is an example of the job-shop problem, but with a twist. After all maintenance is completed on an engine, each engine component's MTBF is compared with other components on the engine. If there is a large disparity among the MTBFs then a component swap may be initiated with another engine in an effort to ensure the MTBFs of the components of a particular engine are similar. This is done so that the engine can have more "time on wing" (TOW) and less time in the shop.

Once the swaps are done, the engine is reassembled and tested as a whole to ensure functionality. This represents a flow-shop problem in that each engine has to have maintenance done first, followed by swapping and then testing. So the problem is a representation of the multi-component scheduling problem.

Figure 33 shows an example of the flow for two engines. Note how the engines flow from one stage to the next and the subcomponents are repaired in a job-shop paradigm.

[1] The MTBF is the mean value of the lengths of time between consecutive failures, under stated conditions, for a stated period in the life of a functional unit. A more simplified MTBF definition for Reliability Predictions can be stated as the average time (usually expressed in hours) that a component works without failure.

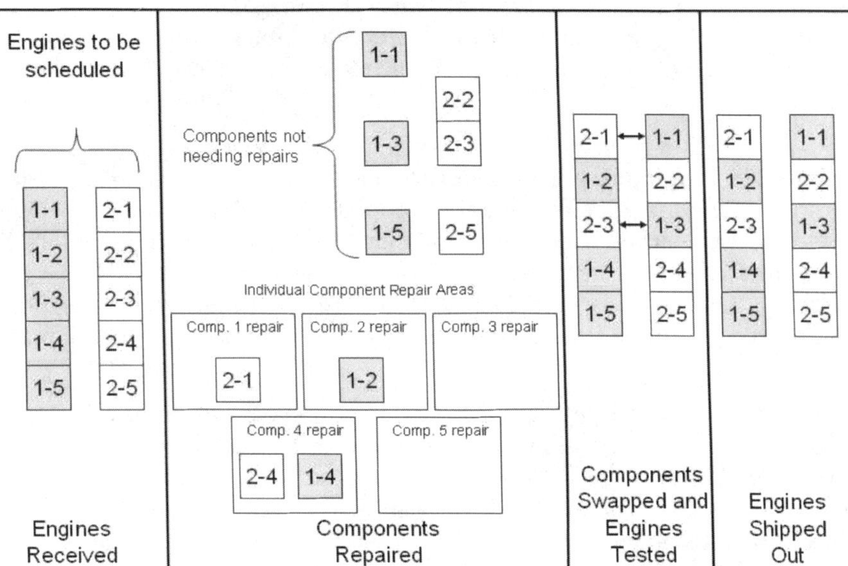

Fig. 33. Example of maintenance flow for two engines. The first number in each block is the engine number and the second number is the component number. So 2-4 refers to the fourth component on engine 2.

Key Objectives of the Problem There are two main optimality objectives. The first objective is to find the schedule that results in repairing the engines in the quickest manner. This is called the makespan. The makespan, which is to be minimized, determines which schedule has a faster process. Our second objective is to attempt to keep the MTBF values within a predetermined range for all engines. This may require a number of component swaps from multiple engines. Consequently, these two objectives conflict. The first objective is to try to get the engines out as quickly as possible, while the second objective slows down the process by swapping components [52].

Ultimately, a solution that takes into account both objectives is needed. A solution can be obtained through a number of different methods. The first method is the weighting method, where each objective is given a predetermined weight, based on it's importance to the decision maker. Another common method is to create a Pareto Front of all the non-dominated solutions and allow the decision makers to pick which solution they deem the best. The first method is an *a priori* weighting method, where the decision maker has no idea concerning the distribution of answers. The second method is a *posterior* weighting method, where the decision maker can see the distribution of possible solutions, and can make a decision based the solutions provided. The Pareto Front method is chosen for reporting solutions. By reporting answers via a Pareto Front, a vector of solutions can be presented. Thus other researchers can compare their results to ours in a straightforward manner. One can also compare various runs and generations to determine how well our algorithm explores and exploits the search space.

GENMOP Particulars The algorithm makes use of four different crossover operators and three different mutation operators. The operators are selected based upon an adaptive probability distribution. If more fit individuals are generated from a certain operator, the algorithm increases the probability of its selection in future generations. This adaptive process is built into the algorithm in an effort to self-tune the algorithm and provide a more hands-off approach.

Operator Selection There are many types of crossover and mutation operators available. For this problem the following crossover operators were chosen:

1) *Whole Arithmetical Crossover*: All genes of x_i and x_r are linearly combined to form chromosomes x_1 and x_2. GenMOP retains x_1 and discards x_2.

2) *Simple Crossover*: One gene is chosen in both x_i and x_r and swapped to form chromosomes x_1 and x_2. GenMOP retains x_1 and discards x_2.

3) *Heuristic Crossover*: Individuals x_i and x_r are combined to form one individual $\ni x_1 = R \cdot (x_r - x_i) + x_r$, where R is a uniform random number between zero and one and the rank of $x_r \leq x_i$.

4) *Pool Crossover*: Randomly chooses genes from individuals in the mating pool and combines them to create x_1.

Note that x_i and x_r are parent chromosomes and x_1 and x_2 are the offspring (children) of the parents.

These operators were chosen for two reasons. First, our chromosome is made up of two different entities, a permutation and a swap section. As such, we couldn't use a permutation based operator on our chromosome unless it were decomposed into two parts, and two separate crossover operators were applied to the problem. It was determined that non-permutation based operators would work better for this problem. The second reason for using these crossover operators is because they have been used in previous experiments, with much success, when dealing with large search landscapes. Since our problem has a very large search space, these operators should provide the exploration and exploitation needed.

The three mutation operators are:

1) *Uniform Mutation*: Chooses a gene existing in the chromosome to reset to a random value within its specified ranges.

2) *Boundary Mutation*: Chooses a gene existing in the chromosome to reset to either its maximum or minimum value.

3) *Non-uniform Mutation*: Chooses a gene to modify by some random value decreases probabilistically, until it equals zero, as the generation number approaches the maximum generations.

These were chosen because they allow varying methods of mutation that should provide for better exploration of the search space.

Chromosome Representation The chromosome representation for this problem encompasses more than just the schedule. Our chromosome represents both the engine precedence listing and the component swaps that should be accomplished. In our initial attempt at this problem, a pseudo-variable chromosome was created. While the chromosome was a constant length, the number of swaps could vary, depending on if a component swap was labelled a zero or not [52]. While this functioned well, a truly variable chromosome is preferred for two reasons. First, it would decrease our search space. Second, it would allow for the dynamic removal and insertion of swaps. The main reason for wanting a variable length chromosome is the uncertainty of how many component swaps each schedule would require in order find the best solutions. Genetic programming (GP) uses variable length chromosomes. In general, GP is more flexible in the number of solutions that it can generate compared to a fixed length chromosome. Since we have no idea of how many swaps each schedule might need, or how many swaps would be needed as the number of engines scheduled is increased, having a variable number of swaps appears to be the best solution.

A fixed number of swaps can be detrimental in two ways. First, if too few swaps were implemented for the problem, then the algorithm would never be able to find all the solutions on the true Pareto Front. Conversely, if we the algorithm implements way too many swap possibilities, then the search space becomes much larger than it needs to be, and the probability of finding the true Pareto Front decreases. The variable length chromosome allows the algorithm to vary the chromosome size and converge on the chromosome length that results in the best results.

The chromosome representation consists of two parts, the schedule and the swaps. As mentioned in Section 3.1, the schedule can be encoded using one of two approaches, direct approach and indirect approach [3, 18].

The first portion of the chromosome representation used in this research is a direct approach that is based on the job-based GA representation [3, 18, 59]. Figure 9 shows an example of this type of representation. This means that order of precedence for scheduling the engines is explicitly listed in which each allele in the first part of the chromosome represents an engine. The first allele listed has precedence over all other engines for component repairs. So if three components need repair, they are scheduled first in the respective backshops. It follows that the last engine listed has its components repaired last. So if engine 3 is listed before engine 5, then engine 3 has priority before engine 5 in all repair operation in both the assembly shops and backshops.

The second portion of the chromosome determines the precedence of the component swaps, the number of swaps, the components to be swapped, and the engines that the components are to come from. Each swap is represented by three alleles. The first two alleles are the two engines to be swapped. These two engine numbers are different from one another. A constraint was added so that the first engine number must always be smaller then the second. This was done in an attempt to keep the search space down. If the smallest engine number can be in either location, then this effectively doubles the number of eligible swaps. The third allele of a swap represents the component to be swapped. Previously, this was set from zero to five to allow for no swap[52]. Since a variable length chromosome is used now, the algorithm only needs to choose between components one to five. This change also reduces our search space. Note that the initial values are all set at random using a uniform distribution. All engines and component swaps are assumed to have similar priorities. If we assumed that earlier engine numbers had a higher priority than later engine numbers, then a different probability distribution, such as an exponential distribution, would be more effective. Figure 34 show a representation of a chromosome for four engines. Note that the first four alleles are for the engine scheduling precedence, while the rest of the alleles represent the swaps.

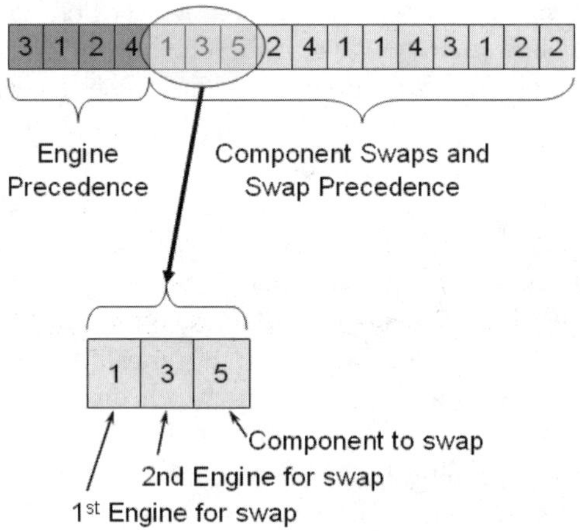

Fig. 34. Example of chromosome representation for four engines.

The most difficult part with the swap portion of the chromosome is determining how many swaps to allow, which is why the variable length is used. An example of why a fixed length chromosome can be bad can be explained using three engines. Suppose that it would take four swaps in order to get the best answer. If you allow for 20 swaps, the search space is increased drastically by the additional 16 swaps that are unnecessary. Conversely, if you only allow for 4 swaps and you need 10, then there is no way that you can ever achieve the best answer because you do not allow for enough swaps. As mentioned earlier, the variable length chromosome overcomes this issue.

One phenomena that has occurred in genetic programming that should be avoided is the tendency of getting very large representations as the number of generations increases [60]. To avoid exceedingly large chromosomes, the problem was constrained to a maximum chromosome size. The size was arbitrarily set based on the number of engines scheduled. The formula for determining the maximum possible chromosome length is found in Equation 1.

$$L^{Max}_{Chromo} = N_e + 3N_e + 15 = 4N_e + 15 \tag{1}$$

where L^{Max}_{Chromo} is the maximum possible length of a chromosome, and N_e symbolizes the number of engines to be scheduled. Essentially, five swaps are always allowed, represented by the 15, and for every engine scheduled, another swap is added to the total. For example, a schedule consisting of ten engines would have a maximum chromosome length of 55 alleles (10 for the schedule, 45 for the swaps).

Figure 35 shows what a small portion of the population might look like with the variable chromosomes. The size of each chromosome is maintained in data array that also maintains the fitness function values for each chromosome.

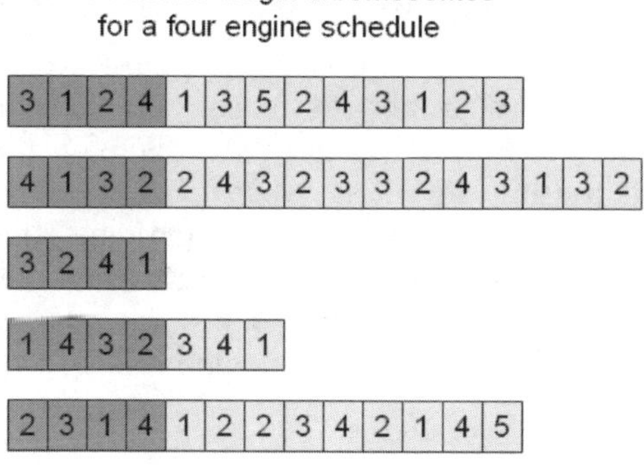

Fig. 35. Examples of possible chromosome representations.

Repair function for the Chromosome The GENMOP algorithm is a real-valued algorithm. As such, the chromosome values were represented as real-values. In order to create the schedule permutation, a repair function is needed in order to effectively compare one allele to another. To do this, the algorithm uses the ceiling function where all real values are rounded up to their nearest integer value. Therefore, a value of 1.234 would be converted to $\lceil 1.234 \rceil = 2$. The algorithm then ensures that the permutation lists every engine only once. If it comes across a duplicate engine, the second engine found is discarded and another random engine is generated. This new engine must be different from all the engines previously listed.

For the swaps, the smallest engine value should be listed in the first allele of each swap. If the circumstance occurs where the first engine listed is larger than the second, the two engines are swapped. If the both engines were identical, the algorithm randomly generates another engine.

On occasion, one of the mutation or crossover operators may cause the values to go beyond the predefined bounds, such as a number larger than the number of engines or a component number of zero, etc. For these occasions, the numbers that exceeded our maximum bounds are replaced with the maximum value allowed. Conversely, numbers that are below the minimum bound are replaced with the minimum value. This was deemed to be a better method instead of randomly creating a new value. This way, the intentions of the initial operation are followed.

One possible problem is using this repair function with the operators provided. The original operators are all real-valued and they weren't developed with permutations in mind. As such, our operators and repair function tend to be more destructive with respect to the permutation. For example, if the first chromosome value is mutated, the change could have an effect on all other values of the permutation. So instead of mutating just one allele of the chromosome, the algorithm could end up mutating many values of the chromosome.

Problem Complexity This problem is much more difficult to solve than a similar sized travelling salesman problem (TSP). The TSP is an NP-complete problem, that has $n!$ possible solutions, where n is the number of cities. For this problem instance, the first part of the chromosome (the engine schedule) is identical to a TSP, but the second part of the chromosome (the engine swap portion) adds a considerable amount of complexity and greatly increases the search space. As a result the complexity of this problem can be calculated as follows:

$$n! * \left(\sum_{i=1}^{n-1} (i) \right) (5)\,(n+5)$$

where n is the number of engines to be scheduled. The $\sum_{i=1}^{n-1}(i)$ is the total number of combinations that two separate engines can be paired. The 5 is the number of components that can be swapped between two paired engines. The $n+$ 5 are the maximum number of possible swaps that can occur for the problem. The problem can be simplified further by making some mathematical substitutions and approximations.

Replacing the summation:

$$n! \left(\frac{(n)(n-1)}{2} \right) (5)\,(n+5)$$

Substituting Stirling's approximation for $n!$:

$$\left(n^n e^{-n} \sqrt{2\pi n} \right) \left(\frac{(n)(n-1)}{2} \right) (5)\,(n+5)$$

So if all the terms are combined in n, the problem is bounded by n^{n+3}.

If the complexities between this problem and an equally sized TSP are compared, the difference can be illustrated numerically. For example, for a 5 city TSP, there are only 120 possibilities, but for a 5 engine problem, there are 60,000 possibilities. Likewise, for a 10 city TSP, the possibilities increase to 3,628,800, and the 10 engine problem increases to 12,247,200,000. So finding a solution for even a small 10 engine problem can be extremely difficult given the large number of possible solutions.

6 Design of Experiments, Testing and Analysis

An objective of this research is to compare the results of a fixed chromosome with the results from our current algorithmic research which incorporates a variable length chromosome. To compare the two approaches, the results of the two fitness functions and the average chromosome length were compared. The new variable chromosome was implemented in such a way that it reduced the search space size and allowed for more swaps than the previous version.

Five engines and ten engines are input in the scheduling process. A variety of initial population sizes, from 10 to 1000 individuals, are tested with a variety of generations, from 10 to 1000. These instances are used in order to compare our results with baseline results, as shown in Tables 1 and 2.

6.1 Analysis of Experiments

One of the goals of this experiment is to determine if our new variable chromosome improves our algorithm execution or if it hinders it. The results of our experiments are compared with those from [52], which used the same algorithm but had a different approach with respect to the chromosome sizing and initialization. The results from [52] are considered the baseline results for the variable length chromosome experiments.

Tables 1 and 2 list the results from our five engine and ten engine experiments. The items in bold are the best mean values for the fitness functions. Note that the variable chromosome always has the lowest makespan while the aggregate swap count results are varied. The most interesting result is the average number of component swaps. In every instance, the new variable chromosome yielded higher chromosome lengths than the baseline representation. This result was probably due to a higher upper bound being set on the number of swaps allowed. Since the initial population is determined in a random fashion, using a uniform distribution to determine the chromosome size, we can calculate the expected value and variance of the number of swaps of the initial population. Equations 2 and 3 show the expected value and variance calculations for a given number of engines to schedule. Note that N_e represents the number of engines the require repair and N_{Fixed_Swaps} is the number of component swaps an engine requires when using a fixed length chromosome.

Table 1. Testing Results for 5 Engines

Number of engines	Generation size	Population Size	Avg/Std Dev of Component Swaps		Avg/Std Dev of Agg. Swap Count		Avg/Std Dev Makespan	
			Baseline Results	Variable Chromosome	Baseline Results	Variable Chromosome	Baseline Results	Variable Chromosome
5	10	10	2.04 / 0.93	5.33 / 3.39	26.23 / 2.01	**25.2** / 1.10	971.5 / 39.4	**947.8** / 56.1
5	10	100	3.43 / 0.93	5.25 / 3.77	**27.6** / 2.68	27.8 / 2.87	941.5 / 78.2	**869.5** / 53.7
5	10	1000	3.61 / 1.10	5.40 / 2.88	**25.4** / 2.01	26.2 / 3.42	921.0 / 88.3	**859.6** / 34.0
5	25	25	2.28 / 1.04	4.75 / 2.92	**25.7** / 1.64	26.5 / 3.29	932.8 / 71.5	**916.8** / 65.8
5	100	10	1.57 / 0.85	10.0 / 0.00	26.6 / 1.29	**24.0** / 2.67	944.6 / 77.2	**833.0** / 22.5
5	100	100	2.80 / 1.16	7.40 / 2.27	25.6 / 1.42	**25.4** / 3.17	904.3 / 79.3	**846.5** / 30.0
5	100	500	3.12 / 0.80	8.00 / 0.00	25.7 / 1.95	**25.0** / 2.65	885.9 / 66.6	**820.0** / 26.5
5	100	1000	3.54 / 0.92	10.0 / 0.00	25.1 / 2.04	**22.0** / 0.00	889.5 / 67.5	**850.0** / 0.00
5	500	1000	3.36 / 1.03	6.21 / 1.69	24.8 / 1.64	**23.4** / 1.56	918.0 / 76.1	**858.5** / 49.6
5	1000	1000	2.94 / 0.75		25.58 / 1.52		886.5 / 68.74	

Table 2. Testing Results for 10 Engines

Number of engines	Generation size	Population Size	Avg/Std Dev of Component Swaps		Avg/Std Dev of Agg. Swap Count		Avg/Std Dev Makespan	
			Baseline Results	Variable Chromosome	Baseline Results	Variable Chromosome	Baseline Results	Variable Chromosome
10	10	10	3.51 / 2.57	11.6 / 2.01	**27.5** / 2.17	28.9 / 2.71	2165.5 / 83.1	**2086.4** / 82.8
10	10	100	5.6 / 2.72	9.42 / 4.42	**27.0** / 2.40	29.7 / 5.05	2148.1 / 50.2	**2023.4** / 136.3
10	10	1000	7.32 / 1.74	12.3 / 1.53	**25.6** / 1.29	28.3 /3.79	2144.9 / 56.4	**1979.7** / 85.9
10	25	25	3.55 / 2.00	11.0 / 1.73	**26.4** / 1.57	28.8 / 3.42	2142.2 / 43.5	**1937.6** / 79.2
10	100	10	3.72 / 1.79	11.4 / 2.19	25.7 / 1.23	**24.3** / 2.69	2139.0 / 29.1	**1832.4** / 50.5
10	100	100	4.01 / 1.97	13.9 / 0.49	26.5 / 1.63	**23.1** / 0.67	2119.1 / 32.7	**1893.4** / 0.67
10	100	500	4.03 / 2.30	9.49 /2.89	25.4 / 1.16	**23.8** / 0.87	2128.8 / 36.0	**1925.9** / 222.4
10	100	1000	4.45 / 1.98	13.5 / 1.12	**24.7** / 0.91	24.8 / 0.43	2133.8 / 53.5	**1678.3** / 106.3
10	1000	1000	4.16 / 2.08	13.9 / 0.65	24.8 / 0.64	**23.7** / 0.55	2105.5 / 4.53	**1638.5** / 54.3

$$E[N_{Fixed_Swaps}] = \frac{(N_e + 5)}{2} \tag{2}$$

$$Var[N_{Fixed_Swaps}] = \frac{(N_e + 5)^2}{12} \tag{3}$$

In the baseline model, every chromosome is set to have one swap per engine scheduled, but a swap could be turned off by placing a zero in for the component number. The component numbers for the initial population are generated using a uniform distribution. Essentially, the number of swaps for in a chromosome is determined via a binomial distribution, with each swap having a $p = 5/6$ probability of occurring. The expected number of swaps for the original chromosome can be found in Equation 4 and the variance is found in Equation 5. Note that N_{Var_Swaps} is the number of component swaps an engine requires when using a variable length chromosome.

$$E[N_{Var_Swaps}] = (N_e)\left(\frac{5}{6}\right) \tag{4}$$

$$Var[N_{Var_Swaps}] = (N_e)\left(\frac{5}{6}\right)\left(\frac{1}{6}\right) \tag{5}$$

Table 3 shows the expected values and variances of the number of swaps, in the initial population, from the two representations. It's interesting to note that in the 10 engine instance, the baseline chromosome population is initially seeded with more swaps, but after all the generations are run, the average chromosome length shrinks, while the variable length chromosome average size increases.

Table 3. Comparison of the number of swaps in the initial population

	Baseline Chromosome		Variable Chromosome	
Number of engines	Expected Value	Variance	Expected Value	Variance
5 Engines	5.0	0.69	4.17	8.33
10 Engines	7.5	1.39	8.33	18.75

Another way to highlight the results is to compare each of the methods using Pareto Front analysis. Figure 36 compares the results of the algorithm implementations using 5 engines and 25 for both the initial population and number of generations. Figure 37 shows the results for the same number of population and generations as listed above, but it show the results with respect to 10 engines. Notice how there is little difference in the 5 engine case, while there is a vast difference between the implementations in the ten engine case. This result is common with all the other instances as well. The trend appears to show that the new GENMOP design is more capable of finding better solutions, especially with respect to the makespan, than the old one. This is most probably caused by our efforts to reduce the search space of the algorithm, through the use of a variable length chromosome and reducing the number of allowable swaps. Figure 38 shows a comparison with more realistic population and generation sizing. Notice how once again our new implementation is much improved over the previous results.

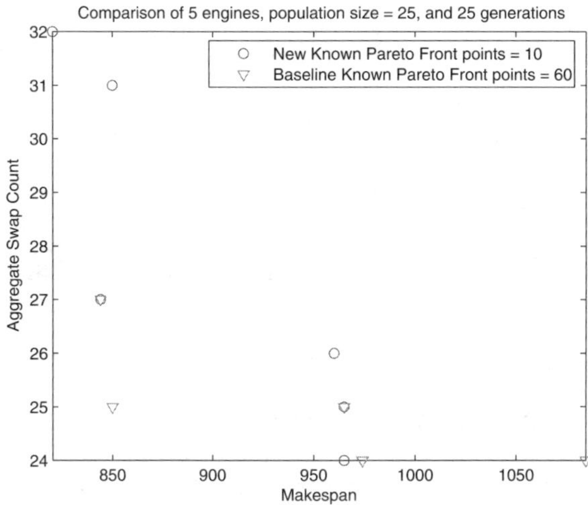

Fig. 36. Comparison of the Pareto front members for the 5 engine instance.

Using an MOEA for the multi-component problem has many advantages, particularly if there is more than one objective a researcher is interested in. For multiple objectives, Pareto front methods are a good way to show the decision maker his options. MOEAs do a good job of finding points along Pareto fronts that are not only concave or convex, but also do well with non-uniform and deceptive problems as well. But MOEAs are only as good as their representation and operators. The key is to balance exploration and exploitation. A proper mix of the two can ensure that the algorithm is capable of finding solutions in both smooth and rugged search landscapes. The MOEA used in this research has proven to be effective in a variety of environments because it is able to provide that good balance.

7 Conclusions

In this chapter five scheduling problems were discussed. These problems led up to the discussion of a new scheduling paradigm, the multi-component scheduling

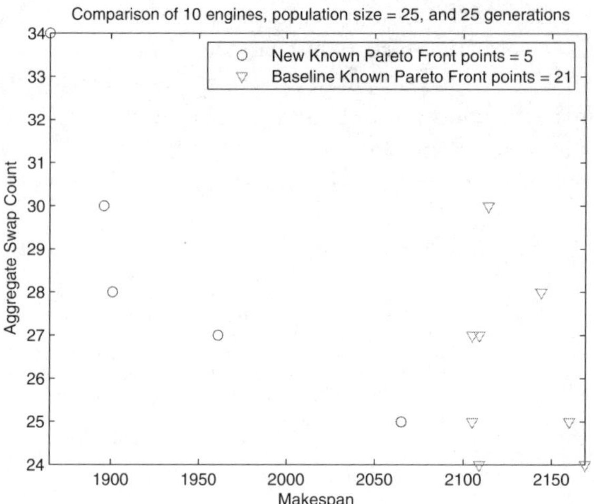

Fig. 37. Comparison of the Pareto front members for the 10 engine instance.

problem. Some specific examples of how EAs have been used with scheduling problems were discussed. Specifically, chromosome representations and various crossover and mutation operators were presented. Then, some examples of the multi-component scheduling problem were given and one of these examples, the engine maintenance scheduling problem, was solved. The GENMOP MOEA is used to optimize the problem. A variable length chromosome is devised in an effort to reduce the search space. The new design performed much better than the baseline when the search space was large, specifically the ten engine instance.

There are several other avenues that can be investigated, given our validated model and MOEA. The one area of research that may increase effectiveness is analyzing the operators currently built into GENMOP and adding various permutation operators that are geared more toward scheduling problems. GENMOP was run with all of its operators available, but some operators may be more useful than others. Also, by adding operators that have been shown to be effective in other scheduling problems, the effectiveness or efficiency of the algorithm may be increased. Analyzing our repair function is another area of research that could pay off big dividends. The current repair method is fairly simple, but it can be rather disruptive. We would like to apply one or two different repair operations and compare how well each one performs. In addition, evaluation of appropriate MOEA quality metrics is to be addressed.

Additionally, incorporating automatic planning systems into the multi-component scheduling problem, is a worthwhile research task. By incorporating that model, it can create another tool that may be useful in solving an additional set of scheduling problems.

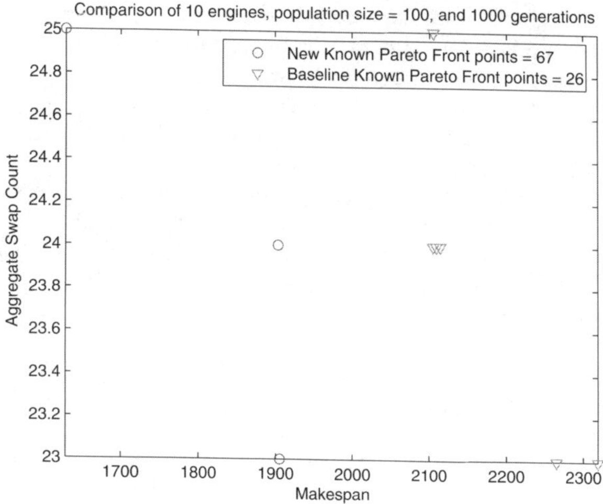

Fig. 38. Comparison of the Pareto front members for the 10 engine instance.

Finally, the search landscapes of the various scheduling algorithms should be studied more in depth. Not much work has been done in this field. By knowing the search landscapes, a researcher can decide what algorithm is best to apply to a particular problem.

References

1. Michael Pinedo. *Scheduling: Theory, Algorithms, and Systems.* Prentice-Hall, Englewood Clifs, New Jersey, 1995.
2. Peter Brucker. *Scheduling Algorithms.* Springer-Verlag New York, Inc., Secaucus, NJ, USA, 1998.
3. Tapan P. Bagchi. *Multiobjective Scheduling by Genetic Algorithms.* Kluwer, Boston, MA, 1999.
4. Vincent T'kindt, Jatinder N.D. Gupta, and Jean-Charles Billaut. Two-machine flowshop scheduling with a secondary criterion. *Comput. Oper. Res.*, 30(4):505–526, 2003.
5. Józef Grabowski and Mieczyslaw Wodecki. A very fast tabu search algorithm for the permutation flow shop problem with makespan criterion. *Comput. Oper. Res.*, 31(11):1891–1909, 2004.
6. Kuo-Ching Ying and Ching-Jong Liao. An ant colony system for permutation flow-shop sequencing. *Computers and Operations Research*, 31(5):791–801, 2004.
7. El-Ghazali Talbi, Malek Rahoual, Mohamed Hakim Mabed, and Clarisse Dhaenens. A hybrid evolutionary approach for multicriteria optimization problems: Application to the flow shop. In *EMO'01: Proceedings of the First International Conference on Evolutionary Multi-Criterion Optimization*, pages 416–428, London, UK, 2001. Springer-Verlag.

8. Jose Antonio Vazquez Rodriguez and Abdellah Salhi. Performance of single stage representation genetic algorithms in scheduling flexible flow shops. In *Congress on Evolutionary Computation (CEC'2005)*, volume 2, pages 1364–1371, Piscataway, New Jersey, September 2005. IEEE Service Center.
9. Ulrich Dorndorf, Erwin Pesch, and Toà Phan-Huy. Solving the Open Shop Scheduling Problem. *Journal of Scheduling*, 4:157–174, 2001.
10. Ching-Fang Liaw. A hybrid genetic algorithm for the open shop scheduling problem. *European Journal of Oprational Research*, 124(1):28–42, 2000.
11. Jesus J. Aguirre-Solis. Tabu Search Algorithm for the Open Shop Scheduling Problem with Sequence Dependent Setup Times. In *Advanced Simulation Technologies Conference*, volume 1. The Society for Modeling and Simulation International, April 2003.
12. Jacek Blażewicz, Erwin Pesch, Malgorzata Sterna, and Frank Werner. Open Shop Scheduling Problems with Late Work Criteria. *Discrete Applied Math*, 134(1-3):1–24, January 2004.
13. C. T. Ng, T. C. E. Cheng, and J. J. Yuan. Concurrent open shop scheduling to minimize the weighted number of tardy jobs. *J. of Scheduling*, 6(4):405–412, 2003.
14. Peter Brucker, Johann Hurink, Bernd Jurisch, and Birgit Wöstmann. A branch & bound algorithm for the open-shop problem. In *GO-II Meeting: Proceedings of the second international colloquium on Graphs and optimization*, pages 43–59, Amsterdam, The Netherlands, The Netherlands, 1997. Elsevier Science Publishers B. V.
15. Ching-Fang Liaw. A tabu search algorithm for the open shop scheduling problem. *Computers and Operations Research*, 26(2):109–126, 1999.
16. Walter H. Kohler and Kenneth Steiglitz. Exact, approximate, and guaranteed accuracy algorithms for the flow-shop problem n/2/f/ f. *Journal of the ACM*, 22(1):106–114, 1975.
17. Mikkel T. Jensen. *Robust and Flexible Scheduling with Evolutionary Computation.* PhD thesis, Department of Computer Science. University of Aarhus, Aarhus, Denmark, October 2001.
18. Runwei Cheng, Mitsuo Gen, and Yasuhiro Tsujimura. A tutorial survey of job-shop scheduling problems using genetic algorithms i: Representation. *Computers & Industrial Engineering*, 30(4):983–997, 1996.
19. Zuobao Wu and M. X. Weng. Multiagent scheduling method with earliness and tardiness objectives in flexible job shops. *IEEE Transactions on Systems, Man, and Cybernetics, Part B*, 35(2):293–301, 2005.
20. Imed Kacem, Slim Hammadi, and Pierre Borne. Pareto-optimality approach for flexible job-shop scheduling problems: Hybridization of evolutionary algorithms and fuzzy logic. *Mathematics and Computers in Simulation*, 60(3-5):245–276, 2002.
21. Takeshi Yamada and Ryohei Nakano. A genetic algorithm applicable to large-scale job-shop problems. In Reinhard Männer and Bernard Manderick, editors, *PPSN*, pages 283–292. Elsevier, 1992.
22. Robert H. Storer, S. David Wu, and Renzo Vaccari. New search spaces for sequencing problems with application to job shop scheduling. *Management Science*, 38(10):1495–1509, 1992.
23. Bryan A. Norman and James C. Bean. A genetic algorithm methodology for complex scheduling problems. *Naval Research Logistics*, 46(2):199–211, 1999.
24. Ulrich Dorndorf and Erwin Pesch. Evolution based learning in a job shop scheduling environment. *Computers and Operations Research*, 22(1):25–40, January 1995.
25. Joseph Adams, Egon Balas, and Daniel Zawack. The shifting bottleneck procedure for job shop scheduling. *Manage. Sci.*, 34(3):391–401, 1988.

26. R. Cheng, M. Gen, and Y. Tsujimura. A tutorial survey of job-shop scheduling problems using genetic algorithms. ii. hybrid genetic search strategies. *Computers & Industrial Engineering*, 37(1-2):51–55, 1999.
27. David E. Goldberg. *Genetic Algorithms in Search, Optimization and Machine Learning*. Addison-Wesley Publishing Company, Reading, Massachusetts, 1989.
28. Lawrence Davis, editor. *Handbook of Genetic Algorithms*. Van Nostrand Reinhold, New York, New York, 1991.
29. Carlos Cotta and José M. Troya. Genetic forma recombination in permutation flowshop problems. *Evolutionary Computation*, 6(1):25–44, 1998.
30. E. Falkenauer and S. Bouffouix. A genetic algorithm for the job shop. In *1991 IEEE International Conference on Robotics and Automation*, pages 824–829, 1991.
31. Shigenobu Kobayashi, Isao Ono, and Masayuki Yamamura. An efficient genetic algorithm for job shop scheduling problems. In Larry J. Eshelman, editor, *ICGA*, pages 506–511. Morgan Kaufmann, 1995.
32. Isao Ono, Masayuki Yamamura, and Shigenobu Kobayashi. A genetic algorithm for job-shop scheduling problems using job-based order crossover. In *International Conference on Evolutionary Computation*, pages 547–552, 1996.
33. Mitsuo Gen, Yasuhiro Tsujimura, and Erika Kubota. Solving job-shop scheduling problems by genetic algorithm. In *Proceedings of the 1994 IEEE International Conference on Systems, Man, and Cybernetics*, volume 2, pages 1577–1582, 1994.
34. Thomas Stützle. An ant approach to the flow shop problem. In *6th European Congress on Intelligent Techniques and Soft Computing*, pages 1560–1564, 1998.
35. Tadahiko Murata and Hisao Ishibuchi. Performance evaluation of genetic algorithms for flowshop scheduling problems. In *International Conference on Evolutionary Computation*, pages 812–817, 1994.
36. Hisao Ishibuchi, Tadashi Yoshida, and Tadahiko Murata. Balance between genetic search and local search in memetic algorithms for multiobjective permutation flow shop scheduling. *IEEE Trans. Evolutionary Computation*, 7(2):204–223, 2003.
37. Tadahiko Murata, Hisao Ishibuchi, and H. Tanaka. Multi-Objective Genetic Algorithm and Its Application to Flowshop Scheduling. *Computers and Industrial Engineering*, 30(4):957–968, September 1996.
38. Hisao Ishibuchi and Tadahiko Murata. Multi-Objective Genetic Local Search Algorithm and Its Application to Flowshop Scheduling. *IEEE Transactions on Systems, Man and Cybernetics*, 28(3):392–403, August 1998.
39. Hisao Ishibuchi and Youhei Shibata. Single-Objective and Multi-Objective Evolutionary Flowshop Scheduling. In Carlos A. Coello Coello and Gary B. Lamont, editors, *Applications of Multi-Objective Evolutionary Algorithms*, pages 529–554. World Scientific, Singapore, 2004.
40. Anant Singh Jain and Sheik Meeran. A state-of-the-art review of job-shop scheduling techniques. *Technical report, Department of Applied Physics, Electronic and Mechanical Engineering, University of Dundee, Dundee, Scotland*, 1998.
41. Christian Bierwirth. A generalized permutation approach to job shop scheduling with genetic algorithms. *OR Spektrum*, 17:87–92, 1995.
42. Monaldo Mastrolilli and Luca Maria Gambardella. Effective neighborhood functions for the flexible job shop problem. *Journal of Scheduling*, 3:3–20, 2000.
43. Dana S. Nau, Tsz-Chiu Au, Okhtay Ilghami, Ugur Kuter, Héctor Muñoz-Avila, J. William Murdock, Dan Wu, and Fusun Yaman. Applications of shop and shop2. *IEEE Intelligent Systems*, 20(2):34–41, 2005.
44. Austin Tate, Brian Drabble, and Richard Kirby. O-plan2: An architecture for command, planning and control. In Mark Fox and Monte Zweben, editors, *Intelligent Scheduling*. Morgan-Kaufmann Publishing, 1994.

45. David E. Wilkins. Can ai planners solve practical problems? *Comput. Intell.*, 6(4):232–246, 1990.
46. Dana S. Nau, Tsz-Chiu Au, Okhtay Ilghami, Ugur Kuter, J. William Murdock, Dan Wu, and Fusun Yaman. Shop2: An htn planning system. *J. Artif. Intell. Res. (JAIR)*, 20:379–404, 2003.
47. Carlos A. Coello Coello, David A. Van Veldhuizen, and Gary B. Lamont. *Evolutionary Algorithms for Solving Multi-Objective Problems.* Kluwer Academic Publishers, New York, May 2002. ISBN 0-3064-6762-3.
48. Kalyanmoy Deb, Samir Agrawal, Amrit Pratab, and T. Meyarivan. A Fast Elitist Non-Dominated Sorting Genetic Algorithm for Multi-Objective Optimization: NSGA-II. In Marc Schoenauer, Kalyanmoy Deb, Günter Rudolph, Xin Yao, Evelyne Lutton, Juan Julian Merelo, and Hans-Paul Schwefel, editors, *Proceedings of the Parallel Problem Solving from Nature VI Conference*, pages 849–858, Paris, France, 2000. Springer. Lecture Notes in Computer Science No. 1917.
49. Eckart Zitzler, Marco Laumanns, and Lothar Thiele. SPEA2: Improving the Strength Pareto Evolutionary Algorithm. In K. Giannakoglou, D. Tsahalis, J. Periaux, P. Papailou, and T. Fogarty, editors, *EUROGEN 2001. Evolutionary Methods for Design, Optimization and Control with Applications to Industrial Problems*, Athens, Greece, September 2001.
50. Carlos M. Fonseca and Peter J. Fleming. An Overview of Evolutionary Algorithms in Multiobjective Optimization. *Evolutionary Computation*, 3(1):1–16, Spring 1995.
51. Mark R. Knarr, Mark N. Goltz, Gary B. Lamont, and Junqi Huang. *In Situ* Bioremediation of Perchlorate-Contaminated Groundwater using a Multi-Objective Parallel Evolutionary Algorithm. In *Congress on Evolutionary Computation (CEC'2003)*, volume 1, pages 1604–1611, Piscataway, New Jersey, December 2003. IEEE Service Center.
52. Mark P. Kleeman and Gary B. Lamont. Solving the Aircraft Engine Maintenance Scheduling Problem Using a Multi-objective Evolutionary Algorithm. In *Evolutionary Multi-Criterion Optimization 2005*, volume 1, pages 782–796, March 2005.
53. Mark P. Kleeman and Gary B. Lamont. Solving the aircraft engine maintenance scheduling problem using a multi-objective evolutionary algorithm. In *GECCO '05: Proceedings of the 2005 workshops on Genetic and evolutionary computation*, pages 196–198, New York, NY, USA, 2005. ACM Press.
54. Traci A. Keller. Optimization of a Quantum Cascade Laser Operating in the Terahertz Frequency Range Using a Multiobjective Evolutionary Algorithm. Master's thesis, Air Force Institute of Technology, Wright-Patterson AFB, OH, June 2004.
55. Zbigniew Michalewicz and Cezary Z. Janikow. Genocop: a genetic algorithm for numerical optimization problems with linear constraints. *Commun. ACM*, 39(12es):223–240, 1996.
56. Zbigniew Michalewicz. Evolutionary computation techniques for nonlinear programming problems. *International Transactions in Operational Research*, 1(2):175, 1994.
57. Mark R. Knarr. Optimizing an In Situ Bioremediation Technology to Manage Perchlorate-Contaminated Groundwater. Master's thesis, Air Force Institute of Technology, Wright-Patterson AFB, OH, March 2003.
58. Traci A. Keller and Gary B. Lamont. Optimization of a Quantum Cascade Laser Operating in the Terahertz Frequency Range Using a Multiobjective Evolutionary Algorithm. In *17th International Conference on Multiple Criteria Decision Making (MCDM 2004)*, volume 1, December 2004.

59. Clyde W. Holsapple, Varghese S. Jacob, Ramakrishnan Pakath, and Jigish S. Zaveri. A Genetics-based Hybrid Scheduler for Generating Static Schedules in Flexible Manufacturing Contexts. *IEEE Transactions on Systems, Man and Cybernetics*, 23:953–972, July-Aug 1993.
60. Jr. Kenneth E. Kinnear. Derivative Methods in Genetic Programming. In T. Bäck, D. B. Fogel, and Z. Michalewicz, editors, *Evolutionary Computation 1: Basic Algorithms and Operators*, pages 103–113. Institute of Physics Publishing, 2000.

Designing Dispatching Rules to Minimize Total Tardiness

Joc Cing Tay and Nhu Binh Ho

Evolutionary and Complex Systems Program (EvoCom)
Nanyang Technological University, School of Computer Engineering,
Blk N4 #2a-32 Nanyang Avenue, Singapore 639798

Summary. We approximate optimal solutions to the Flexible Job-Shop Problem by using dispatching rules discovered through Genetic Programming. While Simple Priority Rules have been widely applied in practice, their efficacy remains poor due to lack of a global view. Composite Dispatching Rules have been shown to be more effective as they are constructed through human experience. In this work, we employ suitable parameter and operator spaces for evolving Composite Dispatching Rules using Genetic Programming, with an aim towards greater scalability and flexibility. Experimental results show that Composite Dispatching Rules generated by our Genetic Programming framework outperforms the Single and Composite Dispatching Rules selected from literature over large validation sets with respect to total tardiness. Further results on sensitivity to changes (in coefficient values and terminals) among the evolved rules indicate that their designs are optimal.

1 Introduction

In today's highly competitive marketplace, a high level of delivery performance has become necessary to satisfy customers. Due to market trends, product orders of low volume, high variety types have been increasing in demand. Hoitomt *et al.* [1] mentions that these products comprise between 50 to 75 % of all manufactured components, thereby making schedule optimization an indispensable step in the overall manufacturing process.

J.C. Tay and N.B. Ho: *Designing Dispatching Rules to Minimize Total Tardiness,* Studies in Computational Intelligence (SCI) **49**, 101–124 (2007)
www.springerlink.com

The Job-Shop Scheduling Problem (JSP) is one of the most popular manufacturing optimization models used in practice [2]. It has attracted many researchers due to its wide applicability and inherent difficulty [3-6]. It is also well known that the JSP is NP-hard [7], hence general, deterministic methods of search are inefficient as the problem size grows larger. The $n \times m$ classical JSP involves n jobs and m machines. Each job is to be processed on each machine in a pre-defined sequence, and each machine processes only one job at a time. In practice, the shop-floor setup typically consists of multiple copies of the most critical machines so that bottlenecks due to long operations or busy machines can be reduced. As such, an operation may be processed on more than one machine having the same function. This leads to a more complex problem known as the Flexible Job Shop Scheduling Problem (FJSP). The extension involves two tasks; assignment of an operation to an appropriate machine and sequencing the operations on each machine. In addition, for complex manufacturing systems, a job can typically visit a machine more than once (known as recirculation). These three features of the FJSP significantly increase the complexity of finding even approximately optimal solutions [8].

The classical JSP and FJSP have been solved by many stochastic local search methods, such as Simulated Annealing [4], Tabu Search [5, 9, 10] and Genetic Algorithms [11-14]. The reported results of applying them show that good approximations of optimality can be found, albeit at the expense of a huge computational cost, particularly when the problem size is large. In practice, dispatching rules have been applied to avoid these costs [15-17]. Although the quality of solutions produced by dispatching rules are no better than the local search methods, they are the more frequently applied technique due to their ease of implementation and their low time complexity. Whenever a machine is available, a priority-based dispatching rule inspects the waiting jobs and selects the job with the highest priority to be processed next. Recently, the introduction of composite dispatching rules (CDR) have been increasingly investigated by the some researchers [18, 19], but typically only for classical JSPs. These rules are the heuristic combination of single dispatching rules that aim to inherit the advantages of the former. The results show that, with careful combination, the composite dispatching rules will perform better than the single ones with regards to the quality of schedules.

In this paper, we investigate the potential use of GP for evolving effective composite dispatching rules for solving the FJSP, with the objective of minimizing total tardiness. The purpose of this research is to find efficient composite dispatching rules that perform better than the dispatching rules presented in literature for solving the same problem. By using a wide training data set, we believe that the evolved CDRs can be applied directed in practice without any modifications. Furthermore, the results of these CDRs could be used as the input to other local search methods in solving FJSP problems, such as Genetic Algorithms [13, 14].

The remainder of this paper is organized as follows. Section 2 gives the formal definition of the FJSP. Section 3 reviews recent related works for solving the JSP and FJSP using dispatching rules and a overview of GP. Section 4 describes our proposed GP framework for evolving CDRs while Section 5 analyzes the performance results of the CDRs obtained with GP. Finally, Section 6 gives some concluding remarks and directions for future work.

2 Problem definition

Similar to the classical JSP, solving the FJSP requires the optimal assignment of each operation of each job to a machine with known starting and ending times. However, the task is more challenging than the classical one because it requires a proper selection of a machine from a set of machines to process each operation of each job. Furthermore, if a job is allowed to recirculate, this will significantly increase the complexity of the system [20]. The FJSP is formulated as follows:

- Let $J = \{J_i\}_{1 \le i \le n}$, indexed i, be a set of n jobs to be scheduled.
- Each job J_i consists of a predetermined sequence of operations. Let $O_{i,j}$ be operation j of J_i.
- Let $M = \{M_k\}_{1 \le k \le m}$, indexed k, be a set of m machines.
- Each machine can process only one operation at a time.
- Each operation $O_{i,j}$ can be processed without interruption on one of a set of machines M_k in a given set $\mu_{i,j} \subset M$ with $p_{i,j,k}$ time units.
- Let C_i and d_i be the completion time and due date of job J_i respectively. The tardiness of this job is calculated by the following formula:
$$T_i = \max\{0, C_i - d_i\}$$
- The objective function T of this problem is to find a schedule that minimizes the sum of tardiness of all jobs (total tardiness problem):
$$T = \min \sum_{i=1}^{n} T_i = \min \sum_{i=1}^{n} \max\{0, C_i - d_i\}$$

Total tardiness is one of the major objectives in production scheduling. A job that is late may penalize the company's reputation and reduce customer satisfaction. Hence, keeping the due dates of jobs under control is one of the most important tasks faced by companies [19].

The FJSP can also be considered to be a Multi Purpose Machine (*MPM*) job-shop [21]. Using the α|β|γ notation of [22], the problem we wish to solve can be denoted by

$$J\ MPM \mid prec\ r_j\ d_j \mid \sum_j T_j$$

where J denotes job-shop problem, *MPM* denotes multi purpose machine, *prec* represents a set of independent *chains* while r_j and d_j represents *release date* and *due date* given to each job respectively; finally, $\sum_j T_j$ represents total tardiness.

In this paper, we shall assume the following:

- All machines are available at time 0.
- Preemption of operations is not allowed.
- Each job has its own release date and due date.

The order of operations for each job is predefined and cannot modified.

3 Previous works

Dispatching rules have received much attention from researchers over the past decades [15-17]. In general, whenever a machine is freed, a job with the highest priority in the queue is selected to be processed on a machine or work center. A comprehensive survey on dispatching rules is by Panwalkar and Wafik [15] and Blackstone *et al.* [16]. Depending on the specification of each rule, it can be classified [15] into:

- Simple Priority Rules
- CDRs
- Weighted Priority Indexes
- Heuristic Scheduling Rules

Simple Priority Rules (SPR) are usually based on a single objective function. They usually involve only one model parameter, such as processing time, due date, number of operations or arrival time. The Shortest Processing Time (SPT) rule is an example of a SPR. It orders the jobs on the queue in the order of increasing processing times. When a machine is freed, the next job with the shortest time in the queue will be removed for processing. SPT has been found to be the best rule for minimizing the mean flow time and number of tardy jobs [17]. The Earliest Due Date (EDD) rule is another example of a SPR where the next job to be processed is the one with the earliest due date. Unfortunately, no SPR performs well across every performance measure such as tardiness or flow time

[23]. To overcome this limitation, CDRs have been studied to combine good features from such SPRs.

There are two kinds of CDRs presented in literature; the first type involves deploying a select number of SPRs at different machines or work centers. Each machine or work center employs a single rule. When a job enters a specific machine or work center, it is processed by the SPR that is preselected for that machine or work center. For instance, Barman [23] applied three different SPRs to solve the flow shop problem corresponding to three work centers. Experimental results show that it obtains better results than a single SPR that is common to all three machines. However, this approach may not be suitable for a shop floor with large number of machines or work centers; and the best independent distribution of single SPRs is difficult to predetermine. Furthermore, it still has the limitation of a localized view. The second type involves applying the composition of several SPRs (otherwise known as a CDR) to evaluate the priorities of jobs on the queue [17]. The latter type is executed similarly to SPRs; when a machine is free, this CDR evaluates the queue and then selects the job with the highest priority. For example, Oliver and Chandrasekharan [17] present five CDRs for solving the JSP. Their results indicate that CDRs are more effective compared to individual SPRs. CDRs inherit the simplicity of SPRs while achieving some scalability as the number of machines increase. Moreover, if well designed, CDRs can solve realistic problems with multiple objectives [8]. However, the challenge is to find a good combination of SPRs to apply to all machines or work centers.

Weighted priority index rules are the linear combination of SPRs with computed weights [18, 19]. Depending on specific business domains, the importance of a job determines it's weight. For instance, consider n jobs with different weights w, each job J_i is assigned weight w_i. The sum of the weighted tardiness as the objective function is given as follows:

$$T = \min \sum_{i=1}^{n} w_i T_i = \min \sum_{i=1}^{n} w_i \times \max\{0, C_i - d_i\}$$

In this paper, weighted priority rules are not considered as they are a generalization of our current formulation of total tardiness where we have assumed instead that all jobs have unit weights (or all jobs are equally important) (see Section 2).

Heuristic rules are rules that depend on the configuration of the system. These rules are usually used together with previous rules, such as SPRs, CDRs or weighted priority index rules. For instance, the heuristic rules may use the expertise of human experience, such as inserting an operation of a job into an idle time slot by visual inspection of a schedule [15].

The results from recent researchers [17, 23] show that CDRs outperform individual SPRs in improving the efficiency of the shop floor. In this work, we focus our attention on finding a computational method to build effective CDRs; one that is based on the composition of fundamental measures rather than on the algebraic combination of SPRs. However, this may be difficult to enumerate manually due to the large parameter and operator space, hence we employ a GP framework.

Genetic programming (GP) [24] belongs to a family of evolutionary computation methods. It is based on the Darwinian principle of reproduction and survival of the fittest. Given a set of functions and terminals and an initial population of randomly generated syntax trees (representing programs), the programs are evolved through genetic recombination and natural selection. GP has been applied to many different problems; from classical tasks, such as function fitting or pattern recognition, to non-trivial tasks that are competitive with significant human endeavours such as designing electrical circuits [25] or antennas [26].

The most important feature that makes GP different from the canonical GA is it's ability to vary the logical structure and size of evolved computer programs dynamically. It can therefore solve more challenging problems that have eluded the canonical GA due to the latter's requirement of a fixed-length chromosome. However, GP has rarely been applied to manufacturing optimization [27]; this is due to the direct permutation property of scheduling where jobs and/or machines can be simply reordered (in the case of JSP) to improve optimality. For instance, the chromosomes presented in [10-14] have fixed lengths, which can be evolved easily by direct permutation. On the other hand, GP uses a tree-based encoding with dynamic length; making it difficult to encode the JSP (for that matter, a FJSP) into a tree-based chromosome. Unlike previous approaches [17-19, 23] where a predefined set of SPRs were combined in advance by human experience, we apply GP to find superior constructions of CDRs which are composed of fundamental terminals (see Table 1). These discovered rules are then used to solve the FJSP directly; the advantage being that the obtained CDRs can solve the FJSPs in shorter computational time as compared to genetic algorithms [10-14].

4 Design of the GP framework

In GP, an individual (ie, computer program) is composed of terminals and functions. Therefore, when applying GP to solve a specific problem, they should be well designed to satisfy the requirements of the current problem.

After evaluating many parameters related to the FJSP, the terminal set and the function set that are used in our algorithm are described as follows.

4.1 Terminal set

In job-shop scheduling, there are many parameters that can effect the quality of results; potentially, all of them can be considered to comprise a dispatching rule. However, in order to create a short and meaningful dispatching rule, only a small and sufficient number of parameters should be evaluated properly. They also help to reduce the search space and improve the efficiency of the algorithm. Based upon the dispatching rules involving due dates in [15-17] and our experimental works, the terminal set proposed in this study is given in Table 1.

Table 1. Terminal Set

Terminal	Meaning
ReleaseDate	Release date of a job (RD)
DueDate	Due date of a job (DD)
ProcessingTime	Processing time of each operation (PT)
CurrentTime	Current time (CT)
RemainingTime	Remaining processing time of each job (RT)
numOfOperations	Number of operations of each job (nOps)
avgTotalProcTime	Average total processing time of each job (aTPT)

In Table 1, *CurrentTime* represents the time when a particular machine is free and starts to select a job to process on its queue. *RemainingTime* corresponds to the elapsed time for the current job to finish. Some previous dispatching rules use total processing time of each job as one of their parameters. However, in FJSP, as an operation of each job can be processed on a set of machines, we normalize the average processing time of each operation with the following formula:

$$\overline{p}_{i,j} = \frac{\sum_{n(\mu_{i,j})} p_{i,j,k}}{n(\mu_{i,j})}$$

where $p_{i,j,k}$ stands for processing time of operation $O_{i,j}$ on machine M_k and $n(\mu_{i,j})$ represents the number of machines that can process $O_{i,j}$.

4.2 Function set

Similar to other applications of GP [24-26] for solving optimization problems, we use four basic operators: addition, subtraction, multiplication, and division for creating a CDR. Furthermore, we employ a well-known Automatically Defined Function (ADF) (proposed by Koza [28]). The ADF is sub-tree which can be used as a function in the main tree. The size of the ADF is varied in the same manner as the main tree. It enables GP to define useful and reusable subroutines dynamically during its run. The results from [28] indicate that GP using ADF outperforms GP without ADF in solving the same optimization problem. The more parameters are used in ADF, the more changes will be needed for GP to evolve good subroutines. However, it can lead to a higher number of generations. We limit the ADF used in our approach to two parameters. The operators used in the ADF are also the four basic operators mentioned above. The operators of the function set in our approach are given in Table 2.

Table 2. Terminal Set

Function	Meaning
+	Addition
-	Subtraction
*	Multiplication
/	Division
ADF(x_1, x_2)	Automatically Defined Function

4.3 Encoding a CDR using a GP chromosome

The obtained results from each generation of GP are a set of computer programs represented as trees. As mentioned in Section 2, the objective in our study is to minimize the total tardiness of the FJSPs. Therefore, we propose a method to form a CDR from the tree-based result of GP. This CDR is then combined with the *least waiting time assignment* [13] to evaluate the total tardiness of the FJSPs. These two processes are described in detail as follows.

To find a suitable machine to process an operation $O_{i,j}$, we apply the *least waiting time assignment* on the set of setting machines that can process $O_{i,j}$. This rule is intended to reduce the workloads of the machines by balancing operations to be assigned. It is calculated by *summing* all the subsequent operations in the waiting list *plus* the remaining processing time on each machine and the processing time of $O_{i,j}$. Therefore, it depends on the total time this operation has to wait to be processed in the worst case, not relying only on its own processing time.

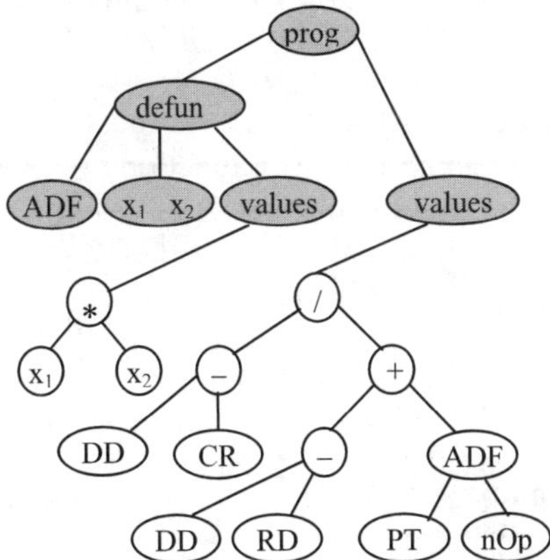

Fig. 1. Example of a GP tree with defined functions and terminals

In determining the proper order of operations on the queue of a particular machine, we use the CDR generated by GP. When a machine is freed, the generated rule is applied directly to the set of operations that are waiting in the queue of the machine. The operation with the highest priority is then selected to be processed on the machine. Figure 1 above gives an example of a dispatching rule tree generated by GP. It shows the overall structure of the generated tree that gives a possible CDR. The left child of *progn* shows the function-defining branch (containing the *defun*). In this case, the ADF function is defined by: $ADF(x_1,x_2)=x_1*x_2$. The right child gives the result-producing branch. This CDR therefore represents the following formula:

$$\frac{(DD-CR)}{(DD-RD)+ADF(PT,nOps)}$$

Since $ADF(x_1,x_2)=x_1*x_2$, we obtain:

$$\frac{(DD-CR)}{(DD-RD)+(PT*nOps)}$$

Any tree in the genomic population of GP that contains our defined functions and terminals can be interpreted as a CDR in the same way. The obtained CDR is then applied to solve a FJSP problem to evaluate its total tardiness.

4.4 GP parameter settings

Through experimentation, the set of parameters used in our GP framework is listed in Table 3.

Table 3. Choice of parameter values

Parameters	Value
Population Size	100
Number of Generations	200
Creation Type	Ramped half and half
Maximum depth for creation	7
Maximum depth for crossover	17
Crossover Probability	100%
Swap Mutation Probability	3%
Shrink Mutation Probability	3%
Number of best rules copy to new generation	5

We implemented *Ramped half and half* to generate the initial population of GP. This method was proposed by Koza [24] and it has been widely used by previous researchers. It divides the initial population into two parts; half of which contains the random generated trees with maximum depth (in this experiment, this value is 7) and the remaining half contains the random generated trees with depth values ranging from one to the maximum depth. In order to keep the best trees that may be destroyed by GP's operators, we sort the current population and copy five of them to the next generation.

5 Experimental results

This section reports and analyses the results of our computational experiments. The system was implemented using C++, running on a 2 GHz PC with 512 MB RAM. We will describe how to generate the test cases that are used to evolve CDRs for minimizing total tardiness objective of FJSP problems. The performance results of the evolved dispatching rules will be compared to some commonly used dispatching rules in literature. Finally, the evolved dispatching rules' sensitive parameters will be discussed.

5.1 Test case generation

Various experiments were conducted to evaluate the efficiency of our proposed algorithms. We categorized these experiments into three classes: FJSP with 100% flexible (FJSP-100), FJSP with 50% of flexibility (FJSP-50), and FJSP with 20% of flexibility (FJSP-20). The FJSP with c% of flexibility means that less than or equal c% of total machines are selected to process an operation. The processing times of each operation was drawn out of U((number of machines)/2, (number of machines)×2), where U refers to a uniform distribution. In practice, an operation can be processed on any of a group of machines that constitute a work center. The variance of these processing times is ideally zero or usually small. Therefore, in our test cases, we set the maximum difference between two operations to be 5 unit times. The release date of each job depends on the number of jobs in a particular test case. If the number of jobs is larger than 50, the release date is drawn out of U[0,40], else it is taken from U[0,20]. Baker [29] proposed a formula to estimate the due date of a job using the TWK-method:

$$d_i = r_i + c \times \sum_{j=1}^{n_i} p_{ij}$$

where r_i and d_i denote release and due dates of job i respectively. p_{ij} presents the processing time of operation O_{ij}, and c denotes the tightness factor of the due date. The higher the value of c, the looser is the job's due date. We adapt this formula to generate due dates of jobs by replacing the parameter p_{iq} with $\overline{p}_{iq}$.

Depending on the tightness of the due date, we separate the samples of each class FJSP-100, FJSP-50, or FJSP-20 into tight, moderate, or loose due dates corresponding to values of c = 1.2, 1.5, and 2. We also generate mixed samples where each sample contains 34% jobs with tight due dates, 33% of jobs with moderate due dates, and the remaining ones with loose due dates. Specifically, the class FJSP-100 holds 9 samples of tight due date, 9 samples of moderate due date, 9 samples of loose due date, and 9 samples of mix due date. Similarly for FJSP-50 and FJSP-20, with 36 samples each. Each training set contains three classes of 108 FJSP problems with different number of jobs, machines and different tightness of jobs. Another five validation sets (with 108x5 FJSP problems) of similar compositions were generated.

In order to understand how our GP framework can adapt to the different conditions of the shop floors for evolving efficient dispatching rules, we divide the experiments into two test samples:

- Test sample 1: varying both the number of jobs and number of machines. Number of jobs and number of machines range from 10 to 200 and 5 to 15,

respectively. This test sample contains 108 training FJSP problems and 108x5 validating FJSP problems.
- Test sample 2: varying the number of jobs and fixing the number of machines. Number of jobs ranges from 20 to 200 and number of machines is fixed at 10. This test sample contains 108 training FJSP problems and 108x5 validating FJSP problems.

The evolved dispatching rules obtained from the test sample 1 are aimed to solve the FJSP problems in the general case of a varying number of jobs and machines while the results of test sample 2 are aimed to solve FJSP problems where the number of machines is unchanged. After training the GP framework with the training set, five best rules were reported. The mean total tardiness of these evolved rules after 500 runs on the validation sets are then reported.

In order to compare the effectiveness of the evolved rules to the human-made rules presented in literature, five frequently used single and composite dispatching rules were selected as benchmarks:

- FIFO (First In First Out): select the job with the earliest coming. This rule is often used in practice since it is simple and easy to implement [16].
- SPT (Shortest Processing Time): select the job with the shortest processing time. This rule is commonly used as a benchmark for minimizing mean flow time and percent of tardy jobs [30].
- EDD (Earliest Due Date): select the job with the earliest due date. This rule is the most popular due date based rule. It is known to be used as a benchmark for reducing maximum tardiness and variance of tardiness [30].
- MDD (Modified Due Date) (= max{$CT+PT_i$, DD_i}): process the jobs in non-decreasing order of MDD. This rule is aimed to minimize total tardiness of jobs [18].
- SL (Slack Time) (= $DD_i - CT - RT_i$): select the job with the minimum slack time. This rule is also used to reduce total tardiness of jobs [17].

Blackstone *et al.* [16] mentions that the study of job shops by analytic techniques, such as queuing theory, becomes extremely complex even for small problems. Therefore, the use of simulation for analyzing dispatching rules is unavoidable. Due to the same difficulty in examining the dispatching rules for solving FJSPs, we also rely on simulation to study the rules' effectiveness. For comparative studies of algorithms in constrained problems, we adopt the approach of [31] in using a one way Analysis of Variances (ANOVA) [32]. The function of ANOVA is based on the ratio of variations. It tests the differences between the means of two or more groups. In this paper, it is used to compare the sample mean of a particular objective for a evolved rule with other sample means (for other rules) that overlap with the former's confidence interval (CI). If an overlap exists, this implies some uncertainty concerning the existance of a performance differential. The values of 99% CI for each sample mean are calculated and presented.

5.2 Test sample 1

The best five dispatching rules that were selected from 5 runs times of GP on the training set of test sample 1 are given in Table 4; where possible, they were simplified algebraically.

Table 4. GP generated dispatching rules for test sample 1

Rule	Expression
Rule_1	aTPT ∗ (CT +RD + PT − 3)+ (CT ∗ PT + RD + nOps) − (nOps ∗ PT + 2PT+CT+1)
Rule_2	(PT+ CT+ RD + 2) ∗ (RT+ PT + aTPT)
Rule_3	CT ∗ aTPT + 5nOps + 3RD
Rule_4	DD ∗ (RD + aTPT + RT + PT)
Rule_5	(aTPT + PT) ∗ (CT + RD) + (DD - RD)

Figure 2 below compares the results of the evolved rules in Table 4 and the five selected dispatching rules for solving different FJSPs. The x-axis represents the dispatching rules while the y-axis represents the average total tardiness of each rule after 500 runs on the five validation test sets.

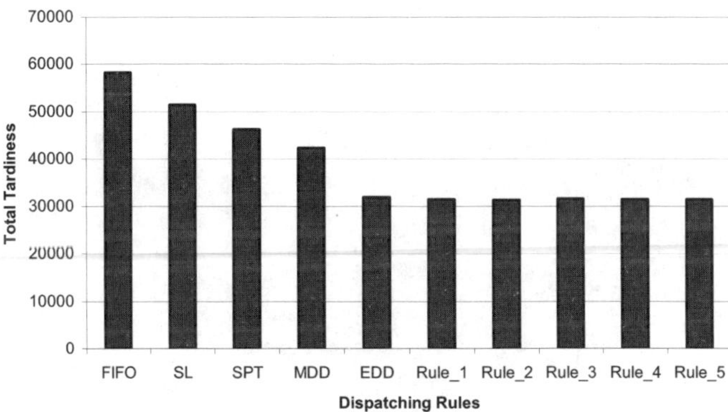

Fig. 2. Performance of dispatching rules on validation test sets in test sample 1

Results from Figure 2 show that the FIFO rule performs poorly in comparison to the others. This is because the due date of jobs are ignored by

FIFO, and therefore the rule does not focus on minimizing total tardiness. The composite dispatching rule SL can obtain slightly better results than FIFO but is still poor in comparison to the remaining rules. Figure 2 indicates that MDD outperforms SL. From the definition of MDD and SL described in Section 5.1, we observe that although these two composite rules contain similar parameters (DD and CT), the gap between the results of the two rules are quite large due to different algebraic combinations of the parameters. This emphasizes that the functions that combine the rules can significantly affect the results. EDD is the best among five rules selected from literature (FIFO, SPT, EDD, MDD, SL) for solving FJSP problems. This could be explained by the presence of the parameter DD in its formula. If the job on the queue is selected by EDD, it has more likely to finish on time since the job with the earliest due date will be selected. Therefore, the total tardiness can be minimized. Although the other rules such as SL or MDD also contain the parameter - due date (DD), EDD obtains almost 50% better results than these rules. This again demonstrates that if an *ineffective* composite dispatching rule is applied to specific problems, it may achieve worse results than the single ones.

We now compare the GP generated rules against the most effective rule (EDD). Figure 3 gives box-plots of the data distribution of EDD and the five GP evolved dispatching rules after 500 runs. For each rule in Figure 3, the box represents the interquartile range which contains the 50% of values. A line across each box denotes the median. Two lines that extend from the box gives the highest and lowest values while the circles denote outliers.

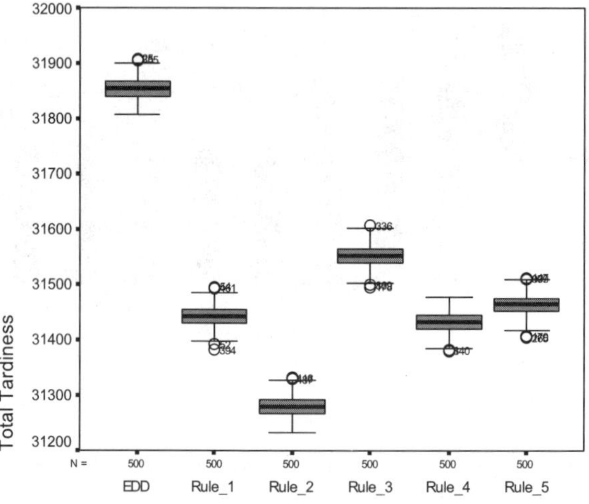

Fig. 3. Data distribution of EDD and *Rule_1* to *Rule_5* after 500 runs

Figure 4 shows in detail the mean total tardiness with 99% CI for each rule. For each rule, the small square in the middle denotes the mean value while two leaves in two sides denote the CI values. It indicates that the results of all evolved rules are much better than by the most effective human-made rule EDD. The best performing rule is the generated rule- *Rule_2* ((PT+ CT+ RD + 2) ∗ (RT+ PT + aTPT)). This is statistically true since its CI does not overlap with the others. It can be considered as the best rule among them to solve total tardiness objective. Although the mean value of total tardiness of *Rule_4* (31432.42) is smaller than the one of *Rule_1* (31442.20), we cannot conclude that Rule_4 is more effective than Rule_1 as their CIs are overlapping. In order to verify if *Rule_4* is really better than *Rule_1* (or not), we apply ANOVA to analyze the data obtained by these rules. Since $F\ ratio = 75.26$ is greater than $F_{critical} = 3.85$, we reject the null hypothesis that the samples are similar. Therefore, the difference between *Rule_4* and *Rule_1* is statistically significant. Jayamohan and Rajendran [30] mentions that the use of both due date information and processing time can lead to good results in minimizing total tardiness. Our five evolved rules present evidence for this conclusion as their formulation contains these parameters. Furthermore, we find that some parameters, such as the total number of operations (nOps) and total processing time of job (aTPT), are ignored or considered insignificant by previous researchers but according to our results, they *do* contribute to reducing mean tardiness. For example, the formula of *Rule_3* suggests that jobs with fewer number of operations have higher priority. In conclusion, the results from this test sample show that the evolved dispatching rules which are formed by the GP framework are very promising in solving the FJSP in general case.

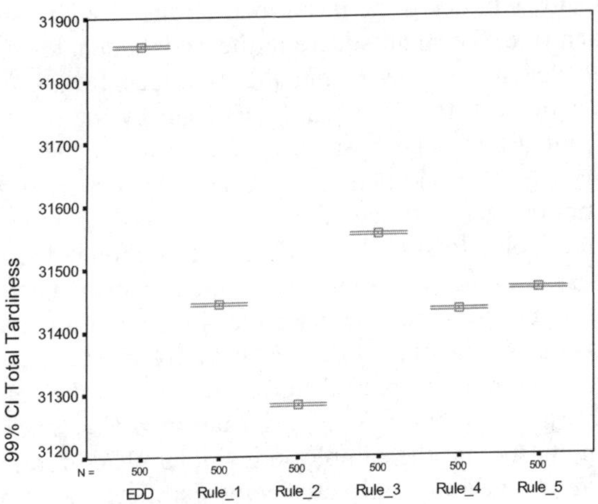

Fig. 4. Mean total tardiness with 99% confidence interval of EDD and *Rule_1* to *Rule_5* after 500 runs

5.3 Test sample 2

Table 5 below represents the best five dispatching rules that were selected from 5 runs times of GP on the training set of test sample 2; where possible, they were simplified algebraically.

Table 5. GP generated dispatching rules for test sample 2

Rule	Expression
Rule_6	3aTPT + (PT/aTPT + 1) * (RD+RT)
Rule_7	6nOps + PT+ CT*(PT+aTPT)
Rule_8	nOps + 9aTPT + 4PT
Rule_9	4aTPT - 2nOps + 3DD + 2PT
Rule_10	DD/aTPT + 2aTPT + PT + DD + RD

Similar to Section 5.2, we compare these evolved rules in Table 5 to five selected rules from literature. The bars on the x-axis from left to right denote FIFO, SL, SPT, MDD, EDD, SL, and *Rule_6 to Rule_10* while the y-axis represents total mean total tardiness after 500 runs. A visual inspection on Figure 5 again demonstrates that when the number of machines is fixed (to 10), FIFO obtains the worst results while EDD obtains the best. In this special case of FJSP customized for a particular shop floor with 10 machines, the order of the rules' performances selected from literature does not change (similar to the results in Figure 2). We conjecture that even with larger validation sets of the types described in Test sample 1 and Test sample 2, the performances of the rules selected from literature are the same regardless of varying number of machines in FJSP problems.

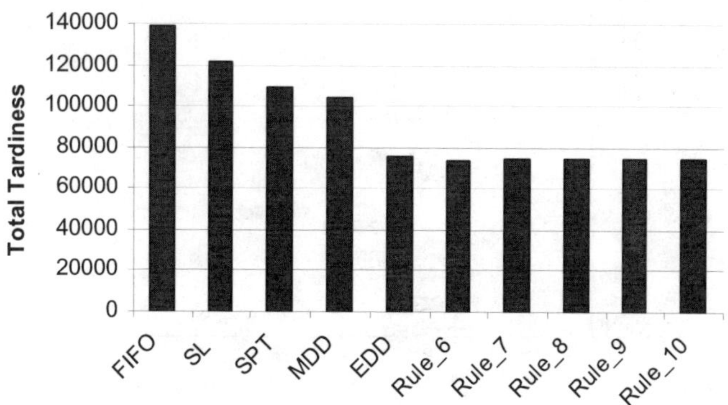

Fig. 5. Performance of dispatching rules on validation test sets in test sample 2

The total tardiness values of the evolved dispatching rules fare better than EDD. Figure 6 and Figure 7 gives the data distribution and the mean total tardiness with 99% CI of EDD and five evolved dispatching rules in Table 5 after 500 runs respectively. The results in these two figures show that the evolved dispatching rules outperform the most effective rule (EDD) among the selected rules from literature. In Figure 7, since the CIs of all the rules do not overlap, we can conclude that *Rule_6* is the most effective rule. The order of the evolved rules' performances decreases from *Rule_6* to *Rule_10*. Similar to the CDRs represented in Table 4, the CDRs in Table 5 are also combined with the same parameters (RD, DD, PT, CT, RT, aTPT, and nOps). The use of both due date information and processing time in their formulas could lead to the effectiveness of the rules [30]. Especially, we believe that the parameters nOps and aTPT contribute to the success of the CDRs as well.

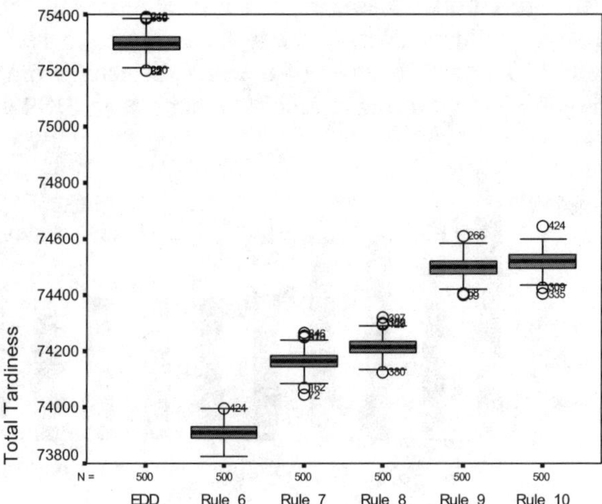

Fig. 6. Data distribution of EDD and *Rule_6* to *Rule_10* after 500 runs

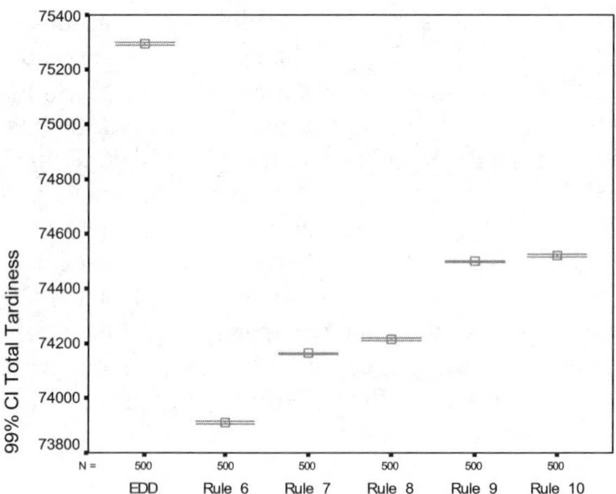

Fig. 7. Mean total tardiness with 99% confidence interval of EDD and *Rule_6* to *Rule_10* after 500 runs

5.4 Sensitivity of parameters

In order to understand why these evolved rules are effective in minimizing total tardiness, we now take a closer look at the combination of their parameters. While single rules consider only one parameter of the shop, the evolved rules employ almost all the important parameters. However, the combination of these parameters plays an essential role to the success of the rule. For instance, the composite rules SL and MDD combine the parameter DD with other parameters CT, PT, and RT but they fail to get better results than the EDD with just one parameter DD (see Figure 2). The parameters aTPT and RD could also be important for solving the problem. They are present in all the rules and contribute mainly to change the priority of one operation to be selected in a queue. For example, *Rule_2* ((PT+ CT+ RD + 2) * (RT+ PT + aTPT)) in Table 4 was constructed with these two terms. The first term operates in favor of release date RD and processing time PT while the second term runs in favor of average total processing time aTPT and remaining time RT. When the release date of a job is small, this means that the job is released early, the first term produces small results. Similarly, when the processing time of the operation is small, the second term produces a small result. Both parameters help to decrease the value of the ratio and assign a high priority to the job. Another example. It is well known that the SPT rule is effective in minimizing the number of tardy jobs [17]. Two terms of this rule also contains PT and aTPT that are in favor of the SPT. Therefore, they also contribute to improve the efficacy of the rule.

For evaluating how good dispatching rules are evolved under the GP framework, we modify *Rule_2* by eliminating or changing slightly the co-efficients of some parameters. The modifications are listed in Table 5 below.

Table 5. Modified Dispatching Rules from *Rule_2*

Rule	Expression	Modification(s) from *Rule_2*
Rule_2	(PT+ CT+ RD + 2) * (RT+ PT + aTPT)	Original Version
Rule_2_1	(PT+ RD+ 2)*(RT+ PT + aTPT)	Removed CT
Rule_2_2	(PT+ CT+ RD + 2)*(PT + aTPT)	Removed RT
Rule_2_3	(PT+ CT+ 20*RD + 2)*(RT+ PT + aTPT)	Changed RD's coefficient from 1 to 20
Rule_2_4	(PT+ CT+ RD + 2)*(RT+ PT + 20*aTPT)	Changed aTPT's coefficient from 1 to 20

In Table 5, *Rule_2_1* and *Rule_2_2* are obtained from *Rule_2* by eliminating CT and RT respectively. By changing the coefficient of RD in *Rule_2* from 1 to 20, we produce *Rule_2_3*. Similarly, *Rule_2_4* is constructed by changing the coefficient of aTPT in *Rule_2* from 1 to 20. They are then applied to solve the FJSP problems in test sample 1. Figure 8 below compares their mean total tardiness with 99% CIs to *Rule_2*'s results after 500 runs.

Figure 8 indicates that although we made small modifications to a small number of parameters of an evolved rule, the results from the obtained rules are much worse than the original one. This implies that the evolved dispatching rules from the GP framework are well designed. It also validates the importance of selecting proper parameters and of the proper algebraic combination of these parameters to construct efficient CDRs. Any changes on the evolved rules could lead to poorer results.

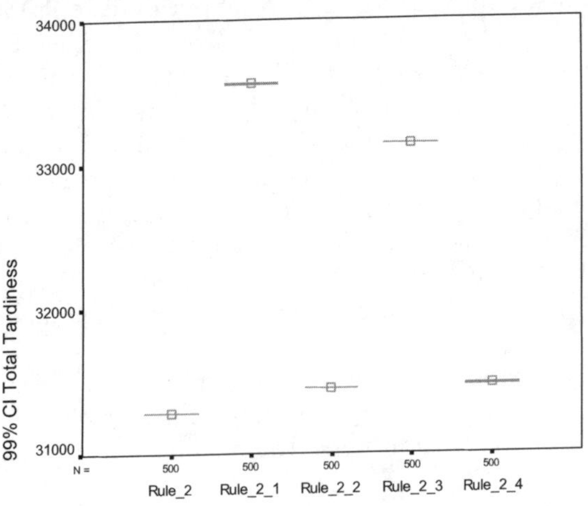

Fig. 8. Mean total tardiness with 99% confidence interval of *Rule_2* and its modified dispatching rules after 500 runs

Generally, the overall experimental results indicate that the evolved rules from our GP framework are more effective than the frequently used dispatching rules in literature. Furthermore, two parameters aTPT and nOps that have received limited study from previous research were found to contribute to the success of evolved CDRs. However, while the importance of selecting proper parameters is one factor to consider when trying to design effective CDRs. We have also proven experimentally that the way to combine these parameters is also crucial. By investigating the potential use of

[11] Kacem, I., Hammadi, S., Borne, P.: Approach by localization and multiobjective evolutionary optimization for flexible job-shop scheduling problems. Ieee T Syst Man Cy C **32** (2002) 1-13

[12] Kacem, I., Hammadi, S., Borne, P.: Pareto-optimality approach for flexible job-shop scheduling problems: hybridization of evolutionary algorithms and fuzzy logic. Math Comput Simulat (2002) 245-276

[13] Ho, N.B., Tay, J.C.: GENACE: an efficient cultural algorithm for solving the flexible job-shop problem. Proceedings of Congress on Evolutionary Computation, Vol. 2 (2004) 1759-1766

[14] Tay, J.C., Wibowo, D.: An effective chromosome representation for evolving flexible job shop schedules. Proceedings of Genetic and Evolutionary Computation (2004) 210-221

[15] Panwalkar, S.S., Iskander, W.: A Survey of Scheduling Rules. Oper Res (1977) 45-61

[16] Blackstone, J.H., Phillips, D.T., Hogg, G.L.: A state-of-the-art survey of dispatching rules for manufacturing job shop operations. Int J Prod Res (1982) 27-45

[17] Holthaus, O., Rajendran, C.: Efficient dispatching rules for scheduling in a job shop. Int J Prod Econ (1997) 87-105

[18] John, J.K., Xiaoming, L.: A Weighted Modified Due Date Rule for Sequencing to Minimize Weighted Tardiness. J Sched (2004) 261-276

[19] Jayamohan, M.S., Rajendran, C.: Development and analysis of cost-based dispatching rules for job shop scheduling. European Journal of Operational Research (2004) 307-321

[20] Pinedo, M.: Scheduling theory, algorithms, and system. Prentice Hall second edition, chapter 2 (2002)

[21] Brucker, P., Jurisch, B., Krämer, A.: Complexity of scheduling problems with multi-purpose machines. Ann Oper Res (1997) 57-73

[22] Graham, R.L., Lawler, E.L., Lenstra, J.K., Kan, A.H.G.R.: Optimization and approximation in deterministic sequencing and scheduling: A survey. Annals of Discrete Mathematics (1979) 287-236

[23] Barman, S.: Simple Priority Rule Combinations: An Approach To Improve Both Flow Time And Tardiness. Int J Prod Res (1997) 2857-2870

[24] Koza, J.: Genetic Programming: on the programming of computers by means of natural selection. MIT Press, Cambrige, MA (1992)

[25] Koza, J.R., Bennett, F.H., Andre, D., Keane, M.A., Dunlap, F.: Automated Synthesis of Analog Electrical Circuits by Means of Genetic Programming. Ieee T Evolut Comput (1997) 109-128

[26] Lohn, J.D., Hornby, G.S., Linden, D.S.: An Evolved Antenna for Deployment on NASA's Space Technology 5 Mission. Proceedings of Genetic Programming Theory Practice (2004)

[27] Dimopoulos, C., Zalzala, A.M.S.: Investigating the use of genetic programming for a classic one-machine scheduling problem. Advances in Engineering Software (2001) 489-498

[28] Koza, J.: Genetic Programming II, Automatic Discovery of Resuable Programs, Chapter 4. MIT Press (1994)

[29] Baker, K.R.: Sequencing Rules and Due-Date Assignments in a Job Shop. Manage Sci (1984) 1093-1104
[30] Jayamohan, M.S., Rajendran, C.: New dispatching rules for shop scheduling: a step forward. Int J Prod Res (2000) 563-586
[31] Quek, H.C., Tay, J.C.: Issues in the Performance Measurement of Constraint Satisfaction Techniques. Artificial Intelligence in Engineering (2000) 281-294
[32] Johnson, R.A.: Statistics: Principles and Methods. John Wiley (2001)

A Robust Meta-Hyper-Heuristic Approach to Hybrid Flow-Shop Scheduling

José Antonio Vázquez Rodríguez and Abdellah Salhi

Mathematical Sciences Department, University of Essex, Wivenhoe Park, Colchester, U. K. {javazq@gmail.com, as@essex.ac.uk}

Summary. Combining meta-heuristics and specialised methods is a common strategy to generate effective heuristics. The inconvenience of this practice, however, is that, often, the resulting hybrids are ineffective on related problems. Moreover, frequently, a high cost must be paid to develop such methods. To overcome these limitations, the idea of using a hyper-heuristic to generate information to assist a meta-heuristic, is explored. The devised approach is tested on the Hybrid Flow Shop (HFS) scheduling problem in 8 different forms, each with a different objective function. Computational results suggest that this approach is effective on all 8 problems considered. Its performance is also comparable to that of specialised methods for HFS with a particular objective function.

1 Introduction

Hybrid Flow Shops (HFS) are manufacturing environments in which a set of jobs must be processed in a series of stages with multiple parallel machines [2]. Given the NP-hard nature, [16], [21], of this problem, to be effective, generic methods rely on the information provided by specialised ones. Often, this information is decided by the objective being optimised. Consequently, the resulting hybrids are not as effective on problems with other objectives. This is a serious shortcoming since, in practice, it is desirable to have solution tools that are reliable on problems with different objectives.

Successful algorithms for HFS use meta-heuristics to schedule the first stage of the shop and a simple constructive procedure to schedule the rest, [29], [28], [34], [15], [26]. Genetic Algorithms (GA) have been used successfully in exploiting this idea. In [23] and [22], for instance, a Random Keys Genetic Algorithm (RKGA) performed well against many specialised heuristics and meta-heuristics such as the problem space-based search method, [24], on HFS problems with sequence dependent setup times. In [15], a GA with a permutation representation of the individuals, and many variants of the crossover operator, also performed well against several heuristics such as ant colonies, tabu search, simulated annealing, other GA's and deterministic methods on

J.A.V. Rodríguez and A. Salhi: *A Robust Meta-Hyper-Heuristic Approach to Hybrid Flow-Shop Scheduling,* Studies in Computational Intelligence (SCI) **49**, 125–142 (2007)

HFS with sequence dependent setup times and machine eligibility. Note that, most of these methods and the ones reviewed in [33], [25] and [32], consider problems with makespan as the optimisation criterion, mainly.

A more recent investigation, [28], studied the performance of the generic combination *GA and a constructive procedure*. The constructive procedure is one of four. Each of these combinations, called Single Stage Representation Genetic Algorithm (SSRGA), was applied to 4 variants of the HFS problem, each in turn having a different objective function. Although the SSRGA variants performed well, there is some variance between the performances. This means that it is not easy to tell which SSRGA will be most suitable for a given problem before carrying out the actual solution.

In this chapter the idea of SSRGA was extended by allowing GA to evolve the heuristic or combination of heuristics to be used to schedule the stages posterior to the first one. The proposed approach uses GA to generate part of the solution to the original problem and also searches the space of heuristics for a combination of some or all of them to generate the remaining components of the solution. In this way, the problem of deciding which SSRGA to use on a specific problem is removed. Moreover, by allowing different heuristics to be used at different stages, the number of usable SSRGA's increases given the different possible combinations of heuristics that can arise, each leading to a different SSRGA variant. The main virtue of the proposed approach is that specialisation occurs during the solution process and not before it. In other words, the specialisation is a function of the instance of the problem in hand rather than the general form of it. Moreover, this allows a high level of portability between related problems.

Heuristics that combine other heuristics are referred to as hyper-Heuristics (HH), [10]. A HH can be seen as a *black box* that takes as input a set of low level heuristics and a problem, [30]. At each decision point, the *black box* selects a low level heuristic and applies it. Note that in HH the high level heuristic acts, exclusively, on the space of low level heuristics, while the latter, are the ones that solve the original problem; see left part of Figure 1.

Hyper-heuristics are becoming popular because they are adaptable, effective on different problems of the same class, and yet relatively easy to implement. This is because most of the effort in implementing them goes into the low level heuristics which themselves are usually easy to implement. Because HH's are adaptable, they have been successfully applied to many problems including personnel scheduling problems such as the sales summit scheduling problem, [10], [11], the project presentation scheduling problem, [13], the nurse scheduling problem, [12], and others, [8], [7]. They have also been applied in manufacturing environments such as job shop, [20], and open shop, [14], scheduling problems and in industrial cutting stock problems, [31]. Meta-heuristics have been adopted as the basis (high level heuristic) of HH. Tabu search was used in [3] and [6]; ant colonies in [4], [5]; and GA's in [14], [20], [9], [19], [17], [18] and [31].

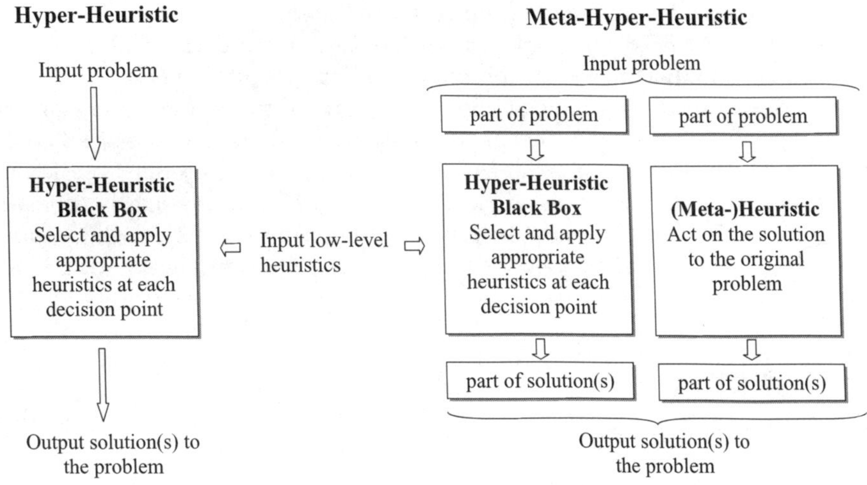

Fig. 1. Left, hyper-heuristics as a black-box model, [30]; right, meta-hyper-heuristic as proposed here

Given that the proposed approach uses an evolutionary method to generate part of the solution to the original problem and also evolves the combination of heuristics to generate the rest, it is a hybrid approach which is both a meta-heuristic and a HH; see right part of Figure 1. From now on we refer to it as the Meta-Hyper-Heuristic-Scheduler (MHHS). MHHS and many variants of the SSRGA are tested on several instances of 8 HFS scheduling problems. Each one of them considers a different objective function. The reported results show that MHHS is superior. The achievements of MHHS compared with those of SSRGA, the latter being a representative of the state-of-the-art in HFS scheduling, suggest that it is a highly competitive solution approach to HFS problems on a wide range of objective functions.

The rest of the chapter is organised as follows. Section 2 presents a formal description of the HFS scheduling problem and the objectives to consider. In Section 3, the low level heuristics, MHHS and SSRGA are explained. Section 4 describes the computational experiments and results. Section 5 discusses the difficulty of predicting the interactions between low level heuristics. Section 6 concludes the chapter.

2 Problem Definition

2.1 Assumptions

A HFS is a manufacturing environment in which a set of n jobs are to be processed in m different stages. All jobs must follow the same processing direction: stage 1 through stage m, with the possibility of skipping one or more stages. At least one of the stages has two or more identical machines in parallel. Any machine can process at most one job at a time and any job is processed on at most one machine at a time. Furthermore, every job is processed on at most one machine in any stage. Preemptions are not allowed, i.e. once the processing of a job has started, it cannot be stopped until it is finished. Using the triplet notation $\alpha\,|\beta|\,\gamma$, see [27], the problem is denoted as $FFc\,|r_{j1}|\,f$, where f is the cost function to optimise and r_{j1} is the release time of job j at the first stage of the shop. The following notation is used through the rest of the paper.

2.2 Notation

- $n =$ number of jobs;
- $m =$ number of stages in the shop;
- $j =$ job index;
- $k =$ stage index;
- $o_{jk} =$ operation of job j to be processed on stage k;
- $O_k = \bigcup\limits_j o_{jk}$;
- $p_{jk} =$ processing time of operation o_{jk};
- $w_j =$ weight of job j;
- $r_{jk} =$ release time of operation o_{jk}; these are calculated while constructing the schedule. Note that r_{j1}, for all j are given as part of the problem;
- $d_j =$ due date of job j;
- $s_{jk} =$ starting time of operation o_{jk};
- $v_{jk} = \sum\limits_{a=k}^{m} p_{ja}$, work remaining for job j at stage k.

2.3 Problem formulation

The following formulation is appropriate when searching in the set of semi-active schedules, i.e., schedules for which an operation can not start earlier without changing the order of processing in any one of the machines, [27]. Let A^{kl} be a set of operations o_{jk} assigned for processing to machine l in stage k. Let S^{kl} be a sequence of the elements in A^{kl} representing the order in which they are to be processed. Let $S^k = \cup_{l=1}^{m_k} S^{kl}$ where m_k is the number of machines at stage k, and $S = \cup_{k=1}^{m} S^k$. Because S^k is the set of the sequences to be followed by the jobs when processed at stage k, S represents a schedule. For S to be feasible the following must hold:

$$\bigcup_{l=1}^{m_k} A^{kl} = O_k \;\forall k \tag{1}$$

Table 1. Objectives to be considered

function	description	where
f_1	$\max_j C_j$	C_j = completion time of j.
f_2	$\sum_j w_j T_j$	$T_j = \max(0, C_j - d_j)$
f_3	$\max_j T_j$	–
f_4	$\sum_j C_j$	–
f_5	$\sum_j w_j U_j$	$U_j = \begin{cases} 1 \text{ if } C_j - d_j > 0 \\ 0 \text{ otherwise} \end{cases}$
f_6	$\sum_j w_j W_j$	$W_j = C_j - s_{j1}$
f_7	$\max_k \left\{ \max_{t=0}^{c_{\max}} \left\{ \sum_j w_j x_{jkt} \right\} \right\}$	$x_{jkt} = \begin{cases} 1 \text{ if } r_{jk} \leq t \text{ and } c_{jk} > t \\ 0 \text{ otherwise} \end{cases}$ $c_{\max} = \max_j C_j$ c_{jk} = completion time of operation o_{jk}
f_8	$\sum_j w_j T_j + \sum_j w'_j E_j$	$E_j = \max(0, d_j - C_j)$

$$\bigcap_{l=1}^{m_k} A^{kl} = \emptyset \; \forall k \tag{2}$$

Let $*$ be a problem instance of the type $FFc|r_{j1}|f_i$ and Ω^*, be the set of all its feasible schedules. The objective is to find $S \in \Omega^*$ such that its incurred cost $f_i(S)$ is minimum, more formally

$$\min_{S \in \Omega^*} f_i(S) \tag{3}$$

f_i can be any of the cost functions described in Table 1.

2.4 Objective functions

The first two functions (f_1 and f_2) of Table 1 are the most common in the literature of HFS. This is because f_1, or maximum completion time, optimises the use of machines in the shop. On the other hand, f_2, or total weighted tardiness, is a good metric of the service quality provided. Both of these functions are regular performance measures, i.e., functions that are nondecreasing in $C_1, \ldots, C_n$, see [27]. The next 3 functions, maximum tardiness (f_3), sum of completion times (f_4), and weighted sum of tardy jobs (f_5), are also regular functions. Functions f_3 and f_5, are concerned with the quality of the service to the client, whereas f_4 with how fast jobs are completed. The sum of weighted waiting times (f_6) measures the total time that the jobs spend in the shop floor, from the moment they start processing in the first stage, until their completion time. Function f_7 is the maximum weighted number of jobs that are in a stage waiting for processing at the same time. Functions f_6 and f_7 can be associated with the cost of inventory of products in

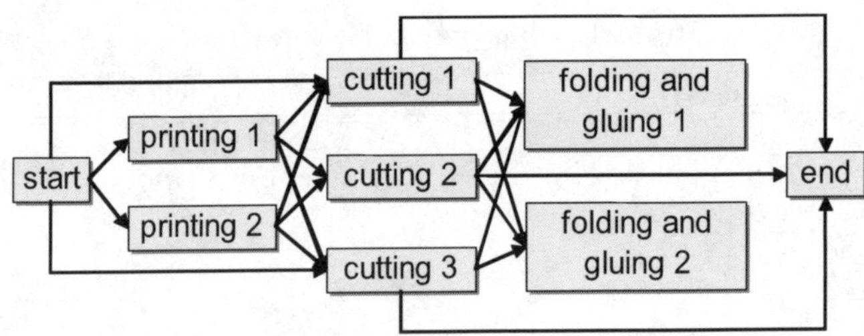

Fig. 2. A cardboard boxes manufacturing process as an example of a HFS

process. Their importance stems from the increasing interest of modern manufacturers in keeping low levels of inventory in process (lean manufacturing). Both of these functions are non-regular. In order to reduce the search to a set of schedules with interesting completion times, MHHS and any of the SS-RGA variants, generate schedules which are semi-active for the first stage of the shop and non-delay for the rest, see definitions in [27]. Function f_8, the weighted earliness and tardiness, is also non-regular. This function is relevant in Just In Time manufacturing systems where delays to meet with the clients' demands in time, and inventories of all sorts, are penalised. Although f_8 has been considered in other contexts, [1], no one has yet considered it in HFS.

2.5 Example of a HFS

As an example of HFS, we consider a production shop that manufactures cardboard boxes. Several kinds are produced, all following the same production flow: printing, cutting, and gluing and folding. First, the sheets are printed with the publicity and information to the client. In the second process, the sheets are cut into the required shape. Finally, the boxes are folded and glued to give them their final form. Figure 2 shows a typical production floor configuration of this kind. As can be seen, there are multiple machines per stage. All products follow the same flow, but, they may skip the processes of printing or folding and gluing.

3 A Meta-Hyper-Heuristic Scheduler and Single Stage Representation Genetic Algorithms

MHHS implements the idea of using GA to schedule the first stage of the shop and the rest through a constructive heuristic. Moreover, it extends this framework by allowing the individuals to encode which heuristic, from a set, to use to schedule stages $2, 3, \ldots, m$. Therefore, in the same solution process, different heuristics may be used to schedule different stages.

Many variants of SSRGA (13), as described in [28], were also implemented. Each of them uses GA and one of the 13 heuristics described next. The same heuristics are in the heuristics repository of MHHS. The rest of this section explains these simple heuristics and provides details of the implementation of MHHS and SSRGA.

3.1 Simple heuristics

Let $O'_k \subseteq O_k$ be a set of operations that: (1) have not been assigned yet and, (2) are ready to be processed at stage k (having been released from the previous stage). Whenever a machine becomes idle, an operation $o_{jk} \in O'_k$ is chosen according to one of the following criteria.

name	description
h_1:	select $o_{jk} \in O'_k$ with the smallest r_{jk} value;
h_2:	select $o_{jk} \in O'_k$ with the smallest p_{jk} value;
h_3:	select $o_{jk} \in O'_k$ with the largest p_{jk} value;
h_4:	select $o_{jk} \in O'_k$ with the smallest $v_{jk} - d_j$ value;
h_5:	select $o_{jk} \in O'_k$ with the largest $v_{jk} - d_j$ value;
h_6:	select $o_{jk} \in O'_k$ with the smallest v_{jk} value;
h_7:	select $o_{jk} \in O'_k$ with the largest v_{jk} value;
h_8:	select $o_{jk} \in O'_k$ with the smallest $w_j p_{jk}$ value;
h_9:	select $o_{jk} \in O'_k$ with the largest $w_j p_{jk}$ value;
h_{10}:	select $o_{jk} \in O'_k$ with the smallest $w_j(v_{jk} - dj)$ value;
h_{11}:	select $o_{jk} \in O'_k$ with the largest $w_j(v_{jk} - d_j)$ value;
h_{12}:	select $o_{jk} \in O'_k$ with the smallest $w_j v_{jk}$ value;
h_{13}:	select $o_{jk} \in O'_k$ with the largest $w_j v_{jk}$ value.

When $O'_k = \emptyset$, o_{jk} is the operation with the smallest release time. Operation o_{jk} is assigned after the last operation assigned to the machine l that allows it the earliest completion time. In all cases ties are broken by preferring smallest job (j) or machine (l) indexes. Hereafter, let us denote as $h_b$ga the SS-RGA variant resulting from the combination of GA and h_b, $b \in \{1, 2, \ldots, 13\}$.

3.2 Representation and evaluation of solutions

For the SSRGA, the adopted representation is a permutation $\Pi = (\pi(1), \pi(2), \ldots, \pi(n))$ where $\pi(i)$ is a job index, and represents the order in which operations are assigned for processing at stage 1. Given an individual Π, it is decoded by assigning the operations $o_{j1} \in O_1$, in the order $\pi(1), \pi(2), \ldots, \pi(n)$, to the machine l in stage 1 that allows them the earliest completion time. The procedure to translate Π into a schedule for stage 1 is as described in Procedure 1.

```
S^{1l} ← ∅, l = 1,...,m_1
for i ← 1,...,|Π| do
  | l ← machine that allows o_{π(i)1} the
  | fastest completion time
  | S^{1l} ← π(i)
endfor
return S^1
```

Procedure 1: $CP(\Pi)$

```
S ← ∅
S^1 ← CP(Π)
S ← S ∪ S^1
for k ← 2,...,m do
  | generate S^k according to h_b
  | S ← S ∪ S^k
endfor
return S
```

Procedure 2: $CP_{SSRGA}(\Pi, h_b)$

The rest of the schedule is build using one of the heuristics (h_b) described in Section 3.1. The evaluation of individuals in SSRGA is as in Procedure 2.

In MHHS, the representation of solutions consists in a permutation Π to schedule the first stage of the shop (as above) and an ordered set, HR, of heuristics to schedule the rest. For instance, an individual (Π', HR') where $\Pi' = (2, 3, 4, 5, 1)$ and $HR' = (\mathrm{h}_1, \mathrm{h}_1, \mathrm{h}_6, \mathrm{h}_7)$, encodes the solution for a 5 job 5 stage shop. The jobs are assigned to the first stage in the order 2, 3, 4, 5, 1; in stages 2 and 3 according to h_1 and in stages 4 and 5 according to h_6 and h_7, respectively. The procedure $CP_{MHHS}(\Pi, HR)$ to evaluate individuals in MHHS is as described in Procedure 3.

```
S ← ∅
S^1 ← CP(Π)
S ← S ∪ S^1
for k ← 2,...,m do
  | generate S^k according to HR_{k-1}
  | S = S ∪ S^k
endfor
return S
```

Procedure 3: $CP_{MHHS}(\Pi, HR)$

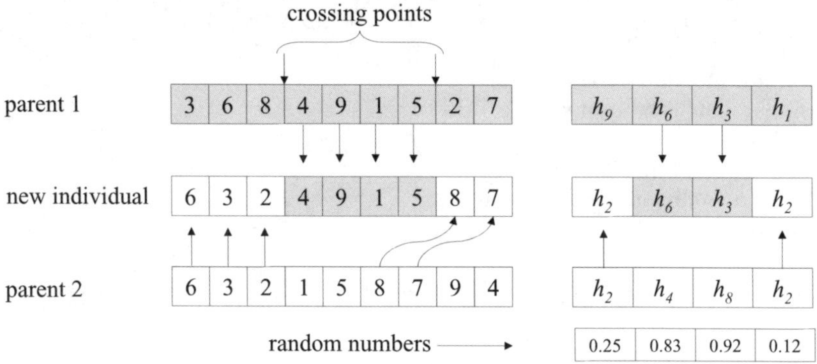

Fig. 3. *2-Point Crossover* assuming that parent 1 is fitter than parent 2.

3.3 Operators

A *2-Point Crossover*, as shown in the left part of Figure 3, was used to recombine Π; two crossover points are randomly selected from parent 1 and the elements between them copied in the same positions in the offspring. The missing jobs are copied from the second parent starting from the beginning of the permutation. In the case of MHHS, each allele in HR is copied with a 0.7 probability from the fittest parent and 0.3 from the other (sec the right part of Figure 3).

As in [15], to mutate an individual, an element in Π is randomly chosen and moved into a new position, the rest of the elements are moved to fill the released spaces as needed. In the case of MHHS, HR remains unchanged.

3.4 General framework

Both MHHS and SSRGA keep a population of individuals, $Pop_{MHHS} = \{(\Pi_1, HR_1), \ldots, (\Pi_N, HR_N)\}$ and $Pop_{SSRGA} = \{\Pi_1, ldots, \Pi_N\}$, respectively. At the initialisation the elements in Pop (for both methods) are generated randomly. In MHHS, each element of the heuristics repository HR_i of every individual i is selected randomly from the set of heuristics presented in Section 3.1. At every generation a new population Pop' of N individuals is created through the crossover operators explained in the previous section. The best individual, S^*, found so far is kept. This is repeated until the stoping condition is met. In algorithmic form the MHHS procedure is as in Procedure 4.

In the case of SSRGA, $CP_{MHHS}(\Pi_i, HR_i)$ is replaced with $CP_{SSRGA}(\Pi_i, \mathrm{h}_b)$.

3.5 Parameter Setting

In order to tune the GA implemented in SSRGA and MHHS, the feasible combinations of the following parameter values were evaluated.

```
Pop ← {(Π_1, HR_1), ..., (Π_N, HR_N)}   // random population
CP_MHHS(Π_i, HR_i)_{i=1}^N
S* ← best individual found so far
while (stoping condition not met) do
  | Pop' ← {(Π'_1, HR'_1), ..., (Π'_N, HR'_N)} // new population
  | CP_MHHS(Π'_i, HR'_i)_{i=1}^N
  | S* ← best individual found so far
endwhile
return S*
```

Procedure 4: $MHHS(FFc|r_{j1}|f_i)$

- crossover probability: 0.6,0.7,0.8,**0.9**,0.95;
- mutation probability: 0.0, 0.01, 0.05, **0.1**, 0.15;
- selection method: **tournament selection**, fitness proportionate selection;
- number of participants in the tournament selection: **2**, 3, 4;
- population size: 25, 50, 75, **100**;
- keeping best individual: **true**, false.
- fitness function: $f_i(S)$, where S is the schedule represented by the individual and f_i, the ith objective function (see Section 2.4).

The stopping condition was set to 10,000 solution evaluations. The best performing combination, shown in bold, was kept for the succeeding experimentation.

4 Computational Results

The SSRGA variants and MHHS were compared on a set of randomly generated instances minimising the objectives described in Section 2.3. All experiments were run on a 3.0 MHz processor with 1.0 Gb of RAM running Windows XP. All implementations were in Java 2 S.E. 5.0.

4.1 Instance generation

1024 instances were generated randomly in a similar fashion as in [15] and [23]. Each of these is a combination of the following variable values.

- $n \in \{20, 40, 60, 80\}$, number of jobs;
- $m \in \{2, 4, 6, 8\}$, number of stages;
- $m_k \in \{U\tilde{}[2,3], U\tilde{}[2,6]\}$, number of machines per stage;
- $p_{jk} \in \{U\tilde{}[50,70], U\tilde{}[10,100]\}$, processing times;
- $r_j = U\tilde{}(0, E(\sum_k p_{jk}))$, release times;
- $d_j \in \{U\tilde{}[0.9D, 1.1D], U\tilde{}[D, 1.5D]\}$, where $D = 1.5E(\sum_k p_{jk})$, due dates;
- $w_j = \{U\tilde{}[2,8]\}$, weights;
- $f \in \{f_1, \ldots, f_8\}$, objective function.

There are $4 \times 4 \times 2 \times 2 \times 1 \times 2 \times 1 \times 8 = 1024$ possible combinations, each a different instance. The number of jobs and stages, which define the size of an instance, are known to have an important impact on the performance of algorithms. Because of this, 4 levels were studied. Different numbers of machines per stage produce instances with different bottleneck criticalities. The release and due dates were generated in proportion to the total expected processing times. This is to avoid to some degree extreme cases where, for instance, the due dates are too easy to meet, or the release times of jobs at the first stage are so expanded that at every iteration there is a single job available, making the instance trivial. D represents the expected processing time plus the expected release time of any job. The due dates were generated in ranges between 0.9 and 1.1 times D for tight instances, and between 1 and 1.5 for extended ones. The objective function of each instance is one of the 8 described in 2.3; therefore, there are 128 instances which consider each function.

Five runs were carried out with each algorithm on every problem instance. The best solution was kept.

4.2 Comparison metrics

The first comparison metric considered, is the average value of the solutions for each of the algorithms on each group of instances (grouped according to the objective they consider; there are 8 different groups with 128 instances in each group). This measure, may be argued, has the limitation of being sensitive to the problem sizes and ranges in which the processing times were generated. Also, it does not distinguish well between the performances of different algorithms, i.e. these average performance values are often close to each other. To remedy this, a second metric is adopted. The latter is a re-scaling of the values obtained by the algorithms to the range 0 to 1, where 0 corresponds to the best value found for a given instance and 1 to that of the worst, as follows:

$$metric_i = \frac{x_i - x_{\min}}{x_{\max} - x_{\min}}, \qquad (4)$$

where $x_{\min}$ and $x_{\max}$ are, respectively, the best and worst solutions found. Given an algorithm i and its returned objective value x_i for a given instance, if this objective value is close enough to the best one overall, i.e., $metric_i$ is close to 0, then this particular algorithm is deemed to be "doing well", and "not so well" otherwise. The only problem is that we need to quantify the idea of "close enough". To this end we choose the value 0.1 to specify just that. This value is based on our observations. It means that when $metric_i \leq 0.1$, algorithm i is "doing well" or its performance is satisfactory.

Table 2. Mean best value of each heuristic on the 128 instances that consider the objective function in the corresponding column

	f_1	f_2	f_3	f_4	f_5	f_6	f_7	f_8
$h_1$ga	1544.8	105976.1	1021.0	49290.3	173.2	92846.3	22.9	111191.7
$h_2$ga	1553.1	109597.0	1042.0	49171.5	172.9	92712.5	25.1	115013.5
$h_3$ga	1555.0	112917.1	1046.6	50376.5	173.8	95644.6	27.8	118900.3
$h_4$ga	1560.4	110841.1	1086.7	49685.0	173.5	93923.9	26.6	116081.3
$h_5$ga	1547.0	111537.3	1014.5	49668.5	174.0	94242.0	26.3	116950.5
$h_6$ga	1564.4	109704.3	1048.5	49256.4	172.9	92893.8	25.2	114909.1
$h_7$ga	1543.9	112695.0	1035.4	50151.6	174.0	94964.5	27.8	118366.0
$h_8$ga	1552.5	116430.7	1042.5	49363.2	173.7	101582.9	32.7	122254.2
$h_9$ga	1553.8	107754.2	1047.4	50040.5	173.4	89430.2	23.3	113474.7
h_{10}ga	1555.2	105858.5	1060.3	49635.4	172.7	88277.3	21.8	111231.5
h_{11}ga	1549.4	118295.6	1027.3	49669.9	174.2	102873.8	34.0	124131.5
h_{12}ga	1560.0	117288.3	1048.0	49471.1	174.0	102093.7	33.6	122813.9
h_{13}ga	1548.8	106895.6	1041.0	49865.4	173.3	88889.0	22.4	112197.8
MHHS	1544.4	105825.6	1014.5	49169.2	172.6	88440.8	21.8	111081.5

4.3 Results

Given the amount of information generated in the described experiments, the results are separated by the objective function and the solution method. These are summarised in Tables 2, 3, 4 and 5. Table 2 is the mean value achieved by each algorithm on the groups of instances corresponding to each function. Table 3 presents the mean value obtained by each method on the *metric* value while Table 4 presents the standard deviations from it. Table 5 displays the number of satisfactory solutions, i.e., metric value ≤ 0.1, as defined in Section 4.2, for each algorithm on each function.

Tables 2, 3 and 5 show that the most effective method, overall, is MHHS. According to the results, MHHS is very competitive when compared with the best of the SSRGA variants on each function. MHHS achieved the best mean value on 6 out of 8 functions (Table 2), the ones on which it did not do well being f_1 and f_6; on these it came second. It obtained the best results according to the metric value (Equation 4), overall, and on 4 of the functions (Table 3). On the other 4, it was second best. Similarly, it obtained the largest or second largest number of successes, as defined in Section 4.2, per function (Table 5) and overall, it obtained a satisfactory solution for more than half of the problems. MHHS also seems to be stable, since it has the smallest standard deviation on the *metric* value (Table 4).

The solution obtained by each algorithm, for every instance, was compared with the one obtained by the rest. The number of times that a given algorithm outperformed, or was outperformed by any of the others, was recorded. The summary of this information is displayed in Figure 4. The bar chart on the left shows the number of times that the algorithm on the x-axis was worse than its competitor. The one on the right, shows the frequency with which

Table 3. Mean *metric* value of each heuristic on the 128 instances that consider the objective function in the corresponding column

	f_1	f_2	f_3	f_4	f_5	f_6	f_7	f_8	Total
$h_1$ga	0.199	0.212	0.222	0.279	0.393	0.294	0.274	0.230	0.263
$h_2$ga	0.473	0.320	0.378	0.186	0.361	0.271	0.295	0.343	0.328
$h_3$ga	0.385	0.578	0.371	0.728	0.450	0.433	0.442	0.579	0.496
$h_4$ga	0.490	0.431	0.757	0.448	0.435	0.329	0.371	0.487	0.468
$h_5$ga	0.281	0.421	0.141	0.473	0.451	0.332	0.355	0.415	0.359
$h_6$ga	0.683	0.304	0.441	0.227	0.374	0.284	0.302	0.359	0.372
$h_7$ga	0.194	0.524	0.298	0.650	0.465	0.383	0.397	0.547	0.432
$h_8$ga	0.411	0.620	0.362	0.304	0.416	0.565	0.537	0.634	0.481
$h_9$ga	0.349	0.335	0.391	0.565	0.410	0.222	0.261	0.388	0.365
h_{10}ga	0.428	0.233	0.477	0.432	0.361	0.135	0.180	0.235	0.310
h_{11}ga	0.359	0.721	0.276	0.452	0.484	0.696	0.611	0.730	0.541
h_{12}ga	0.536	0.670	0.405	0.360	0.465	0.628	0.595	0.679	0.542
h_{13}ga	0.270	0.301	0.333	0.512	0.426	0.213	0.237	0.318	0.326
MHHS	0.174	0.195	0.161	0.217	0.317	0.166	0.194	0.175	0.200

Table 4. Standard deviation of the *metric* values of each heuristic on the 128 instances that consider the objective function in the corresponding column

	f_1	f_2	f_3	f_4	f_5	f_6	f_7	f_8	Total
$h_1$ga	0.297	0.267	0.291	0.253	0.362	0.217	0.339	0.286	0.297
$h_2$ga	0.338	0.254	0.298	0.268	0.345	0.246	0.311	0.269	0.303
$h_3$ga	0.354	0.308	0.267	0.348	0.337	0.324	0.360	0.316	0.346
$h_4$ga	0.372	0.254	0.366	0.262	0.365	0.258	0.335	0.265	0.335
$h_5$ga	0.295	0.264	0.260	0.255	0.403	0.257	0.327	0.228	0.307
$h_6$ga	0.373	0.226	0.304	0.260	0.359	0.239	0.308	0.270	0.324
$h_7$ga	0.305	0.306	0.255	0.306	0.369	0.292	0.345	0.305	0.340
$h_8$ga	0.342	0.298	0.276	0.253	0.356	0.373	0.404	0.305	0.348
$h_9$ga	0.325	0.316	0.299	0.282	0.384	0.282	0.349	0.330	0.336
h_{10}ga	0.343	0.277	0.302	0.256	0.343	0.231	0.322	0.291	0.321
h_{11}ga	0.325	0.320	0.278	0.236	0.386	0.402	0.435	0.319	0.379
h_{12}ga	0.365	0.312	0.267	0.263	0.366	0.390	0.425	0.285	0.356
h_{13}ga	0.306	0.287	0.267	0.255	0.371	0.295	0.337	0.316	0.319
MHHS	0.261	0.250	0.288	0.303	0.350	0.286	0.313	0.246	0.292

the algorithm in the x-axis showed a better result. These bar charts show that MHHS is, overall, superior.

Even though h_1 is the simplest of the heuristics, on the whole, the second best algorithm is $h_1$ga. This is, perhaps, because h_1 allows GA to have a higher impact on the solution than the rest of the heuristics; given that it schedules the jobs in stages $k > 1$, by considering their release times only. On the other hand, heuristics h_b for $b > 1$, reduce the influence of GA by prioritising the jobs in stages $k > 1$ by different criteria than their release times. However,

Table 5. Number of times that each heuristic obtained a *metric* value ≤ 0.1 on the 128 instances that consider the objective function in the corresponding column

	f_1	f_2	f_3	f_4	f_5	f_6	f_7	f_8	Total
$h_1$ga	78	68	66	43	44	31	58	63	451
$h_2$ga	25	28	28	76	45	40	45	26	313
$h_3$ga	41	11	26	13	29	29	37	11	197
$h_4$ga	34	10	16	10	38	36	41	5	190
$h_5$ga	46	17	92	12	40	31	40	10	288
$h_6$ga	17	25	21	56	46	36	46	19	266
$h_7$ga	84	16	36	7	37	32	43	12	267
$h_8$ga	34	11	26	23	41	28	40	5	208
$h_9$ga	44	44	27	11	43	70	72	42	353
h_{10}ga	32	64	19	14	44	89	94	63	419
h_{11}ga	34	8	42	9	38	26	40	10	207
h_{12}ga	27	10	22	21	38	28	38	5	189
h_{13}ga	58	44	29	10	39	69	75	48	372
MHHS	82	70	90	73	56	88	88	75	622

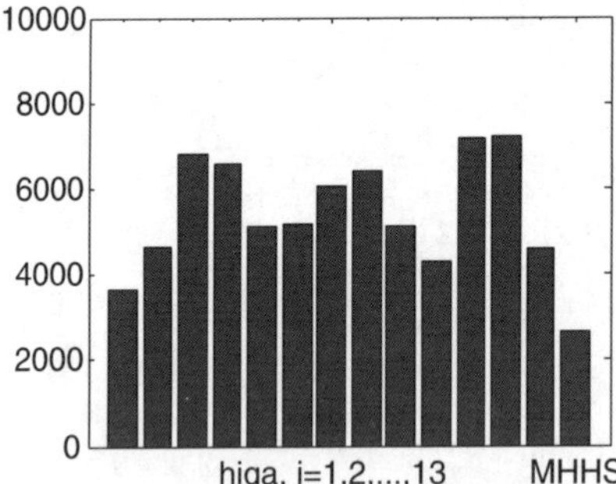

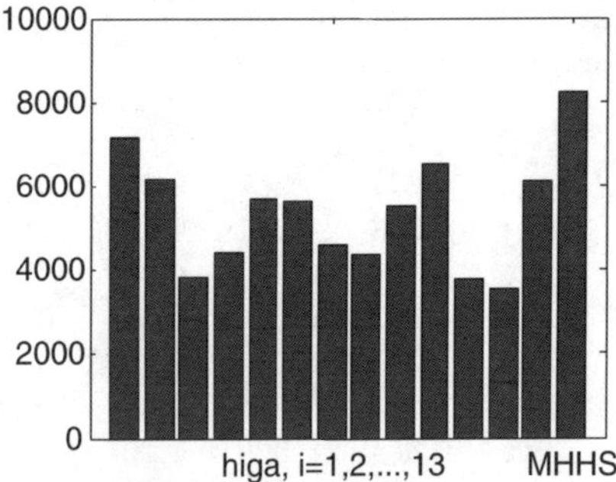

Fig. 4. Frequency on which the algorithm in the horizontal was outperformed (left) or outperformed (right) other algorithms.

the evidence presented suggests that MHHS finds situations in which using these other criteria is convenient.

Note that, in all the studied algorithms, the sorting of operations at stages posterior to the first one, dominates the complexity of the evaluation of individuals. The evaluation of a solution, then, has a theoretical run-time of $O(mn \log n)$. All the algorithms required from 0.3 seconds, for the smallest instances, to 6.1 seconds, for the largest ones, without significant differences between the heuristics.

Table 6. f_1 value (makespan) obtained by each of the algorithms in the first column for the corresponding instance

	instance 1	*instance 2*	*instance 3*	*instance 4*	*instance 5*
$h_1$ga	1549	1636	2265	1493	2437
$h_7$ga	1534	1654	2269	1494	2437
$h_4$ga	1542	1650	2326	1514	2496
$h_6$ga	1550	1663	2319	1531	2458
MHHS{h_1,h_7}	1532	1651	2265	1496	2471
MHHS{h_4,h_6}	1532	1641	2307	1512	2529
MHHS{h_1,h_4}	1538	1691	2269	1495	2486
MHHS{h_1,h_6}	1541	1652	2274	1493	2467
MHHS{h_7,h_4}	1532	1647	2265	1495	2469
MHHS{h_7,h_6}	1533	1638	2265	1494	2444
MHHS{h_1,h_4,h_7}	1544	1645	2269	1498	2430
MHHS{h_1,h_6,h_7}	1535	1681	2265	1493	2460
MHHS{h_1,h_4,h_6}	1542	1679	2265	1494	2454
MHHS{h_4,h_6,h_7}	1550	1649	2270	1495	2434
MHHS{h_1,h_4,h_6,h_7}	1532	1631	2265	1493	2423

5 Discussion

The results of the previous section provide evidence to support the hypothesis that a HH framework such as MHHS can provide useful support to a generic algorithm such as GA. But, as already mentioned, some of the algorithms performed poorly and it seems reasonable to exclude their corresponding heuristic from MHHS, leaving just those that are good. However, often, it is not easy to predict how beneficial the interactions between heuristics would be. Sometimes, individually poor heuristics perform well when combined, and good ones, do not. To illustrate this, the following experiment was carried out. The simple heuristics in the best two SSRGA variants ($h_1$ga and $h_7$ga) and in the worst two ($h_4$ga and $h_6$ga) on solving f_1, were used to form different variants of MHHS. They are denoted MHHS{.,.,...}, for instance, MHHS{h_4,h_1} is one that combines h_4 and h_1. All the possible variants (11 of them) were run 5 times on 5 randomly selected instances from the ones explained in Section 4.1. Table 6 presents the best result of the 5 runs for each heuristic.

In most cases the combinations are successful, although not always. Perhaps the most interesting case is the one of *instance 2*. In this case, the combination of the two worst heuristics, h_4 and h_6, obtained better results than the combination of the best two, h_1 and h_7. This is made even more striking by the fact that $h_1$ga and $h_4$ga are the best two SSRGA's for this instance, i.e. *instance 2*, and MHHS{h_1, h_4} obtained the worst result. Surprisingly, MHHS{h_6, h_7} did relatively well. The difference in performance between the MHHS with two heuristics and with three is not clear. However, without any doubt, the best results were obtained by MHHS{h_1, h_4, h_6, h_7} which constantly obtained the best results and discovered two new upper bounds for *instance 2* and *instance 5*. Even though these observations are not conclusive, it is clear that the problem of how to select heuristics and their combinations in a HH framework is not a trivial one. Note that a systematic exploration of combinations of heuristics is not practical since their number increases exponentially with the number of heuristics and stages in the shop. The evolutionary approach is, therefore, appropriate.

6 Conclusion

We have presented a meta-hyper-heuristic method for the solution of Hybrid Flow Shop (HFS) scheduling problems. Its performance is overall superior to that of many variants of the Single Stage Representation Genetic Algorithm (SSRGA) on many instances of the HFS problem in 8 forms, each with a different objective function. This superiority is importantly concerned with the fact that MHHS works equally well on the HFS when different objectives are considered, contrary to the SSRGA variants which may perform well on some objective functions but not on others. It is important to state that the present study shows that MHHS is on the whole superior to the state-of-the-art SSRGA for the problem of interest.

Even though hyper-heuristics are effective by themselves, our findings suggest that they can be hybridised with meta-heuristics to lead to even more effective methods, such as MHHS. Although this method has been tested on HFS only, it is generic and may prove to be efficient on other intractable optimisation problems.

The discussion of Section 5 is far from over. Further work is being carried out on several issues, in particular, the interaction of heuristics in hyper-heuristic frameworks.

Acknowledgements: This work is supported by CONACYT grant 178473.

References

1. K. R. Baker and G. D. Scudder. Sequencing with earliness and tardiness penalties: A review. *Operations Research*, 38:22–36, 1990.
2. Shaukat A. Brah. *Scheduling in a Flow Shop with Multiple Processors.* PhD thesis, University of Houston, 1988.
3. E. Burke and E. Soubeiga. Scheduling nurses using a tabu-search hyperheuristic. In *1st Multidisciplinary International Conference on Scheduling: Theory and Applications (MISTA 2003)*, 2003.
4. Edmund Burke, Graham Kendall, Ross O'Brien, D. Redrup, and E. Soubeiga. An ant algorithm hyper-heuristic. In *Proceedings of The Fifth Metaheuristics International Conference (MIC 2003)*, 2003.

5. Edmund Burke, Graham Kendall, Dario Landa Silva, Ross O'Brien, and Eric Soubeiga. An ant algorithm hyperheuristic for the project presentation scheduling problem. In *Proceedings of the Congress on Evolutionary Computation (CEC 2005)*, pages 2263–2270. IEEE press, 2005.
6. Edmund K. Burke, Graham Kendall, and Eric Soubeiga. A tabu-search hyperheuristic for timetabling and rostering. *Journal of Heuristics*, 9:451–470, 2003.
7. Konstantin Chakhlevitch and Peter Cowling. Choosing the fittest subset of low level heuristics in a hyper-heuristic framework. In G.R. Raidl and J. Gottlieb, editors, *EvoCOP 2005, LNCS 3448*, pages 23–33, Berlin Heidelbergh, 2005. Springer-Verlag.
8. Peter Cowling and Konstantin Chakhlevitch. Hyperheuristics for managing a large collection of low level heuristics to schedule personnel. In *Proceedings of Congress on Evolutionary Computation (CEC2003)*, pages 1214–1221. IEEE, 2003.
9. Peter Cowling, Graham Kendall, and Limin Han. An investigation of a hyperheuristic genetic algorithm applied to a trainer scheduling problem. In *Proceedings of Congress on Evolutionary Computation (CEC2002)*, pages 1185–1190. IEEE, 2002.
10. Peter Cowling, Graham Kendall, and Eric Soubeiga. A hyperheuristic approach to scheduling a sales summit. In E. K. Burke and W. Erben, editors, *LNCS 2079, Practice and Theory of Automated Timetabling III : Third International Conference, PATAT 2000*, pages 176–190. Springer-Verlag, 2000.
11. Peter Cowling, Graham Kendall, and Eric Soubeiga. A parameter-free hyperheuristic for scheduling a sales summit. In *Proceedings of 4th Metahuristics International Conference (MIC 2001)*, pages 127–131, 2001.
12. Peter Cowling, Graham Kendall, and Eric Soubeiga. Hyperheuristics: A robust optimisation method applied to nurse scheduling. In *Proceedings of Parallel Problem Solving from Nature Conference, 7th International Conference, LNCS 2439*, pages 851–860. Springer-Verlag, 2002.
13. Peter Cowling, Graham Kendall, and Eric Soubeiga. Hyperheuristics: A tool for rapid prototyping in scheduling and optimisation. In S. Cagoni, J. Gottlieb, E. Hart, M. Middendorf, and R. Günther, editors, *LNCS 2279, Applications of Evolutionary Computing : Proceedings of Evo Workshops 2002*, pages 1–10. Springer-Verlag, 2002.
14. Hsiao-Lan Fang, Peter Ross, and Dave Corne. A promising hybrid GA/Heuristic approach for open-shop scheduling prblems. In A. Cohn, editor, *11th European Conference on Artificial Intelligence (ECAI 94)*, pages 590–594. John Wiley & Sons, Ltd., 1994.
15. Rubén Ruíz García and Concepción Maroto. A genetic algorithm for hybrid flow shops with sequence dependent setup times and machine elegibility. *European Journal of Operational Research*, 169:781–800, 2006.
16. J. N. D. Gupta. Two-stage hybrid flow shop scheduling problem. *Operational Research Society*, 39:359–364, 1988.
17. Limin Han and Graham Kendall. Guided operators for hyper-heuristic genetic algorithm. In *Proceedings of The 16th Australian Conference on Artificial Intelligence (AI'03), LNAI 2903*, pages 807–820. Springer-Verlag, 2003.
18. Limin Han and Graham Kendall. An investigation of a tabu assisted hyperheuristic genetic algorithm. In *Proceedings of Congress on Evolutionary Computation (CEC2003)*, pages 2230–2237. IEEE, 2003.

19. Limin Han, Graham Kendall, and Peter Cowling. An adaptive length chromosome hyperheuristic genetic algorithm for a trainer scheduling problem. In *Proceedings of the 4th Asia-Pacific Conference on Simulated Evolution and Learning (SEAL'02)*, pages 267–271, 2002.
20. Emma Hart and Peter Ross. A heuristic combination method for solving job-shop scheduling problems. In *Lecture Notes in Computer Sciences (1498)*, pages 845–854. Springer-Verlag, 1998.
21. J. A. Hoogeveen, J. K. Lenstra, and B. Veltman. Preemptive scheduling in a two-stage multiprocessor flow shop is NP-hard. *European Journal of Operational Research*, 89:172–175, 1996.
22. M. E. Kurz, M. Runkle, and S. Pehlivan. Comparing problem-based-search and random keys genetic algorithms for the SDST FFL makespan scheduling problem. working paper, 2005.
23. Mary E. Kurz and Ronald G. Askin. Scheduling flexible flow lines with sequence dependent set-up times. *European Journal of Operational Research*, 159:66–82, 2003.
24. V. Jorge Leon and Balakrishnan Ramamoorthy. An adaptable problem space based search method for flexible flow line scheduling. *IIE Transactions*, 29:115–125, 1997.
25. Richard Linn and Wei Zhang. Hybrid flow shop scheduling: A survey. *Computers & Industrial Engineering*, 37:57–61, 1999.
26. Ceyda Oguz and M. Fikret Ercan. A genetic algorithm for hybrid flow shop scheduling with multiprocessor tasks. *Journal of Scheduling*, 8:323–351, 2005.
27. Michael Pinedo. *Scheduling Theory, Algorithms and Systems.* Prentice Hall, 2002.
28. José Antonio Vázquez Rodríguez and Abdellah Salhi. Performance of single stage representation genetic algorithms in scheduling flexible flow shops. In *Congress on Evolutionary Computation (CEC2005)*, pages 1364–1371. IEEE Press, 2005.
29. Funda Sivrikaya Serifoglu and Gunduz Ulusoy. Multiprocessor task scheduling in multistage hybrid flow shops: A genetic algorithm approach. *Journal of the Operational Research Society*, 55:504–512, 2004.
30. Eric Soubeiga. *Development and Application of Hyperheuristics to Personnel Scheduling.* PhD thesis, School of Computer Science and Information Technology, The University of Nottingham, 2003.
31. Hugo Terashima-Marín, Armando Morán-Saavedra, and Peter Ross. Forming hyper-heuristics with GAs when solving 2d-regular cutting stock problems. In *Proceedings of Congress on Evolutionary Computation CEC(2005)*, pages 1104–1110. IEEE Press, 2005.
32. A. Vignier, J. C. Billaut, and C. Proust. Les problèmes d'ordonnancement de type flow-shop hybride: état de l'art. *Operations Research*, 33:117–183, 1999.
33. Hong Wang. Flexible flow shop scheduling: Optimum, heuristics and artifical intelligence solutions. *Expert Systems*, 22:78–85, 2005.
34. Bagas Wardono and Yahya Fathi. A tabu search algorithm for the multi-stage parallel machines problem with limited buffer capacities. *European Journal of Operational Research*, 155:380–401, 2004.

Hybrid Particle Swarm Optimizers in the Single Machine Scheduling Problem: An Experimental Study

Leticia Cagnina[1], Susana Esquivel[1] and Carlos A. Coello Coello[2]

[1] Lab. de Investigación y Desarrollo en Inteligencia Computacional (LIDIC)**
Universidad Nacional de San Luis
Ejército de los Andes 950 - 5700 - San Luis - Argentina
[2] CINVESTAV-IPN (Evolutionary Computation Group)
Departamento de Ingeniería Eléctrica, Sección Computación
Av. IPN No. 2508, Col. San Pedro Zacatenco, México D.F. 07360, México

Summary. Although Particle Swarm Optimizers (PSO) have been successfully used in a wide variety of continuous optimization problems, their use has not been as widespread in discrete optimization problems, particularly when adopting non-binary encodings. In this chapter, we discuss three PSO variants (which are applied on a specific scheduling problem: the Single Machine Total Weighted Tardiness): a Hybrid PSO ($HPSO$), a Hybrid PSO with a simple neighborhood topology ($HPSO_{neigh}$) and a new version that adds problem-specific knowledge to $HPSO_{neigh}$ ($HPSO_{kn}$). The last approach is used to guide the blind search that PSO usually does and reduces its computational cost (measured in terms of the objective function evaluations performed). It is also shown that $HPSO_{kn}$ obtains good results with a lower computational cost, when comparing it against the other PSO versions analyzed, and with respect to a classical PSO approach and to a multirecombined evolutionary algorithm (*MCMP-SRI-IN*), which contains specialized operators to tackle single machine total weighted tardiness problems.

1 Introduction

Particle Swarm Optimization (PSO) is a bio-inspired heuristic that was proposed by James Kennedy and Russell Eberhart [16]. PSO is a population-based stochastic heuristic that simulates the flight of a flock of birds. In PSO, each particle in the swarm (i.e., the population) is a possible solution within the multidimensional search space. Such a particle has some properties such as a position (within the search space), a velocity of exploration which is constantly updated, and a record of its past behavior. Each particle evaluates

** The LIDIC is supported by the Universidad Nacional de San Luis and the ANPCyT (National Agency to Promote Science and Technology).

L. Cagnina et al.: *Hybrid Particle Swarm Optimizers in the Single Machine Scheduling Problem: An Experimental Study*, Studies in Computational Intelligence (SCI) **49**, 143–164 (2007)

its relative position with respect to a goal (fitness) at every iteration and it adjusts its own velocity using the best position that it has found so far and the best position reached by any particle in its neighborhood (or in the swarm, if no neighborhood topology is adopted). Then, the velocity is used to update the position of each particle. The update is done using the following equations:

$$vel_{ij} = w * vel_{ij} + c_1 * r_1 * (p_{ij} - part_{ij}) + c_2 * r_2 * (p_{gj} - part_{ij}) \quad (1)$$

$$part_{ij} = part_{ij} + vel_{ij} \quad (2)$$

where vel_{ij} is the velocity of the particle i in the dimension j, w is the inertia factor [15] whose goal is to balance global exploration and local exploitation, c_1 and c_2 are the personal and social learning factors, r_1 and r_2 are two random numbers in the range (0,1), p_{ij} is the best position reached by the particle i and p_{gj} is the best position reached by any particle in the neighborhood (or swarm).

PSO was originally designed to work in continuos search spaces, and the specialized literature reports a significant amount of research that makes evident the great search capabilities of PSO in such type of search spaces. However, the use of *PSO* in discrete search spaces is relatively scarce, particularly when non-binary encodings (e.g., permutations) are adopted (see for example [24, 31, 12, 22]).

The authors recently proposed a hybrid PSO approach, which was called *HPSO* [8]. *HPSO* incorporates a *random keys* representation [4] for the particles and a dynamic mutation operator similar to the one used in evolutionary algorithms. The use of the *random keys* encoding allows to represent permutations using real numbers. This, in turn, allows us to use *PSO* with real numbers instead of having to rely on more complex encodings to represent a permutation of integers. In further work by the authors, $HPSO_{neigh}$ was introduced [7]. This approach adds to *HPSO* a local neighborhood (known as *circle topology* [25]) to each particle.

In this chapter, we propose a new *PSO* variant, which we call $HPSO_{kn}$. This algorithm is an extended version of $HPSO_{neigh}$, which incorporates problem-specific knowledge to guide the search.

The three previously indicated *PSO* approaches are used to solve a hard combinatorial optimization problem called *Total Weighted Tardiness Scheduling* (TWT). To the authors' best knowledge, this chapter constitutes only the third reported attempt to use PSO in scheduling (the two other attempts are reported in [38] and [8]).

The main goal of this chapter is to show the performance of our proposed $HPSO_{kn}$ using some instances of the TWT problem in single machine environments. We also aim to compare the results produced by the new algorithm against those obtained with the classical *PSO*, the *HPSO*, the $HPSO_{neigh}$, and a multirecombined evolutionary algorithm (*MCMP-SRI-IN*) [14] that was specially designed for dealing with the problem of our interest. Such an approach also adopts the knowlegde insertion concept (adopted in this chapter) that consists of incorporating in the population three seeds generated with other traditional heuristics.

The remainder of the chapter is organized as follows. In Section 2, the scheduling problem of our interest is properly defined. In Section 3, we briefly review the previous related work. Section 4 describes the PSO algorithms adopted for our experimental study, including our new proposed approach. Section 5 contains a description of our experimental design, including the parameters settings adopted. Our results are shown and discussed in Section 6. Finally, our conclusions and some possible paths for future research are provided in Section 7.

2 Single Machine Total Weighted Tardiness Problem

The single machine scheduling model is the simplest of all possible machine environments and it is a special case of more complicated machine environments. This model was selected because the results obtained for it provide the basis to develop heuristics for more complex machine environments. In this work only the deterministic model is analyzed.

The term *machine* is used to specify any resource that will process an assignment. In the single machine system just one resource is available; thus, only one job can be processed by the machine at any time. Each job or task consists of one or more operations (sub-tasks).

The objective function or criterion selected to evaluate the quality of the schedule was the Total Weighted Tardiness (TWT) because it is important in a wide range of production activities. In this problem, the jobs or assignments that have to be processed are characterized by several elements:

- *Processing time (**p**)*, the amount of time the job needs the resource to complete its task. It includes a setup and a knock-down time;
- *Weight (**w**)*, a value indicating the importance of the job with respect to the other jobs in the system. It represents a factor of priority, that is, what job should be chosen (among all the available jobs) to be processed next;
- *Due date (**d**)*, in which the job should finish and free the resource. It denotes the date the job is promised to be delivered to the customer.

Assuming the deterministic model and that the system consists of a set of n jobs ($j = 1, \ldots, n$) to be processed without preemption in a single machine, each job j has its own p_j (processing time), w_j (weight) and d_j (due date). For a given processing order of all jobs, the earliest *completion time* C_j can be defined like the time the job j uses from the moment in which it enters the system and until it leaves the system. Also, for each job j the *tardiness* T_j is defined like the maximum value among zero and the completion time minus the due date: $T_j = max\{0, C_j - d_j\}$. Then, the TWT problem consists of finding an appropriate processing order of the jobs with the purpose of minimizing the number of weighted tardy jobs, that is, to minimize the *Total Weighted Tardiness*:

$$\sum_{j=1}^{n} w_j T_j$$

over the n jobs in the system.

3 Previous Related Work

The single machine total weighted tardiness problem is an NP-hard [27] scheduling problem. The TWT problem has been tackled by a number of exact methods such as Branch and Bound [37, 21, 33], where some schedules are discarded because they exceed the objective function value set as a bound. A competitive technique in this context is dynamic programming [37, 23], which constructs all possible sets of jobs and recursively obtains a solution. The problem with these two approaches (branch & bound and dynamic programming) is the exponential growth and the considerable computer resources (computational time and memory requirements) that they require as the size of the problem grows.

Several enumerative methods have also been proposed, such as those that use dominance rules to restrict the search for the optimal solution [23] and those that characterize adjacent jobs in the optimal sequence [36]. An experimental study of these methods might be found in [37].

Some schedule construction heuristics have also been proposed to tackle this problem. These heuristics generate good, but not necessarily optimal solutions. For example, some authors have proposed dispatching rules to build a solution by fixing a job in a position in the sequence at each step of the process. There are a lot of rules widely used for the TWT problem. Comparisons between *weighted shorted processing time* (WSPT), *earliest due date* (EDD), *modified cost over time* (MCOVERT) and *apparent urgency* (AU) might be found in [34]. Additionally, an experimental study of this sort of heuristic may be found in [2]. The apparent tardiness cost (ATC) was proposed and tested in [40]. Then, in [10], the same rule was tested with other dispatching rules in job and flow shops, showing its effectiveness in minimizing the average tardiness.

A dominance rule for the most general case of total weighted tardiness problem is presented in [1], showing the sufficient condition for local optimality and how it generates schedules that cannot be improved by adjacent job interchanges.

There are other useful methods, such as the method of interchanges. Such interchanges require an initial sequence over which the change will take place. If the changed solution is better than the non-changed one, the method keeps it; otherwise, the changed solution is discarded. When the solution cannot be improved, the interchanges stop and the process returns the sequence solution. Comparisons among several heuristics (including interchanges) might be found in [34]. Some results indicate that the pairwise interchange methods are very good for this problem.

There exist several local search algorithms that propose to solve the TWT problem using insertion and swap movements to find a good schedule. These heuristics compute the neighborhood of a solution through movements of jobs in the sequence. For example, an exponentially sized "dynasearch" that swaps positions within the neighborhood in polynomial time is described in [35], where every swap is a single movement. The authors of that paper showed that their results were the best known so far in terms of both solution quality and computational time. In [13], the common swap neighborhood is extended with generalized pairwise interchanges, showing how effective are the neighborhoods for some scheduling problems. An enhanced dynasearch swap neighborhood was developed in [20], precisely by adding generalized pairwise interchanges. A fast and efficient algorithm is presented in [17], which combines the insertion, swap and twist neighborhoods; its searching process takes $O(n^2)$ time.

Metaheuristics offer a good compromise between computational effort and solution quality. In the case of TWT, a number of metaheuristics have been applied to its solution, including simulated annealing, tabu search, genetic algorithms, ant colony optimization and, more recently, particle swarm optimization.

SA and TS are advanced local search techniques. SA uses a parameter named *temperature* for changing the probability of moving from one point within search space to another one [26]. This technique is based on a thermodynamic analogy: "start heating a row of materials to a fusing state for growing a crystal. Then reduce the temperature T until the crystal structure is frozen. But if the cooling is done quickly, bad things might occur (irregularities in the crystal structure, for instance, and the level of energy trapped is higher than a perfect crystal structured)". The state can be looked as a feasible solution, ground state as an optimal solution, temperature control parameter T, and the energy as the evaluation function. In the process, the T parameter used to influence the search of a better value, is updated periodically. Usually T starts with a high value (doing the procedure similar to a purely random search) and gradually decreases its value. In each iteration the best value is updated. The process is executed until some external condition is reached. SA approaches for the TWT problem are stated in [29, 34, 21].

TS, instead has a memory, which forces the algorithm to explore new areas without visiting previous ones [19]. The solutions examined recently become "tabu" (forbidden) points to select as a new solution and are stored in a list H. The process is structurally similar to that of SA. It returns an accepted solution which needs not be better. The acceptance is based on the previous history of the search H. The process makes a new movement in the search space only when the search is stuck in a local optimum, although SA does not have this condition. In [21], TS was applied to solve the TWT problem.

Genetic Algorithms (GAs) are a particular type of Evolutionary Algorithm (EA) which normally adopt a binary encoding for their individuals. GAs are based in the "survival of the fittest" principle from Darwin's evolutionary theory. GAs choose the fittest individuals to recombine, aiming to increase the fitness of all the population over time. GAs use operators such as selection, mutation and crossover to create a new population. Comparisons of methods that include GAs might be found in [14]. In that work, the authors presented a competitive GA to solve the TWT. This GA uses problem-specific knowledge which is inserted with the aim of removing some of the "blindness" at the search traditionally performed by a GA. This GA outperformed other evolutionary algorithms in the TWT, which showed the efficacy of using problem-specific knowledge.

The Ant Colony Optimization (ACO) is a paradigm inspired by the trail following behavior observed in colonies of real ants. ACO was applied to TWT in [30], in which a pheromone summation evaluation was adopted for the probability of transition, and a specific heuristic was tailored for the TWT. Better results were presented in [28] and [6]. The latter introduced local search which is combined with the constructive phase obtaining an algorithm that uses heterogeneous colonies of ants.

Particle Swarm Optimizer (PSO) is a population-based stochastic heuristic which is inspired in the flight patterns of a flock of birds, in which a group (called the "swarm") follows a leader. As indicated before, PSO has been scarcely applied to scheduling problems. In [38], there is a comparative study between PSO, ACO and Iterative Local Search algorithm in the TWT problem. In [8], the authors proposed to adopt the random keys encoding for the individuals combined with a dynamic mutation operator. In [8], results are compared with respect to conventional heuristics and with respect to an evolutionary algorithm [14] that was fairly competitive at that time. In both cases, results indicated that PSO is a promising heuristic to tackle the TWT problem.

4 Improved Hybrid PSO Algorithms for the TWT Problem

In this section, the three PSO variants adopted in this chapter (i.e., $HPSO$, $HPSO_{neigh}$, and $HPSO_{kn}$) are described. However, we first present the pseudocode of the classical PSO algorithm (see Figure 1), because it will serve as the basis for all the other algorithms.

As can be seen in Figure 1, once the swarm, the velocities of each particle and the particle best memory are initialized (lines 2 to 4), the swarm is evaluated and the *leader* (the best particle of the swarm or the best in the neighboorhod, if appropriate) is selected (line 5). Then, at each iteration,

```
1. InitializeSwarm(Part)
2. InitializeVelocities(v)
3. Copy(Part, PartBests)
4. EvaluateParticles(Part, ObjectiveFunction)
5. Remember Leader of Swarm
6. do
7.     UpdateVelocities(v)
8.     UpdatePositionParticles(Part)
9.     EvaluateParticles(Part, ObjectiveFunction)
10.    UpdateParticleMemory(PartBests) if appropriate
11.    SelectNewLeader
12. while (¬termination)
```

Fig. 1. General outline of the classical *PSO* Algorithm

the velocities and positions of the particles are updated using equations (1) and (2) defined in Section 1 (lines 7 and 8). After the update process takes place, each particle is evaluated at its new position (line 9). If the new particle is better than its personal best position (line 10), then this last one is accordingly updated, i.e. $PartBest_i$ is set to $Part_i$.

4.1 *HPSO* Algorithm Description

As indicated before, there is sufficient evidence of the good performance of the *PSO* algorithm in continuous search spaces. The main motivation for the development of the *HPSO* algorithm was to preserve such efficiency when dealing with discrete optimization problems. Thus, we decided to adopt the random keys encoding proposed in [4] so that we could preserve a real-numbers encoding when dealing with permutations. The main idea of the random keys encoding is to adopt a set of randomly generated real numbers, which are then sorted and decoded in such a way that their position in the sequence is interpreted as a permutation position. In the scheduling problem studied, each particle is an n-dimensional vector and each dimension (a real number with two digits of precision) corresponds to a job. The components are randomly generated when the algorithm starts within the range $(0, 1)$. Then, the particle is transformed into a schedule by sorting those values in ascending order. Let's illustrate this with an example: for a nine-job problem, let's assume that we have the particle vector <0.23, 0.08, 0.97, 0.96, 0.32, 0.55, 0.18, 0.87, 0.99>. If we sort this list of real numbers in ascending order, we have the following sequence: <0.08, 0.18, 0.23, 0.32, 0.55, 0.87, 0.96, 0.97, 0.99>. Now, from this sorted list, we extract the mapping that we need: the first value (0.08) corresponds the the integer 1, the second value (0.18), corresponds to the integer 2, and so on. Going back to the original (unsorted) list of real numbers, the permutation that it encodes can be obtained by replacing the integers that

```
1. InitializeSwarm(Part)
2. InitializeVelocities(v)
3. Copy(Part, PartBests)
4. EvaluateParticles(Part, ObjectiveFunction)
5. Remeber Best Leader of Swarm
6. do
7.     CalculateProbability Mutation(p_mut)
8.     UpdateVelocities(v)
9.     UpdatePositionParticle(Part)
10.    EvaluateParticles(Part, ObjectiveFunction)
11.    UpdateParticleMemory(PartBests) if appropriate
12.    MutateSwarm(Part)
13.    EvaluateParticles(Part, ObjectiveFunction)
14.    UpdateParticleMemory(PartBests) if appropriate
15.    SelectNewLeader
16. while (¬termination)
```

Fig. 2. General outline of the *HPSO* Algorithm

we produced from the sorted list. So, we have the following schedule: <3 1 8 7 4 5 2 6 9>. This is thus the permutation evaluated to determine the objective function value of this particle. It is worth noting, however, that due to the redundancy of the representation, many random key vectors may result in the same schedule. So, with the aim of maintaining diversity in the population, we adopted a dynamic mutation operator.

The mutation operator is applied to change the value of a component of a particle, with a probability *pm* varying between *max_pm* and *min_pm*, which depends on the total number of cycles *max_cycles* and the *current_cycle*.

$$pm = max_pm - \frac{max_pm - min_pm}{max_cycle} \times current_cycle \tag{3}$$

where *max_pm* and *min_pm* are the maximum and minimum values that *pm* can take, *max_cycle* is the total number of cycles that the algorithm will iterate, and *current_cycle* is the current cycle in the iterative process. In this way, mutation is more frequently applied at the beginning of the search process and its application decreases as the number of iterations increases. The particle is updated only if the objective function value of the new particle is better than the objective function value prior to applying mutation. Figure 2 displays the pseudocode for the *HPSO* approach.

The differences between *HPSO* and the *PSO* algorithm are described in Figures 1 and 2, and are expressed in lines 7, 12, 13, and 14, where the *HPSO* algorithm includes the mutation operator and the re-evaluation of the swarm to see if each mutated particle is better than its ancestor; if this is the case, then the best position memory is updated.

4.2 $HPSO_{neigh}$ Algorithm Description

As was observed in [8], $HPSO$ converges to a local optimum in some difficult instances of the TWT, which causes stagnation in the search. In order to avoid this problem, $HPSO$ was improved through the use of a neighborhood circle topology (see Figure 3). In this topology, each particle is influenced both by the best value found by the particle itself and by the best value found in the neighborhood so far (neighborhood leader).

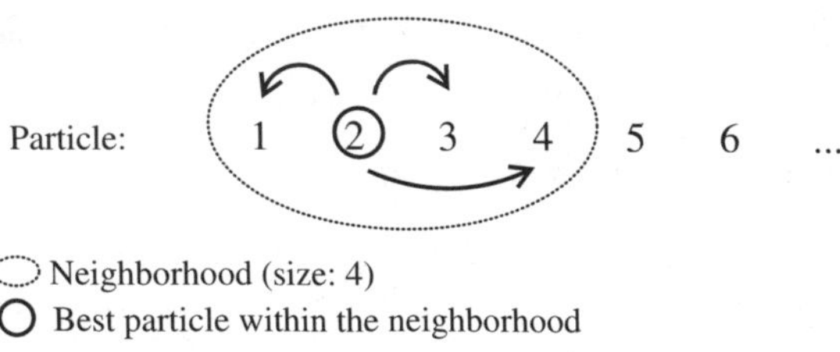

Fig. 3. Graphical illustration of the circle topology adopted by the $HPSO_{neigh}$ algorithm

For example, if we have a swarm with 6 particles and the neighborhood size is 4, then the following neighborhoods are considered: 0 1 2 3, 1 2 3 4, 2 3 4 5, 3 4 5 0, 4 5 0 1, and 5 0 1 2 (the numbers indicating the particle index). Then, each particle is influenced by the performance of the leader of a smaller group instead of being influenced by the performance of the best global leader (i.e., of the complete swarm). Figure 4 presents the pseudocode of the $HPSO_{neigh}$ algorithm. In line 7 the neighborhood of any $part_i$ is composed by the particles whose index are in the interval $[i, i + neighborhood_size - 1]$ if $i + neighborhood_size - 1 < number_particles$. Otherwise, the neighborhood of any $part_i$ consisting of particles whose index are in the interval $[i, neighborhood_size - 2]$ ($\forall i = 1, \ldots, number_particles$).

Besides the inclusion of the neighborhood handler it is important to note that $HPSO_{neigh}$ differs from $HPSO$ in that the former does the particle processing asynchronously, whereas the last one does such processing synchronously. In the asynchronous update, the neighbors on one side of the particle to be adjusted have been updated, while the neighbors on the other side have not. In the synchronous update, the leader is the same for all the particles; therefore, they can be updated in parallel [9].

The algorithms presented in this work were implemented following these criteria since there is prior empirical evidence of the efficiency of these types of processing [18, 9].

```
InitializeSwarm(Part)
InitializeVelocities(v)
Copy(Part, PartBests)
do
    for i = 1 to number_particles do
        CalculateProbabilityMutation(p_mut)
        Search the leader in the neighborhood of part_i
        UpdateVelocity(v_i)
        UpdateParticle(part_i)
        EvaluateParticle(part_i, ObjectiveFunction)
        UpdateParticleMemory(part_i, PartBest_i) if appropriate
        MutateParticle(part_i)
        EvaluateParticle(part_i)
        UpdateParticleMemory(part_i, PartBests_i) if appropriate
    end
while (¬termination)
```

Fig. 4. General outline of the $HPSO_{neigh}$ Algorithm

4.3 $HPSO_{kn}$ Algorithm Description

To improve the previous approach ($HPSO_{neigh}$), we inserted problem-specific knowledge through three seeds generated by three good heuristics: *Rachamadagu and Morton Heuristic* (R&M), *Covert* and *Montagne Heuristic* [32] whose principal property is not only the quality of the results, but also to give an ordering of the jobs (schedule) close to the optimal sequence.

The Rachamadagu and Morton Heuristic, provides a schedule according to the following expression:

$$\pi_j = (w_j/p_j)[exp\{-(S_j)^+/kp_{av}\}] \tag{4}$$

where $S_j = [d_j - (p_j + Ch)]$ is the slack of job j at time Ch and Ch is the total processing time of the jobs already scheduled, k is a parameter of the method (usually $k = 2.0$) and p_{av} is the average processing time of the jobs competing for top priority. In this heuristic, jobs are scheduled one at a time and every time a machine becomes free, a ranking index is computed for the remaining ones. The job with the hightest ranking index is selected to be processed.

The Covert Heuristic, works in a similar way to R&M in cases of a single resource (our case), but applies instead the expression:

$$\pi_j = (w_j/p_j)\{1 - (S_j)^+/kp_j\} \tag{5}$$

The Montagne Heuristic, for its part, uses the following equation:

$$\pi_j = (w_j/p_j)[1 - (d_j)\sum_{i=1}^{n} p_i] \quad (6)$$

This equation does not consider the slack factor, but the due date of every job (d_j) and the sum of all the processing time (p_i).

Several other heuristics previously proposed for the TWT problem in the specialized literature were also tested (using the PARSIFAL package [32]), but we found the three above heuristics to be the most effective and therefore our choice. Needless to say, all of these heuristics are representative of the state-of-the-art in this problem.

As the seed values for each of these three heuristics are very close from each other (in most cases, the Euclidean distance among them is less than one unit in objective function value), we hypothesized that if we put them together, they would influence each other and, slowly, they would also influence the other solutions. That was the reason why we decided to introduce the three seeds within the initial population of particles. Note however, that different positions of the population were adopted for the insertion in each case (see Figure 5). The R&M seed is inserted randomly within the first third of the population, the Montagne seed in the second third, and Covert in the last third of the population (this was done considering the positions of the particles within the storage structure). In that way, each particle is forced to be influenced by some of these good permutations. In some cases, the particles located in the limit of each range might be influenced by two seeds. However, the final value will be the result of the influence of the best of them. Figure 6 shows the pseudocode for our $HPSO_kn$ algorithm.

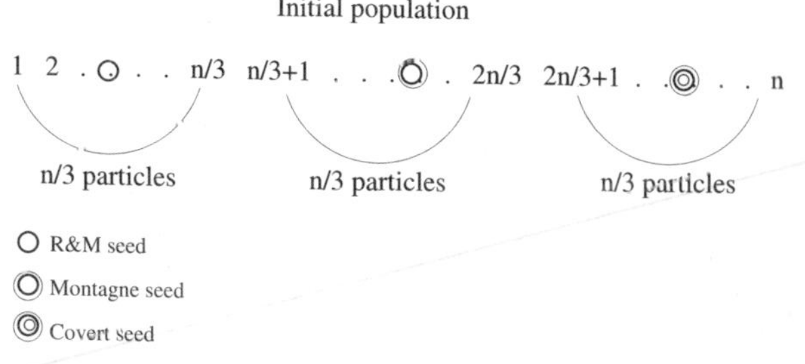

Fig. 5. Graphical illustration of the way in which the three types of seeds (produced by the three heuristics adopted) are inserted in the population

Finally, we will proceed to briefly describe the evolutionary algorithm used to compare our results. The MCMP-SRI-IN [14] approach considers the mating of an evolved individual (the *stud*) with both random and seed immigrants. The process for creating offspring is the following. From the old population, the stud is selected by means of proportional selection and inserted into the mating pool. A number of n_1 parents in the mating pool is completed

```
InitializeSwarm(Part)
InitializeVelocities(v)
Copy(Part, PartBests)
// Seeds Insertion
s = rnd(0, number_particles/3)
CopySeed(seed_{R&M}, part_s)
s = rnd(number_particles/3 + 1, 2 * number_particles/3)
CopySeed(seed_{Covert}, part_s)
s = rnd(2 * number_particles/3 + 1, number_particles)
CopySeed(seed_{Montagne}, part_s)
do
    for i = 1 to number_particles do
        CalculateProbabilityMutation(p_{mut})
        Search the leader in the neighborhood of part_i
        UpdateVelocity(v_i)
        UpdateParticle(part_i)
        EvaluateParticle(part_i, ObjectiveFunction)
        UpdateParticleMemory(part_i, PartBest_i) if appropriate
        MutateParticle(part_i)
        EvaluateParticle(part_i)
        UpdateParticleMemory(part_i, PartBests_i) if appropriate
    end
while (¬termination)
```

Fig. 6. General outline of the $HPSO_{kn}$ Algorithm

both with randomly created individuals (the "random immigrants") and with "seed immigrants". The stud mates every other parent. The couples undergo crossover (partial mapped crossover) and $2\times(n_2-1)$ offspring are created. The best of these offspring is stored in a temporary children pool. The crossover operation is repeated n_1 times, for different cut points each time, until the children pool is full. Finally, the best offspring created from n_2 parents and n_1 crossover operations is inserted into the new population. Figure 7 displays this process.

5 Experimental Design

As indicated before, the goal of the work reported here was to determine the performance of different PSO optimizers when used to solve the total weighted tardiness problem in single machine environments. As indicated before, even with this relatively simple formulation, this model leads to an optimizacion problem that is NP-hard.

The algorithms were tested on twenty instances of 40 and 50 jobs, which were extracted from the OR-Library [5]. The numbering of the problems are

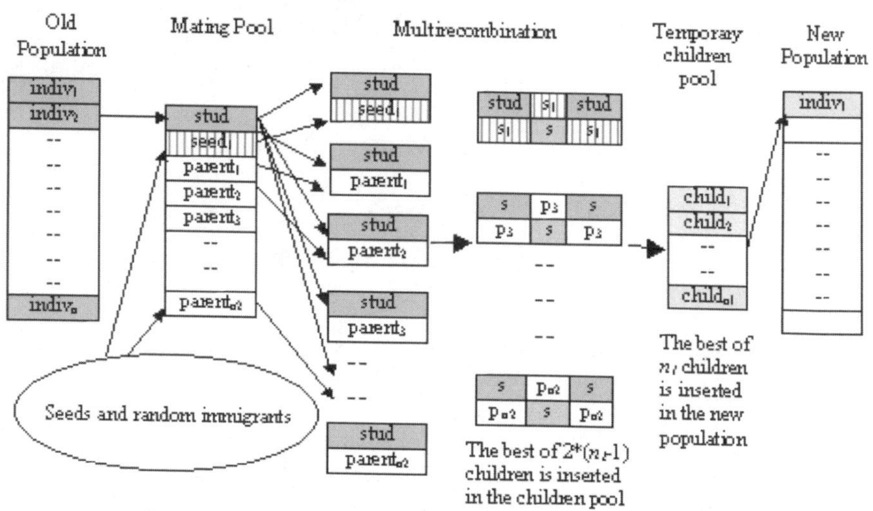

Fig. 7. General outline of $MCMP - SRI - IN$ approach

not consecutive because each one was randomly selected from different groups. The tardiness factor, which is an instance parameter that controls the number of tardy jobs, is harder for those with a higher identifier number. That means that a higher identifier number of instances involves a greater number of tardy jobs.

As it is well-known for researchers working with metaheuristics, the parameters setting of the technique is a very important issue that deserves special attention. Thus, we conducted some preliminary experiments in order to determine the most suitable values for the PSO approaches considered in our study. The values of w (inertia factor), c_1 and c_2 (personal and social learning factors, respectively) were defined following the suggestions from van den Bergh [39]. Analogously, the neighborhood size was fixed between the 8% and 10% of the total swarm size. The values adopted for these parameters in all the experiments conducted are shown in Table 1. The swarm size was set proportional to the permutation length, as suggested by Clerc [11]. 30 independent runs were performed in each experiment. The maximum number of iterations was fixed as follows: $HPSO$ 6000 (40 jobs) and 9000 (50 jobs); $HPSO_{neigh}$ and $HPSO_{kn}$ 50000 (40 jobs) and 65000 (50 jobs). These values were empirically derived after an exhaustive series of experiments. Initially, $HPSO$ ran for the same number of cycles as the other approaches, but its performance did not improve. Thus, as a consequence, we decided to reduce its total number of iterations.

For $HPSO_{neigh}$ and $HPSO_{kn}$, it was neccesary to determinate the values for the mutation probability (pm). This parameter depends of two values:

Table 1. Parameter settings for the *PSO* algorithms considered

Parameters	$HPSO$	$HPSO_{neigh}$	$HPSO_{kn}$
Inertia factor	0.3	0.5	0.5
Learning factors	1.3	1.5	1.5
Neighbor-hood size	-	4	4

min_pm and *max_pm* which, in our case, were fixed to 0.1 and 0.4, respectively.

Additionally, the parameter settings for $MCMP-SRI-IN$ were taken from [14] and are the following: the evolutionary algorithm ran for 200 generations with a population size of 100 individuals. The crossover probability was 0.65 and the mutation probability was 0.05. The algorithm performed 14 crossover operations on each pair of parents and it used 16 parents to recombine. The number of seed was 3 (generated with R&M, Covert, and Modified R&M heuristics).

5.1 Performance Metrics

To compare the algorithms, the following performance metrics were chosen:

- **Best**: It indicates the best value found by an algorithm.
- μ**Best**: It is the mean objective value obtained from the best found particles throughout all runs.
- σ**Best**: It is the standard deviation of the objective values corresponding to the best found particles throughout all runs with respect to μ**Best**.
- (σ/μ)**Best**: This coefficient of variation is calculated as the $\sigma Best$ and $\mu Best$ ratio. It represents the desviation as a percentage of the $\mu Best$ value. The closer this value is to zero, the higher the robustness of the results obtained by an algorithm.
- **Mean Evaluations (ME)**: It is the mean number of evaluations necessary to obtain the best value of the objective function found throughout the runs performed.
- **Hit Ratio (HR)**: It is the percentage of runs where the algorithm reaches the best known values for each test function.

6 Analysis of Results

In this section, we present the results obtained for the algorithms compared as well as a brief discussion of them. First, we present the results obtained by the classical *PSO*, which are displayed in Tables 2 and 3 for instances of

40 and 50 jobs, respectively, where **IN** denotes the problem instance number and the **Best Known Values**, were taken from the OR-Library [5].

Table 2. *PSO* performance for problem instances of 40 jobs

IN	Best Known Value	Best	σ/μ	ME	HR
1	913	913	0.4631	320	0.03
6	6955	8708	0.1363	3220	0.00
11	17465	20652	0.1324	120002	0.00
19	77122	81184	0.0501	85233	0.00
21	77774	81057	0.0583	125512	0.00
26	108	108	0.8715	240	0.03
31	6575	9832	0.1789	135522	0.00
41	57640	63311	0.0643	2445	0.00
46	64451	67088	0.0570	289874	0.00
51	0	661	0.5225	47877	0.00
56	2099	2779	0.2827	586588	0.00
66	65386	75419	0.0617	298854	0.00
71	90486	93072	0.0510	147455	0.00
76	0	0	1.8088	200954	0.70
91	47683	57484	0.0706	568852	0.00
96	126048	130657	0.0333	75665	0.00
101	0	0	0.0000	1552	1.00
106	0	0	0.0000	2544	1.00
116	46770	56139	0.0872	185587	0.00
121	122266	128107	0.0581	299847	0.00

From the results shown in Tables 2 and 3, it can be seen that the classical *PSO* is unable to reach the best known values in almost all the instances. This is indicated by the zero values for the HR metric, except for instances 101 and 106 in the case of 40 jobs. This is the reason by which in the remainder of this section, only the results for $HPSO$, $HPSO_{neigh}$, $HPSO_{kn}$ and $MCMP-SRI-IN$ are discussed.

Tables 4 and 5 summarize the best objective function values found by the *PSO* variants and by the evolutionary algorithm for problem instances with 40 jobs and 50 jobs, repectively. Observing these values in both tables, we can see that the $HPSO$ algorithm has, for some instances, the worst performance (marked with boldface). This is due to the fact that each particle in the swarm is attracted towards the position of the global best particle, which leads to a stagnation of the algorithm in a local optimum. In the case of 40 jobs, $HPSO_{neigh}$ and $HPSO_{kn}$ converge to the same best values, and both algorithms outperform to $MCMP-SRI-IN$ in instance 21. The results for the 50 jobs problems (Table 5) show that in instance 6, $HPSO_{kn}$ obtains the worst best value (but yet it is closer to the best known value). For instances

Table 3. *PSO* performance for problem instances of 50 jobs

IM	Best Known Value	Best	σ/μ	ME	HR
1	2134	2259	0.22623	98558	0.00
6	26276	29241	0.08538	568856	0.00
11	51785	53844	0.08870	866585	0.00
19	89299	99698	0.06118	248898	0.00
21	214546	221119	0.03595	25442	0.00
26	2	10	1.11608	17452	0.00
31	9934	14426	0.16843	34252	0.00
41	123893	124855	0.02765	27784	0.00
46	157505	167009	0.46806	89552	0.00
51	0	0	0.7619	131905	0.03
56	1258	1258	0.19776	57884	0.10
66	76878	76991	0.01623	25995	0.00
71	150580	151322	0.02495	69899	0.00
76	0	0	1.0000	27741	0.20
91	9298	39787	0.02778	98778	0.00
96	77909	187222	0.00991	33541	0.00
101	0	0	0.9935	37787	0.50
106	0	0	0.9000	47785	0.35
116	35727	38544	0.03077	78448	0.00
121	8315	79884	0.02304	35884	0.00

Table 4. *Best* metric values for TWT 40 jobs problem size

IN	Best Known Value	$HPSO$	$HPSO_{neigh}$	$MCMP-SRI-In$	$HPSO_{kn}$
1	913	913	913	913	913
6	6955	6955	6955	6955	6955
11	17465	17465	17465	17465	17465
19	77122	77122	77122	77122	77122
21	77774	77774	77774	77774	77774
26	108	108	108	108	108
31	6575	6575	6575	6575	6575
41	57640	57640	57640	**57876**	57640
46	64451	**64459**	64451	64451	64451
51	0	0	0	0	0
56	2099	2099	2099	2099	2099
66	65386	**65402**	65386	65386	65386
71	90486	**90523**	90486	90486	90486
76	0	0	0	0	0
91	47683	47683	47683	47683	47683
96	126048	126048	126048	126048	126048
101	0	0	0	0	0
106	0	0	0	0	0
116	46770	**46771**	46770	46770	46770
121	122266	**122304**	122266	122266	122266

Table 5. *Best* metric values for TWT 50 jobs problem size

IN	Best Known Value	$HPSO$	$HPSO_{neigh}$	$MCMP-SRI-In$	$HPSO_{kn}$
1	2134	2134	2134	2134	2134
6	26276	26276	26276	26276	**26281**
11	51785	51785	51785	51785	51785
19	89299	**89308**	**89308**	89299	89299
21	214546	**214585**	**214744**	**214555**	**214555**
26	2	2	2	2	2
31	9934	9934	9934	9934	9934
44	123893	**124261**	123893	123893	123893
46	157505	**157536**	157505	157505	157505
51	0	0	0	0	0
56	1258	1258	1258	1258	1258
66	76878	**76948**	76878	76878	76878
71	150580	**150667**	150580	150580	150580
76	0	0	0	0	0
91	89298	**89543**	**89323**	**89448**	**89474**
96	177909	**178007**	177909	177909	177909
101	0	0	0	0	0
106	0	0	0	0	0
116	35727	**35830**	**35728**	35727	35727
121	78315	**78396**	78315	78315	78315

19, 21, 91, and 116, $HPSO_{kn}$ is the algorithm with the best performance, and specially in instance 21 where none of the algorithms reaches the best know values, $HPSO_{kn}$ obtains the same value than $MCMP-SRI-IN$. As a conclusion, we can say that except for some instances (marked with boldface), all the algorithms find the best known values. In fact, even when these values are not reached, $HPSO_{kn}$ and $MCMP-SRI-IN$ converge to very similar values.

Nevertheless, it is important to analyze these results in more details, by using other performance metrics such as Hit Ratio and the mean number of evaluations that each algorithm has to perform to find the best value.

Figure 8 shows the analysis of the Hit Ratio metric. In this case, we can see that $HPSO_{neigh}$ finds the best known values approximately 70% of the time for the case of 40 jobs and around 50% of the time for the case of 50 jobs. In contrast, $HPSO_{kn}$ reaches the best known values in approximately the 80% and 70% of the runs for the 40 and 50 jobs instances, respectively. Also, we can observe that the results obtained with $MCMP-SRI-IN$ are slighly better than those found by $HPSO_{kn}$, although none of the algorithms finds the best known values for all the instances in all the runs. With the previous observations in mind, we can conclude that $HPSO_{kn}$ is superior to $HPSO_{neigh}$ and its results are comparable to those obtained by $MCMP-SRI-IN$ (which can be seen as an evolutionary algorithm that has been carefully tailored for the problem being solved in this study).

Figure 9 shows the cost measured as the mean number of evaluations that an algorithm performs to reach the best known values. In this case, $HPSO_{kn}$ performs, on average, a lower number of evaluations when compared with $HPSO_{neigh}$ and $MCMP-SRI-IN$, a difference that becomes even higher

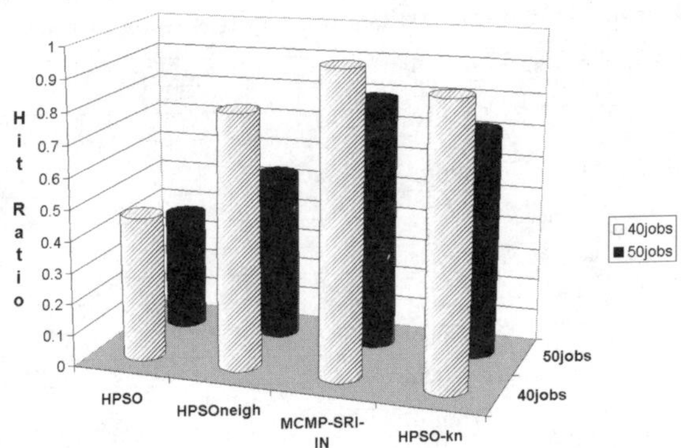

Fig. 8. Performance evaluation with respect to the Hit Ratio metric

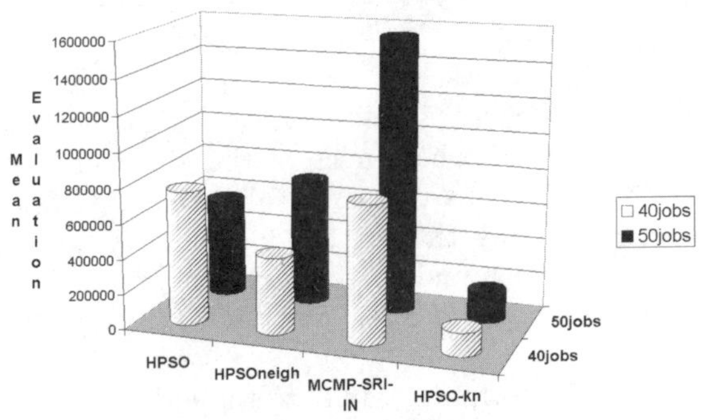

Fig. 9. Performance evaluation with respect to the Mean Evaluation metric

when the problem size is increased. This difference was somehow expected (with respect to $HPSO_{neigh}$) due to the guided search that the $HPSO_{kn}$ performs. The idea of including knowledge about the problem in the algorithm is not new, since it has been successfully applied in the past in several evolutionary algorithms [3].

Table 6. $(\sigma/\mu)Best$ mean values obtained by the PSO variants compared

Problem Size	HPSO	HPSO_neigh	HPSO_kn
40	0.003825	0.002880	0.001950
50	0.003320	0.001565	0.000100

In Table 6, we show the mean values over all the coefficients of variation of the best values calculated for all the instances, for the different PSO variants for the two instances studied (40 jobs and 50 jobs). These values are grouped around the mean. Although not all the coefficient values were equal to zero, they are very close, which suggests robustness of the algorithms with respect to the results that they found.

7 Conclusions and Future Work

In this chapter, three improved PSO variants were presented to deal with permutation problems. To determine the performance of the algorithms studied, the weighted tardiness scheduling on the single machine environments problem was selected as a case of study. $HPSO$ is a hybridized PSO in the sense that a suitable representation and a dynamic mutation operator were adopted to make it more competitive in sequencing problems. However, we saw that this approach in which the global leader is always followed, is prone to converge to a local optimum, causing a premature convergence of the algorithm.

As a way of dealing with this drawback, we proposed an approach called $HPSO_{neigh}$, which incorporates a simple neighborhood topology, so that each particle is only influenced by the best local particle in its neighborhood. This modification allowed that the algorithm could find all the best known values for the 40 jobs problem size and increased the number of instances in which the algorithm found the best known values for the instances of 50 jobs (instances 19, 21, 44, 46, 66 and 71). A further modification was introduced, which consisted of the incorporation of specific domain knowledge by means of the inclusion of seeds (generated with another heuristic) in the swarm. This new version was named $HPSO_{kn}$. All these algorithms were compared among themselves and with respect to $MCMP-SRI-IN$, which is an evolutionary algorithm specially tailored for the problem of interest and which also uses the inclusion of knowledge through seeds. Although $HPSO_{neigh}$, $HPSO_{kn}$ and $MCMP-SRI-IN$ found objective values which are similar, $HPSO_{kn}$ and $MCMP-SRI-IN$ exceeded widely to $HPSO_{neigh}$ in the number of runs in which they reached the best known values as was shown with the Hit Ratio values. In spite of that, the cost (measured in the number of evaluations performed to reach the best known values) of $HPSO_{kn}$ is fairly smaller than the one required by $MCMP-SRI-IN$ and also (as expected) is about a 50% lower than the cost of $HPSO_{neigh}$. We believe that these preliminary results are good enough to consider $HPSO$ variants as a promising approach for scheduling problems. Thus, we are convinced that this topic deserves further study.

As part of our future work, we are considering different possibilities. The first one is to minimize the redundancy of the encoding currently adopted by exploring alternative encodings. Second, we aim to study the effect of incorporating and adapting other operators which have been typically used with evolutionary algorithms to solve permutations problems. Finally, it is of great relevance for us the study of the behavior of our proposed approach in much larger instances of this problem (between 100 and 200 jobs).

Acknowledgments

The third author gratefully acknowledges support from CONACyT project number 45683-Y.

References

1. M. S. Akturk and M. B. Yildirim. A New Dominance Rule for the Total Weighted Tardiness Problem. *Production Planning and Control*, 10(2):138–149, 1999.
2. B. Alidaee and K. R. Ramakrishnan. A computational experiment of covert_au class of rules for single machine tardiness scheduling problem. *Computers and Industrial Engineering*, 30(2):201–209, 1996.
3. Thomas Bäck, David B. Fogel, and Zbigniew Michalewicz. *Handbook of Evolutionary Computation*. IOP Publishing LTD and Oxford University Press, 1997.
4. James C. Bean. Genetics and random keys for sequencing and optimization. *ORSA Journal on Computing*, 6(2):154–160, 1994.
5. J. E. Beasley. OR Library, Scheduling: Weighted Tardiness. http://people.brunel.ac.uk/~mastjjb/jeb/info.html.
6. M. Den Besten, T. Stutzle, and M. Dorigo. Ant colony optimization for the total weighted tardiness problem. In *Proc. of PPSN-VI: Sixth International Conference on Parallel Problem Solving from Nature*, volume LNCS 1917, pages 611–620, 2000.
7. L. Cagnina and S. Esquivel. Particle swarm optimization para un problema de optimización combinatoria. In *Memorias del X Congreso Argentino de Ciencias de la Computación*, pages 1847–1855, La Matanza, Buenos Aires, Argentina (in Spanish), 2004. http://www.lidic.unsl.edu.ar/publicaciones/info_publicacion.php?id_publicacion=199.
8. L. Cagnina, S. Esquivel, and R. Gallard. Particle swarm optimization for sequencing problems: a case study. In *Proceedings of the 2004 IEEE Congress on Evolutionary Computation (CEC'2004)*, pages 536–541, Portland, Oregon, USA, 2004.
9. A. Carlisle. *Applying The Particle Swarm Optimization to Non-Stationary Environments*. PhD thesis, Auburn University, USA, December 2002.
10. K. Caskey and R. L. Storch. Heterogeneous dispatching rules in job and flow shops. *Production Planning and Control*, 7:351–361, 1996.
11. M. Clerc. Discrete particle swarm optimization illustred by the traveling salesman problem, 2000. http://www.mauriceclerc.net.
12. Carlos A. Coello Coello, Erika Hernández Luna, and Arturo Hernández Aguirre. Use of particle swarm optimization to design combinational logic circuits. In

Andy M. Tyrell, Pauline C. Haddow, and Jim Torresen, editors, *Evolvable Systems: From Biology to Hardware. Proceedings of the 5th International Conference, ICES 2003*, pages 398–409, Trondheim, Norway, 2003. Springer, Lecture Notes in Computer Science Vol. 2606.

13. F. Della Croce. Generalized pairwise interchanges and machine scheduling. *European Journal Operations Research*, 83:310–319, 1995.
14. M. De San Pedro, D. Pandolfi, A. Villagra, M. Lasso, G. Vilanova, and R. Gallard. Adding problem-specific knowledge in evolutionary algorithms to solve wt scheduling problems. In *Memorias del VIII Congreso Argentino de Ciencias de la Computación*, pages 343–353, Buenos Aires, Argentina (in Spanish), 2002. http://www.lidic.unsl.edu.ar/publicaciones/info_publicacion.php?id_publicacion=198.
15. R. Eberhart and Y. Shi. A modified particle swarm optimizer. In *International Conference on Evolutionary Computation, IEEE Service Center*, Anchorage, AK, Piscataway, NJ, 1998.
16. R. C. Eberhart and J. Kennedy. A new optimizer using particle swarm theory. In *6th International Symposium on Micro Machine and Human Science (Nagoya, Japan)*, pages 39–43, Piscataway, NJ., 1995. IEEE Service Center.
17. O. Ergun and J. B. Orlin. A fast algorithm for searching insertion, swap, and twist neighborhoods for the single machine total weighted tardiness problem. In *Working Paper, Operations Research Center, MIT*, 2004.
18. Susana C. Esquivel and Carlos A. Coello Coello. On the Use of Particle Swarm Optimization with Multimodal Functions. In *Proceedings of 2003 Congress on Evolutionary Computation (CEC'2003)*, volume 2, pages 1130–1136, Piscataway, NJ., December 2003. IEEE Press.
19. Fred Glover and Manuel Laguna. *Tabu Search.* Kluwer Academic Publishers, Boston, Massachusetts, 1997.
20. A. Grosso, F. Della Croce, and R.Tadei. An enhanced dynasearch neighborhood for the single-machine total weighted tardiness scheduling problem. *Operations Research Letters*, 32:68–72, 2004.
21. C. N. Potts H. A. J. Crauwels and L. N. Van Wassenhove. Local search heuristics for the single machine total weighted tardiness scheduling problem. In *INFORMS Journal on Computing*, volume 10(3), pages 341–350, 1998.
22. X. Hu and R. Eberhart. Swarm intelligence for permutation optimization: a case study on n-queens problem. In *Proceeding of the IEEE Swarm Intelligence Symposium*, pages 243–246, Indianapolis, Indiana, USA, 2003.
23. A. H. G. Rinnooy Kan, B. J. Lageweg, and J. K. Lenstra. Minimizing total costs in one-machine scheduling. *Operations Research*, 23(3):908–927, 1975.
24. J. Kennedy and R. Eberhart. A discrete binary version of particle swarm algorithm. In *Proceedings of the World Multiconference on Systemics, Cybernetics and Informatics*, pages 4104–4109, Piscataway, NJ, 1997.
25. James Kennedy and Russell Eberhart. *Swarm Intelligence.* Morgan Kaufmann Publishers, 2001.
26. S. Kirkpatrick, C. D. Gellatt, and M. P. Vecchi. Optimization by Simulated Annealing. *Science*, 220(4598):671–680, 1983.
27. J. K. Lenstra, A. H. G. Rinnooy Kan, and P. Brucker. Complexity of machine scheduling problem. In P. L. Hammer, E. L. Johnson, B. H. Korte, and G. L. Nemhauser, editors, *Studies in Integer Programming, volume I of Annals of Discrete Mathematics*, pages 343–362. North-Holland, The Netherlands, 1977.

28. Y. C. Liang. *Ant colony optimization approach to combinatorial problems.* PhD thesis, Department of Industrial and Systems Engineering, Auburn University, 2001.
29. T. E. Matsuo, C. J. Suh, and R. S. Sullivan. A controlled search simulated annealing method for the single machine weighted tardiness problem. *Annals of Operations Research*, 21:95–108, 1989.
30. D. Merkle and M. Middendorf. An ant algorithm with a new pheronome evaluation rule for total tardiness problem. In *Proc. of EvoWorkshops 2000: Real-World Applications of Evolutionary Computing*, volume LNCS 1803, pages 287–296, 2000.
31. C. Mohan and B. Al-Kazemi. Discrete particle swarm optimization. In *Proceeding of the Workshop on Particle Swarm Optimization*, Indianapolis, IN, 2001.
32. T. Morton and D. Pentico. *Heuristic Scheduling Systems.* Wiley series in Engineering and Technology management, John Wiley and Sons, 1993.
33. C. N. Potts and L. N. Van Wassenhove. A branch and bound algorithm for the total weighted tardiness scheduling problem. In *Operations Research*, volume 33 number 2, pages 363–377, 1985.
34. C. N. Potts and L. N. Van Wassenhove. Single machine tardiness sequencing heuristics. *IIE Transactions*, 23(4):346–354, 1991.
35. C. N. Potts R. K. Congram and S. Van de Velde. An iterated dynasearch algorithm for the single-machine total weighted tardiness scheduling problem. *INFORMS Journal on Computing*, 14:52–67, 2002.
36. R. M. V. Rachamadugu. A note on weighted tardiness problem. *Operations Research*, 35:450–452, 1987.
37. C. N. Potts T. S. Abdul-Razaq and L. N. Van Wassenhove. A survey of algorithms for the single machine total weighted tardiness scheduling problem. In *Discrete Applied Mathematics*, volume 26, pages 235–253, 1990.
38. M. Fatih Tasgetiren, Mehmet Sevkli, Yun-Chia Liang, and Gunes Gencyilmaz. Particle Swarm Optimization Algorithm For Single Machine Total Weighted Tardiness Problem. In *Proceedings of the 2004 IEEE Congress on Evolutionary Computation (CEC'2004)*, pages 1412–1419, Portland, Oregon, USA, 2004.
39. Frans van den Bergh. *An Analysis of Particle Swarm Optimization.* PhD thesis, Faculty of Natural and Agricultural Science, University of Petroria, Pretoria, South Africa, November 2002.
40. A. P. J. Vepsalainem and T. E. Morton. Priority rules for job shops with weighted tardiness costs. *Management Science*, 33:1035–1047, 1987.

An Evolutionary Approach for Solving the Multi-Objective Job-Shop Scheduling Problem

Kazi Shah Nawaz Ripon[1], Chi-Ho Tsang[1], and Sam Kwong[2]

Department of Computer Science, City University of Hong Kong
83 Tat Chee Avenue, Kowloon, Hong Kong
[1]{ripon, wilson}@cs.cityu.edu.hk [2]cssamk@cityu.edu.hk

Summary. In this chapter, we present an evolutionary approach for solving the multi-objective Job-Shop Scheduling Problem (JSSP) using the Jumping Genes Genetic Algorithm (JGGA). The jumping gene operations introduced in JGGA enable the local search process to exploit scheduling solutions around chromosomes, while the conventional genetic operators globally explore solutions from the population. During recent decades, various evolutionary approaches have been tried in efforts to solve JSSP, but most of them have been limited to a single objective, which is not suitable for real-world, multiple objective scheduling problems. The proposed JGGA-based scheduling algorithm heuristically searches for near-optimal schedules that optimize multiple criteria simultaneously. Experimental results using various benchmark test problems demonstrate that our proposed approach can search for the near-optimal and non-dominated solutions by optimizing the makespan and mean flow time. The proposed JGGA based approach is compared with another well established multi-objective evolutionary algorithm (MOEA) based JSSP approach and much better performance of the proposed approach is observed. Simulation results also reveal that this approach can find a diverse set of scheduling solutions that provide a wide range of choice for the decision makers.

1 Introduction

The objective of scheduling is to allocate resources efficiently, such that a number of tasks can be completed economically within given hard or soft constraints. In essence, scheduling can be considered as a searching or optimization problem, where the goal is to find the best possible schedule. Among all of the scheduling

K.S. Nawaz Ripon et al.: *An Evolutionary Approach for Solving the Multi-Objective Job-Shop Scheduling Problem,* Studies in Computational Intelligence (SCI) **49**, 165–195 (2007)
www.springerlink.com

problems, the JSSP is one of most challenging one. It is widely found in industry, and it is often considered to include many general scheduling problems that exist in practice. The complexity of JSSP increases with the number of constraints imposed and the size of search space employed. Except for some highly restricted special cases, the JSSP is an *NP*-hard problem, and it has also been considered as the worst of the worst among combinatorial problems [1]. These days, many real-world JSSPs include a larger number of jobs and machines as well as additional constraints and flexibilities, all of which in turn further increases the complexity of JSSP. Exact methods, such as the branch-and-bound method or dynamic programming, are always computationally expensive to use when searching for an optimum scheduling solution with a large search space. To overcome this difficulty, it is more reasonable to achieve near-optimal solutions instead. Stochastic search techniques such as Evolutionary Algorithms (EAs) can be efficiently used to find such solutions. The EAs differ from other conventional optimization algorithms in that they search among and evolve from a population of solutions, rather than a single solution. EAs have proved very powerful and robust for solving many significant single or multiple objective problems.

Various classes of scheduling problems have been investigated over the years, and many different methods have been developed for solving them. Yet, most of the research conducted in the field of scheduling has concerned a single objective, and principally the optimization of makespan. By contrast, real-life scheduling problems are multi-objective by nature and they require the decision maker to consider a number of criteria before arriving at any conclusion. A solution that is optimal with respect to a certain given criterion might be a poor candidate for where another is paramount. Hence, the trade-offs involved in considering several different criteria provide useful insights for the decision maker. Surprisingly, however, research that takes multiple criteria into account has been scarce, particularly when compared to the research in the area of single criterion scheduling. There have been only a few attempts to tackle the multi-objective JSSP [2,3,4]. The goal of multi-objective job-shop scheduling is to find as many different potential schedules as possible, each of which is near-optimal and is not dominated by consideration of a particular objective. Some performance measures used frequently in this field include makespan, mean flow time, and mean tardiness. The makespan is defined as the maximum completion time of all jobs, mean flow time is the average of the flow times of all jobs, and the mean tardiness is defined as the average of tardiness of all jobs. In an attempt to address multiple objectives simultaneously in this work, we apply makespan and mean flow time as the objectives of our scheduling algorithm.

The Jumping Genes Genetic Algorithm (JGGA) [5,6] is a very recent MOEA. It imitates a jumping gene phenomenon that was discovered by Nobel Laureate McClintock in her work on the corn plants. In this work, an extended JGGA is proposed to search for the Pareto-optimal schedules in static JSSP. The jumping gene operations proposed in JGGA exploit scheduling solutions around the chromosomes, while the general genetic operators explore solutions from the population globally using multiple objective functions. The central idea behind the jumping operations is to provide a local search capability that will make it possible to fine-tune the scheduling solutions during evolution. In previous research [5,6], the JGGA has performed robustly in searching the non-dominated solutions, taking into consideration both convergence and diversity. This is significant because it is very important to obtain both converged and diverse scheduling solutions in the Pareto-optimal set. Converged solutions guarantee that the most near-optimal schedules that consider multiple criteria will be found. The diverse solutions, in particular the extreme solutions, are useful when selecting the best compromise schedule from among the non-dominated solutions, according to the specific objectives required in different production orders or customer demands. To justify our approach, we also compared our proposed approach with another well established MOEA (NSGAII [7]) based approach, and found that the JGGA-based scheduling approach is able to produce more non-dominated and near-optimal solutions that provide the decision makers with enough alternatives to find out suitable solution.

This chapter is organized as follows. Section 2 describes the scheduling and the JSSP. The literature review of various scheduling approaches, in particular the multi-objective genetic algorithms for JSSP, are presented in Section 3. The advantages of applying JGGA in JSSP are highlighted at the end of Section 3. Section 4 briefly describes the overview of the original JGGA. Detailed description of the proposed extended JGGA in solving JSSP is discussed in Section 5. To demonstrate the performance of JGGA, experimental results are presented and analyzed in Section 6, followed by the conclusions in the final section.

2 Description of Scheduling and Job-Shop Scheduling Problem

2.1 Scheduling Problem

Scheduling can loosely be described as the allocation of shared resources (machines, people etc) efficiently over time to competing activities (jobs, tasks, etc) such that a certain number of goals can be achieved economically and a set of given constraints can be satisfied. In general, the construction of a schedule is an optimization problem of arranging time, space and (often limited) resources simultaneously [8]. The constraints can be classified as hard and soft. Hard constraint must not be violated under any circumstances. The solutions that satisfy such constraints can be called feasible. For the soft constraint, it is desirable to satisfy as many soft constraints as possible, but if one of them is violated, a penalty is applied and the solution is still considered to be feasible. In practice, scheduling can be considered as a search problem where it is required to search for any feasible schedule or as an optimization problem where it is desirable to search for the best feasible schedule. In practical problems, it is not easy to express the conditions that make a schedule more preferable than another and to incorporate this information in an automated system. Furthermore, the combinatorial nature of these problems leads to explore huge search spaces and for that human involvement is often inevitable to guide the search towards promising regions.

The shop scheduling is one of the most challenging scheduling problems. It can be classified into four main categories: (i) single-machine scheduling, (ii) flow-shop scheduling, (iii) job-shop scheduling, and (iv) open-shop scheduling. Single-machine scheduling is the simplest shop scheduling problem, in which there is only one machine available and arriving jobs require services from this machine. In flow-shop scheduling, jobs are processed on multiple machines in an identical sequence. Job-shop scheduling is a general case of flow-shop scheduling in that the sequencing of each job through the machines is not necessarily identical. An open-shop scheduling is similar to a job-shop scheduling except that a job may be processed on the machines in any sequence the job needs. In other words, there is no operationally dependent sequence that a job must follow.

This chapter focuses on solving the JSSP since it is widely found in the industry and is often considered to be representative of many general scheduling problems in practice.

2.2 Job-Shop Scheduling Problem (JSSP)

The JSSP is commonly found in many real-world applications such as industrial production and multi-processor computer systems. A job-shop scheduling involves processing of the jobs on several machines without any "series" routing structure. The challenge here is to determine the *optimum sequence* in which the jobs should be processed in a way that one or more performance measure, such as the total time to complete all the jobs, the average mean flow time, or the mean tardiness of jobs from their committed dates, is minimized. A classical n job, m machine JSSP consists of a finite set $\{J_j\}_{1\leq j\leq n}$ of n independent jobs or tasks that must be processed in a finite set $\{M_k\}_{1\leq k\leq m}$ of m machines. The problem can be characterized as follows:

- each job $j \in J$ must be processed by every machine $k \in M$;
- the processing of job J_j on machine M_k is called the operation O_{jk};
- operation O_{jk} requires the exclusive use of machine M_k for an uninterrupted duration t_{jk}, its processing time;
- each job consists of an operating sequence of x_j operations (technological sequence of each job);
- O_{jk} can be processed by only one machine k at a time (disjunctive constraint);
- each operation, which has started, runs to completion (non-preemption condition);
- each machine performs operations one after another (resource/capacity constraint);

Table 1 shows an example of *6*x*6* job-shop scheduling benchmark problem [9]. In this example, the Job-1 is processed by Machine-3 for 1 time unit, and it is also processed by Machine-1 for 3 time units, and so forth.

Table 1. A 6x6 job-shop scheduling benchmark problem [9]

Job-n	(k,t)	(k,t)	(k,t)	(k,t)	(k,t)	(k,t)
Job-1	3,1	1,3	2,6	4,7	6,3	5,6
Job-2	2,8	3,5	5,10	6,10	1,10	4,4
Job-3	3,5	4,4	6,8	1,9	2,1	5,7
Job-4	2,5	1,5	3,5	4,3	5,8	6,9
Job-5	3,9	2,3	5,5	6,4	1,3	4,1
Job-6	2,3	4,3	6,9	1,10	5,4	3,1

The JSSP is often considered as *n*x*m* minimum-makespan optimization problem, due to the fact that minimum-makespan is the simplest criterion which directly corresponds to a good schedule. However, it is almost impossible to optimize all the above mentioned criteria, because they are often conflicting.

2.3 Complexity of JSSP

The complexity of JSSP increases with the number of constraints imposed and the size of search space employed. Except for some highly restricted special cases, the JSSP is an *NP*-hard problem and finding an exact solution is computationally intractable. For example, a small *10*x*10* (10 jobs, 10 machines and 100 operations) scheduling problem proposed by Muth and Thompson [9] in 1963 remained unsolved until two decades later. In addition, the JSSP has also been considered as a hard combinatorial optimization problem, which is also one of the worst members in that class [1]. Even a simple version of the standard job-shop scheduling is *NP*-hard if the performance measure is the makespan and $m > 2$. For the standard JSSP, the size of search space is $(n!)^m$, and for this reason, it is computationally infeasible to try every possible solution. This is because the required computation time increases exponentially with the problem size. In practice, many real-world JSSPs have a larger number of jobs and machines as well as additional constraints and flexibilities, which further increase its complexity.

3 Related Works

3.1 Traditional and Heuristic Approaches for JSSP

The JSSP has been extensively studied over the past forty years. A wide variety of approaches have been proposed in many diverse areas, such as operations research, production management and computer engineering. Traditionally, the exact methods such as the branch-and-bound algorithm [10], the time orientation approach [11] and the Lagrangian relaxation with dynamic programming [12] have been successfully applied to solve small JSSPs. However, for today's large JSSPs with complex search spaces, it is computationally intractable for them to obtain an exact optimal schedule within a reasonable time. Because obtaining the exact optimal solution for large JSSPs is non-trivial, it is desirable to obtain as many as near-optimal or possibly optimal solutions in polynomial time, which can be later judged by human experts. Many meta-heuristic techniques have been proposed in the literature to search for near-optimal scheduling solutions in a reasonable amount of processing time. The meta-heuristic approaches include the Simulated Annealing (SA) [13], the Tabu Search (TS) [14], the Genetic Algorithms (GA) [15,16], a hybrid of SA and GA [17], the Ant Colony Optimization (ACO) algorithm [18], the Particle Swarm Optimization (PSO) algorithm [19], and the like. Recently, hybrid heuristics have been a vital topic in the fields of both computer science and operations research, and it is often the case that local search is incorporated into evolutionary approaches in order to improve the results obtained with these methods. Such methods are sometimes called memetic algorithms. These approaches include the local search [20], the shifting bottleneck approach [21], the guided local search with shifting bottleneck approach [22], constraint propagation algorithm [23], parallel greedy randomized adaptive search procedure (GRASP) [24], and the like. Comprehensive surveys of the general JSSPs are found in [25,26].

3.2 Evolutionary Algorithms for JSSP

EAs are the stochastic search algorithms inspired by the natural selection and survival of the elitist in the biological world. The goal of EAs is to search for global near-optimal solutions from the fitness functions using exploration and exploitation methods. These methods include biologically inspired mating, mutation and selection operations. EAs differ from other conventional optimization algorithms in that EAs evolve and search from a population of solutions rather than a single solution. They have proved to be very powerful and robust for solving manynontrivial single or multiple objective problems. Furthermore, they often allow objectives and constraints to be easily modified. The work related to the GA in JSSPs is focused and discussed in this chapter. An in-depth survey of the application of the GA in JSSPs is found in [27].

The GA, proposed by Holland [28] in the 1970s has been successfully applied to solve many combinatorial optimization problems, including scheduling. Unlike many heuristic approaches, the GA is more general and abstract for different independent problem domains. And, this superiority of GA comes from its structural formulation. For example, the GA simultaneously finds the best solution from many points of the search domain rather than analyzing one domain point at a time, and thus avoiding getting stuck in local optima. To evaluate solutions, GA uses the objective function rather than auxiliary knowledge such as derivative functions. In addition, the GA is not only effective in performing global searches, but it is also flexible enough to hybridize with other domain-dependent heuristics or local search techniques to solve specific problems. Because exact methods typically take exponential time and because many heuristic approaches can only find suboptimal solutions for large JSSPs, the EAs, in particular the GA, becomes a more popular approach for solving JSSPs. The application of the GA in JSSPs was introduced by Davis [29] in 1985; since then, many GA-based job-shop scheduling approaches have been proposed. The literature already shows that GA-based approaches can often achieve more robust and better performance than many traditional and heuristic approaches applied in JSSP [30].

3.3 Importance of Multi-Objective Job-Shop Scheduling

In many real-world JSSPs, it is often necessary to optimize several criteria, such as the length of a schedule or the utilization of different resources simultaneously. In general, minimization of makespan is often used as the optimization criterion in single objective JSSP. However, the minimizations of lateness, tardiness, flow time, machine idle time, and such others are also the important criteria in JSSP. As discussed in [31], makespan may not be the only commercial interest in scheduling, since it is unusual to have a scheduling problem that has a fixed pre-determined ending, which doesn't change unexpectedly and in which all details are known at the beginning. Some other objectives, such as mean flow time or tardiness are also important like the makespan. It is desirable to generate many near-optimal schedules considering multiple objectives according to the requirements of the production order or customer demand. Then, the production manager can selectively choose the most demanding schedule from among all of the generated solutions for specific order or customer. On the other hand, if multiple objectives conflict with each other, then the production manager does not need to omit any required objective before the assistance of multi-objective scheduler. For the explanation of the correlations of different objective functions used in scheduling, the reader may refer to [32] for details. Based on the principle of multi-objective optimization, obtaining an optimal scheduling solution that satisfies all of the objective functions is almost impossible, due to the conflicting nature of objective functions where improving one objective may only be achieved when worsening another objective. However, it is desirable to obtain as many different Pareto-optimal scheduling solutions as possible, which should be non-dominated, converged to, and diverse along the Pareto-optimal front with respect to these multiple criteria. The description of the Pareto-optimal optimization and the application of multi-objective GA in JSSP are discussed as follows.

3.4 Multi-Objective Genetic Algorithms for JSSP

Multi-objective optimization has been studied by many researchers and industrial practitioners in various areas due to the multi-objective nature of many real-world problems. In general, many practical problems have no optimal solution satisfying all the objectives simultaneously, as the objectives may conflict with each other. However, there exists a set of equally efficient, non-dominated, admissible, or non-inferior solutions, known as the Pareto-optimal set [33]. The goal of multi-objective optimization is to search a set of Pareto-optimal solutions. Without the loss of generality, an unconstrained multi-objective optimization problem can be formally formulated as follows. The goal is to minimize $z = f(x)$ where $f(x) = (f_1(x), f_2(x), \ldots, f_k(x))$ and $k \geq 2$ is the number of objective functions $f_i : \Re^n \mapsto R$, subject to $x \in X$. A solution $x^* \in X$ is called Pareto optimal if there is no $x \in X$ such that $f(x) < f(x^*)$. If x^* is Pareto optimal, $z^* = f(x^*)$ is called (globally) non-dominated. The set of all Pareto optima is called the Pareto optimal set, and the set of all non-dominated objective vectors is called the Pareto front. Searching an approximation to either the Pareto optimal set or the Pareto front is called the Pareto optimization. From the results of Pareto optimization, the human decision makers can choose the suitable compromise solutions. In general, multi-objective optimization approaches can be broadly classified into three categories as shown below. Note that for these optimization approaches, applying weighting coefficients, priorities, goal values and dominance of solutions are the commonly used methods to formalize the preference articulation.

(i) A *priori* articulation of preferences: The decisions are made before searching, and the individual distinct objectives are combined as a single objective prior to optimization. An example of its application in multi-objective scheduling can be found in [34].

(ii) A *posteriori* articulation of preferences: The search is made before making decisions, and decision makers choose the trade-off solution from Pareto-optimal set by inspection. An example of its application in multi-objective scheduling can be found in [2].

(iii) *Progressive* articulation of preferences: Both the search and decisions making are integrated at interactive steps. Decision makers provide partial preference information to the optimization algorithm such that the algorithm generates better solutions according to the received information. An example of its application in multi-objective scheduling can be found in [35].

Since population-based EAs are capable of generating and evolving multiple elements of the Pareto-optimal set simultaneously in a single run and are less susceptible to the shape and continuity of the Pareto front, the EAs have been extensively developed for multi-objective optimization. In this work, we consider solving JSSP using multi-objective EA, in particular multi-objective GA. Although the advantages and good performance of multiobjective GA in many combinatorial optimization problems have been demonstrated in the literature [36], their applications on JSSP are still considered limited and mostly dominated by the unrealistic single objective GA. In fact, the impact of scheduling research on real-world problems is limited due to the single objective nature of many proposed scheduling algorithms. A significant work on multi-objective evolutionary scheduling can be found in [2], and an early example of a practical application of multiobjective GA in scheduling is available in [29]. It should be noted that many proposed multi-objective optimization approaches for JSSP are based on a *priori* articulation of preferences, in which multiple objectives are combined into a single scalar objective using weighted coefficients [37]. As the relative weights of the objectives are not exactly known and cannot be pre-determined by users, the objective function that has the largest variance value may dominate the multi-objective evaluation, which in turn produces inferior non-dominated solutions as well as poor diversity of the solutions. Therefore, as the weighting of objective functions cannot be determined, it is essential to apply a *posteriori* articulation of preferences and to present all of the Pareto-optimal scheduling solutions to the decision makers in advance. Then, the decision makers can choose the most compromising schedule from the Pareto-optimal set. These trade-off solutions, considering more than one criterion, are particularly useful for decision makers since improving one of the multiple criteria may unexpectedly worsen the other criterion. In the literature, there are only a few GA based multi-objective scheduling approach utilizing *posteriori* articulation of preferences are available [4,38], but they mainly focused on flow-shop scheduling.

3.5 Advantages of JGGA for JSSP

In this work, an extended JGGA [6] is proposed to search for Pareto-optimal schedules in static JSSP. The JGGA, which is developed based on NSGAII [7], optimizes multiple objectives using a posteriori articulation of preference based approach. The detailed methodology proposed in JGGA is discussed in Sections 4 and 5. The main reasons to advocate for using JGGA to solve JSSP are summarized as follows.

It is well known that GA is not very effective for fine-turning the solutions that are already close to the optimal solution, as the crossover operator may not be sufficient enough to generate feasible schedules. Hence it is necessary to integrate some local search strategies in GA for enhancing the Pareto-optimal solutions. Such hybridization is often called Genetic Local Search (GLS). The rationale behind the hybridization is that GA is used to perform global exploration among the population, while local search strategies are used to perform local exploitation around the chromosomes. As discussed in several papers [20,39,40], the hybridization of local search strategies and GA always provides better performance than that obtained by GA alone. In addition, it should be noted that as the length of the chromosome increases with the problem size of JSSP, the multi-objective GA might suffer from premature convergence, due to the long chromosomes in a large search space. This is due to the fact that the genes cannot be excited consistently during the entire evolutionary process, despite the assistance of crossover and mutation [6]. To remedy this drawback, the jumping gene operations proposed in JGGA offer a local search capability in order to exploit solutions around the chromosomes, while the usual genetic operators globally explore solutions from the population using multiple objective functions. Since the genes in JGGA can jump from one position to another either within their own or to another chromosome under multiple stresses, the central idea to incorporate the jumping operations in this approach is to provide a local search capability to fine-tune the scheduling solutions during evolution. The following section contains a detailed discussion of this issue in relation to the JGGA.

As discussed in the previous section, the applications of multi-objective GA in JSSP are still considered to be limited, and the solutions to JSSP are still dominated by the unrealistic single-objective GA. For these reasons, the JGGA is proposed in order to optimize multiple objectives simultaneously, such as the makespan and the mean flow time, during the evolution. In previous work [6], the JGGA has performed robustly in searching non-dominated solutions that take into consideration both convergence and diversity. Obtaining converged and diverse scheduling solutions in the Pareto-optimal set is very important, for the reasons discussed above.

4 Jumping Genes Genetic Algorithm (JGGA)

In this section, the initial implementation of JGGA that was proposed for multi-objective evolutionary optimization is discussed. Detailed description of JGGA can be found in [5,6]. JGGA is a relatively new MOEA that imitates a jumping gene phenomenon discovered by Nobel Laureate McClintock during her work on the corn plants. The main feature of JGGA is that it only has a simple operation in which a transposition of gene(s) is induced within the same or another chromosome in the GA framework.

4.1 Overview of JGGA

The jumping genes (also called transposons) effect was discovered by Noble Laureate McClintock [41]. She observed that not all chromosomes, but one of them (number 9) tended to be broken from generation to generation and the place of fracture is always the same. She also discovered that there are non-autonomous transposable elements called Dissociation (Ds), which can transpose (jump) from one position to another within the same or to a different chromosome under the presence of autonomous transposable elements called Activator (Ac). Ds itself cannot jump unless Ac activates it. However, Ac can jump by itself. According to the experimental observation, jumping genes can move around the genome in two ways, *cut and paste* and *copy and paste*. The former means a piece of DNA is cut and pasted somewhere else. The later means the genes remain at the same location while the message in the DNA is copied into RNA and then copied back into DNA at another place in the genome. Whichever the process of transposition, the jumping genes in one place can affect the genetic behavior at other places in the genome. The transposition phenomenon applied in JGGA is discussed below.

4.2 Computational Jumping Gene Paradigm

As nature tends to be opportunistic rather than deterministic, the jumping process is neither streamlined nor can it be planned in advance. Therefore, the behavior of jumping genes is similar to many other genetic operations that operate on the basis of opportunity. To incorporate the jumping genes paradigm into an EA framework, a new operation – jumping gene transposition is introduced after the selection process (before the crossover process). The non-dominated sorting strategy, crowding-distance mechanism and elitism strategy used in JGGA are the same as used in NSGAII.

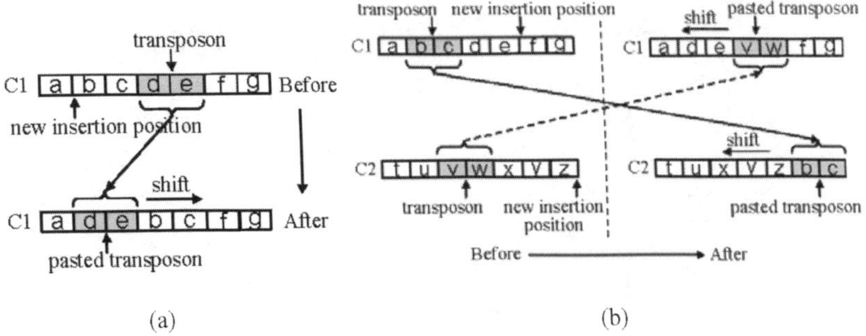

Fig. 1. Cut and paste transposition: (a) same chromosome; (b) different chromosome

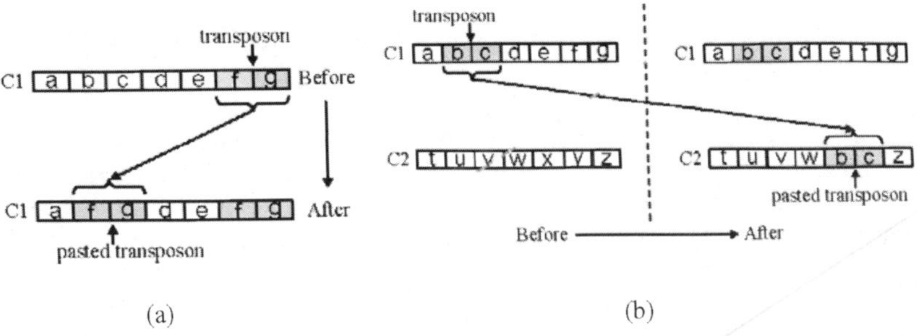

Fig. 2. Copy and paste transposition: (a) same chromosome; (b) different chromosome

The implementation of JGGA is such that each chromosome has some consecutive genes which are selected as a transposon. The number of transposons in a chromosome can be greater than one and the length of each transposon can be more than one unit (e.g. one bit for binary code, one integer for integer code, etc). The locations of the transposons are assigned randomly, but their contents can be transferred within the same or even to different chromosomes in the population pool. The actual implementation of the cut and paste operation is that the element is cut from the original site and pasted into a new site (Fig. 1). In the case of copy and paste, the element replicates itself, with one copy of it inserted into a new site, while the original one remains unchanged at the same site (Fig. 2). The jumping operators are chosen randomly on the basis of opportunity like other genetic operations. Also, the transpositions made within the same chromosome or to a different chromosome are chosen randomly and there is no restriction to the chromosome choice. The flowchart of a complete evolutionary cycle of JGGA is shown in Fig. 3.

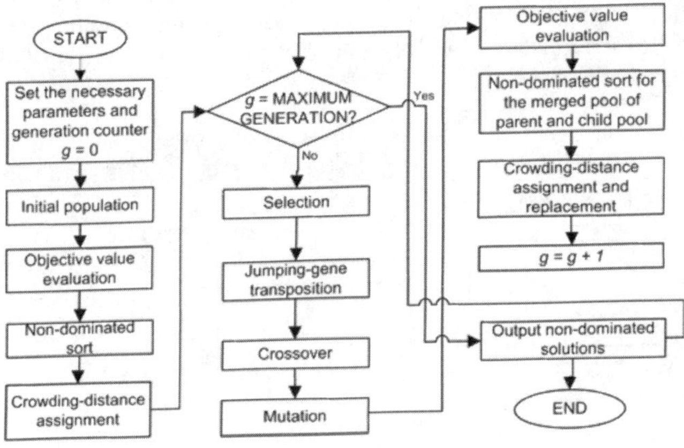

Fig. 3. Flowchart of JGGA

4.3 Consequence of Jumping Genes on Multi-Objective Functions

Every conventional genetic operator employs only vertical transmission of genes from generation to generation (i.e. from parent to children). However, the jumping operators introduce a kind of horizontal transmission. This type of gene transmission is a lateral movement of genes within a chromosome or even to other individuals. The effects of the two jumping operators are described below.

The cut and paste operator opens a path through the landscape of a larger number of possible chromosome changes. It is a more efficient strategy for creating and trying out new genes [41]. Furthermore, in the context of cut and paste transposons and host interaction; the environment that natural selection must be concerned with consists of not only the physical environment and other species, but also the microenvironment of the chromosome itself [5]. In the case of a copy and paste operation, a transposon can not only jump, but will also carry information for shaping the genome. These movements create places in the genome where stretches of DNA can pair and exchange information, and genes with similar sequences of the family do recombine. These repetitive sequences that are present throughout the genome may enable the exchange of information between unrelated genes by recombination [41]. Hence, a copy and paste operator eventually benefits the phenotypic shaping of chromosomes. Moreover, an increased probability of recombination at repeat sequences provides a more focused strategy for genetic exploration rather than wandering the vast landscape with random change.

It is well-known that the genes which can jump in a genome are largely dependent upon the environmental conditions particularly under stress. When the genome senses the stress, genes jump [41]. This is a better way of exploration and exploitation than only the use of Pareto-optimal solutions themselves. The JGGA makes the best use of this phenomenon when multiple stresses are induced. In the evolutionary process, selection plays an important role in the exploration versus exploitation issue. A strong selective pressure can lead to premature convergence and weak selective pressure can make the search ineffective. The JGGA is capable of exerting appropriate selection pressure on the chromosomes because the new genetic operators can perform a horizontal transformation, particularly when multiple stresses are induced. As a result, it creates more chances to achieve better convergence and diversity, as well as avoiding premature convergence. The success and performance of JGGA as an MOEA have been discussed in [5,6].

5 Multi-Objective Evolutionary Job-Shop Scheduling using JGGA

5.1 Chromosome Representation and Population Initialization

Chromosome formulation is a key issue in designing efficient GAs for heavily constrained JSSPs because different methods for representing parameters (genes) in scheduling solutions (chromosomes) create different search spaces and different difficulties for genetic optimization operators. In [27], the authors classify all the chromosome formulation methods into two major approaches: direct representation and indirect representation. In indirect representation, the chromosome encodes a sequence of preferences. These decision preferences can be heuristic rules or simple ordering of jobs in a machine. After that a schedule builder is required to decode the chromosome into a schedule. In a direct representation, the schedule itself is directly encoded onto the chromosome and thereby eliminating the need for a complex schedule builder. At the same time, applying simple genetic operators on direct representation string often results in infeasible schedule solutions. For this reason, domain-specific genetic operators are required. Representation schemes in these two major approaches are critically reviewed in [27] for ease of implementation in GA, choices of crossover and mutation operators, and their advantages and disadvantages.

In our work, indirect representation incorporated with a schedule builder is applied. The JGGA is implemented with an un-partitioned operation based representation where each job integer is repeated *m* times (*m* is the number of machines). This representation was employed by Bierwirth [42] and mathematically known as “permutation with repetition”. By scanning the permutation from left to right, the *k*-th occurrence of a job number refers to the *k*-th operation in the technological sequence of this job as depicted in Fig. 4. In this representation, it is possible to avoid the schedule operations whose technological predecessors have not been scheduled yet. Therefore, any individual can be decoded to a feasible schedule. However, two or more different individuals may be decoded to an identical schedule.

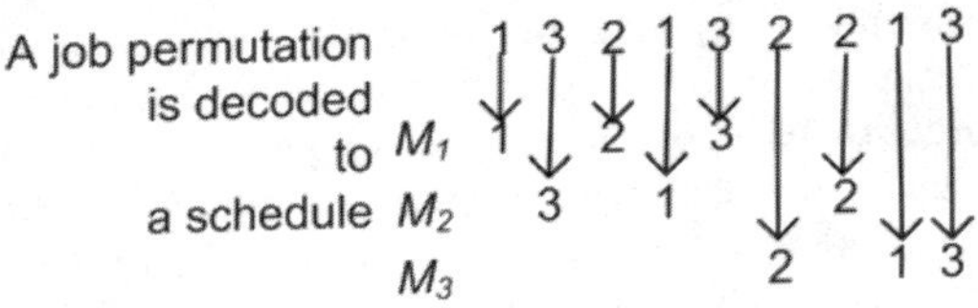

Fig. 4. Permutation with repetition approach for a 3x3 JSSP

The advantage of such a scheme is that it requires a very simple schedule builder because all the generated schedules are legal. And the number of possibilities explored by genotype is measured as [42]:

$$\frac{M_1 + M_2 + M_3 + \ldots\ldots\ldots + M_m}{!M_1 \times !M_2 \times !M_3 \times \ldots\ldots \times !M_n} \tag{1}$$

where M_1, M_2, $\ldots M_m$ denote the number of machines the jobs 1, 2 and 3 would visit in the entire schedule.

5.2 Schedule Builder

In indirect representation, the chromosome contains an encoded schedule and a scheduler builder is used to transform the chromosomes into a feasible schedule. The schedule builder is a module of the evaluation procedure and should be chosen with respect to the performance-measure of optimization. It is well-established that the minimization of makespan plays the major role in converting the chromosomes into a feasible schedule [40]. Computational experiments discussed in [43] show that the genetic minimum-makespan in JSSP improves by the use of a powerful schedule builder.

A schedule is called semi-active when no operation can be started earlier without altering the operation sequences of any machine. The makespan of a semi-active schedule may often be reduced by shifting an operation to the left without delaying other jobs, which is called the permissible left shift. When no such shifting can be applied to a schedule, it can be called an active schedule. The third type of schedule is called non-delay schedule, in which no machine is idle, if an operation is ready to be processed. The set of non-delay schedules is a subset of active schedules, which is the subset of semi-active schedule. For regular performance measures, it can be shown that for any problem an optimal active schedule exists [40]. However it has also been demonstrated that some problems have no optimal non-delay schedule. For this reason, when solving scheduling problems involving regular performance measures, usually only the set of active schedules are searched. Since searching for active schedules brings a huge reduction of the search space while still ensuring that an optimal schedule can be found, it is safe and efficient to limit the search space to the set of all active schedules. In fact, an active schedule builder performs a kind of local-search, which can be used to introduce heuristic-improvement into genetic search.

One of the efficient approaches to generate an active schedule builder is the Giffler & Thompson algorithm [44]. In this work, we employed a variant of hybrid Giffler & Thompson Algorithm proposed by Varela et al [45]. We made slight modification to this algorithm in order to fit with the representation of chromosome and to produce active schedules only. This algorithm is given below (Algorithm 1) where S is the schedule being constructed. The set A is used to hold the set of schedulable operations, where an operations o is said to be schedulable if it has not been scheduled yet.

Algorithm 1. Hybrid Giffler and Thompson

1. Set $S = \{\ \}$;
2. Let $A = \{o_{j1} \mid 1 \leq j \leq N\}$;

while $A \neq \emptyset$ **do**

3. $\forall o_i \in A$ let $st(o_i)$ be the lowest starting time of i, if scheduled now;
4. Let $o_k \in A$ such that $st(o_k) + du(o_k) \leq st(o) + du(o),\ \forall o \in A$; where $du(o)$ is the processing time for operation o. (if two or more operations are tied, pick the leftmost operation in the chromosome);
5. Set M^* is the machine that is to process o_k;
6. Let $B = \{\ o \in A \mid$ it is to process on machine M^* and $st(o) < st(o_k) + du(o_k)\}$;
7. Let $o_t \in B$ such that $st(o_t) \leq st(o),\ \forall o \in B$;
8. Select $o^* \in B$ such that o^* is the leftmost operation in the chromosome and add o^* to S with starting time $st(o^*)$;
9. Let $A = A \backslash \{o^*\} \cup \{SUC(o^*)\}$; where $SUC(o)$ is the next operation to o in its job if any exists;

end while

5.3 Jumping Operations

The chromosome representation in this extended JSSP is different from that of conventional JGGA. Hence, the direct application of the original jumping operators may create illegal schedule. This problem gets more serious in the case of copy and paste, which copies some number of consecutive genes from one place to another while remaining in the original positions. As in the original chromosome representation, the number of jobs must be equal to the number of machines on which it will be processed; the resulting chromosome will produce infeasible solution. As a result, some problem specific jumping operators are required. In this extended JSSP, we classify the jumping operators based on the number of participating parent chromosomes. If it is applied with one chromosome, the concept is relative easy. Two gene positions are selected randomly. Then the same number (random) of consecutive genes is selected and they change their positions. This procedure is described in Fig. 5:

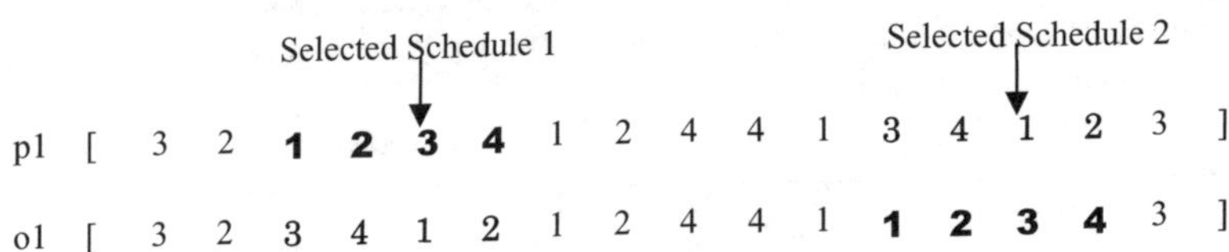

Fig. 5. Jumping operation within single chromosome

We follow the concept of "partial schedule exchange crossover" [46] to implement jumping operations between two different chromosomes. At first, partial schedules from both chromosomes are selected randomly. A partial schedule is identified with the same job in the first and last positions of the selected portion. The following steps describe this operation in case of a 4x4 JSSP.

Step 1: Randomly pick one partial schedule (a substring of random length) in the first parent. Let it be the job *4* located at position 6 as shown in Fig. 6.

Step 2: Find the next-nearest job *4* in the same parent **P1**, which is in position 9 and the partial schedule 1 is (4 1 2 4).

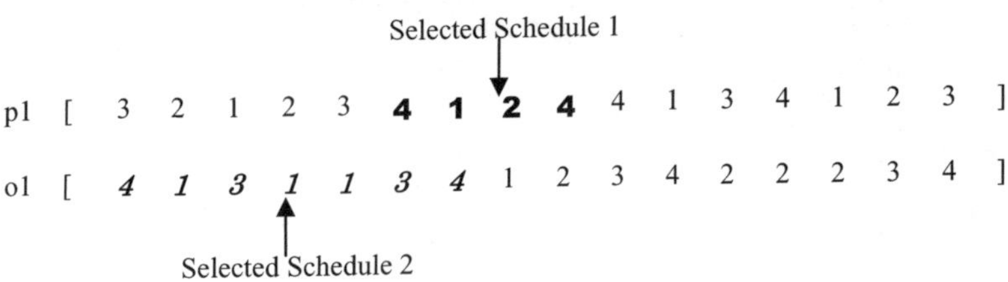

Fig. 6. Selecting partial schedule

Step 3: Identify the partial schedule 2 in parent **P2** in the same way like partial schedule 1, provided that the first gene must be job *4*. Thus the second partial schedule is (4 1 3 1 3 4), as shown in Fig. 6.

Step 4: Exchange these two partial schedules to produce offspring **o1** and **o2**, as shown in Fig. 7.

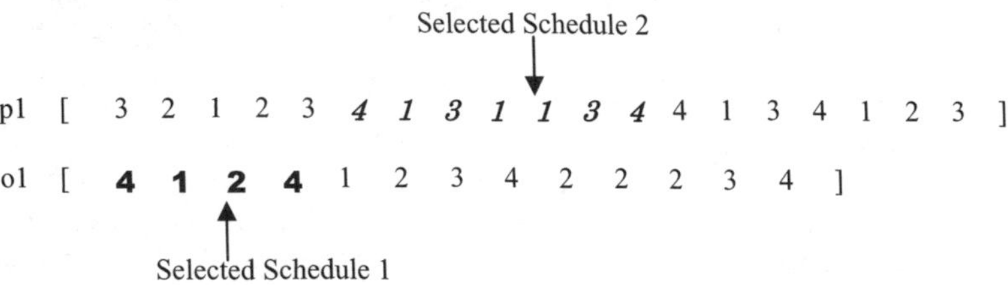

Fig. 7. Exchange of partial schedules

Usually, the partial schedules being exchanged contain a different number of genes, and the offspring may not include or may have operations in excess of those required for each job. So, the offspring may be illegal. As shown in Fig. 8, offspring **o1** gains extra genes 3, 1, 1 and 3, while it lost gene 2; on the other hand **o2** lost 3, 1, 1, 3; and gained 2. The following steps are required to perform repair works to convert the chromosomes into a legal schedule.

Partial Schedule1	4 1 2 4	Partial Schedule2	4 1 3 1 1 3 4
Replaced in p1 by	4 1 3 1 1 3 4	Replaced in p2 by	4 1 2 4

Fig. 8. Missed/exceeded genes after exchange

Step 5: In offspring **o1**, the extra genes (3, 1, 1 and 3) are deleted without disturbing the inserted partial schedule, and the missing gene(s) (here 2) are to be inserted in the position immediately after the newly acquired partial schedule, as shown in Fig. 9.

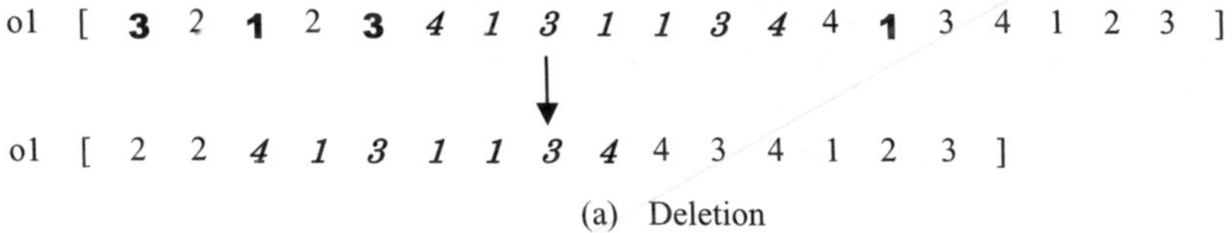

(a) Deletion

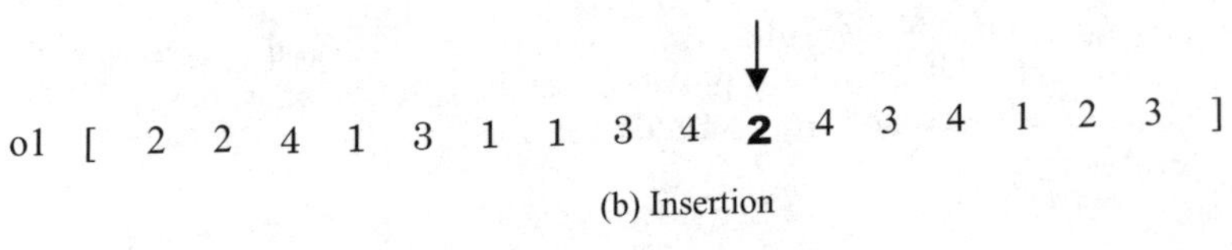

(b) Insertion

Fig. 9. Repair work for **o1**

Step 6: The same procedure is repeated for insertion and deletion in case of **o2**.

5.4 Crossover Operation

One of the unique and important techniques involved in GA is crossover, which is regarded as the backbone of GA. It intends to inherit nearly half of the information of the two parent solutions to one or more offspring solutions. Provided that the parents keep different aspects of high quality solutions, crossover induces a good chance to find the better offspring. The permutation with repetition chromosome representation allows each job to be repeated exactly the number of times equal to machine on which it will be processed and the crossover operator has to respect this repetition-structure of a certain permutation. Since it is difficult for the simple one-point or two-point crossover to maintain this constraint, these types of crossover cannot be used in this representation. A large number of crossover operators have been proposed in the literature, due to the need for designing specialist crossover operations to use with permutation representations. The details of crossover operators specifically designed for ordering applications can be found in [47].

In this approach, we applied the Generalized Order Crossover (GOX) operator [48] for performing the crossover. In GOX, the donor parent contributes a substring of length normally in the range of one third to half of the length of the chromosome string. The idea behind choosing such a length of substring is that the offspring will inherit almost the same amount of information from both the parents. And this choice is important because implanting a crossover-string into the receiver chromosome requires a preparation which usually causes some loss of information [42]. The operations in the receiver that correspond to the substring are located and deleted, and the child is created by making a cut in the receiver at the position where the first operation of the substring used to be and inserting the donator substring there. This has been exemplified in Fig. 10 for a *3*x*3* problem. The chosen substring has been underlined in the donator and the child, while the operations deleted in the receiver have been marked by bold-face.

donator	3	1	2	1	2	3	2	3	1
receiver	3	**2**	1	**2**	**1**	1	**3**	3	2
child	3	2	1	2	3	1	1	3	2

Fig. 10. GOX crossover (the substring does not wrap around the end-points)

When the donator substring wraps around the boundary points of the donator chromosome, a different and simplified procedure is followed. The operations in the receiver that correspond to the operations in the donor substring are still deleted, and the donor substring is inserted in the receiver in the same positions, it occupies in the donor (at the ends). The procedure is exemplified in Fig. 11.

donator	3	2	2	1	1	3	2	3	1
receiver	**2**	**3**	1	3	1	**1**	**3**	2	2
child	**3**	**2**	1	3	1	2	2	**3**	**1**

Fig. 11. GOX crossover (the substring wraps around the boundary)

5.5 Mutation Operation

Mutation is the second main transformation operators in an EA. It is the principal source of variability in evolution and it provides and maintains diversity in a population. Normally, after crossover, each child produced by the crossover undergoes mutation with a low probability. Here, we followed the 'job-pair exchange mutation operator' [46]. In our approach, two non-identical jobs are picked randomly and then they exchange their positions as shown in Fig. 12. As a result, all of the resulting schedules are legal and no repair is necessary.

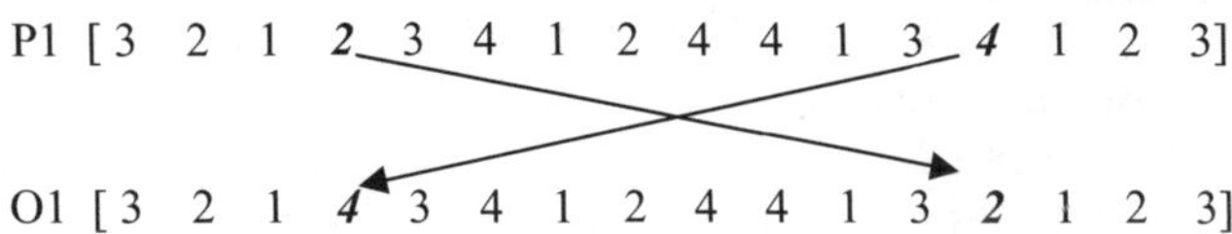

Fig. 12. The job-pair exchange mutation

In this approach, both mutation and jumping operators (in the case of within a single chromosome) seem to be doing nearly the same tasks. This might create some confusion about using an additional mutation operator. Basically, the distinction among these operators lies in the extent of genes allowed on each operator. If only one gene is used, it should be called a mutation operation. The basic idea behind using an additional mutation operator is to make use of multiple crossover/mutation operations to increase the population diversity by introducing some sort of adaptation mechanisms. In most cases, this concept seems to be better than classical GAs for optimization problems, particularly in optimization of hard unimodal and multimodal problems [49].

Fig. 13 presents the complete evolutionary cycle of the proposed multi-objective JSSP algorithm.

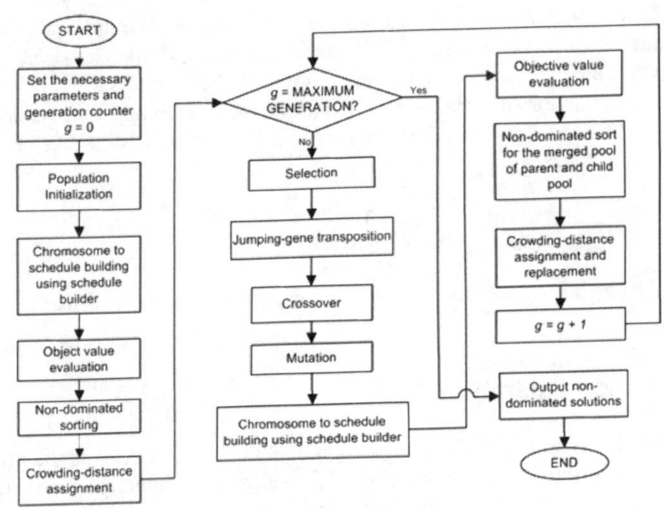

Fig. 13. Complete evolutionary cycle for multiobjective JSSP

6 Experimental Results

6.1 Benchmark Problems

Computational experiments are carried out to investigate the performance of our proposed multi-objective evolutionary job-shop scheduling algorithm based on JGGA. To evaluate how JGGA performs with respect to solution quality, we run the algorithm on various benchmark data. The first three well-known benchmark problems, known as *mt06, mt10* and *mt20,* formulated by Muth and Thompson [9] are commonly used as test beds to measure the effectiveness of a certain method. The *mt10* and *mt20* problems have been the good computational challenges for a long time. However, it is no longer a computational challenge now. Indeed, the *mt10* problem has been referred as "notorious", because it remained unsolved for over 20 years. The *mt20* problem has also been considered as quite difficult. Applegate and Cook proposed a set of benchmark problems called the "ten tough problems" as a more difficult computational challenge than the *mt10* problem, by collecting difficult problems from literature, some of which still remain unsolved [50]. The 'ten tough problems' consist of *abz7, abz8, abz9, la21, la24, la25, la27, la29, la38,* and *la40*. The *abz* problems are proposed by Adams in [21]. The *la* problems are parts of 40 problems *la01-la40* originated from [51]. Table 2 shows the problem size, best lower bound of these problems, and whether an optimal solution is known or not. The problem data and the lower bounds information are taken from the OR-library [52].

Table 2. Benchmark Problems

Instance Data	Number of Jobs	Number of Machines	Lower Bound	Optimal Solution
mt06	6	6	55	known
mt10	10	10	930	known
mt20	20	5	1165	known
abz7	20	15	654	known
abz8	20	15	635	known
abz9	20	15	656	known
la21	15	10	1040	unknown
la24	15	10	935	known
la25	15	10	977	known
la27	20	10	1235	unknown
la29	20	10	1120	unknown
la38	15	15	1184	unknown
la40	15	15	1222	known

6.2 Experimental Evaluation and Discussions

Till now, almost all of the evolutionary JSSP algorithms try to optimize single criterion only. In this work, the makespan is considered as the first objective. The mean flow time, as the second objective, continues to be very important since it assists to select the appropriate one when many algorithms proposed in the literature have reached the same makespan for many instances. To evaluate our proposed algorithm, first we perform the experiments in a single objective (makespan) context to justify its capability in optimizing makespan. The experiments are conducted using 100 chromosomes and run for 100 generations. After that, we showed its performance as a multi-objective evolutionary JSSP approach by optimizing makespan and mean flow time, which are to be simultaneously minimized.

(1) Makespan = $\max[C_i]$ where C_i is the completion time of job i.

(2) Mean flow time $= \frac{1}{n}\sum_{i=1}^{n} C_i$ where C_i is the completion time of job i.

We also compare our proposed algorithm with another well-known MOEA (NSGAII) based JSSP algorithm to justify the proposed one. For both of these two algorithms, the experiments were conducted using 100 chromosomes and 150 generations. The probabilities of crossover and mutation were 0.9 and 0.3 respectively. For JGGA, the probability of jumping operations was 0.6. Using the same setting, each benchmark problem was tested for thirty times with different seeds. Then all the final generations were combined and a non-dominated sorting [7] was performed to constitute the final non-dominated solutions.

6.2.1 Single Objective Context. The values provided in Table 3 show the makespan of the best schedules obtained in case of *mt* problems by some GA-based scheduling algorithms. The column labeled sGA is based on the GA using the simple mutation that swaps two positions of two random jobs [53], where as SGA is based on simple GA proposed by Nakano and Yamada [30]. LSGA and GTGA indicate the GA-based job-shop scheduling algorithm incorporating local search [53] and the GA based on GT crossover [15], respectively. From this Table, it can be easily found that the proposed JGGA-based scheduling algorithm is capable of producing near-optimal schedules in most of the problems. For *mt06* and *mt10* problems it achieves the lower bound. In case of *mt20*, JGGA cannot obtain the lower bound, however it outperforms the other contested algorithms; although some other heuristically based scheduling algorithms may achieve these lower bounds.

Table 3. Comparison with some GA based algorithms

Instance Data	L.B	sGA	GTGA	LSGA	SGA	JGGA
mt06	55	<u>55</u>	<u>55</u>	<u>55</u>	<u>55</u>	<u>55</u>
mt10	930	994	<u>930</u>	976	965	<u>930</u>
mt20	1165	1247	1184	1209	1215	1180

We also perform experiments to compare our proposed JSSP algorithm with some heuristic evolutionary scheduling approaches. Table 4 summarizes the makespan performance in comparison with these methods for the ten tough problems. The column headings Nowi indicates the best performance of the Tabu search proposed in [14], CBSA+SB indicates the Critical Block Simulated Annealing with shifting bottleneck heuristics [43]. Aart, Kopf and Appl indicate the Simulated Annealing results proposed in [54], GA performance in [55] and [50], respectively. The MSFX is based on GA using Multi-Step Crossover Fusion [43]. The LSGA and sGA are the same as above. Note that, this Table is partially cited from [43] and the parameter settings for the proposed JGGA based JSSP algorithm is the same as above. The comparative results indicate that our proposed algorithm finds the near-optimal solutions in case of five out of ten problems and optimal solutions can be found in three of them. However other algorithms can also find the same makespan like the proposed one in some problems, but not in large numbers like the proposed one. An exception is the performance of CBSA+SB, which performs very well. In fact, it achieves the best results in seven test problems. Despite that it should be mentioned that CBSA+SB is designed for single objective only, and the main goal of our proposed algorithm is to find the trade-off solutions for multi-objective JSSP, which is very rare in literature. While considering this, the overall performance of JGGA is very promising for all the problems. Moreover, for those cases where the proposed algorithm fails to achieve the known best results, it performs consistently and achieves very close to the near-optimal solutions.

Table 4. Comparison of the makespan obtained by various evolutionary and heuristic methods

Instance data	L.B	Nowi	sGA	CBSA +SB	LSGA	Aarts	Kopf	Appl	MSFX	JGGA
abz7	654	-	-	665	-	668	672	668	678	665
abz8	635	-	-	675	-	670	683	687	686	685
abz9	656	-	-	686	-	691	703	707	697	694
la21	1040	1047	1180	1046	1114	1053	1053	1053	1046	1046
la24	935	939	1077	935	1032	935	938	935	935	935
la25	977	977	1116	977	1047	983	977	977	977	977
la27	1235	1236	1469	1235	1350	1249	1236	1269	1235	1235
la29	1120	1160	1195	1154	1311	1185	1184	1195	1166	1156
la38	1184	1196	1435	1198	1362	1208	1201	1209	1196	1199
la40	1222	1229	1492	1228	1323	1225	1228	1222	1224	1225

6.2.2 Multi-Objective Context. Multi-objective optimization differs from single objective optimization in many ways [33]. For two or more objectives, each objective corresponds to a different optimal solution, but none of these trade-off solutions is optimal with respect to all objectives. Thus, multi-objective optimization does not try to find one optimal solution but all trade-off solutions. Apart from having multiple objectives, another fundamental difference is that multi-objective optimization deals with two goals. The first goal is to find a set of solutions as close as possible to the Pareto-optimal front. The second goal is to find a set of solutions as diverse as possible.

Table 5 shows the performance statistics of the evolutionary JSSP algorithms based on JGGA and NSGAII in the context of makespan and mean flow time. The parameter settings are the same as in the single objective context. The results shown in the Table indicate that both JGGA and NSGAII based algorithms perform well in achieving near-optimal solutions. However, the JGGA based approach clearly outperforms the other in terms of *mt10*, *abz7*, *abz8*, *la21*, *la24*, *la25*, *la27*, and *la29*. Only in the case of *mt06*, NSGAII outperforms JGGA. In other cases, the solutions produced by both algorithms are non-dominated to each other.

Table 5. Results of test problems

Instance		Makespan		M-Flow Time		Spread
		Best	Average	Best	Average	(S)
mt06	JGGA	55	59.1538	47	48.7692	0.00216
	NSGAII	55	59.0952	46	48.8571	0.19162
mt10	JGGA	930	1009.067	796	844.833	0.71937
	NSGAII	930	1022.6	805	842	1.69952
mt20	JGGA	1180	1347.533	819	906.2	1.84535
	NSGAII	1184	1270.25	815	873.25	0.7
abz7	JGGA	665	716.222	598	617.667	0.74684
	NSGAII	667	729.428	606	620.143	0.46291
abz8	JGGA	685	735.833	600	640.333	0.53124
	NSGAII	686	736	606	638.667	1.16451
abz9	JGGA	694	718.8	574	599.6	0.65145
	NSGAII	690	736.857	585	595.166	0.67679
la21	JGGA	1046	1120.273	885	908.7	1.663
	NSGAII	1046	1090.75	892	904	3.91265
la24	JGGA	935	989.111	792	829.556	1.31383
	NSGAII	935	966.5	811	823.333	0.2357
la25	JGGA	977	1024.667	784	809.555	1.0633
	NSGAII	977	1009.667	793	800.5	2.03214
la27	JGGA	1235	1333.714	1072	1096.83	1.13665
	NSGAII	1235	1330.444	1073	1093.37	1.20253
la29	JGGA	1156	1210.1	952	1032.6	0.48968
	NSGAII	1160	1214.11	959	991.444	0.97182
la38	JGGA	1199	1250.917	964	1028.33	3.75653
	NSGAII	1196	1267.333	1005	1011.5	4.13693
la40	JGGA	1225	1316.182	1076	1144.09	2.03899
	NSGAII	1228	1299.25	1073	1107	4.13927

To illustrate the convergence and diversity of the solutions, non-dominated solutions of the final generation produced by JGGA and NSGAII for the test problems *mt10*, *abz8*, *la29* and *la40* are presented in Fig. 14. From these Figures, it can be observed that the final solutions are well spread and converged. In particular, the solutions produced by the proposed approach are more spread than that of NSGAII and for this reason it is capable of finding extreme solutions. It can be further justified by the values of Space (*S*) metric [56] given in Table 5. An algorithm having a smaller S is better in terms of diversity. It can be easily inferred from the Table that JGGA is able to find more diverse solutions than NSGAII. However, the ability of producing diverse solution has some side effects, which can be well described in conjunction with Table 5. From this Table, it can be found that, in most cases, the gap between the average and best values of the solutions produced by JGGA is more than that of the solutions produce by NSGAII. This, in turn, indicates that NSGAII produces solutions that are relatively stable. However, certainly our proposed approach produces more near-optimal and trade-off solutions to fulfill the main goals of multi-objective scheduling implementation.

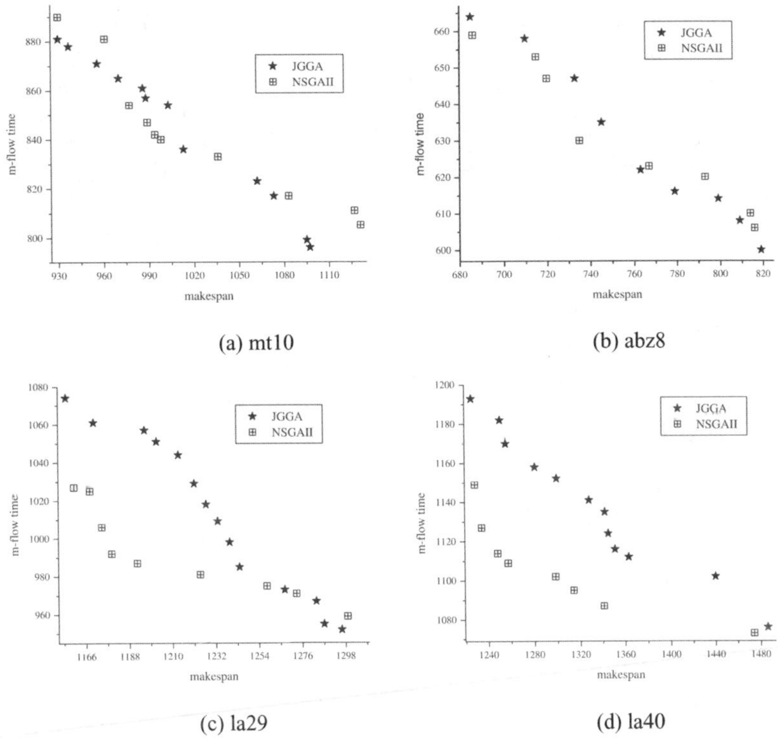

Fig. 14. Final Pareto-optimal front

In general, the proposed JGGA-based multi-objective JSSP algorithm is able to achieve consistent near-optimal scheduling solutions which are both well spread and converged. As compared to NSGAII, the JGGA produced much more non-dominated solutions for most of the test problems. The plots of the obtained non-dominated solutions per generations in a single run for the test problems *mt20*, *abz7*, *la21* and *la25* presented in Fig. 15 justify this. Apart from *mt20* and *la27*,

JGGA finds more non-dominated solutions than NSGAII for all the problems. To summarize the result, the proposed algorithm is capable of producing near-optimal and non-dominated solutions, which are also the lower bounds in many cases. The simulation results clearly show that our proposed approach is able to find a set of diverse solutions, which are also close to the Pareto-optimal front.

The properties exhibited by the solutions act upon the basic properties of jumping gene operations. Jumping gene adaptation is used to enhance the exploitation by incorporating local search in GA. Hence it can find a set of well diverse solutions along the Pareto front of a converged population. From the experimental results presented in [5,6], it can be grasped that the main strength of the jumping gene operators is its ability to produce diverse solutions. It should be noted that jumping gene operations are beneficial to the convergence also. As a result, it is able to find extreme solutions. However, this phenomenon in turn causes some deficient affects. In some rare cases, the solutions are too diverse. Hence, it achieves the extreme solutions but the number of non-dominated solutions per generation is not satisfactory enough. This problem can be tackled by applying some hybrid jumping gene approach that can balance between the diversity and convergence. In future, we wish to employ hybrid approach to observe the results.

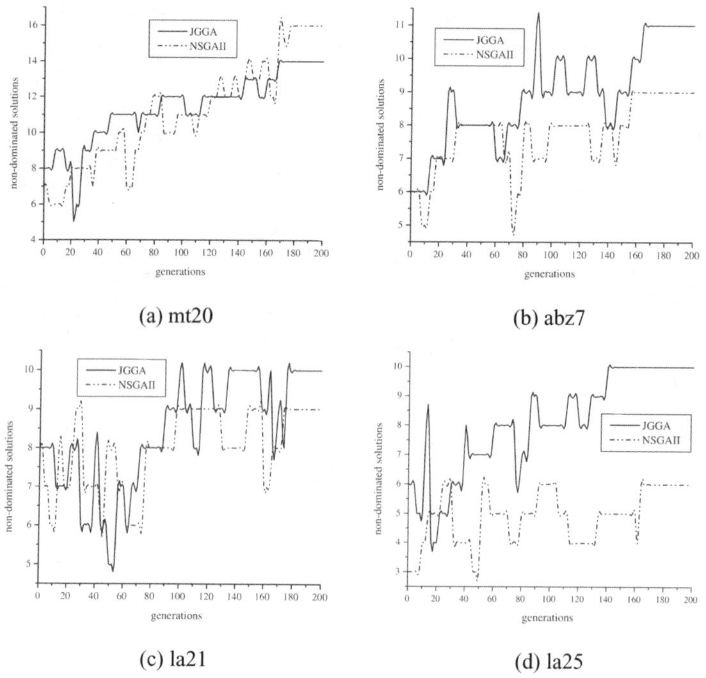

Fig. 15. Non-dominated solutions per generations (only for a single run)

7 Conclusions

As the job-shop scheduling problem becomes ever more sophisticated, due to the increasing number of jobs and machines involved, as well as additional objectives, constraints, and flexibilities, there is a growing demand for a robust and intelligent job-shop scheduler. Drawing inspiration from the jumping gene phenomenon discovered by Nobel Laureate McClintock, the JGGA is a very recent MOEA that has been demonstrated to perform robustly in solving combinational optimization problems. In this work, we have presented a JGGA-based evolutionary scheduling approach for solving the multi-objective JSSP. The JGGA searches for the Pareto-optimal schedules in the static JSSP. The jumping gene operations proposed in JGGA exploit scheduling solutions around the chromosomes while the general genetic operators in JGGA globally explore solutions from the population, in which makespan and mean flow time are applied as the multiple objectives.

The experimental results demonstrate that, as compared to other existing heuristically based evolutionary scheduling approaches, the JGGA can produce an overall strong performance for all of the applied benchmark problems related to the makespan in the context of single-objective optimization. For the multi-objective optimization, the results show that the JGGA is capable of finding a set of solutions that are both diverse and close to the Pareto-optimal front. In particular, the near-optimal solutions are also the lower bound found in many cases. The comparative results with another well established MOEA-based scheduling algorithm reveal that the main strength of JGGA is its ability to produce diverse solutions, in particular the extreme solutions, while maintaining the consistency and convergence of the trade-off non-dominated solutions. In general, the JGGA fulfills the goals of multi-objective scheduling optimization by taking into consideration convergence and diversity, as well as the local search strategy.

References

1. Garey, M., Johnson, D., Sethi, R.: The Complexity Of Flow Shop And Job Shop Scheduling. Maths Ops Res., Vol. 1 (1976) 117-129
2. Bagchi, T.P.: Multiobjective Scheduling by Genetic Algorithms, Kluwer Academic Publishers, Boston/Dordrecht/London (1999)
3. Garen, J.: A Genetic Algorithm for Tackling Multiobjective Job-Shop Scheduling Problems. In: Gandibleux X., Sevaux, M., Sörensen, K., T'kindt, V. (eds): Metaheuristics for Multiobjective Optimisation. Lecture Notes in Economics and Mathematical Systems, Springer, Berlin, Vol. 535 (2004) 201-219
4. Bagchi, T.P.: Pareto-optimal Solutions for Multi-objective Production Scheduling Problems. In: Int. Conf. on Evolutionary Multi-Criteria Optimization, LNCS 1993 (2001) 458-471

5. Chan, T.M., Man, K.F., Tang, K.S., Kwong, S.: A Jumping Gene Algorithm for Multiobjective Resource Management in Wideband CDMA Systems. Computer Journal, Vol. 48, No. 6. (2005) 749-768
6. Man, K.F., Chan T.M., Tang, K.S., Kwong, S.: Jumping Genes in Evolutionary Computing. In: Thirtieth Annual Conf. of the IEEE Industrial Electronics Society, Busan, Korean (2004) 1268-1272
7. Deb, K., Pratap, A., Agarwal, S., Meyarivan, T.: A Fast and Elitist Multiobjective Genetic Algorithm: NSGA-II. IEEE Trans. Evolutionary Computation, Vol. 6, No. 2. (2002) 182-197
8. Landa Silva, J.D., Burke, E.K., Petrovic S.: An Introduction to Multiobjective Metaheuristics for Scheduling and Timetabling. In: Gandibleux X., Sevaux, M., Sörensen, K., T'kindt, V. (eds): Metaheuristics for Multiobjective Optimisation. Lecture Notes in Economics and Mathematical Systems, Springer, Berlin, Vol. 535 (2004) 91-129
9. Muth, J.F., Thompson, G.L.: Industrial Scheduling. Prentice-Hall, Englewood Cliffs, N.J. (1963)
10. Bucker, P., Jurish B., Sievers, B.: A Branch and Bound Algorithm for the Job-shop Scheduling Problem. Discrete Applied Mathematics, Vol. 49. (1994) 105-127
11. Martin, P.D.: A Time-Oriented Approach to Computing Optimal Schedules for the Job-Shop Scheduling Problem. Ph.D. Thesis, School of Operations Research and Industrial Engineering, Cornell University, NY, USA. (1996)
12. Chen, H., Chu, C., Proth, J.M.: A More Efficient Lagrangian Relaxation Approach to Job-shop Scheduling Problems, In: IEEE Int. Conf. on Robotics and Automation. (1995) 496-501
13. Steinhöfel, K., Albrecht, A., Wong, C.K.: Fast Parallel Heuristics for the Job Shop Scheduling Problem. Computers & Operations Research, Vol. 29. (2002) 151-169
14. Nowicki E., Smutnicki, C.: A Fast Taboo Search Algorithm for the Job Shop Scheduling Problem. Management Science. Vol. 42. (1996) 797-813
15. Yamada T., Nakano, R.: A Genetic Algorithm Applicable to Large-scale Job Shop Problems. In: Second Int. Conf. on Parallel Problem Solving from Nature (PPSN-II), North-Holland, Amsterdam. (1992) 281-290
16. Pérez, E., Herrera F., Hernández, C.: Finding Multiple Solutions in Job Shop Scheduling by Niching Genetic Algorithm. J. Intelligent Manufacturing. Vol. 14. (2003) 323-339
17. Wang L., Zheng, D.Z.: An Effective Hybrid Optimization Strategy for Job-shop Scheduling Problems. Computers & Operations Research. Vol. 28. (2001) 585-596
18. Blum, C., Sampels, M.: An Ant Colony Optimization Algorithm for Shop Scheduling Problems. J. Mathematical Modelling and Algorithms, Vol. 3. (2004) 285-308
19. Ge., H.W., Liang, Y.C., Zhou, Y., Guo, X.C.: A Particle Swarm Optimization-based Algorithm for Job-shop Scheduling Problems. Int. J. Computational Methods, Vol. 2, No. 3. (2005) 419-430

20. Vaessens, R.J.M., Aarts E.H.L., Lenstra, J.K.: Job Shop Scheduling by Local Search. INFORMS J. Computing, Vol. 8. (1996) 302-317
21. Adams, J., Balas, E., Zawack, D.: The Shifting Bottleneck Procedure for Job Shop Scheduling. Management Science, Vol. 34. (1988) 391-401
22. Balas, E., Vazacopoulos, A.: Guided Local Search with Shifting Bottleneck for Job Shop Scheduling. Management Science, Vol. 44. (1998) 262-275
23. Brinkkötter W., Brucker, P.: Solving Open Benchmark Problems for the Job Shop Problem. J. Scheduling, Vol. 4. (2001) 53-64
24. Aiex, R.M., Binato S., Resende, M.G.C.: Parallel GRASP with Path-relinking for Job Shop Scheduling. Parallel Computing, Vol. 29. (2003) 393-430
25. Blazewicz, J., Domschke, W., Pesch, E.: The Job Shop Scheduling Problem: Conventional and New Solution Techniques, European J. Operations Research, Vol. 93. (1996) 1-33
26. Jain, A., Meeran, S.: Deterministic Job-shop Scheduling: Past, Present and Future. European J. Operations Research, Vol. 113. (1999) 390-434
27. Cheng, R., Gen, M., Tsujimura, Y.: A Tutorial Survey of Job-shop Scheduling Problems using Genetic Algorithms – I: Representation. Computers and Industrial Engineering, Vol. 30, No. 4. (1996) 983-997
28. Holland, J.H.: Adaptation in Natural and Artificial Systems. University of Michigan Press, Ann Arbor. (1975)
29. Davis, L.: Job-shop Scheduling with Genetic Algorithm. In: First Int. Conf. on Genetic Algorithms and Their Applications, Pittsburgh, PA, USA, Lawrence Erlbaum. (1985) 136-140
30. Nakano, R., Yamada, T.: Conventional Genetic Algorithm for Job-shop Problem. In: Fourth Int. Conf. on Genetic Algorithms, San Diego, CA, Morgan Kaufmann, San Mateo, CA. (1991) 474-479
31. Hart, E., Ross P., Corne, D.: Evolutionary Scheduling: A Review. Genetic Programming and Evolvable Machines, Vol. 6. (2005) 191-220
32. Hapke, M., Jaszkiewicz, A., Kurowski, K.: Multi-objective Genetic Local Search Methods for the Flowshop Problem. In: Advances in Nature-Inspired Computation: The PPSN IV Workshops, PEDAL, University of Reading, UK. (2002) 22-23
33. Deb, K.: Multi-objective Optimization using Evolutionary Algorithms. John Wiley & Sons. (2001)
34. Allahverdi, A.: The Two- and M-machine Flowshop Scheduling Problem with Bicriteria of Makespan and Mean Flowtime. European J. of Operational Research, Vol. 147. (2003) 373-396
35. Hapke, M., Jaszkiewicz A., Słowiński, R.: Interactive Analysis of Multiple-criteria Project Scheduling Problems. European J. of Operational Research, Vol. 107. (1998) 315-324
36. Fonseca, C.M., Fleming, P.J.: Genetic Algorithms for Multiobjective Optimization: Formulation, Discussion and Generalization. In: Forrest, S. (ed): Fifth Int. Conf. on Genetic Algorithms, San Mateo, California, UIUC, Morgan Kaufmann Publishers. (1993) 416-423
37. T'kindt V., Billaut, J.C.: Multicriteria Scheduling: Theory, Models and Algorithms. Springer. (2006)

38. Murata, T., Ishibuchi, H., Tanaka, H.: Multi-objective GA and Its Applications to Flowshop Scheduling. Computers and Industrial Engineering, Vol. 30, No. 4. (1996) 957-968
39. Jain, A., Meeran, S.: A State-of-the-art Review of Job-shop Scheduling Techniques. Technical Report, University of Dundee. (1998)
40. Yamada, T., Nakano, R.: Scheduling by Genetic Local Search with Multi-step Crossover. In: Voigt, H.-M., Ebeling, W., Rechenberg I., Schwefel, H.-P. (eds): Parallel Problem Solving from Nature – PPSN IV, LNCS 1141, Springer. (1996) 960-969
41. Caporale, L.H.: Jumping Genes. In: Darwin in the Genome: Molecular Strategies in Biological Evolution. McGraw-Hill, New York. (2003) 145-153
42. Bierwirth, C.: A Generalized Permutation Approach to Job Shop Scheduling with Genetic Algorithms, OR Spektrum. (1995) 87-92
43. Yamada, T.: Studies on Meta Heuristics for Jobshop and Flowshop Scheduling Problems. PhD. Thesis, Kyoto University, Japan. (2003)
44. Giffler, B., Thompson, G.: Algorithms for Solving Production Scheduling Problems. Operations Research, Vol 8, No 4. (1960) 487-503
45. Varela, R., Serrano, D., Sierra, M.: New Codification Schemas for Scheduling with Genetic Algorithms. In: Mira J., Álvarez, J.R. (eds): IWINAC 2005, LNCS 3562 (ISBN: 3-540-26319-5), Springer-Verlag. (2005) 11–20
46. Gen, M., Tsujimura, Y., Kubota, E.: Solving Job-Shop Scheduling Problem Using Genetic Algorithms, In: Sixteenth Int. Conf. on Computers and Industrial Engineering. (1994) 576-579
47. Poon P., Carter, N.: Genetic Algorithm Crossover Operators for Ordering Applications. Computers and Operations Research, Vol. 22. (1995) 135–147
48. Bierwirth, C., Matfield, D.C., Kopfer, H.: On Permutation Representation for Scheduling Problems. In: Parallel Problem Solving from Nature, Vol. 4. (1996) 310-318
49. Spirov, A.V., Kazansky, A.B.: Jumping Genes-Mutators Can Rise Efficacy of Evolutionary Search. In: Genetic and Evolutionary Computation Conference, New York, USA. (2002) 561–568
50. Applegate, D., Cook, W.: A Computational Study of the Job-shop Scheduling Problem. ORSA J. Computing, Vol. 3, No. 2. (1991) 149–156
51. Lawrence, S.: Resource Constrained Project Scheduling: An Experimental Investigation of Heuristic Scheduling Techniques (Supplement). Technical report, Graduate School of Industrial Administration, Carnegie Mellon University. (1984)
52. OR Library. URL: http://mscmga.ms.ic.ac.uk
53. Ombuki, B., Ventresca, M.: Local Search Genetic Algorithms for Job Shop Scheduling Problem. Technical Report No. CS-02-22, Brock University, Canada. (2002)
54. Aarts, E.H.L., Van Laarhoven, P.J.M., Lenstra, J.K., Ulder, N.L.J.: A Computational Study of Local Search Algorithms for Job Shop Scheduling. ORSA J. Computing, Vol. 6, No. 2. (1994) 118–125

55. Mattfeld, D.C., Kopfer, H., Bierwirth, C.: Control of Parallel Population Dynamics by Social-like Behavior of GA-individuals. In: Parallel Problem Solving from Nature, Vol. 866 (1994) 16-25
56. Schott, J.R.: Fault Tolerant Design Using Single and Multi-Criteria Genetic Algorithms. Master's Thesis, Department of Aeronautics and Astronautics, Massachusetts Institute of Technology, Boston, MA. (1995)

Multi-Objective Evolutionary Algorithm for University Class Timetabling Problem

Dilip Datta[1], Kalyanmoy Deb[1], and Carlos M. Fonseca[2]

[1] Indian Institute of Technology Kanpur, Kanpur - 208 016, India, {ddatta,deb}@iitk.ac.in
[2] Universidade do Algarve, Campus de Gambelas, 8000-117 Faro, Portugal, cmfonsec@ualg.pt

Abstract. After their successful application to a wider range of problems, in recent years evolutionary algorithms (EAs) have also been found applicable to many challenging problems, like complex and highly constrained scheduling problems. The inadequacy of classical methods to handle discrete search space, huge number of integer and/or real variables and constraints, and multiple objectives, involved in scheduling, have drawn the attention of EAs to those problems. Academic class timetabling problem, one of such scheduling problems, is being studied for last four decades, and a general solution technique for it is yet to be formulated. Despite multiple criteria to be met simultaneously, the problem is generally tackled as single-objective optimization problem. Moreover, most of the earlier works were concentrated on school timetabling, and only a few on university class timetabling. On the other hand, in many cases, the problem was over-simplified by skipping many complex class-structures. The authors have studied the problem, considering different types of class-structures and constraints that are common to most of the variants of the problem. NSGA-II-UCTO, a version of NSGA-II (an EA-based multi-objective optimizer) with specially designed representation and EA operators, has been developed to handle the problem. Though emphasis has been put on university class timetabling, it can also be applied to school timetabling with a little modification. The success of NSGA-II-UCTO has been presented through its application to two real problems from a technical institute in India.

1 Introduction to Class Timetabling Problem

Preparation of an academic class timetable is a typical real-world scheduling problem that appears to be a tedious job in every academic institute once or twice a year. The problem involves the arrangement of classes (lectures), students, teachers and rooms at a fixed number of time-slots, respecting certain restrictions - for example, no class, student, teacher or room can be engaged more than once at a time, and many more. An effective timetable is crucial for the satisfaction of educational requirements, and the efficient utilization of human and space resources. In case of a school, the students are generally grouped in fixed classes, and rooms are also fixed for conducting the classes, which make the problem

D. Datta et al.: *Multi-Objective Evolutionary Algorithm for University Class Timetabling Problem,* Studies in Computational Intelligence (SCI) **49**, 197–236 (2007)

somewhat simpler. The only task left is to schedule the classes at suitable time-slots. However, a university usually offers a range of optional courses, for which the number of students in a class is not known in advance. Hence, the classes are to be scheduled to suitable rooms and time-slots only after the enrollment of students. Moreover, the existence of compound classes, such as multi-slot, split or combined, makes the problem more complicated. Traditionally, the scheduling of a class timetable is made through manual processes, involving the labour of several days of several persons. Such a process is based on *trial and hit* only, and hence, a valid solution is not guaranteed. Even if a valid solution is found, it is likely to miss far better solutions. These uncertainties motivated for the scientific study of the problem, and to develop an automated solution technique for it. Based on various such studies, the problem can be categorized as below:

1. The class timetabling problem belongs to the class of combinatorial optimization problem[3] (Papadimitriou and Steiglitz 1982; Melicio et al. 2004).
2. It is also a variant of general *resource allocation problem*[4] (Papadimitriou and Steiglitz 1982).
3. Most of its variants are highly constrained optimization problems, and their worst case complexity is NP-complete (Even et al. 1976; Cooper and Kingston 1995).

The timetabling problem is important not only because every institute faces it once or twice a year, but also because it is just one of the many existing scheduling problems. An effective solution technique to the problem could be applied to other scheduling problems (Abramson 1991). The problem drew the attention of the researchers in the early 60's with the study of Gotlieb (1962), who formulated a class-teacher timetabling problem by considering that each lecture contained one group of students, one teacher, and any number of time-slots which could be chosen freely. Schaerf (1996, 1999) surveyed that most of the early techniques for automated timetabling were based on *successive augmentation*, where a partial timetable was filled in lecture by lecture until either all lectures were scheduled or no further lecture could be scheduled without violating constraints. In another survey, Abramson (1991) reported the general techniques applied to the problem in the past, such as network flow analysis (Greko 1965), random number generator (Fujino 1965), integer programming (Lawrie 1969; Tripathy 1984), and linear algorithm (Akkoyunlu 1973). In addition to these, worth mentioning methods are *exact method-based heuristic algorithm* (de Werra 1971), and *graph coloring theory* (Neufeld and Tartar 1974). However, the classical techniques are not fully capable to handle the large number of integer and/or real variables and constraints, involved in the huge discrete search space of the timetabling problem. These inadequacy of classical techniques have drawn the attention of the researchers towards the non-classical techniques. Worth mentioning non-classical

[3] In a combinatorial optimization problem, the search space is discrete and finite, and the quality of a solution is measured by arbitrary functions.

[4] A resource allocation problem means the execution of a given set of tasks, utilizing available resources, subject to certain constraints - such as priority of a task over another, deadlines, and so on.

techniques, that were/are being used to the problem, are genetic algorithms (Colorni et al. 1990, 1992; Abramson and Abela 1992), neural network (Looi 1992), and tabu search algorithm (Costa 1994). However, compared to other non-classical methods, the widely used are the genetic/evolutionary algorithms (GAs/EAs). The reason might be their successful implementation in a wider range of applications. Once the objectives and constraints are defined, GAs appear to offer the ultimate *free lunch* scenario of good solutions by evolving without a problem solving strategy (Al-Attar 1994). In the GA, used by Abramson and Abela (1992) to school timetabling problem, a solution is likely to loose or duplicate a class under crossover operator. A repairing mechanism, in the form of a mutation operator, was used by them to fix up such lost or duplicated classes. Piola (1994) applied three evolutive algorithms to school timetabling problem, and showed their capability to tackle highly constrained combinatorial problems, such as timetabling problem. Lima et al. (2001) used a GA to university class timetabling problem with the objective to facilitate the conciliation of the students' class hours with their work time. Srinivasan et al. (2002) applied an EA to university class timetabling problem. However, they over-simplified the problem by considering only a single batch of the university, and infinite number of rooms for scheduling classes. Blum et al. (2002) studied university class timetabling problem using a GA, where attempt is made to reduce the complexity by initializing a solution through a set of sequential heuristic rules. Bufé et al. (2001) used a hybrid EA to school timetabling problem, where a deterministic heuristic-based timetable builder is used for generating feasible solutions. The EA, used by Rossi-Doria and Paechter (2003) to university class timetabling problem, evolves the choice of the heuristics for generating a feasible solution.

2 Aim of the Present Study

The class timetabling problem involves several criteria that must be satisfied simultaneously, such as compliance with regulations, proper utilization of resources, and satisfaction of people's preferences (Silva et al. 2004). However, the problem is generally tackled as single-objective optimization problem by combining multiple criteria into a single scalar value. In most of the cases, cited in Sect.1, minimization of weighted sum of constraint violation was used as the only objective function. Only in recent years, a few multi-objective optimization techniques have been proposed to tackle the problem. Carrasco and Pato (2001) used a bi-objective GA to school timetabling problem for minimizing violation of soft constraints from two competitive perspective of teachers and classes. Filho and Lorena (2001) also tackled the school timetabling problem using bi-objective model of constructive EA, which uses the well-known schemata theory (Goldberg (1989). They modelled the problem as clustering of teachers and classes, and minimized their conflicts. Desef et al. (2004) used another bi-objective tabu search to German primary schools. They chose minimization of weighted sum of idle time-slots and the extents to which the daily classes were finished, where higher priority was given to the earlier one in one objective function, and the latter in the other objective function. However, the total number of techniques for handling the class timetabling problem, whether single or multi-objective, is still relatively scarce. Moreover, most of the earlier works were concentrated on school timetabling, and only a few on university class timetabling. On the other hand, a timetabling problem, particularly in universities and in many higher secondary schools, may contain different types of classes, such as multi-slot, split, combined, grouped, etc., which make the problem more complex and constrained. However, only a limited number of researchers considered one or more such class-structures (Abramson 1991; Melicio et al. 2004; Piola 1994; Rudová and Murry 2002; Carrasco and Pato 2001).

The above mentioned shortfalls depict the fact that a lot of research is still required to obtain some improved general solution techniques for class timetabling problem. A general technique should not be over-simplified by leaving different aspects to users. Instead of that, it should include all possible aspects which can easily be simplified as per the requirement of individual user. Owing to such requirements, the current authors felt the necessity for studying the problem, considering different types of class-structures and constraints that are common to most of the variants of the problem. In this regard, NSGA-II-UCTO, a heuristic-based multi-objective evolutionary optimization technique, has also been developed to handle the problem. It employs NSGA-II (Deb 2001; Deb et al. 2002), an EA-based multi-objective optimizer, with specially designed representation and EA operators for class timetabling problem. Though emphasis was put on universities, NSGA-II-UCTO can also be applied to schools with a little modification. It was applied by the authors to two highly constrained real problems from NIT-Silchar, a technical institute in India. The problems were considered under different cases of EA operators and parameters, and NSGA-II-UCTO was found successful in every instance.

3 Types of Academic Courses

In a class timetabling problem, classes[5] of a course[6] are events, to which resources (rooms and time-slots) are to be assigned in such a way that all the events are scheduled using the available resources. Before assigning the resources effectively to the events, we need to know the types and natures of the events. Throughout the academic career, generally one has to pass through three types of courses, namely *Simple Course*, *Compound Course* and *Open Course*.

1. **Simple Course:** A simple course is one which is offered to a particular batch[7] only, and compulsory to all the students of the batch. In other words,

[5] A class is a meeting of students and teacher in a room for a lecture.

[6] A course is a subject to be studied, for example, *Theory of Optimization* or *Introduction to Numerical Methods.*

[7] A batch means, for example, first year BSc in Physics, or third semester of Electronics Engineering.

the allocation of students to a class is fixed, and the classes are disjoint, i.e., no two classes at a time have any common student. Moreover, such a class is scheduled at a single time-slot only (single-slot class). A school timetable is generally composed of simple courses only.

2. **Compound Course:** It is also compulsory to all the students of a batch. The class-structure of a compound course may be *multi-slot, split* or *combined.* The usual duration of a time-slot is around 30-60 minutes, which is not sufficient for many classes, for example, laboratory classes. Hence, such a class is generally spanned over two or more consecutive time-slots, and known as a *multi-slot class.* Sometimes the students of a class are split into two or more groups for better teacher-students interaction, or due to the limited resources required for conducting the class, for example tutorial and laboratory classes. Such a class is known as a *split class.* When one group of students of a batch attends a split class of one course, the other groups of students of the batch may attend the split classes of other courses, if any. On the other hand, sometimes a single course, or a few similar courses is (are) offered to different batches. If the students of two or more batches can be accommodated in a single room, classes of such courses may be combined to teach together. These three class-structures of compound courses complicate the scheduling of a class timetable. Without these class-structures, a compound course reduces to a simple course. The compound courses are found in universities, and also in many higher secondary schools. However, only a limited number of authors have reported about one or more such class-structures. Abramson (1991) considered multi-slot classes. Melicio et al. (2004), Piola (1994), and Rudová and Murry (2002) discussed about combined classes. Carrasco and Pato (2001) considered multi-slot and combined classes.
3. **Open Course:** As the name implies, an open course is a flexible/optional course. Unlike a simple or compound course, an open course is offered to more than one batch. A student can randomly opt his/her required number of open courses from a given set of such courses, offered by different departments. This flexibility restricts the scheduling of open courses to be done only after the students' course registration, so that no two classes, to be attended by a student, are scheduled at the same time. It also complicates the scheduling of classes as the availability of suitable time-slots and rooms is drastically reduced with increasing number of students and courses. Open courses are generally offered only in universities and higher secondary schools. When offered to a single batch, an open course reduces to a compound or simple course according to its class-structure. A special case of open courses is the *group courses,* which are generally known as *elective courses.* A batch may have some sets of elective courses, and a student has to opt any one course from a given set. For example, a student can choose any one course on computer programming from three available choices of *Programming in C*, *Programming in C++* and *Programming in Java.* That is, the students of the batch are divided in the classes of those courses, and the classes are scheduled in different rooms, but at the same time. It is to be noted that

group classes are not the split classes. In group classes, a student has to attend the classes of one course only, whereas in split classes, a student has to attend the classes of all courses in rotation.

4 University Class Timetable as a Multi-Objective Optimization Problem

As stated earlier, a class timetabling problem involves the arrangement of classes, students, teachers and rooms at a fixed number of time-slots, respecting certain constraints. Constraints in a class timetabling problem can be classified as *hard constraints* and *soft constraints*. Hard constraints must be satisfied by a solution of the problem, whereas soft constraints are desired to be satisfied, but not essential. A solution is feasible/valid if it satisfies all the hard constraints, and a feasible solution is the best one if it satisfies all the soft constraints also (Silva et al. 2004). However, it is very tough, or may even impossible to satisfy all the soft constraints. This complexity requires the scheduling of a class timetable to be treated as a solution over hard constraints, and optimization over soft constraints (Rudová and Murry 2002). However, since the large number of variants of the problem differ from each other, based on the types of universities, and their distinct class-structures and constraints, it is very difficult to formulate a general problem that will serve the requirements of all universities. Many researchers attempted to generalize it by considering only simple class-structures. However, a general technique should not be over-simplified by leaving different aspects to users. Instead of that, it should include all possible aspects which can easily be simplified as per the requirements of individual users. Since the numbers and types of constraints also vary from university to university, they should be classified very carefully in order to generalize the problem. Instead of considering the case of a particular university, only those constraints, that must be satisfied in most of the universities, should be brought under the category of hard constraints. All other constraints can be categorized as soft constraints, which should be expressed explicitly. That is, soft constraints should not be made an integral part of the problem, so that the addition or deletion of any such constraints does not affect the formulation, or solution of the main problem. Based on these assumptions, a model university class timetabling problem has been designed by the current authors, the definition and formulation of which are described in the following sub-sections.

4.1 Common Hard Constraints

Only the following six types of hard constraints were identified by the authors, that are common to most of the variants of the class timetabling problem, and must be satisfied by a solution to accept it as a valid one:

1. A student should have only one class at a time.
2. A teacher should have only one class at a time.

3. A room should be booked only for one class at a time (a set of combined classes may be treated as a single class).
4. Only one class of a course should be scheduled on a day.
5. A class should be scheduled only in a specific room, if required, otherwise in any room which has sufficient sitting capacity for the students of the class. Due to the requirement of some extra facilities, such as laboratory apparatus, many classes may need to be scheduled only in specific rooms.
6. A class should be scheduled only at a specific time-slot, if required. Due to many reasons, such as non-availability of part-time teachers in peak hours of the day, or involvement of senior teachers in administrative works, some classes may need to be scheduled only at specific time-slots.

4.2 Common Soft Constraints

Soft constraints do not represent any physical conflict, but preferences, which are encouraged to be fulfilled whenever possible (Melicio et al. 2004; Bufé et al. 2001; Carrasco and Pato 2004). A solution is valid even if it does not satisfy the soft constraints, but not so good as one satisfying them. The quality of a solution is inversely proportional to the amount of violated soft constraints (Piola 1994). Though the imposition of excess soft constraints would produce a greater preferred timetable, but it will increase the computational complexity of a problem. Hence, the number of such constraints should be as less as possible. The authors considered only the following three types of soft constraints which are imposed in many universities:

1. A student should not have any free time slot between two classes on a day.
2. A teacher should not be booked at consecutive time-slots on a day (except in multi-slot classes).
3. A smaller class should not be scheduled in a room which can be used for a bigger class.

The soft constraint (1) implies a compact timetable, whereas the constraint (2) conflicts with it, and seeks a well-spread timetable. The constraint (3) takes care of proper utilization of rooms. As already assumed that the soft constraints should be defined explicitly in a problem, more such constraints can be imposed without any loss of generality of the problem. However, the imposition of a constraint will force a solution to respect it, and hence, the solution may get altered by the imposition of each soft constraint. A few other soft constraints, considered in different universities, are: (i) preferences of teachers on time-slots or class rooms, if any, should be respected (Melicio et al. 2004), (ii) in case of an institute with multiple campuses, a teacher should not be engaged in more than one campus on a day, (iii) classes to be attended by a student should be well distributed over the week (Bufé et al. 2001), (iv) classes to be attended by a student should be in the same room, or in rooms of shorter distances to avoid him/her from running long distances to attend the classes (Daskalaki et al. 2004), (v) a timetable should be compact enough for having long-spanned free hours, or one or two free days for both teachers and students (Bufé et al. 2001; Gaspero and Schaerf 2002), (vi) free time-slots, if any, should be grouped at the beginning or end of the day (Piola 1994), and so on.

4.3 Common Objective Functions

Once the class timetabling problem has been identified as an optimization problem, we need one or more objectives to optimize. However, an objective function in this problem is just an arbitrary measure of the quality of a solution (Abramson and Abela 1992). Hence, the choice of objectives varies from university to university. Most of the researchers treated the problem as a single-objective optimization problem, and took the minimization of total constraints violation as the only objective function (Abramson and Abela 1992; Blum et al. 2002; Lima et al. 2001; Piola 1994). However, various choices of objectives, such as a compact timetable, minimum number of consecutive classes of teachers and so on, lead the scheduling of a class timetable to a multi-objective optimization problem. However, a very limited number of researchers considered multiple objectives in the problem (Paquete and Fonseca 2001; Silva et al. 2004). Carrasco and Pato (2001) used a bi-objective model to school timetabling problem for minimizing violation of soft constraints from two competitive perspective of teachers and classes. Filho and Lorena (2001) also tackled the school timetabling problem using bi-objective model where teachers and classes are clustered, and their conflicts are minimized. Desef et al. (2004) used another bi-objective model to German primary schools. Both of their objective functions are the minimization of weighted sum of idle time-slots and the extents to which the daily classes are finished, where higher priority is given to the earlier one in one objective function, and the latter in the other objective function. The current authors also considered a bi-objective model. The two conflicting soft constraints, constraints (1) and (2) as mentioned in Sect.4.2, were taken as the two objective functions for minimizing during the optimization process, i.e,. the objective functions are:

1. Minimize the average number of weekly free time-slots between two classes of a student.
2. Minimize the average number of weekly consecutive classes of a teacher.

In these objective functions, classes only of the same day, not of different days, are considered in finding the free time-slots of a student, or consecutive classes of a teacher. Objectives in a class timetabling problem are nothing but functions of the soft constraints imposed on it. Hence, without any loss of generality of the problem, different objectives can be incorporated by imposing different soft constraints on the problem.

4.4 Mathematical Formulation

After defining the objective and constraint functions, the scheduling of university class timetable, as a multi-objective optimization problem, can now be expressed mathematically as below:

Let,
- D Number of days/week.
- T Number of time-slots/day.
- R Number of rooms.
- S Number of students.
- M Number of teachers.
- C Number of courses.
- E Number of events (classes).

$d_{e,i}$ = 1 if the event e is scheduled on day i, else 0.
$t_{e,j}$ = 1 if the event e is scheduled at time-slot j, else 0.
$r_{e,k}$ = 1 if the event e is scheduled in room k, else 0.
$p_{e,s}$ = 1 if the event e is attended by student s, else 0.
$\mu_{e,m}$ = 1 if the event e is taught by teacher m, else 0.
$l_{e,c}$ = 1 if the event e belongs to course c, else 0.

– Objective functions:
  1. Minimize the average number of weekly free time-slots between two classes of a student:

$$f_1 \equiv \frac{1}{\mathrm{S}} \sum_{s=1}^{\mathrm{S}} \sum_{i=1}^{\mathrm{D}} \sum_{j=t^{\mathrm{o}}_{s,i}}^{t^{\mathrm{e}}_{s,i}} I\left(\sum_{e=1}^{\mathrm{E}} d_{e,i}.t_{e,j}.p_{e,s} = 0\right), \tag{1}$$

  where $t^{\mathrm{o}}_{s,i}$ and $t^{\mathrm{e}}_{s,i}$ are, respectively, the first and last time-slots of classes of student s on day i. The *indicator function*[8] I shows 0 engagement of student s at time-slot j on day i, i.e., a free time-slot for the student.
  2. Minimize the average number of weekly consecutive classes of a teacher:

$$f_2 \equiv \frac{1}{\mathrm{M}} \sum_{m=1}^{\mathrm{M}} \sum_{i=1}^{\mathrm{D}} \sum_{j=1}^{\mathrm{T}} I_1(A_1 > 0).I_2(A_2 > 0).I_3(A_3 = 0), \tag{2}$$

  where $A_1 = \sum_{e=1}^{\mathrm{E}} d_{e,i}.t_{e,j-1}.\mu_{e,m}$, $A_2 = \sum_{e=1}^{\mathrm{E}} d_{e,i}.t_{e,j}.\mu_{e,m}$ and $A_3 = \sum_{e=1}^{\mathrm{E}} d_{e,i}.t_{e,j-1}.t_{e,j}.\mu_{e,m}$. The indicator functions I_1 and I_2 confirm the engagement of teacher m at both of time-slots (j-1) and j on day i. The third indicator function I_3 shows the engagement of the teacher at those time-slots for different events. It (I_3) has been used just to assure that multi-slot classes are not taken as consecutive classes.

[8] **Indicator Function:** A function which returns either 1 or 0, respectively, depending on whether its argument is true or false. Suppose Ω is a set with typical element ω, and let A be a subset of Ω. Then the indicator function of A, denoted by $I_A(\cdot)$, is defined by (Murison 2000)

$$I_A(\omega) = \begin{cases} 1 \text{ if } \omega \in A \\ 0 \text{ if } \omega \notin A. \end{cases}$$

- Subject to hard constraints:
  1. Number of classes to be attended by a student at a time:

$$g_{(s-1)\mathrm{TD}+(i-1)\mathrm{T}+j} \equiv \sum_{e=1}^{\mathrm{E}} d_{e,i}.t_{e,j}.p_{e,s} \leq 1, \tag{3}$$
$$s = 1,..,\mathrm{S}; i = 1,..,\mathrm{D}; \text{ and } j = 1,..,\mathrm{T},$$

     Total number of constraints in this category is STD.
  2. Number of classes to be taught by a teacher at a time:

$$g_{\mathrm{STD}+(m-1)\mathrm{TD}+(i-1)\mathrm{T}+j} \equiv \sum_{e=1}^{\mathrm{E}} d_{e,i}.t_{e,j}.\mu_{e,m} \leq 1, \tag{4}$$
$$m = 1,..,\mathrm{M}; i = 1,..,\mathrm{D}; \text{ and } j = 1,..,\mathrm{T},$$

     It is to be noted that a set of combined classes, to be taught together by a teacher, has been taken as a single class (event). Total number of constraints in this category is MTD.
  3. Number of classes in a room at a time:

$$g_{(\mathrm{S+M})\mathrm{TD}+(k-1)\mathrm{TD}+(i-1)\mathrm{T}+j} \equiv \sum_{e=1}^{\mathrm{E}} d_{e,i}.t_{e,j}.r_{e,k} \leq 1, \tag{5}$$
$$i = 1,..,\mathrm{D}; j = 1,..,\mathrm{T}; \text{ and } k = 1,..,\mathrm{R},$$

     Total number of constraints in this category is RTD.
  4. Number of classes of a course on a day:

$$g_{(\mathrm{S+M+R})\mathrm{TD}+(c-1)\mathrm{D}+i} \equiv \sum_{e=1}^{\mathrm{E}} d_{e,i}.l_{e,c} \leq 1, \tag{6}$$
$$c = 1,..,\mathrm{C}; \text{ and } i = 1,..,\mathrm{D},$$

     Total number of constraints in this category is CD.
  5. Room where a class is to be scheduled:
     (a) Type of the room:

$$g_{(\mathrm{S+M+R})\mathrm{TD}+\mathrm{CD}+2e-1} \equiv \sum_{k=1}^{\mathrm{R}} r_{e,k}.I(k \in \rho_e) = 1, \quad e = 1,..,\mathrm{E}, \tag{7}$$

     where ρ_e is the set of specific rooms for event e. The indicator function I confirms the allotment of an event in a room from its set of specific rooms. When no room is specified for an event, any general room becomes specific for that event.
     (b) Capacity of the room:

$$g_{(\mathrm{S+M+R})\mathrm{TD}+\mathrm{CD}+2e} \equiv \sum_{s=1}^{\mathrm{S}} p_{e,s} \leq \sum_{k=1}^{\mathrm{R}} r_{e,k}.h_k, \quad e = 1,..,\mathrm{E}, \tag{8}$$

     where h_k is the sitting capacity of room k.

Total number of constraints in this category is 2E.

6. Time-slot at which a class is to be scheduled:

$$g_{(\mathrm{S+M+R})\mathrm{TD+CD}+2E+e} \equiv \sum_{j=1}^{\mathrm{T}} t_{e,j}.I(j \in \tau_e) = 1, \quad e = 1,..,\mathrm{E}, \tag{9}$$

where τ_e is the set of specific time-slots for event e. The indicator function I confirms the allotment of an event at a time-slot from its set of specific time-slots. Total number of constraints in this category is E.

- Soft constraints:
  1. Total number of weekly free time-slots between two classes of students:

$$sc_1 \equiv \sum_{s=1}^{\mathrm{S}} \sum_{i=1}^{\mathrm{D}} \sum_{j=t^{o}_{s,i}}^{t^{e}_{s,i}} I\left(\sum_{e=1}^{\mathrm{E}} d_{e,i}.t_{e,j}.p_{e,s} = 0\right) = 0, \tag{10}$$

  where notations are the same as in (1).

  2. Total number of weekly consecutive classes teachers:

$$sc_2 \equiv \sum_{m=1}^{\mathrm{M}} \sum_{i=1}^{\mathrm{D}} \sum_{j=1}^{\mathrm{T}} I_1(A_1 > 0).I_2(A_2 > 0).I_3(A_3 = 0) = 0, \tag{11}$$

  where notations are the same as in (2).

  3. Capacity of a class, and that of the room where the class is scheduled:

$$sc_{2+\sum(\mathrm{E}-e)+e'} \equiv \sum_{s=1}^{\mathrm{S}} p_{e',s} > \sum_{k=1}^{\mathrm{R}} r_{e,k}.h_k, \text{ if } \sum_{s=1}^{\mathrm{S}} p_{e',s} > \sum_{s=1}^{\mathrm{S}} p_{e,s}, \tag{12}$$

  where $e = 1,..,\mathrm{E}$ and $e' = e+1,..,\mathrm{E}$. Maximum number of constraints in this category is $\frac{\mathrm{E(E-1)}}{2}$. This constraint is invalid if the capacity of any available room is greater than the number of students of any event (class) which is yet to be scheduled.

From the above formulation, it is seen that the considered university class timetabling problem is composed of (S+M+R)TD+CD+3E hard constraints and $2+\frac{\mathrm{E(E-1)}}{2}$ soft constraints, where S, M, R, T, D, C and E represent, respectively, the total numbers of students, teachers, rooms, time-slots/day, days/week, courses, and classes (events). These constraints cause classical optimization methods difficult to be applied on this problem. Since the soft constraints of (10) and (11) have already been taken into account by the objective functions, the main optimization problem can be defined by the objective functions and hard constraints only, i.e., by (1)-(9). The last soft constraint, given by (12), will be taken into account by the heuristic approach addressed in Sect.5. In case of a class timetable without any open course, (1), (3) and (10) can also be expressed in terms of batches, instead of students. Since a batch doesn't have any open course, the batch is free/engaged at a time-slot means all the students of the batch are also free/engaged at that time-slot. Hence, computational time can be reduced by expressing these quantities in terms of batches, because the number of batches in a timetabling problem is always less than the number of students.

5 Heuristic Approach to University Class Timetabling Problem

Any numerical optimization process starts with an initial guess to the desired solution. The process becomes computationally less expensive, as well as faster, if the initial guess is feasible, or near feasible. In many cases, better solutions can be expected with such an initial guess than with an arbitrary one. For a class timetabling problem, however, it is very difficult to obtain an initial feasible solution, or may even impossible to accomplish it (Melicio et al. 2004). The job may be an easy one with simple classes, such as single-slot classes without any restriction on the assigned rooms and time-slots. The size of the search space is the same for all classes, and hence, they can be scheduled randomly in any rooms and time-slots. For example, in case of a timetable with 5 days per week, 7 time-slots per day, and 10 rooms per time-slot, there are total of 350 possible slots[9] (= 5 days × 7 time-slots × 10 rooms), and a class can be scheduled in any one of them. However, any restriction, imposed on a class, will reduce the size of the search space for that class. Say, the *Physics Laboratory class* is to be scheduled only in *Physics Laboratory.* Then the size of the search space for this class would be 35 ($=5\times7\times1$). If it happens to be a 2-slot class, it cannot be scheduled at the last time-slot of a day, and the size of the search space will be reduced to 30 ($=5\times6\times1$). Again, if it is to be taught by a part-time teacher, who will not be available before 6-th time-slot, the size of the search space will further be reduced to 5 ($=5\times1\times1$). As a result, this class may not be possible to schedule without violating any constraint, if it comes at the last during the scheduling. This is because, the 6-th time-slot of all the days, only where this class can be scheduled, might have already been assigned to some other classes. Hence, in order to get a feasible or near feasible solution, random scheduling of classes may not work. Rather classes should be scheduled in some *order* - based on the size of the search space, and/or other complexities of a class. The developers of EAs also cannot expect an EA to generate initial feasible, or near feasible, solutions from nothing, i.e., without being armed with any primitive guideline. Such expectations may require hours or days to evolve solutions, which even may not be of acceptable quality. Hence, at least, in case of large and complex problems, like class timetabling problem, an EA needs be guided by some *Heuristic Approach*[10] (HA in short) to generate initial solutions (Al-Attar 1994). Many

[9] A slot does not mean a time-slot, but a location under a time-slot where a class can be scheduled. For example, if there are 10 rooms, there will be 10 slots under a single time-slot.

[10] An approach without a formal guarantee of performance is known as *Heuristic Approach.* It relates or uses a problem-solving technique, in which the most appropriate

authors have suggested, and also presented many such HAs for class timetabling problem (Bufé et al. 2001; Carrasco and Pato 2001; Lewis and Paechter 2004). Howover, the detail behind the development of such an approach is hardly reported in literature. Hence, it is required to study different types of problems, and see how such an approach can be developed. In this regard, an experiment on three virtual problems was performed by the authors, considering different types of classes, as addressed in Sect.3, and/or imposing restrictions on the random use of rooms and time-slots. Both random and some ordered searches were performed under different conditions. Due to space limit, the detail of the experiment could not be reported here. It was found from the experiment that a class with specific time-slots is the most complex one. The next complex one is a multi-slot class. However, classes with specific rooms, group classes and split classes do not follow any common pattern of complexity. This might be due to their comparatively lower complexities, which, even in many cases, are comparable with those of simple single-slot classes. It was also observed, in most of the cases, that the ordered search can reduce the complexities of the problems by using less resources than the random search does. Hence, the experiment clearly depicts the need of using an ordered search for obtaining a feasible or near feasible solution with optimum use of resources. Therefore, based on the experiment, and the size of the search space for a class, a sequential heuristic approach (HA) has been developed that has the capacity to produce an acceptable timetable. It is economic in terms of resources, and also computationally inexpensive. In this approach, the events are first sorted in descending order of their complexities, and then resources are assigned to them in order. The detail of the approach is as below:

1. All classes are first sorted in the following order:
   (a) Ascending order of number of specific time-slots. If not mentioned, this number for a class is the number of total available time-slots.
   (b) If numbers of specific time-slots are equal, descending order of number of time-slots per class.
   (c) If numbers of specific time-slots as well as numbers of time-slots/class are equal, ascending order of number of specific rooms. If not mentioned, this number for a class is the number of total available rooms.
   (d) If numbers of specific time-slots, numbers of time-slots/class as well as numbers of specific rooms are equal, preference to group classes over other classes.
   (e) If numbers of specific time-slots, numbers of time-slots/class as well as numbers of specific rooms are equal, preference to split classes over other classes (group classes are not supposed to be split).
   (f) Since sitting capacity of a room also plays an important role, similar classes, under each of the above five steps, may be sorted in descending order of number of students in each class.

one, out of several solutions found by alternative methods, is selected for successive problems.

2. Once the sorting of classes is over, they may be considered in order to assign resources to them, respecting as many hard constraints as possible. If classes were not sorted in Step (1f) as per the numbers of students in them, a class may be scheduled in an arbitrary capable room, and then an exhaustive search may be performed among the remaining rooms for finding a suitable smaller one. This will avoid the possibility of occupation of an over-size room by a smaller class which, as per the sorting under the Steps (1a) to (1e), may come before a bigger class. This will also satisfy the soft constraint (3) of Sect.4.2.

In the above HA, classes with specific time-slots, and multi-slot classes have been ordered as per the results of the conducted experiment. This ordering is also obvious from the sizes of the search spaces for those classes. Because, the number of specific time-slots for a class, in general, would be much smaller than the number of possible time-slots for a multi-slot class. Hence, the size of the search space for a class with specific time-slots is smaller than that for a multi-slot class. Since no fixed pattern was found in case of classes with specific rooms, group classes, and split classes, those have been ordered only on the basis of increasing sizes of their search spaces. Although the sizes of the search spaces for a split single-slot class and a simple single-slot class are equal, the former one has been preferred because of its relatively complex nature over the latter one.

5.1 Few Characteristics of Class Timetabling Problem

As the available resources for any problem is limited, based on the above experiment, the following characteristics of class timetabling problem have been observed:

1. Random scheduling of events may lead to the improper use of resources, and as a result, many later events may be left unscheduled due to the non-availability of proper resources,
2. Even if the events are scheduled in order, many resources may be required to leave unused, i.e., sufficient events may not be available to use all the resources, and
3. Amount of unused resources increases with the increasing complexities of the events.

6 Evolutionary Chromosome and Operators for University Class Timetabling Problem

Since the decision variables (classes) of class timetabling problem seek only positive integer values of different parameters of the problem, such as the serial numbers of time-slots and rooms where classes are to be scheduled, or the serial numbers of teachers and students to whom the classes are assigned, the scheduling of class timetable becomes a pure integer programming problem. It also belongs to the class of combinatorial optimization problems (Melicio et al. 2004). Moreover, as seen in Sect.4, it is a multi-objective optimization problem. Though there exists a number of classical algorithms for handling such problems (Rao 1996), most of them suffer from computational complexity, and/or multiple objective functions. Among non-classical algorithms, an evolutionary algorithm (EA) is one which can successfully handle integer variables, multiple objective functions, as well as combinatorial optimization problems (Deb 2001). Moreover, in the literature also, EAs are found as the widely used non-classical methods to the class timetabling problem. The current authors also used a specially designed multi-objective EA to this problem. The developed EA, known as NSGA-II-UCTO, along with chromosome representation and EA operators (crossover and mutation), are addressed in detail in the following subsections.

6.1 Chromosome Representation

The performance of an EA, or the quality of a solution from an EA, is greatly influenced by the chromosome representation. Hence, this representation should be as simple and intuitive as possible. In case of class timetabling problem, a chromosome represents a class timetable, and each of its genes describes a component of the timetable. In spite of the matrix-pattern of representation of an actual class timetable, as shown in Table 1, different authors use different types of direct or indirect chromosome representations to suit their algorithms. The direct chromosome representation usually encodes an actual timetable as a vector of classes to which time-slots and rooms are to be assigned, or as a vector of time-slots at which classes and rooms are to be scheduled. The indirect representation, on the other hand, encodes a set of instructions, and then a *timetable builder* is used to produce a timetable from those instructions (Burke et al. 1995). Fang (1994) used a direct representation where the length of a chromosome is equal to the number of classes times the number of parameters to be assigned to a class. The position of a gene, in each set of genes equal to the number of classes, represents the serial number of a class, and the value of the gene represents a parameter, such as a time-slot, a room or a teacher. One of the main reasons for using such representations is to apply directly the standard EA operators to the chromosomes. However, the major disadvantage of this type of direct representation is that the EA operators are likely to produce infeasible solutions which are required to be repaired by some *repair operator* (Bufé et al. 2001; Rossi-Doria and Paechter 2003). Bufé et al. (2001), and Rossi-Doria and Paechter (2003) used indirect chromosome representation, and preserved the feasibility of solutions under the action of EA operators. Bufé et al. (2001) used parametric chromosome representation, containing a permutation of events, and then a deterministic heuristic-based timetable builder was used to generate a

Table 1. Example of a class timetable.

	09-10	**10-11**	**11-12**	**.....**
R1	Phy-I	Math-I	Chem-I	.....
R2	Math-II	Chem-II	Phy-II	.....
R3	Chem-III	Phy-III	Math-III	.....
...	...	...	...	.....

feasible solution from the permutation. A chromosome, used by Rossi-Doria and Paechter (2003), is composed of two rows, each of length equal to the number of events, representing different heuristics to choose feasible events and resources, respectively. In another indirect representation, Abramson and Abela (1992) represented a chromosome by time-slots, where a time-slot is composed of number of tuples. Each tuple is identified by its label, and the fields of a tuple contain a valid class, teacher and room. The drawback of this representation is that the crossover operator may not be able to give correct set of tuples to the offspring (new solutions), which they termed as *label replacement problem*, and proposed a *label replacement algorithm*, in the form of a mutation operator, to correct such problems.

A direct chromosome representation, which looks like the representation of an actual class timetable (as shown in Table 1), was used by the current authors. A chromosome is a two-dimensional matrix, and represents a class timetable. Each column of the matrix represents a time-slot, and a row represents a room. That is, the chromosome is a vector of time-slots, and a time-slot, which represents a gene of the chromosome, is a vector of rooms. Then, the value of each cell of the matrix represents the event(s), scheduled in the corresponding room and time-slot. Hence, mathematically a chromosome can simply be expressed as:

$$T = (T_1, T_2, ..., T_j, ..., T_t),$$

and $T_j = (R_1, R_2, ..., R_k, .., R_r)^{\mathrm{T}}$,

where,

T_j = j-th time-slot (gene),
R_k = k-th room,
t = total number of time-slots, and
r = total number of rooms.

The proposed chromosome representation is also shown in tabular form in Table 2. Since the chromosome directly represents an actual timetable, the advantage of this representation is obvious. This is one of the best chromosome representations for any topology optimization of structures (Datta and Deb 2005),

Table 2. Chromosome representation.

<table>
<tr><th>R/T</th><th>T_1</th><th>T_2</th><th>T_3</th><th>..</th><th>T_j</th><th>..</th><th>T_t</th></tr>
<tr><td>R_1</td><td>C20</td><td>C11</td><td>C39</td><td>...</td><td>C05</td><td>...</td><td>C16</td></tr>
<tr><td>R_2</td><td>C33</td><td>C21</td><td>C15</td><td>...</td><td>C40</td><td>...</td><td>C12</td></tr>
<tr><td>R_3</td><td>C01</td><td colspan="2">C35</td><td>...</td><td>C07</td><td>...</td><td>C08</td></tr>
<tr><td>..</td><td>...</td><td>...</td><td>...</td><td>...</td><td>...</td><td>...</td><td>C27</td></tr>
<tr><td>R_k</td><td>C13</td><td>C02</td><td>C14</td><td>...</td><td>C22</td><td>...</td><td>C38</td></tr>
<tr><td>..</td><td>...</td><td>...</td><td>...</td><td>...</td><td>...</td><td>...</td><td>C18</td></tr>
<tr><td>R_r</td><td>C06</td><td>C04</td><td>C17
C36</td><td>...</td><td>C28</td><td>...</td><td>C31</td></tr>
</table>

or circuits synthesis (Mesquita et al. 2002). Two major advantages of this representation are:

1. It is very easy to control the possibility of scheduling, at a particular time-slot, more than one class in a room, or to a teacher, student or batch. To check any such possibility, it is not required to go through the entire timetable, but only the column of the matrix, representing that time-slot, is sufficient.
2. Compound and group classes can easily be scheduled and checked. For example, Table 2 shows the scheduling of 2-slot class $C35$ in room R_3 at time-slots T_2 and T_3, and combined classes $C17$ and $C36$ in room R_r at time-slot T_3.

Carrasco and Pato (2001) used a similar representation. However, they handled only combined and multi-slot classes. Lewis and Paechter (2004) also used a similar representation in examination timetabling problem. Instead of the columns, they took each cell of the matrix as a gene of the chromosome, which is similar with the vector representation of events.

6.2 Crossover Operator (XVRA)

A crossover operator trades sets of genes between two chromosomes, allowing the sets to exploit particularly beneficial portions of a search space. An example of a traditional *single-point* crossover operator on two chromosomes (Parent-1 and Parent-2) is shown in Fig.1. This crossover operator may produce very useful offspring in many optimization problems, including the scheduling in unrestricted/ less-restricted search space (Corne et al. 1994; Fang 1994). However, a complex scheduling, like class timetabling problem in highly restricted search space, may even fail under such a crossover operator. The reason behind it is clear from Fig.1. If the positions and values of the genes were to represent, respectively, classes and time-slots (or rooms), the offspring are not acceptable as two classes have been scheduled to the same time-slot (or room). On the other hand, if the positions and values of the genes were to represent, respectively, the time-slots (or rooms) and classes, then also the offspring are not acceptable as the same class is scheduled at two different time-slots (or rooms), and some classes are left unscheduled. As mentioned in Sect.6.1, these are the reasons why researchers fail to prevent the feasibility of offspring generated by the standard crossover operator, and need some kind of repair operator, or specially designed mutation operator, to repair an infeasible offspring (Abramson and Abela 1992; Lewis and Paechter 2004). Due to high complexity of class timetabling problem, it is very difficult to get a feasible solution for it, perhaps even impossible to accomplish it. Hence, it was suggested by Melicio et al. (2004), and Lewis and Paechter (2004) that the main objective

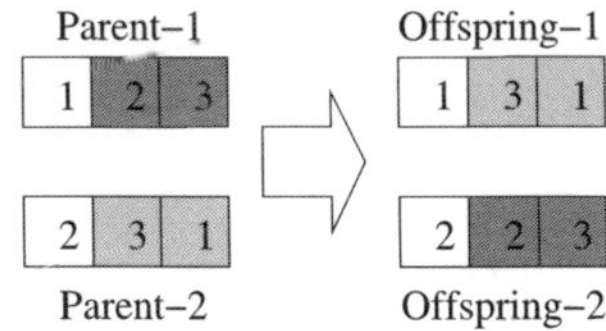

Fig. 1. Traditional 1-point crossover.

for this kind of problems should be to solve over hard constraints, and optimize over soft constraints, remaining in feasible search space only. For preserving the feasibility of solutions under crossover operators, Lewis and Paechter (2004) proposed four specialized and problem specific crossover operators for university class timetabling problem, namely sector-based crossover, day-based crossover, student-based crossover, and conflicts-based crossover operators. Earlier Burke et al. (1995) used a hybrid crossover operator, which they called *heuristic hybrid recombination operator*, for producing feasible offspring. However, they handled examination timetabling problems, for which they could use unlimited number of time-slots with the aim of reducing it by incorporating it into their objective function. They also proposed eight different selection heuristics for generating feasible offspring under crossover operator.

Since the traditional crossover operator, in scheduling problems, is likely to produce infeasible offspring, these problems need some computationally expensive repair operator to regain the feasibility. To eliminate such expensive repair operator, research has been aimed at designing suitable crossover operators that should be able to produce feasible offspring directly. Such a crossover operator, known as *Crossover for Valid Resource Allocation* (XVRA), has been developed by the current authors to allow valid (feasible) resources to events. It is similar with the sector-based or day-based crossover operator, proposed by Lewis and Paechter (2004). In XVRA, the first offspring is generated from the first parent assisted by the second parent. Similarly, the second offspring is generated from the second parent assisted by the first parent. A similar model was used by Okabe et al. (2005) also. Unlike in traditional crossover operator, where a complete offspring is generated from two parents, a portion of an offspring in XVRA is generated by the parents, and the remaining portion is completed by the heuristic approach (HA) of Sect.5. Though the sizes of the search spaces, exploited from the parents, get reduced, the size of the search space for introducing the remaining information by the HA is increased. In class timetabling problem, this reduces the possibility of getting stuck in scheduling complex classes, if any, where more than one room at a time-slot (for group classes), a room for consecutive time-slots (for multi-slot classes), or a specific time-slot/room may be required. In generating the first offspring using XVRA, few genes, $\{T_i\}$, are randomly selected from the first parent, and transferred the information in them to the corresponding genes of the offspring. Then few more genes, $\{T_j \mid T_j \neq T_i, \forall i, j\}$, are randomly selected from the second parent, and transferred the information in them, which are still required by the offspring, to the corresponding genes of the offspring. Finally, the HA is used to introduce the remaining information required by the offspring. Generation of the first offspring, using XVRA, is also shown in Fig.2 with the help of a small example. 10 classes (*C1-C10*) are to be scheduled at 10 time-slots (T1-T10) in 5 days with 2 time-slots per day, and 3 rooms (R1-R3) per time-slot. Time-slots T1-T2 and T5-T6 have been chosen randomly from the first parent, and classes at them have been transferred directly to the first offspring. Similarly, T3-T4 and T9-T10 have been chosen randomly from the sec-

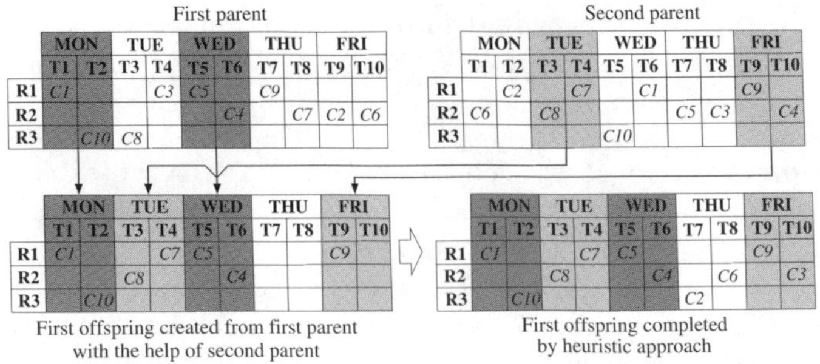

Fig. 2. Generation of the first offspring using XVRA.

ond parent. Since $C7$ and $C8$ of T3-T4 are still required by the offspring, they have also been transferred directly. Classes at T9-T10 are $C4$ and $C9$. Since $C4$ has already been scheduled at T6 of the offspring, only $C9$ has been transferred. $C2$, $C3$ and $C6$ could not be transferred from any of the parents. Hence, the HA has been applied to schedule those classes to the offspring. This completes the generation of the first offspring. Similarly, the second offspring can be generated from the second parent, assisted by the first parent. It is to be noted that no attempt has been made in any parent of Fig.2 to transfer the classes of T7-T8 to the offspring. As mentioned above, this has been done deliberately to increase the size of the search space for successfully scheduling the remaining classes by the HA. Though XVRA has been developed for generating feasible offspring, provision has been kept to accept infeasible offspring also, so that the operator would not require excessive guidance to generate an offspring. In that case, during transferring information from the parents to the offspring, or scheduling remaining classes by the HA, one or more hard constraints may be relaxed.

6.3 Mutation Operators (MFRA, MIRA, MRRA & MSIS)

In binary representation of chromosomes, the mutation operator simply alters the value of a randomly chosen gene from 0 to 1, or from 1 to 0. However, this technique is not applicable in any other form of chromosome representation, and some specific functions are required for such mutation operators. The most suitable function in case of class timetabling problem is to swap the classes of two slots, and/or to shift a class to an empty slot. Since it is tough to obtain feasible solutions for class timetabling problem, and crossover operators are likely to lose their feasibility, researchers generally put more attention in designing different mutation operators. The current authors also designed three mutation operators, namely MFRA, MIRA and MRRA, for mutating a solution. MFRA and MIRA are objective-specific, and perform, respectively, exploration and exploitation of a search space. While MRRA is for both exploring and exploiting a search space. Based on a problem in hand, these mutation operators can be tested to find which one would perform better in case of a particular objective function. Then the selected objective-specific operators can be applied simultaneously for optimizing all the objective functions together. Besides MFRA,

MIRA and MRRA, the authors designed one more mutation operator, namely MSIS. When any constraint is relaxed, MSIS is used for steering an infeasible solution towards the feasible region.

1. **Mutation for Fresh Resource Allocation (MFRA):** It explores a search space by assigning fresh resources to events. That is, only shifting of a class to a free slot is allowed in class timetabling problem. Preliminary observation reveals that it should work in reducing the number of free time-slots of a student (objective function f_1, given by (1)). A similar mutation operator was used by Carrasco and Pato (2001). However, MFRA may not work well, or even fail, in a problem where the number of overall free slots is very limited.
2. **Mutation for Interchanging Resource Allocation (MIRA):** It exploits a search space by interchanging the resources allotted to events. In case of class timetabling problem, it only swaps classes between two slots. MIRA is expected to reduce the number of consecutive classes of teachers (objective function f_2, given by (2)).
3. **Mutation for Reshuffling Resource Allocation (MRRA):** MRRA is just a combination of MFRA and MIRA, i.e., allotment of resources to events are reshuffled, where an event may be re-allotted new resources, or interchanged with that of another event. In case of class timetabling problem, it may swap classes at two slots, or shift a class to an empty slot. However, since the HA of Sect.5 schedules a class in a minimally suitable room, it is worthless to attempt here again to change the room of a class. Hence, MRRA may be allowed to change only the time-slot of a class. Since the operator is allowed to both exploring and exploiting a search space, it cannot be predicted if it gets biased to any specific objective function. In MRRA, two time-slots are first chosen randomly, and then the classes in each room at those time-slots are swapped one by one, provided no non-relaxed constraint is violated. If any multi-slot class appears in any room, the range of time-slots are expanded accordingly. Similarly, number of rooms to be handled at a time is expanded if any group class appears in any room. If the classes of any room cannot be swapped due to the violation of any non-relaxed constraint, attempt may be made for that room by considering another random pair of time-slots. The working procedure of MRRA is also shown in Fig.3 with an example. The problem contains the scheduling of 10 courses (*C1-C10*). *C8* has one 2-slot class per week. *C1* and *C2* are group courses, each having one single-slot class. All other courses have 2 single-slot classes in each. The randomly chosen time-slots are T3 and T9. Classes in room R1 at those time-slots are *C7* and *C10*, respectively (Fig.3(a)), which have been swapped without violating any constraint (Fig.3(b)). Since the class in R2 at T3 is *C8*, which is a 2-slot class, the range of T3 and T9 have been expanded to T4 and T10, respectively, and then the classes have been swapped. Because of the violation of the constraint under (6) (same class more than once a day), which would happen with *C3* on Tuesday if it were shifted from T9 to T3 in R3, another pair of time-slots, T1 and T8, have

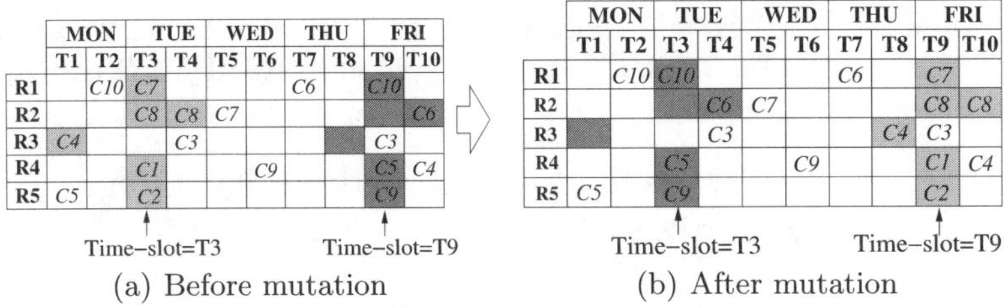

(a) Before mutation (b) After mutation

Fig. 3. Mutation for Reshuffling Resource Allocation (MRRA).

been chosen randomly. Then *C4* has been shifted from T1 to T8, at which R3 was empty. Since the group classes of *C1* and *C2* apprear at T3 in R4 and R5, respectively, R4 and R5 have been considered together. Then the classes of those rooms at T3 and T9 have been swapped. MRRA has been designed in such a way that it can preserve the feasibility of a solution during mutation. However, like in the case of XVRA, provision has been kept here also to accept an infeasible offspring by relaxing one or more hard constraints. Since MRRA changes the time-slots of classes, it is likely to work in taking care of student and/or teacher-clash constraints, given by (3) and (4), respectively, if these were relaxed. MRRA can be simplified easily to obtain the working procedure of the other two mutation operators, MFRA and MIRA.

4. **Mutation for Steering Infeasible Solution (MSIS):** MSIS has been designed to take care of violation of the relaxed constraints, if any. It is used only when there is some constraint violation, with an expectation of steering an infeasible solution towards the feasible region. MSIS is not explicitly a repairing mechanism, but mutation is performed only in those slots, for the classes of which any constraint is violated. Such a slot is first identified by an exhaustive search, and then the class(es) of the slot is(are) interchanged with that of another slot chosen randomly, provided no non-relaxed constraint is violated. An example of the working procedure of MSIS is shown in Fig.4. As shown in Fig.4(a), class *C1* was scheduled twice on Monday at time-slots

	MON		TUE		WED		THU		FRI	
	T1	T2	T3	T4	T5	T6	T7	T8	T9	T10
R1	C1			C7	C5		C2			
R2			C8			C4		C6	C8	
R3		C1							C9	

(a) Before mutation

	MON		TUE		WED		THU		FRI	
	T1	T2	T3	T4	T5	T6	T7	T8	T9	T10
R1	C2			C7	C5		C1			
R2			C8		C8	C4		C6		
R3		C1							C9	

(b) After mutation

Fig. 4. Steering infeasible solution towards feasible region using MSIS.

T1 and T2 in rooms R1 and R3, respectively, which violated the constraint under (6) (same class more than once a day). Hence, T7 has been chosen randomly where *C2* was scheduled in R1. Then *C1* and *C2* of R1 have been swapped (Fig.4(b)). Again, if both *C8* and *C9*, scheduled at T9 in R2 and R3, respectively, were to be attended by the same students, student-clash constraint under (3) would be violated. Hence, T5 has been chosen randomly where R2 was empty. Then *C8* has been shifted in R2 at T5 (Fig.4(b)).

7 NSGA-II-UCTO: NSGA-II as University Class Timetable Optimizer

Once the chromosome representation, and EA operators have been designed, an EA is required where these can be incorporated for optimizing a problem. Since multiple objectives are to be achieved in class timetabling problem, a multi-objective EA is required to tackle the problem. A number of such EAs, differing in one or more aspects from each other, have been developed, and implemented in different types of problems. Widely accepted among those are MOGA (Fonseca and Fleming 1993), NPGA (Horn et al. 1994), NSGA (Srinivas and Deb 1994), SPEA (Zitzler and Thiele 1999), and NSGA-II (Deb 2001; Deb et al. 2002). *Non-dominated Sorting Genetic Algorithm*-II (NSGA-II) has been selected by the current authors for optimizing university class timetabling problem, and it has been named as NSGA-II-UCTO (*NSGA-II as University Class Timetable Optimizer*). The reason for selecting NSGA-II among its counterparts is its successful implementation in a wider range of problems, which lead it to be crowned as the fast breaking paper in engineering in February, 2004 (THOMSON 2004). The salient features of NSGA-II-UCTO are as given below:

1. The chromosome representation, addressed in Sect.6.1, is used to form an EA population of N solutions.
2. The heuristic approach, addressed in Sect.5, is used to initialize the solutions of the population.
3. *Crowded Tournament Selection Operator* (Deb 2001) is used to form a *mating pool* (Deb 1995) of N solutions from the population. It is done by randomly selecting two solutions from the population, and sending a copy of the best one, based on ranks and crowding distances (Deb 2001), to the mating pool. The process is continued until the mating pool is filled up with N solutions. The mating pool is later used by EA operators for generating offspring.
4. The crossover operator XVRA, addressed in Sect.6.2, is used for generating a new population of N offspring.
5. One or more mutation operator(s) from MFRA, MIRA and MRRA, addressed in Sect.6.3, is(are) used for mutating the offspring of the new population.
6. Next, if any constraint violation is allowed, the mutation operator MSIS, addressed in Sect.6.3, is used for steering an infeasible offspring, if any, towards the feasible region.
7. Both the populations, obtained so far, are combined to form a combined population of 2N solutions.
8. Based on ranks and crowding distances, the best N solutions from the combined population are picked up to form a single population.
9. Steps (3)-(8) are repeated for required number of generations.
10. Result obtained after the required number of generations is accepted as the optimum result.

7.1 Capabilities of NSGA-II-UCTO

All the noticeable features of university class timetabling problem have been incorporated in NSGA-II-UCTO. It can be applied to any institute, particularly in India, with slight modification in objective and constraint functions, and/or input data files. The included provisions are:

1. The provisions, that can be availed through input data files only, are:
   (a) Compound courses, having multi-slot, combined and split classes. If not present, the number of any such course can simply be made zero.
   (b) Open courses, including group courses. Identities of students in a batch are not required, if no open course is offered to that batch.
   (c) Choices for specific rooms and time-slots for conducting classes.
   (d) Status of constraint functions, whether violation will be allowed or not.
2. Addition/deletion of any constraint can be made through subroutines, coded in C programming language.
3. It can also be applied to school timetabling problem with slight modification. Since the major requirements in this problem are to start classes from the very first time-slot of a day, and no free time-slots between two classes on a day, objective function f_1, given by (1), can be used as a hard constraint for the latter case. In addition to that, one more hard constraint can be defined to look after the starting of classes from the very first time-slot of a day. Then it can be handled as a single-objective (f_2) optimization problem, or one or more new objective function(s) can also be defined.

8 Application of NSGA-II-UCTO

NSGA-II-UCTO was tested on a number of virtual and real problems, and it was found successful in every case. Its application to two real problems from National Institute of Technology - Silchar (NIT-Silchar), a technical institute in India, will be addressed in detail in this section. Moreover, different features of EAs, such as attainment and summary attainment surfaces, and tuning of EA operators and parameters, will be presented through these problems.

In addition to Masters and PhD programmes, NIT-Silchar offers five under-graduate degree programmes in the disciplines (batches) of Civil, Mechanical, Electrical, Electronics and Computer Engineering. These under-graduate programmes are of four-year (eight-semester) duration, and the annual intake capacities are 50, 60, 45, 30 and 30, respectively. There are two sessions in a year: odd semester and even semester. Odd semester covers the classes of first, third, fifth and seventh semesters, while even-semester covers those of second, fourth, sixth and eighth semesters. To allow for change of discipline at the end of first year, the first and second semesters consist of common compulsory courses. Since the numbers of students in Electronics and Computer Engineering are comparatively less, these two disciplines are treated as one in these two semesters. The scheduling problems of odd and even semesters, for those five disciplines, were considered by the current authors to study using NSGA-II-UCTO. The problems have been named, in short, as NITS1 and NITS2, respectively. The reasons behind the selection of these two problems are their complex class-structure and high requirements. Moreover, as discussed in Section 4.4, NITS1 can be tackled by expressing (1) and (3) in terms of batches, instead of students. It will be shown shortly how drastically the number of constraints can be reduced from such replacement of students by batches. Both NITS1 and NITS2 contain all types of courses, addressed in Sect.3. Laboratory classes span over two to four consecutive time-slots. Most of the tutorial and laboratory classes are split into two sections. Moreover, many classes are either combined or grouped. Even many classes are to be taught by multiple teachers. On the other hand, many batches are fully packed over the week, and students of such batches hardly get 6 or 7 free time-slots in a week of total 35 time-slots. Moreover, there are very limited numbers of rooms and teachers for conducting the classes. In terms of the number of time-slots, weekly engagement of teachers varies from 3 to 18 in a week. There are total 522 classes in NITS1, and these span over 583 time-slots. Similarly, the number of classes in NITS2 is 506, spanning over 559 time-slots. Detail of classes in both of NITS1 and NITS2 are given, respectively, in Tables 3 and 4 (NITS 2005), where *Smp, Splt, Cmb* and *Opn* denote simple, split, combined and open classes, respectively. These classes are required to be scheduled in 5

Table 3. Classes in NITS1.

Classes	Number of Classes				# of
	Smp	Splt	Cmb	Opn	Slots
1-Slot	254	98	36	62	432
2-Slot	15	50	–	–	130
3-Slot	05	02	–	–	21
Total	**274**	**150**	**36**	**62**	**583**

Table 4. Classes in NITS2.

Classes	Number of Classes				# of
	Smp	Splt	Cmb	Opn	Slots
1-Slot	238	98	42	66	423
2-Slot	06	46	–	–	104
3-Slot	–	08	–	–	24
4-Slot	–	02	–	–	08
Total	**244**	**154**	**42**	**66**	**559**

days/week, where each day has 7 time-slots with a recess after the 4-th time-slot. Available resources (teachers and class-rooms) were modified slightly, from the actual one, to show the scheduling using optimum resources. The considered number of rooms (including laboratories) in NITS1 is 37, and that in NITS2 is 35. The numbers of teachers in NITS1 and NITS2 are 89 and 63, respectively. Hence, as per the formulation of Sect.4.4, the numbers of hard constraints in the problems are given in Table 5, where S, M, R, T, D, C and E represent the numbers of students, teachers, rooms, time-slots/day, days/week, courses, and classes, respectively. There are limited number of open courses in NITS1: only one or two set(s) of *departmental elective courses* in seventh semester. Classes of these courses are again grouped, and a student has to attend only one class of a set. Hence, by treating a set of group courses as a single course, the problem can be tackled as one without any open course. Hence, as mentioned in Sect.4, computational time for NITS1 can be reduced by expressing (1) and (3) in terms of

Table 5. Numbers of hard constraints in NITS1 and NITS2.

Problem	S	M	R	T	D	C	E	No. of Constraints (S+M+R)TD+CD+3E
NITS1	860	89	37	7	5	154	522	36,846
NITS2	860	63	35	7	5	143	506	35,763

batches, instead of students. In that case, replacing 860 students by 19 batches in Table 5, the number of hard constraints in NITS1 is reduced from 36,846 to 7,411. For the ease of computer programming, an original course with lecture, tutorial and laboratory classes, was divided into applicable sub-courses, for example, Physics lecture course, Physics tutorial course and Physics laboratory course. This was done because the structures of those classes vary from each other. In NSGA-II-UCTO, the HA needs two large numbers as the maximum numbers of attempts for two searches: one for finding suitable time-slot and room for a class, and the other for completing the initialization of a solution. These numbers are arbitrary. Based on the execution of a number of computational tests, the former one was taken equal to the total number available slots, and the latter one as 1000. If the HA fails to initialize a solution in those maximum numbers of attempts, the solution is replaced by already initialized one. In the crossover operator, initially attempt was made to transfer 40% of genes (randomly selected) from each parent to an offspring. In that case, when feasibility of a solution was required to be preserved, the HA frequently failed to complete offspring in the specified maximum number of attempts. Hence, only 20% of genes from each parent was transferred to an offspring. If the HA fails to complete an offspring in the specified maximum number of attempts, it is replaced by one of the best solutions from the previous population. Each mutation operator also needs a large number as the maximum number of attempts for finding a pair of time-slots, at which the classes of a room are to be swapped/shifted. This number was taken arbitrarily as the number of the total time-slots in a week.

When applied, NSGA-II-UCTO was found successful in every considered case of NITS1 and NITS2. The obtained results are presented here through one or more of the following plot(s):

1. Final non-dominated trade-off solutions (*Pareto Front*),
2. Computational times,
3. Comparison of results obtained from the use of different combinations of EA operators, and/or parameters,
4. Comparison of results obtained from single and multi-objective cases,
5. Attainment and summary attainment surfaces of Pareto fronts, etc.

8.1 Selection of Mutation Operators

As discussed in Sect.6.3, before going to solve a problem, it is required to find objective-specific mutation operator(s) which would best suit to the problem in hand. Then the selected operator(s) can be applied simultaneously in NSGA-II-UCTO for optimizing all the objective functions together. In this regard, the three mutation operators, MFRA, MIRA and MRRA, were applied to NITS2 as the only operator for generating new solutions. Each operator was applied, separately to each objective function, under single-objective optimization for different values of mutation probability (p_m) and random seed[11] (r_s). Three values of p_m and ten values of r_s were considered in the range of (0,1). Then each value of p_m, separately under all the ten values of r_s, was used to each of the three mutation operators. Single-objective optimization problems were solved using NSGA-II-UCTO only by keeping all objective functions, other than the required one, constant at zero. In each case, NSGA-II-UCTO was executed for 1000 generations with a population of 50 solutions, and then the means and standard deviations of the objective functions, f_1 and f_2, were computed. The obtained results for f_1 and f_2 are shown in Tables 6 and 7, respectively. f_1

Table 6. Performance of MFRA, MIRA and MRRA on f_1 of NITS2.

Random	MFRA			MIRA			MRRA		
Seed	**0.01**	**0.45**	**0.90**	**0.01**	**0.45**	**0.90**	**0.01**	**0.45**	**0.90**
0.010	3.931	3.846	3.280	5.137	5.074	5.043	4.060	3.634	3.502
0.050	4.520	3.971	3.003	5.095	5.134	5.127	4.260	3.683	3.737
0.125	4.681	4.135	3.673	5.178	5.178	5.178	4.751	3.667	3.270
0.232	4.500	3.806	3.613	5.310	5.137	5.110	4.796	3.700	3.663
0.345	4.538	3.752	3.822	4.592	4.370	4.463	4.371	3.696	3.663
0.463	4.091	3.606	3.462	4.844	4.844	4.758	4.380	3.398	3.560
0.587	3.880	3.328	2.857	5.123	4.928	5.074	4.594	3.456	2.972
0.712	4.620	3.920	3.317	5.320	5.241	5.206	4.186	3.308	2.836
0.852	4.150	3.934	3.862	5.271	5.076	5.033	4.465	4.043	3.494
0.999	4.733	3.945	3.499	5.251	5.230	4.902	4.720	3.776	3.145
Mean f_1	**4.364**	**3.824**	**3.439**	**5.112**	**5.021**	**4.989**	**4.458**	**3.636**	**3.384**
Std. Dev.	**0.303**	**0.213**	**0.313**	**0.218**	**0.247**	**0.216**	**0.239**	**0.198**	**0.296**

and f_2, along with their means, under each case are also shown in Fig.5(a) and Fig.5(b), respectively. Means are shown in the figures by filled-circles. Though it was expected that MFRA would perform better on f_1, and MIRA on f_2, the obtained results revealed that the improvement of both f_1 and f_2 were much better under MRRA only. Hence, MRRA was accepted as the best one among all the three mutation operators. The performance of MIRA was the worst, while that of MFRA was comparable with that of MRRA in many cases. Moreover, MFRA slightly outperformed MRRA in case of f_1 under $p_m = 0.01$. The best performance of MRRA might be due to the fact that it explores, as well as

[11] A *random seed* is a user-specified random number in the range of (0,1). In NSGA-II-UCTO, it is used for generating random integers as the serial numbers of class-rooms and time-slots where classes are to be scheduled.

Table 7. Performance of MFRA, MIRA and MRRA on f_2 of NITS2.

Random Seed	MFRA			MIRA			MRRA		
	0.01	**0.45**	**0.90**	**0.01**	**0.45**	**0.90**	**0.01**	**0.45**	**0.90**
0.010	0.905	0.825	0.778	1.095	1.048	1.032	0.825	0.714	0.762
0.050	0.873	0.651	0.730	0.936	0.952	0.905	0.841	0.635	0.667
0.125	0.952	0.825	0.778	1.048	1.048	0.984	0.873	0.778	0.714
0.232	1.000	0.873	0.794	1.048	1.032	0.921	0.921	0.809	0.746
0.345	0.968	0.921	0.841	1.143	0.984	0.984	0.936	0.809	0.746
0.463	0.857	0.603	0.667	0.984	0.936	0.968	0.905	0.651	0.603
0.587	0.809	0.794	0.730	1.095	0.984	0.952	0.809	0.682	0.746
0.712	0.905	0.794	0.714	1.016	1.000	0.968	0.825	0.651	0.651
0.852	0.921	0.794	0.682	1.032	1.063	1.032	0.825	0.746	0.762
0.999	0.889	0.841	0.714	1.079	1.016	1.000	0.952	0.809	0.682
Mean f_2	**0.908**	**0.792**	**0.743**	**1.048**	**1.006**	**0.975**	**0.871**	**0.729**	**0.708**
Std. Dev.	**0.053**	**0.091**	**0.051**	**0.057**	**0.040**	**0.040**	**0.050**	**0.067**	**0.052**

(a) Performance on f_1

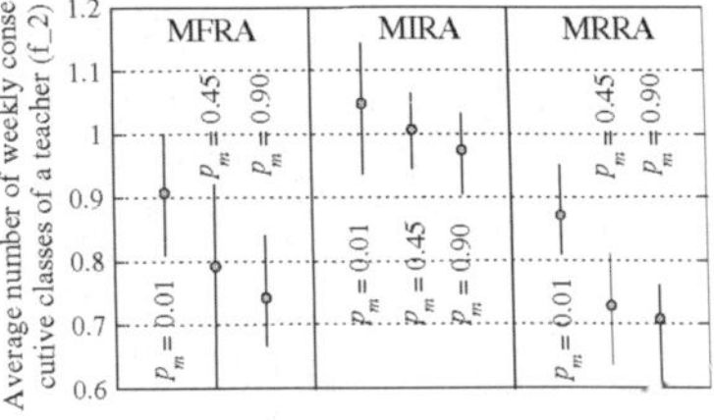

(b) Performance on f_2

Fig. 5. Performance of MFRA, MIRA and MRRA on f_1 and f_2 of NITS2.

exploits, the search space, whereas other two operators perform either of these two searches. Two more important observations from this experiment are that

1. NSGA-II-UCTO depends on user-defined mutation probabilities. This is clear from any row of Tables 6 and 7, which shows the objective functions for different mutation probabilities under the same random seed.
2. NSGA-II-UCTO also depends on random seeds (in other words, initial solutions). This can be observed from any column of Tables 6 and 7, which shows the objective functions for different random seeds under the same mutation probability.

To overcome the dependency of NSGA-II-UCTO on user-defined p_m, evolving or self-adapting p_m (Anastasoff 1999; Smith and Fogarty 1996) can be adopted. This is well acceptable as wide evidence of the evolution of mutation rates is found in nature also (Lund 1994). In uncertain and changing environments, the evolved mutation rates must be able to adapt to new situations. In EA problems, therefore, it is not required to fix p_m for the whole evolution process, but can be left free to evolve itself. In NSGA-II-UCTO, evolving p_m have been encoded genotypically in the chromosomes. Each chromosome contains an additional gene as its p_m. Initially, p_m is assigned a random number in the range of (0,1), and thereafter, allowed to evolve to p'_m. p_m is evolved to p'_m using *polynomial mutation of real number* (Deb 2001) with 100% probability. If a chromosome gets mutated at p'_m, then p'_m is stored as p_m in the chromosome. Since NSGA-II-UCTO is also dependent on initial solutions, a problem may be tested under different values of r_s, and then the best solution can be accepted from the multiple alternatives.

8.2 NITS1 and NITS2 with Relaxed Hard Constraints

As the first case, NITS1 and NITS2 were considered with relaxed batch/student and teacher-clash constraints, given by (3) and (4), respectively. The crossover probability (p_c), mutation probability (p_m) and random seed (r_s) were chosen as 0.90, 0.01 and 0.125, respectively. Then NSGA-II-UCTO was executed with 50 solutions in the population. In case of NITS1, though all the initial solutions were infeasible, the population was filled by feasible solutions just in 285 generations. However, the whole population at that stage was filled by the copies of a single solution with $f_1 = 8.015$ and $f_2 = 0.640$. After that, NSGA-II-UCTO was unable to generate any new solution till 2000 generations. Hence, the execution was stopped. The solutions, obtained so far, are shown in Fig.6(a). In case of NITS2

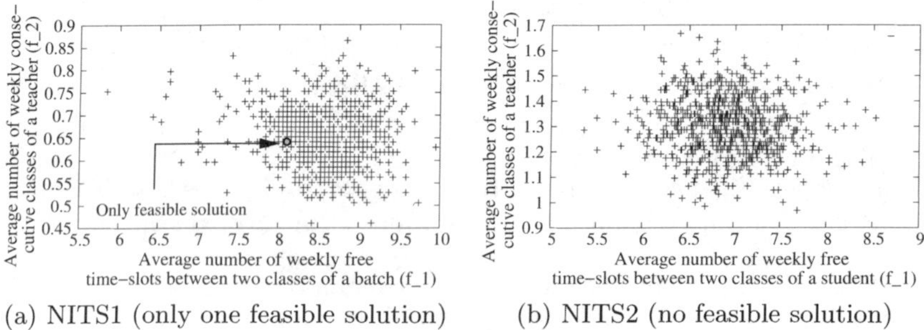

(a) NITS1 (only one feasible solution) (b) NITS2 (no feasible solution)

Fig. 6. NITS1 and NITS2 with relaxed student/batch and teacher-clash constraints ($p_c = 0.90$, $p_m = 0.01$ and $r_s = 0.125$).

also, the HA could not initialize any feasible solution. NSGA-II-UCTO was also unable to produce even a single feasible solution in 5000 generations. However, some improvements in student and teacher-clash constraints were noticed during the optimization process. Hence, the possibility of obtaining feasible solutions can not be denied if the execution was continued for more generations. However, execution of NSGA-II-UCTO up to 5000 generations itself was very huge (which took 40 hours 47 minutes 7 seconds in Linux environment in a Pentium IV machine with 1.0 GB RAM and 2.933 GHz processor). Hence, its execution over that, without any guarantee, would be computationally uneconomic. The solutions, obtained till 5000 generations, are shown in Fig.6(b).

8.3 NITS1 and NITS2 Maintaining Feasibility of Solutions

Being not succeeded by relaxing constraints, NITS1 and NITS2 were solved again maintaining the feasibility of solutions. The EA parameters were kept the same as in Sect.8.2, i.e., $p_c = 0.90$, $p_m = 0.01$ and $r_s = 0.125$. The attempt was successfull at this time. The obtained results of NITS1, after 5000 generations, are shown in Fig.7. Though the attempt was successful, it is observed

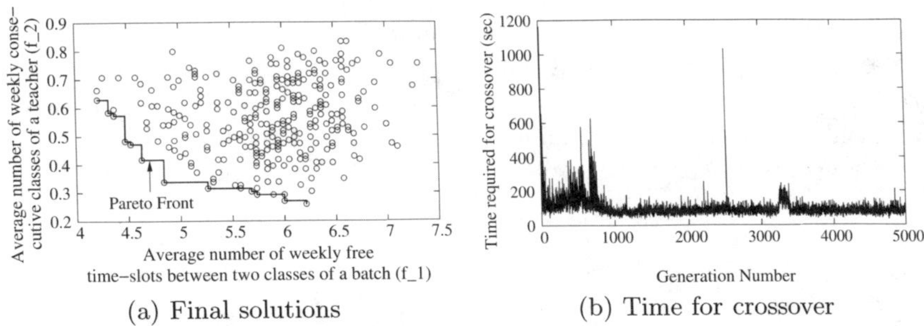

(a) Final solutions (b) Time for crossover

Fig. 7. Optimized solutions of NITS1 ($p_c = 0.90$, $p_m = 0.01$ and $r_s = 0.125$).

from Fig.7(b) that the success was obtained at the cost of huge computational time, required by the crossover operator. Total execution time was 136 hours 52 minutes 42 seconds. The HA took 6 minutes 53 seconds to initialize the 50 solutions of the population. Each function evaluation took only around 1 second. One mutation was completed in less than 1 second. Total data printing time was 7 minutes 15 seconds. Rest of the computational time was required in crossover, most of which was exhausted by the HA in completing the offspring. The same situation was arisen in case of NITS2 also, results of which are shown in Fig.8. Total execution time was 19 hours 59 minutes 36 seconds. As shown in Fig.8(b),

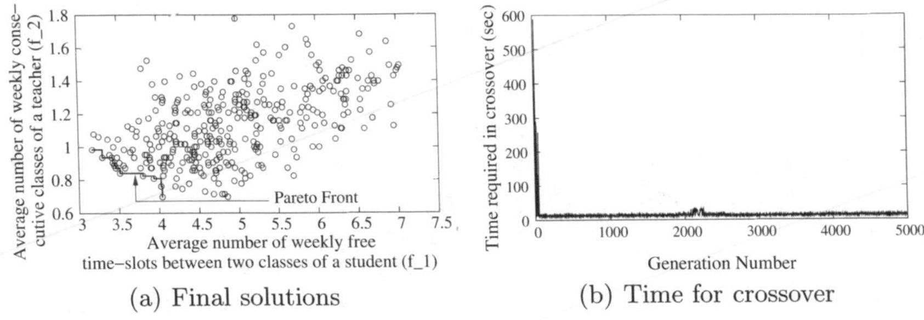

(a) Final solutions (b) Time for crossover

Fig. 8. Optimized solutions of NITS2 ($p_c = 0.90$, $p_m = 0.01$ and $r_s = 0.125$).

here also most of the execution time was required by the crossover operator.

Initially it took huge time, 11-588 seconds per crossover till 49 generations, and thereafter, it was stable at 9-34 seconds till the completion of the execution. The reason of taking excess time till 49 generations was that the HA frequently failed to complete the offspring, and hence, it was required to make the specified maximum number of attempts before replacing an incomplete offspring by one of the best solutions from the previous generation.

8.4 NITS1 and NITS2 Using Only Mutation Operator (MRRA)

Since the crossover operator (XVRA) was found too expensive in terms of computational time, an experiment was made using only mutation operator (MRRA). Since the whole responsibility for generating new solutions was imposed on MRRA only, p_m was set very high at 0.90. r_s was kept the same, i.e., $r_s = 0.125$. The obtained results of NITS1 are shown in Fig.9. Interesting enough, the total

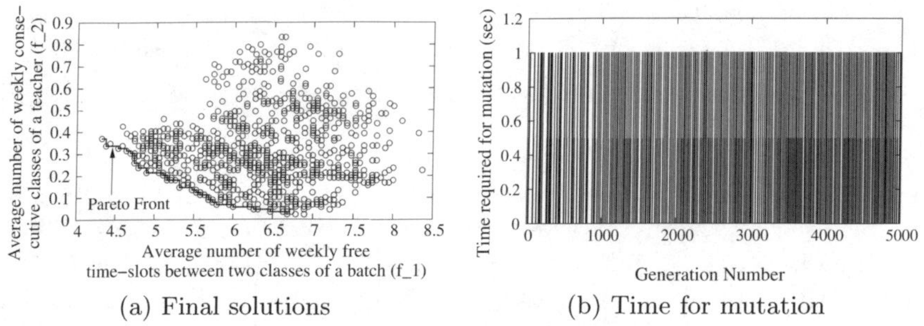

(a) Final solutions (b) Time for mutation

Fig. 9. Optimized solutions of NITS1 using only MRRA ($p_m = 0.90$ and $r_s = 0.125$).

execution time in this case came down to 1 hour 19 minutes 17 seconds only. The number of search points was also increased from 284 to 926. Moreover, much better result was obtained than that in the previous case, shown in Fig.7, i.e., only MRRA performed better than that of combined XVRA and MRRA. This was a very good outcome, which needed further confirmation whether it is a general case or just for the particular case in hand. Since it was revealed in Sect.8.1 that NSGA-II-UCTO depends on both user-defined p_m and r_s, multiple runs under different p_m and r_s were required for this confirmation. However, instead of taking few selective p_m, which would again depend on the choice of an individual user, the test was performed under evolving p_m. In the experiment in hand, the *polynomial probability distribution index* (Deb 2001) for mutating the evolving p_m was set at 30. Then NITS1 was solved again under nine different values of r_s, and the use of only MRRA was found outperforming the use of combined XVRA and MRRA in all the nine cases. The compared Pareto fronts are shown in Fig.10. Hence, it can be concluded that the mutation operator (MRRA) alone performs well on NITS1 than that of the combined crossover (XVRA) and mutation (MRRA) operators.

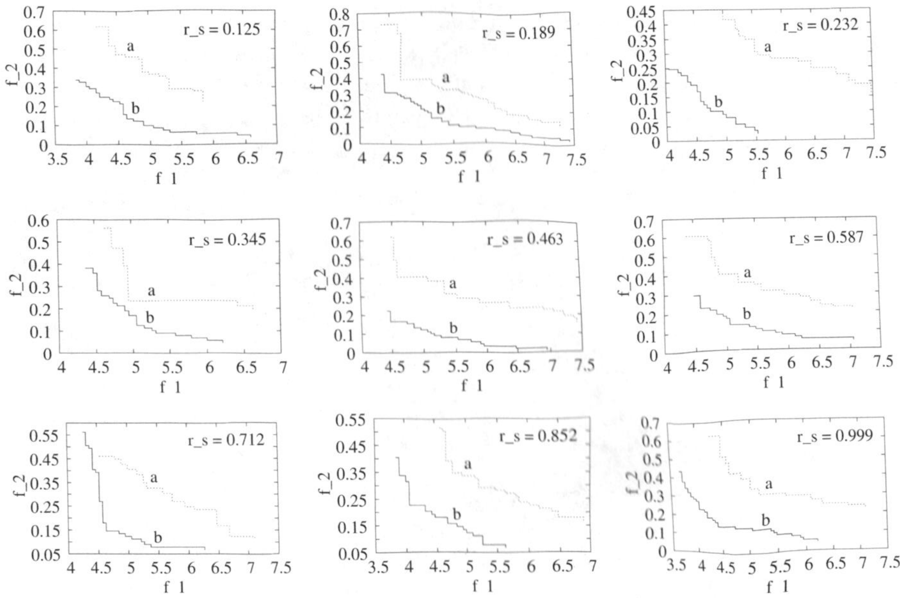

Fig. 10. Comparison of Pareto fronts of NITS1, obtained under different initial solutions (r_s) and evolving p_m. Curve (a): combined XVRA ($p_c = 0.90$) and MRRA, and curve (b): only MRRA.

The results of NITS2, when solved using only MRRA, are shown in Fig. 11. The total execution time in this case was only 14 hours 23 minutes 30 seconds,

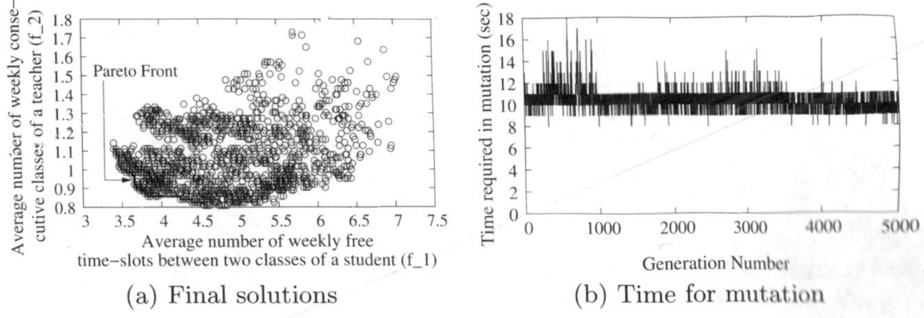

(a) Final solutions (b) Time for mutation

Fig. 11. Optimized solutions of NITS2 using only MRRA ($p_m = 0.90$ and $r_s = 0.125$).

out of which each mutation took 8-18 seconds. That is, in NITS2 also, the time required by the use of only MRRA is less than that required by the use of combined XVRA and MRRA (Fig. 8). The number of search points was also increased from 357 to 1228. However, unlike in the case of NITS1, the solution quality of NITS2 was better under the combined XVRA and MRRA. Before making any conclusion on this outcome, here also multiple runs were required to check whether it is a general case or just a particular one. In this regard, NITS2 was also solved nine times with evolving p_m under different values of r_s. The compared Pareto fronts are shown in Fig.12. It is observed in Fig.12 that the

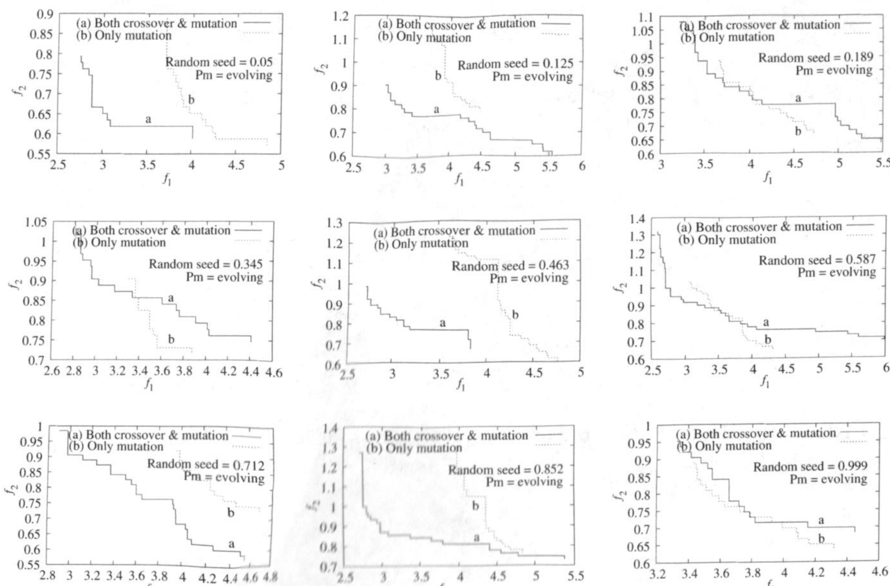

Fig. 12. Comparison of Pareto fronts of NITS2, obtained under different initial solutions (r_s) and evolving p_m. Curve (a): combined XVRA ($p_c = 0.90$) and MRRA, and curve (b): only MRRA.

performance of only MRRA was comparable with that of combined XVRA and MRRA. However, the overall performance of the latter was much better than the former one. Hence, it can be concluded that, unlike in the case of NITS1, the combined crossover (XVRA) and mutation (MRRA) operators perform well on NITS2 than that of the mutation operator (MRRA) alone.

8.5 *Attainment* and Summary Attainment Surfaces (AS & SAS)

When evaluating the performance of a stochastic optimizer, it is sometimes desirable to express the performance in terms of the quality attained in a certain number of sample runs. In multi-objective evolutionary optimization, the outcome of a run is measured as an *attainment surface* (AS) in k-dimensional space, where k is the number of objective functions. A surface formed by joining the points, representing the non-dominated solutions, is known as an attainment surface (e.g., the Pareto front of Fig.7(a)). It divides the objective space into two regions: one is dominated, and the other is not dominated by the obtained solutions (Fonseca and Fleming 1996; Knowles 2005). Multiple attainment surfaces, obtained from multiple runs of an optimizer, can be superimposed, and interpreted probabilistically for visualizing the performance of a run of the optimizer. For this, the region, bounded by all the attainment surfaces, is divided by equal number of non-crossing surfaces, known as *summary attainment surfaces* (SAS) (Knowles 2005). SAS are denoted in percentage, and indicate the performance-level of a solution. For example, the 50% SAS identifies the region of objective space, which is dominated by half of the given SAS. Similarly, the 0% SAS identifies the region not dominated by any given SAS, whereas the 100% SAS identifies the region dominated by every given SAS. Thus, the 0% SAS visualizes the best-case performance, while the 100% SAS visualizes the worst-case performance. An experiment was performed on NITS1 and NITS2 for obtaining such AS and SAS. Since NSGA-II-UCTO performed well on NITS1 using only MRRA, the AS and SAS for the problem were also obtained using only MRRA. Similarly, the AS and SAS for NITS2 were obtained using the combined XVRA and MRRA, since they showed better performance on the problem than that of MRRA alone. It can be observed from the plots, shown in Sect.8.2 to 8.4, that the ranges of the objective functions (f_1 and f_2) are different, for which the progress of one objective function is incomparable with the other one. Hence, these were normalized in [0,1] in AS and SAS, using the following relations:

$$\bar{f}_{1,i} = \frac{f_{1,i} - f_1^{\min}}{f_1^{\max} - f_1^{\min}} \quad \text{and} \quad \bar{f}_{2,i} = \frac{f_{2,i} - f_2^{\min}}{f_2^{\max} - f_2^{\min}} \,, \tag{13}$$

where $f_j^{\min}$ and $f_j^{\max}$ $(j=1,2)$ are, respectively, the minimum and maximum values of f_j in the combined solutions of Pareto fronts, whose AS and SAS are required. $\bar{f}_{1,i}$ and $\bar{f}_{2,i}$ are, respectively, normalized values of $f_{1,i}$ and $f_{2,i}$ in those solutions.

NITS1 was solved separately with $p_m = 0.90$ and evolving p_m, under eleven different values of r_s in each case. The obtained AS and SAS for both the cases are shown in Fig.13 and Fig.14, respectively. It is observed from the AS of Fig.13(a) and Fig.14(a) that the Pareto fronts followed a common pattern in all the cases. Theoretically, all fronts had to be identical. However, the slight variation on the actual fronts depicts the dependency of NSGA-II-UCTO on various EA parameters, such as p_m or r_s. This is clear from the AS and SAS of Fig.14 with evolving p_m, which are more uniform than those with fixed p_m, shown in Fig.13. The comparison of 0%, 50% and 100% SAS from both the cases are also shown in Fig.14(c). Another important observation from these comparison of SAS is that solutions with evolving p_m are more diverged than those with fixed p_m.

Two cases were considered for obtaining the AS and SAS for NITS2 also: one with $p_c = 0.90$ and $p_m = 0.010$, and the other with $p_c = 0.90$ and evolving p_m. In each case, the problem was solved under nine different values of r_s. The obtained results for both the cases are shown in Fig.15 and Fig.16, respectively. The comparison of 0%, 50% and 100% SAS from both the cases are also shown

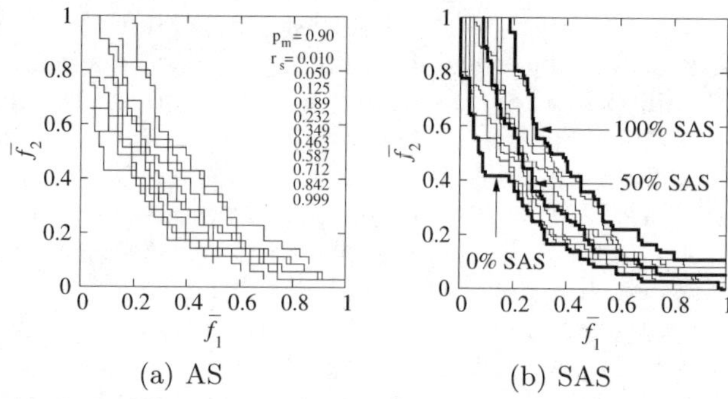

Fig. 13. Attainment and Summary Attainment Surfaces for NITS1 using only MRRA with $p_m = 0.90$.

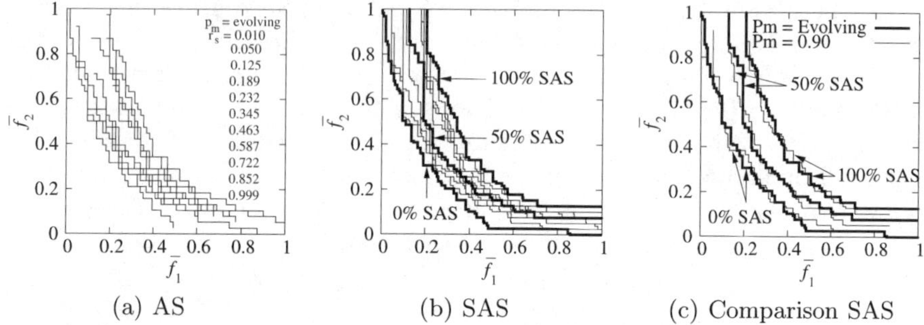

Fig. 14. Attainment and Summary Attainment Surfaces for NITS1 using only MRRA with evolving p_m, and comparison of SAS with those shown in Fig.13.

in Fig.16(c), where it is observed that significant improvements in SAS were obtained from the use of evolving p_m. However, the solutions were more diverged under fixed p_m.

8.6 Comparison of Results of NITS1 from Single and Multi-Objective Optimization

Presently timetables of NITS1 and NITS2 are prepared manually by *trial and hit* method, where a feasible solution is never achieved. As a result, a few teacher-clash is always allowed in the timetables, and then it becomes the responsibility of a teacher to run to others to adjust his/her clashed-classes. Hence, the multi-objective results, produced by NSGA-II-UCTO, could not be compared with any existing solution. Therefore, it was tested how much the multi-objective results of NITS1 could be supported by those of single-objective optimization. It is seen in literature that most of the earlier works on class timetabling problem were based on single-objective optimization. Hence, the results of single-objective optimization of NITS1 were chosen in the present work to compare its results of multi-objective optimization. The comparison was made by considering two cases of NITS1: one with $p_m = 0.90$ and $r_s = 0.125$, and the other

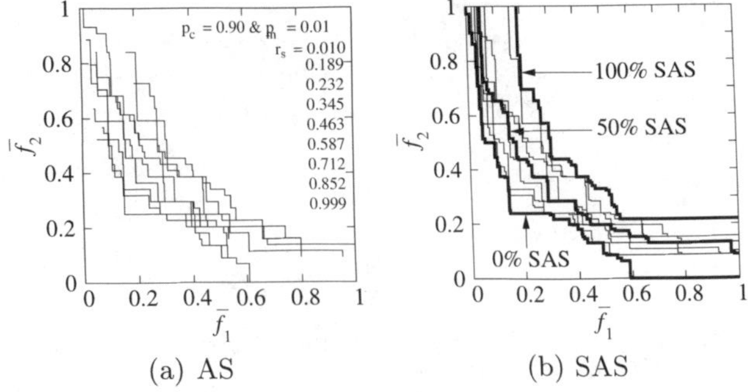

(a) AS (b) SAS

Fig. 15. Attainment and Summary Attainment Surfaces for NITS2 using combined XVRA and MRRA with $p_c = 0.90$ and $p_m = 0.010$.

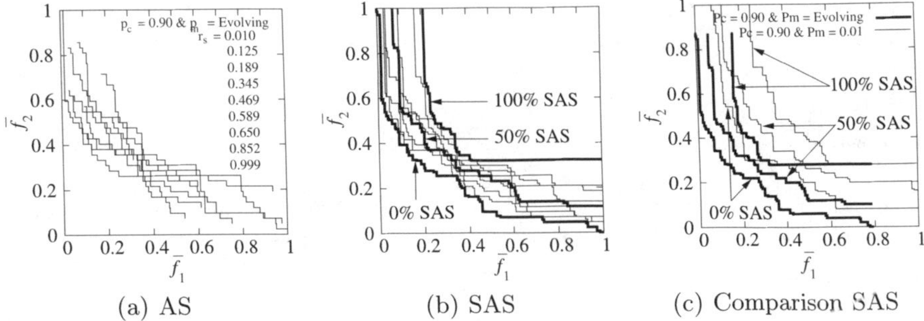

(a) AS (b) SAS (c) Comparison SAS

Fig. 16. Attainment and Summary Attainment Surfaces for NITS2 using XVRA and MRRA with $p_c = 0.90$ and evolving p_m, and comparison of SAS with those shown in Fig.15.

with evolving p_m and $r_s = 0.125$. In each case, NITS1 was solved thrice, under single-objective optimization, for three different objective functions of f_1, f_2 and $(f_1 + f_2)$. As mentioned in Sect. 8.1, single-objective problems were solved using NSGA-II-UCTO only by keeping all objective functions, other than the required one, constant at zero. The obtained results are shown in Figure 17. Points A, B and C, respectively, represent the independently minimized f_1, f_2 and $(f_1 + f_2)$ with fixed $p_m = 0.90$, and A', B' and C' represent those with evolving p_m. Curve D is the Pareto front obtained from multi-objective optimization with fixed $p_m = 0.90$, and curve D' is that with evolving p_m. As seen in the figure, in case of the fixed p_m, point A was marginally better than any point on the curve D, while point B was dominated by the curve. Point C, which indirectly optimized both f_1 and f_2, was also dominated by the curve D. Since point C was obtained by *weighted-sum method* (Deb 2001) with equal weightage to both f_1 and f_2, possibly it could be improved by properly tuning the weightages of the objectives. On the other hand, the results of single-objective optimization, under evolving p_m, were quite different. A' and C' were deteriorated from their earlier values (A and C, respectively), while some improvement was noticed in B'. However, no significant change was observed in the curve D', obtained from

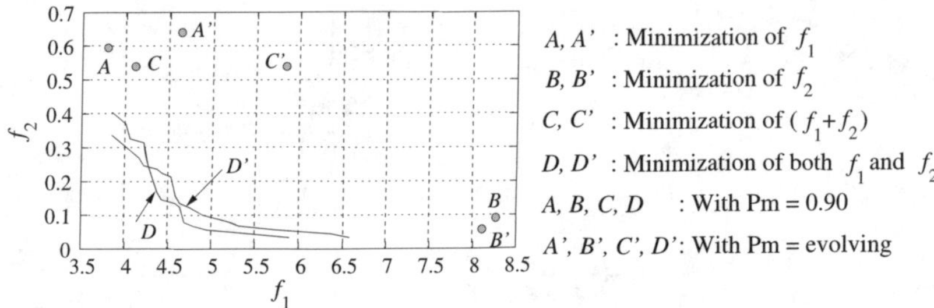

Fig. 17. Optimized solutions of NITS1 under single-objective optimization (Points A, B and C, and A', B' and C'), and multi-objective optimization (Curves D and D'). A, B, C and D are with $p_m = 0.90$ and $r_s = 0.125$, and A', B', C' and D' are with evolving p_m under $r_s = 0.125$.

multi-objective optimization under evolving p_m. Combining all the results, it is seen that only point A, under single-objective optimization with fixed p_m, was slightly better than any point of multi-objective optimization. The value of f_1 at A was 3.789, whereas its minimum value in both of D and D' was equal to 3.842.

It can be concluded from this experiment that, in case of class timetabling problem, better results can be obtained from multi-objective optimization over any single-objective optimization.

9 Conclusion

Though the academic class timetabling problem is being studied for more than four decades, a general solution technique for it, considering different aspects of its variants, is yet to be formulated. Despite multiple criteria to be met simultaneously, the problem is generally tackled as single-objective optimization problem. Moreover, most of the earlier works were concentrated on school timetabling, and only a few on university class timetabling. On the other hand, in many cases, the problem was over-simplified by skipping many complex class-structures. Hence, NSGA-II-UCTO has been developed by the current authors, as an attempt to overcome those shortfalls, with the following salient features:

1. It is a multi-objective optimizer,
2. It is directly applicable to university class timetabling problem. It can also be applied to school timetabling with a little modification,
3. Different types of classes can be handled through input datafiles only,
4. Choices for rooms and time-slots for conducting classes, and status of constraint handling, can also be made through input datafiles only, and
5. Addition/deletion of any constraint can be made through subroutines, coded in C programming language.

NSGA-II-UCTO has been applied to two highly constrained real problems from a technical institute in India, and it has been found successful under different combinations of EA operators and parameters. However, it has been observed that NSGA-II-UCTO is dependent on user-defined mutation probabilities, as well as on initial solutions.

1. Dependency of NSGA-II-UCTO on user-defined mutation probabilities could be sorted out to some extent by using evolving/self-adapted mutation probabilities.
2. Since it depends on initial solutions also, multiple runs may be performed with different initial solutions, and then the best solution can be captured from the multiple alternatives.

On the other hand, out of the considered two problems, NSGA-II-UCTO has been found performing well on one with mutation operator alone, while on the other with combined crossover and mutation operators. Hence, though the use of crossover operator has been found to be expensive in terms of computational time, a problem may be tested under both the cases, if time is not a big factor.

Acknowledgement: This work was partially supported by an Indo-Portuguese scientific and technical cooperation grant from DST, India, and GRICES, Portugal.

References

Abramson, D.: Constructing school timetables using simulated annealing: sequential and parallel algorithms. Management Science **37(1)** (1991) 98–113

Abramson, D., Abela, J.: A parallel genetic algorithm for solving the school timetabling problem. In Proceedings of 15 Australian Computer Science Conference, Hobart, (1992) 1–11

Akkoyunlu, E. A.: A linear algorithm for computing the optimum university timetable. The Computer Journal **16(4)** (1973) 347–350

Al-Attar, A.: White Paper: A hybrid GA-heuristic search strategy. AI Expert, USA (1994)

Anastasoff, S. J.: Evolving mutation rates for the self-optimisation of genetic algorithms. Lecture Notes in Computer Science, Springer-Verlag, London **1674** (1999) 74–78

Blum, C., Correia, S., Dorigo, M., Paechter, B., Rossi-Doria, O., Snoek, M.: A GA evolving instructions for a timetable builder. In Proceedings of the Practice and Theory of Automated Timetabling (PATAT) (2002) 120–123

Bufé, M., Fischer, T., Gubbels, H., Häcker, C., Hasprich, O., Scheibel, C., Weicker, K., Weiker, N., Wenig, M., Wolfangel, C.: Automated solution of a highly constrained school timetabling problem - preliminary results. EvoWorkshops-2001, Como, Italy (2001) 431–440

Burke, E., Elliman, D., Weare, R.: Specialised recombinative operators for timetabling problems. In Proceedings of the AISB (AI and Simulated Behaviour) Workshop on Evolutionary Computing (1995) 75–85

Carrasco, M. P., Pato, M. V.: A multiobjective genetic algorithm for the class/teacher timetabling problem. In Proceedings of the Practice and Theory of Automated Timetabling (PATAT-2000), Lecture Notes In Computer Science, Springer **2079** (2001) 3–17

Carrasco, M. P., Pato, M. V.: A comparison of discrete and continuous neural network approaches to solve the class/teacher timetabling problem. European Journal of Operational Research **153(1)** (2004) 65–79

Colorni, A., Dorigo, M., Maniezzo, V.: Genetic algorithms and highly constrained problems: The time-table case. In Proceedings of the first International Workshop on Parallel Problem Solving from Nature (PPSN-1, 1990), Lecture Notes in Computer Science (1991), Springer **496** (1990) 55–59

Colorni, A., Dorigo, M., Maniezzo, V.: A genetic algorithm to solve the timetable problem. Tech. rep. 90-060 revised, Politecnico di Milano, Italy (1992)

Cooper, T. B., Kingston, J. H.: The complexity of timetable construction problems. In Proceedings of Practice and Theory of Automated Timetabling (PATAT-95), Lecture Notes in Computer Science (1996), Springer-Verlag **1153** (1995) 283–295

Corne, D., Ross, P., Fang, H-L.: Fast practical evolutionary timetabling. Lecture Notes in Computer Science 865, Springer-Verlag (Evolutionary Computing AISB Workshop, Leeds, UK) (1994) 251–263

Costa, D.: A tabu search algorithm for computing an operational timetable. European Journal of Operational Research **76(1)** (1994) 98–110

Daskalaki, S., Birbas, T., Housos, E.: An integer programming formulation for a case study in university timetabling. European Journal of Operational Research, **153** (2004) 117–135

Datta, D., Deb, K.: Design of optimum cross-sections for load-carrying members using multi-objective evolutionary algorithms. In Proceedings of International Conference on Systemics, Cybernetics and Informatics (ICSCI), Hyderabad, India **1** (2005) 571–577

de Werra, D.: Construction of school timetables by flow methods. INFOR - Canadian Journal of Operations Research and Information Processing **9** (1971) 12–22

Deb, K.: Optimization for Engineering Design-Algorithms and Examples. Prentice-Hall of India Pvt. Ltd., New Delhi, India (1995)

Deb, K.: Multi-Objective Optimization using Evolutionary Algorithms. John Wiley & Sons Ltd, Chichester, England (2001)

Deb, K., Agarwal, S., Pratap, A., Meyarivan, T.: A fast and elitist multi-objective genetic algorithm: NSGA-II. IEEE Transactions on Evolutionary Computation **6(2)** (2002) 182–197

Desef, T., Bortfeldt, A., Gehring, H.: A tabu search algorithm for solving the timetabling problem for German primary schools (Abstract). In Proceedings of the Practice and Theory of Automated Timetabling (PATAT) (2004) 465–469

Even, S., Itai, A., Shamir, A.: On the complexity of timetable and multicommodity flow problems. SIAM Journal of Computation **5(4)** (1976) 691–703

Fang, H-L.: Genetic algorithms in timetabling and scheduling. PhD Thesis, Department of Artificial Intelligence, University of Edinburgh (1994)

Filho, G. R., Lorena, L. A. N.: A constructive evolutionary approach to school timetabling. In Proceedings of First European Workshop on Evolutionary Computation in Combinatorial Optimization (EvoCOP-2001) (2001) 130–139

Fonseca, C. M., Fleming, P. J.: Genetic Algorithms for Multiobjective Optimisation: Formulation, discussion and generalization. In Proceedings of the fifth International Conference on Genetic Algorithms. S. Forrest, ed. Morgan Kaufmann, San Mateo (1993) 416–423

Fonseca, C. M., Fleming, P. J.: On the performance assessment and comparison of stochastic multiobjective optimizers. In Proceedings of 4th International Conference on Parallel Problem Solving from Nature (PPSN)-IV, Lecture Notes in Computer Science, Springer-Verlag (1996) 584–593

Fujino, K.: A preparation for the timetable using random number. Information processing in Japan **5** (1965) 8–15

Gaspero, L. D., Schaerf, A.: Multi-neighbourhood local search for course timetabling. In Proceedings of the Practice and Theory of Automated Timetabling (PATAT) (2002) 128–132

Goldberg, D. E.: Genetic Algorithms in Search, Optimization, and Machine Learning. Addison-Wesley (1989)

Gotlieb, C. C.: The construction of class-teacher timetables. In Proceedings of IFIP Congress, North-Holland Pub. Co., Amsterdam (1962) 73–77

Greko, B.: School scheduling through capacitated network flow analysis. Swed. Off. Org. Man., Stockholm (1965)

Horn, J., Nafpliotis, N., Goldberg, D. E.: A Niched Pareto Genetic Algorithm for Multiobjective Optimization. In Zbigniew Michalewicz (ed.): Proceedings of the first IEEE Conference on Evolutionary Computation **1** (1994) 82–87

Knowles, J.: A summary-attainment-surface plotting method for visualizing the performance of stochastic multiobjective optimizers. IEEE Intelligent Systems Design and Applications (ISDA-2005) (2005) 552–557

Lawrie, N.: An integer programming model of a school timetabling problem. The Computer Journal **12** (1969) 307–316

Lewis, R., Paechter, B.: New crossover operators for timetabling with evolutionary algorithms. In A. Lofti (Ed.) 5th International Conference on Recent Advances in Soft Computing (RASC) **5** (2004) 189–195

Lima, M. D., de Noronha, M. F., Pacheco, M. A. C., Vellasco, M. M. R.: Class scheduling through genetic algorithms. IV Workshop do Sistema Brasileiro de Technologia de Informação (SIBRATI), Poli/USP-Säo Paulo (2001)

Looi, C.: Neural network methods in combinatorial optimization. Computers and Operations Research **19(3/4)** (1992) 191–208

Lund, H. H.: Adaptive approaches towards better GA performance in dynamic fitness landscapes. Technical Report, Aarhus University, Daimi, Denmark (1994)

Melicio, F., Caldeira, J. P., Rosa, A.: Two neighbourhood approaches to the timetabling problem. In Proceedings of the Practice and Theory of Automated Timetabling (PATAT) (2004) 267–282

Mesquita, A., Salazar, F. A., Canazio, P. P.: Chromosome representation through adjacency matrix in evolutionary circuits synthesis. In Proceedings of the 2002 NASA/DOD Conference on Evolvable Hardware (EH'02) (2002) 102–109

Murison, B.: Indicator Functions. http://mcs.une.edu.au/ stat354/notes/node16.html (October, 2000)

Neufeld, G. A., Tartar, J.: Graph coloring conditions for the existence of solutions to the timetable problem. Communications of the ACM **17(8)** (1974) 450–453

NITS: National Institute of Technology - Silchar. http://www.nits.ac.in (2005)

Okabe, T., Jin, Y., Sendhoff, B.: A new approach to dynamics analysis of genetic algorithms without selection. In Proceedings of Congress on Evolutionary Computation, Edinburgh (2005) 374–381

Papadimitriou, C. H., Steiglitz, K.: Combinatorial Optimization - Algorithms and Complexity. Prentice-Hall of India Private Limited, New Delhi (1982)

Paquete, L. F., Fonseca, C. M.: A study of examination timetabling with multiobjective evolutionary algorithms. In 4th Metaheuristics International Conference (MIC-2001), Porto (2001) 149–154

Piola, R.: Evolutionary solutions to a highly constrained combinatorial problem. In Proceedings of IEEE Conference on Evolutionary Computation (First World Congress on Computational Intelligence), Orlando, Florida **1** (1994) 446–450

Rao, S. S.: Engineering Optimization-Theory and Practice. New Age International (P) Ltd, India (1996)

Rossi-Doria, O., Blum, C., Knowles, J., Sampels, M., Socha, K., Paechter, B.: A local search for the timetabling problem (Abstract). In Proceedings of the Practice and Theory of Automated Timetabling (PATAT) (2002) 124-127

Rossi-Doria, O., Paechter, B.: An hyperheuristic approach to course timetabling problem using an evolutionary algorithm. The first Multidisciplinary International Conference on Scheduling: Theory and Applications (MISTA) (2003)

Rudová, H., Murry, K.: University course timetabling with soft constraints. In Proceedings of the Practice and Theory of Automated Timetabling (PATAT) (2002) 73–89

Schaerf, A.: Tabu search techniques for large high-school timetabling problems. In Proceedings of thirteenth National Conference of the American Association for Artificial Intelligence (AAAI-1996), AAAI Press/MIT Press (1996) 363–368

Schaerf, A.: A survey of automated timetabling. Artificial Intelligence Review **13(2)** (1999) 87–127

Silva, J. D. L., Burke, E. K., Petrovic, S.: An introduction to multiobjective metaheuristics for scheduling and timetabling. Metaheuristic for Multiobjective Optimisation, Lecture Notes in Economics and Mathematical Systems-Springer **535** (2004) 91–129

Smith, J., Fogarty, T. C.: Self adaptation of mutation rates in a steady state genetic algorithm. In Proceedings of the third IEEE Conference on Evolutionary Computation, IEEE Press, Piscataway, NJ (1996) 318–323

Srinivas, N., Deb, K.: Multiobjective optimization using Nondominated Sorting in Genetic Algorithms. Journal of Evolutionary Computation **2(3)** (1994) 221–248

Srinivasan, D., Seow, T. H., Xu, J. X.: Automated time table generation using multiple context reasoning for university modules. In Proceedings of IEEE International Conference on Evolutionary Computation (CEC) (2002) 1751–1756

THOMSON: ISI Essential Science Indicators: Special Topics - Fast Breaking Papers. http://www.esi-topics.com/fbp/fbp-february2004.html (2004)

Tripathy, A.: School timetabling - A case in large binary integer linear programming. Management Science **30(12)** (1984) 1473–1489

Zitzler, E., Thiele, L.: Multiobjective evolutionary algorithm: A comparative case study and the Strength Pareto Approach. IEEE Transactions on Evolutionary Computation **3(4)** (1999) 257–271

Metaheuristics for University Course Timetabling

Rhydian Lewis[1], Ben Paechter[1], Olivia Rossi-Doria[2]

[1]Centre for Emergent Computing,
Napier University, Edinburgh EH10 5DT, Scotland.
[2]Dipartimento di Matematica Pura ed Applicata,
Univesita' degli Studi di Padova, via G. Belzoni 7, 35131 Padua, Italy.

Summary. In this chapter we consider the NP-complete problem of university course timetabling. We note that it is often difficult to gain a deep understanding of these sorts of problems due to the fact that so many different types of constraints can ultimately be considered for inclusion in any particular application. Consequently we conduct a detailed analysis of a benchmark problem version that is slightly simplified, but also contains many of the features that make these sorts of problems "hard". We review a number of the algorithms that have been proposed for this particular problem, and also present a detailed description and analysis of an example algorithm that we show is able to perform well across a range of benchmark instances.

1 Introduction to Timetabling

Timetables are ubiquitous in many areas of daily life such as work, education, transport, and entertainment: it is, indeed, quite difficult to imagine an organized and modern society coping without them. Yet in many real-world cases, particularly where resources (such as people, space, or time) are not overly in abundance, the problem of constructing workable and attractive timetables can often be a very challenging one, even for the experienced timetable designer. However, given that these timetables can often have large effects on the day-to-day lives of the people who use them, timetable construction is certainly a problem that we should try to solve as best we can. Additionally, given that timetables will often need to be updated or completely remade (e.g. school timetables will often be redes-

R. Lewis et al.: *Metaheuristics for University Course Timetabling,* Studies in Computational Intelligence (SCI) **49**, 237–272 (2007)

igned at the beginning of each academic year; bus timetables will need to be modified to cope with new road layouts and bus stops, etc.), their construction is also a problem that we will have to face on a fairly regular basis.

1.1 Timetabling at Universities

In this chapter we will be concerning ourselves with the problem of constructing timetables for universities. The generic university-timetabling problem may be summarised as the task of assigning events (lectures, exams, etc.) to a limited set of timeslots, whilst also trying to satisfy some constraints.

Probably the most universally encountered constraint for these problems is the *event-clash* constraint: if one or more persons are required to attend a pair of events, then these events must not be assigned to the same timeslot. However, beyond this simple example, university timetabling problems, in general, are notorious for having a plethora of different problem definitions in which any number of different constraints can be imposed. These constraints can involve factors such as room facilities and capacities, teacher and student preferences, physical distances between venues, the ordering of events, the timetabling policies of the individual institution, plus many more. (Some good surveys on constraints can be found in [9, 12, 22, 41].) Some problem definitions may even directly oppose others in their criteria for what makes a good timetable. For example, some might specify that we want timetables where each member of staff is given one day free of teaching per week (e.g. [20]). Others, however, might discourage or disallow this. Obviously, which constraints are imposed, as well as the relative importance that each one has, depends very much on each individual university's preference. This, on the whole, makes it difficult to formulate meaningful and universal generalisations about timetabling in general.

One important feature that we do know, however, is that timetable construction is NP-complete in almost all variants [46]. Cooper and Kingston [21], for example, have shown a number of proofs to demonstrate that NP-completeness exists for a number of different problem interpretations that can often arise in practice. This, they achieve, by providing transformations from various well-known NP-complete problems (such as graph-colouring, bin-packing, and three-dimensional matching) to a number of different timetabling problem variants. Even, Itai, and Shamir [28] have also shown a method of transforming the NP-complete 3-SAT problem into a timetabling problem.

Of course, this general NP-completeness implies that whether we will be able to obtain anything that might be considered a workable timetable in any sort of reasonable time will depend very much on the nature of the problem instance being tackled. Some universities, for example, may have timetabling requirements that are fairly *loose*: perhaps there is an abundance of rooms or some extra teaching staff. In these cases, maybe there are many good timetables within the total search space, of which one or more can be found quite easily. On the other hand, some university's requirements might be much more demanding, and maybe only a small number of workable timetables – or perhaps none – may exist. (It should also be noted that in practice, the combination of constraints that are imposed by timetabling administrators could often result in problems that are *impossible* to solve unless some of the constraints are relaxed.) Thus, in cases where "harder" problems are encountered, there is an implicit need for powerful and robust heuristic search methods. Some excellent surveys of these can be found in [9, 11, 14, 15, 17, 41].

When looking at the timetabling problem from an operations research point-of-view, the constraints that are imposed on a particular problem tend usually to be classified as either *hard* or *soft*.[1] Hard constraints have a higher priority than soft, and are usually mandatory in their satisfaction. Indeed, timetables are usually only considered feasible if *all* of the hard constraints have been satisfied. Soft constraints, on the other hand, are those that we want to obey only if possible, and more often than not will describe what it is for a timetable to be *good* (with regards to the timetabling policies of the university, as well as the experiences of the people who will have to use it). As can be imagined, most real-world timetabling problems will have their own particular idiosyncrasies, and while this has resulted in a rich abundance of different timetabling algorithms, it also makes it very difficult to compare and contrast them. However, as Schaerf [46] points out, this situation is perfectly understandable given that many people will often be more interested in solving the timetabling problems of their own university rather than spending time comparing results with others.

It is widely accepted, however, that timetabling problems within universities can be loosely arranged into two main categories: exam timetabling problems and course timetabling problems. In reality, and depending on the university involved, both types of problem might often exhibit similar characteristics (both are usually likely to require a satisfaction of the event-clash constraint, for example), but one common and generally acknowl-

[1] A good review of the many different sorts of constraints that can be encountered in real-world timetabling problems can be found in [22].

edged difference is that in exam timetabling, multiple events can be scheduled into the same room at the same time (providing seating-capacity constraints are not exceeded), whilst in course timetabling, we are generally only allowed one event per room, per timeslot. A second common difference between the two can also sometimes concern issues with the timeslots: course timetabling problems will generally involve assigning events to a fixed set of timeslots (e.g. those occurring in exactly one week) whilst exam-timetabling problems may sometimes allow some flexibility in the number of timeslots being used (see for example [8, 10, 23, 27]).

1.2 Chapter Overview

In this chapter we will primarily concern ourselves with university course timetabling. We will, however, also refer to exam timetabling research when and where it is useful to do so. (Readers more interested in the latter are invited to consult some good texts presented by Burke *et al.* [12, 13], Thompson and Dowsland [49], Cowling *et al.* [24], and Carter [14-16].)

The remainder of this chapter is set out as follows: in the next section we will review some of the most common forms of timetabling algorithm apparent in the literature, and will discuss some possible advantages and disadvantages of each. Next, in section 3, we will give a definition and history of the particular timetabling problem that we will be studying here, and will include a survey of some of the best works proposed for it. In section 4, we will then describe an example algorithm for this problem and will provide a short experimental analysis. Finally, section 5 will conclude the chapter.

2 Dealing with Constraints

When attempting to design an algorithm for university timetabling, one of the most important issues that needs to be addressed is the question of how the algorithm proposes to deal effectively with both the hard constraints and the soft constraints. A survey of the literature indicates that most metaheuristic timetabling algorithms (of which there are many) will generally fall into one of three categories:

1. **One-Stage Optimisation Algorithms**: where a satisfaction of both the hard and soft constraints is attempted simultaneously (e.g. [20, 22, 26, 45]).

2. **Two-Stage Optimisation Algorithms**: where a satisfaction of the soft constraints is only attempted once a feasible timetable has been found (e.g. [5, 18, 19, 31, 32, 49]).
3. **Algorithms that allow Relaxations**: Violations of the hard constraints are disallowed from the outset by relaxing some other feature of the problem. Attempts are then made to try and satisfy soft constraints whilst also giving consideration to the task of eliminating these relaxations (e.g. [8, 10, 27, 37]).

Looking at category (1) first, algorithms of this type generally allow the violation of both hard and soft constraints within the timetable, and the aim is to then search for a timetable that has an adequate satisfaction of both. Typically, the algorithm will attempt this by using some sort of weighted sum function, with violations of the hard constraint usually being given much higher weightings than the soft constraints. For example, in [22] Corne, Ross, and Fang use the following evaluation function: given a problem with k types of constraint, where the penalty weighting associated with constraint i is w_i, and where $v_i(tt)$ represents the number of constraint violations of type i in a timetable tt, quality can be calculated using the formula:

$$f(tt) = 1/\left(1 + \sum_{i=1}^{k} w_i v_i(tt)\right) \quad (1)$$

In [22], the authors use this evaluation method in conjunction with an evolutionary algorithm, although, one large advantage of this method is that it can, of course, be used with any reasonable optimisation technique (see [26] and [45], for example). Another immediate advantage of this approach is its flexibility: any sensible constraint can be incorporated into the algorithm provided that an appropriate penalty weighting is stipulated in advance (thus indicating its relative importance compared to others).

However, this sort of approach also has some disadvantages. Some authors (e.g. Richardson *et al.* [38]) have argued that this sort of evaluation method does not work well in problems that are sparse (i.e. where only a few solutions exist in the search space). Also, even though the choice of weights in the evaluation function will often critically influence the algorithm's navigation of the search space (and therefore its timing implications and solution quality), there does not seem to be any obvious methodology for choosing the best ones. Some authors (e.g. Salwach [44]) have also noted that a weighted sum function can be problematic, because it can cause a discontinuous fitness landscape, where small changes to a candidate solution can actually result in overly large changes to its fitness.

With regards to timetabling problems, however, it is worth noting that some researchers have tried to circumvent some of these problems by allowing penalty weightings to be altered dynamically during the search. For example, in order to penalise hard constraint violations in his tabu search algorithm for school timetabling, Schaerf [45] defines a weighting value w, which is initially set to 20. However, at certain points during the search, the algorithm is able to increase w when it is felt that the search is drifting into search-space regions that are deemed too infeasible. Similarly, w can also be reduced when the search consistently finds itself in feasible regions.

The operational characteristics of two-stage optimisation algorithm for timetabling (category (2)) may be summarised as follows: in stage-one, the soft constraints are generally disregarded and only the hard constraints are considered for optimisation (i.e. only a feasible timetable is sought). Next, assuming feasibility has been found, attempts are then made to try and minimise the number of the soft constraint violations, using techniques that only allow feasible areas of the search space to be navigated[2].

Obviously, one immediate benefit of this technique is that it is no longer necessary to define weightings in order to distinguish between hard and soft constraints (we no longer need to directly compare feasible and infeasible timetables), meaning that a number of the problems inherent in the use of penalty weightings no longer apply. In practical situations, such a technique might also be very appropriate where finding feasibility is the primary objective, and where we only wish to make allowances towards the soft constraints if this feasibility is definitely not compromised. (Indeed, use of a one-stage optimisation algorithm in this situation could be inappropriate in many cases because, whilst searching for feasibility, the weighted sum evaluation function would always be taking the soft constraints into account. Thus, by making concessions for the soft constraints, the search could suffer the adverse effect of actually being led away from attractive (i.e. 100% feasible) regions of the search space.)

One of the major requirements for the two-stage timetabling algorithm to be effective, however, is that a reasonable amount of movement in the feasible-only search space must be achievable. If the feasible search space of the problem is convex and constitutes a reasonable part of the entire search space, then this may be so. However, if we are presented with a

[2] This could be achieved using neighbourhood operators that always preserve feasibility (c.f. [49]); by using some sort of repair mechanism to convert infeasible individuals into feasible ones (e.g. [32]); or by immediately rejecting any infeasible candidate solutions that crop up during the search. (In evolutionary computation, the latter is sometimes known as the "death penalty" heuristic [36].)

non-convex feasible search space, then searches of this kind could turn out to be extremely inefficient because it might simply be too difficult for the algorithm to explore the search space in any sort of useful way. (In these cases, perhaps a method that allows the search to take "shortcuts" across infeasible parts of the search space might be more promising.)

Lastly, whether this technique will be appropriate in a practical sense also depends largely on the users' requirements. If, for example, we are presented with a problem instance where feasibility is very difficult or seemingly impossible to achieve, then an algorithm of this form will never end up paying any consideration to the soft constraints. In this case, users may prefer to be given a solution timetable in which a suitable compromise between the number of hard and soft constraint violations has been achieved (suggesting that, perhaps one of the other two types of algorithm might be more appropriate).

Looking now at category (3), some authors have shown that good timetabling algorithms can also be achieved through the use of more specialised methodologies whereby various constraints of the problem are relaxed in order to try and facilitate better overall searches. For example, in their evolution-based algorithm for exam timetabling, Burke, Elliman, and Weare [8, 10] do not allow the direct violation of any of the problem's hard constraints; instead, they choose to open up new timeslots for events that cannot be feasibly placed into any existing timeslot. The number of timeslots being used by a candidate timetable then forms part of the evaluation criteria. In addition to this, the authors also define an opposing soft constraint that specifies that exams for individual students must be spread out (in order to avoid situations where students are required to sit exams in consecutive timeslots). Because a reasonable satisfaction of this type of constraint will usually rely on there first being an adequate number of timeslots available, the overall aim of the algorithm is to find a suitable compromise between the two objectives.

A second example of this type of approach is provided by Paechter *et al.* in [37]. Here, the authors describe a memetic approach for course timetabling in which an evolutionary algorithm is supplemented by a local-search routine that aims to improve each timetable. In this approach, a constructive scheme is also used and, rather than break any hard constraints, events that cannot be feasibly assigned are left to one side unplaced. Soft constraint violations are also penalised through the use of weightings that can be adjusted by the user during the search.

One interesting aspect of this approach is the authors' use of sequential evaluation: when comparing two candidate timetables, the algorithm deems the one with the least number of unplaced events as the fitter. However, ties are broken by looking at the penalties caused by each of the timetable's soft constraint violations. Thus many of the problems encountered when judging a timetable's quality through a single numerical value alone (as is the case with category (1)) can be avoided. Note, however, that this method of evaluation is only useful for algorithms where it is sufficient to know the ordering of a set of candidate solutions, rather than a quality score for each (in this case, the authors use binary tournament selection with their evolutionary algorithm); it is thus perhaps less well suited to other types of optimisation methods.

Concluding this section, it should be clear to the reader that the question of how to deal with both the hard and soft constraints in a timetabling problem is not always easily answerable, yet it is certainly something that we have to effectively address if automated timetabling is to be considered a worthwhile endeavour. As we have noted at various points, the issue of meeting the user's timetabling requirements (whatever these might be) often constitutes an important part in this decision. Indeed, it would seem reasonable to assume that perhaps this is the most important issue, considering that solutions to practical problems will inevitably have to be used by real people. However, it is, of course, also desirable for the algorithm to be fast, reliable and robust whenever possible.

3 The UCTP and the International Timetabling Competition

In the previous two sections, we mentioned that a difficulty often experienced in automated timetabling is that it is not always easy to compare and contrast the performance of different timetabling algorithms. Indeed, many authors often only report results from experiments with their own university's timetabling problem and fail to provide comparisons to others. (In many of these situations we also, of course, have no way of determining if the problem instances used in the experiments were actually "hard" or not, although what actually constitutes a "hard" timetabling instance is still not completely understood). These difficulties in making algorithm comparisons are in contrast to many other problems faced in operations research (such as the travelling salesperson problem and the bin-packing problem) where we often have standardised problem definitions, together with an abundance of different problem instance libraries available for benchmarking algorithms[3].

---

[3] See, for example, http://www.research.att.com/~dsj/chtsp/index.html or http://www.wiwi.uni-jena.de/Entscheidung/binpp/index.htm for libraries of TSP and bin packing problems respectively.

However, over the past few years a small number of instance sets have become publicly available. In 1996 for example, Carter [16] published a set of exam-timetabling problem instances taken from twelve separate educational establishments from various parts of the world[4]. A number of different studies have now used these in their experimental analyses [8, 16, 23, 48, 49]. More recently, a number of problem instances for course timetabling have also been made publicly available [1, 2]. It will be these particular collections of problem instances that we will focus our studies upon in this chapter.

3.1 Origins of this Problem Version

The so-called University Course Timetabling Problem (UCTP) was originally used by the Metaheuristics Network[5] – a European Commission funded research project – in 2001-2, but was also subsequently used for the International Timetabling Competition in 2002 [1], of which further details will be given later. The problem, which was formulated by the authors, is closely based on real-world problems, but is also simplified slightly. Although, from the outset, we were not entirely happy about using a simplified problem, we had a number of reasons for doing this. Firstly the problem was intended for research purposes, particularly with regards to analysing what actually happens in algorithms that are designed to solve the problem. (Real problems are often too complicated and messy to allow researchers to properly study these processes.) Secondly, the large number of hard and soft constraints usually found in real-world problems often makes the process of writing code (or updating existing programs to be suitable) a long and arduous process for timetabling researchers. Thirdly, many of the constraints of real-world problems are idiosyncratic and will often only relate to specific institutions, and so their inclusion in a problem will not always be instructive when trying to learn about timetabling in general.

The UCTP therefore offers a compromise: a variety of real world aspects of timetabling are included, yet for ease of scientific investigation, many of the messy fine-details found in practical problems have been removed.

[4] Download at http://www.or.ms.unimelb.edu.au/timetabling/atdata/carterea.tar

[5] http://www.metaheuristics.org/

3.2 UCTP Problem Description

A problem instance for the UCTP consists of a set E of n events to be scheduled into a set of timeslots T and a set of m rooms R, each that has an associated seating capacity. We are also given a set of students S each attending some subset of E. Pairs of events are said to *conflict* when one or more students are required to attend them both. Finally, we are given a set of features[6] F. These are *satisfied* by rooms and *required* by events. In order for a timetable to be *feasible*, every event $e \in E$ must be assigned to a room $r \in R$ and timeslot $t \in T$ (where $|T| \leq 45$, to be interpreted as five days of nine timeslots), such that the following hard constraints are satisfied:

H1. No student is required to attend more than one event at any one time (or, in other words, conflicting events should not be assigned to the same timeslot);

H2. All of the features required by an event are satisfied by its room, which must also have an adequate seating capacity;

H3. Only one event is put in any room in any timeslot (i.e. no double booking of rooms).

Note that the presence of H1 above makes the task of finding a feasible timetable similar to the well-known NP-hard graph colouring problem. In order to convert one problem to the other, each individual event is considered a node, and edges are then added between any pair of nodes that represent conflicting events. In very basic timetabling problem formulations (e.g. [40]), the task is to then simply colour the graph with as many colours as there are available timeslots. (Indeed, graph colouring heuristics are often used in timetabling algorithms [13, 16, 35, 49]).

However, as we demonstrate in fig. 1, in the case of this UCTP, the presence of H2 and H3 add extra complications because we must now also ensure that for any given timeslot (i.e. colour class) there are adequate and appropriate rooms available. From a pure graph colouring perspective, this means that many feasible colourings might still represent infeasible timetables[7].

---

[6] In the real world, these features might be things such as audio equipment, computing facilities, wheelchair access, etc.

[7] Note that the presence of the rooming constraints provides us with a lower bound to the underlying graph colouring problem, because a feasible solution can never use less than $\lceil n/m \rceil$ colours (timeslots).

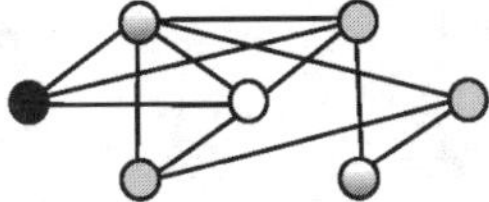
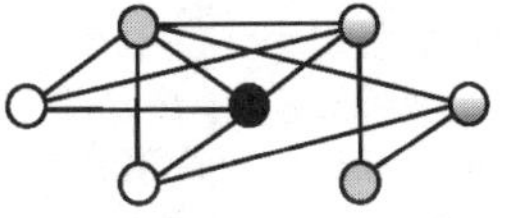

Fig. 1. In this example, both graphs have been coloured optimally. However, in the case of our timetabling problem, if only 2 rooms were available per timeslot then the left graph could never represent a feasible timetable because one of the timeslots would have 3 events assigned to it. The right solution, on the other hand, might represent a feasible timetable, providing that each event can also be granted the rooming features and seating capacity that they require.

In addition to the hard constraints outlined above, in this problem there are also three soft constraints. These are as follows:

S1. No student should be required to attend an event in the last timeslot of a day;

S2. No student should sit more than two events in a row;

S3. No student should have a single event in a day.

Note that each soft constraint is slightly different (indeed, this was done deliberately): violations of S1 can be checked with no knowledge of the rest of the timetable, violations of S2 can be checked when building the timetable, and, lastly, violations of S3 can only be checked once all events have been assigned to the timetable.

Formally, we work out the number of soft constraint violations in the following way. For S1, if a student has a class on the last timeslot of the day, we count this as one penalty point. Naturally, if there are s students in this class, we consider this as s penalty points. For S2, if one student has three events in a row we give one penalty point. If a student has four events in a row we count this as two, and so on. Note that adjacent events occurring over two separate days are not counted as a violation. Finally, each time we encounter a student with a single event on a day, we count this as one penalty point (two for two days with single events etc.). Our soft constraint evaluation function is simply the total of these three values.

We consider a timetable to be *perfect* if it is feasible (i.e. has no hard constraint violations) and if it contains no soft constraint violations.

3.3 Initial Work and the International Timetabling Competition

Rossi-Doria *et al.* conducted one of the first studies using this timetabling problem in 2002 [43]. Here, the authors used five different metaheuristic techniques (namely, evolutionary algorithms, ant colony optimisation, iterated local search, simulated annealing, and tabu search) to produce five separate algorithms for the UCTP. In order to facilitate a fair comparison of these algorithms (the main objective of their study), all used the same solution representation and search landscape. In some cases satisfaction of both the hard and soft constraints was attempted simultaneously (in the case of the evolutionary algorithm, for example, a weighted sum function was used to give higher penalties for hard constraint violations). Others, such as the iterated local search and simulated annealing algorithms, used a two-stage approach. Upon completing a comparison of these five metaheuristic algorithms, two interesting conclusions were offered by the authors:

- "The performance of a metaheuristic, with respect to satisfying hard constraints and soft constraints may be different;"
- "Our results suggest that a hybrid algorithm consisting of at least two phases, one for taking care of feasibility, the other taking care of minimising the number of soft constraint violations, is a promising direction."

Following this work, the International Timetabling Competition [1] was organised and run in 2002-3. The idea of this competition was for participants to design algorithms for this timetabling problem, which could then be compared against each other using a common set of benchmark instances and a fixed execution time limit[8]. Upon the close of the competition, the participant whose algorithm was deemed to perform best across these instances (and checked against a number of unseen instances only available to the organisers) was awarded a prize. The exact criteria for choosing the winner can be found on the competition web site [1].

The twenty problem instances used for the competition consisted of between 200 and 300 students, 350 to 440 events, and 10 or 11 rooms. As usual, the number of timeslots was fixed at 45. Additionally, in 13 of the 20 instances the number of events n was equal to the number of rooms multiplied by 40. This means that, because all instances were ensured to have at least one perfect solution[9], optimal solutions to these instances had

[8] The execution time limit was calculated for a participant's computer by a program that measured various characteristics of that computer during execution. The effectiveness of this benchmarking program was later verified by running the best competition entries on a single standard machine.

[9] In fact, we actually know that there are at least $5! = 120$ perfect timetables, because we note that the soft constraints do not actually span across different days. Thus, we can permute the days of a perfect timetable, and it will still have no soft constraint violations.

to have 40 timeslots completely filled with events (as, obviously, perfect solutions would not have any events assigned to the five end-of-day timeslots.)

Another important aspect of the competition was the way in which timetables were chosen to be evaluated. The official rules of the competition stated that timetable quality would only be measured by looking at the number of soft constraint violations: if a timetable contained any hard constraint violations, used any extra timeslots, or had any unplaced events, then it would immediately be considered worthless. Participants were then only allowed to submit an entry to the competition if their algorithms could find feasibility on all twenty instances. Given this rule, and also taking into consideration the conclusions of Rossi-Doria *et al.* [43] quoted above, it is perhaps, unsurprising that many of the entrants to this competition therefore elected to use the two-stage timetabling approach mentioned in section 2. Another consequence of the evaluation scheme was that the problem instances were chosen so that feasibility was relatively easy to find.

The competition, which ended in March 2003, eventually saw a total of 21 official entries, plus 3 unofficial entries (the latter were not permitted to enter the competition because they were existing members of the Metaheuristics Network). The submitted algorithms used a variety of techniques including simulated annealing, tabu search, iterated local search, ant colony optimisation, some hybrid algorithms, and heuristic construction with backtracking. The winning algorithm was a two-stage, simulated annealing-based algorithm by Philipp Kostuch of Oxford University. Details of this, plus many of the others mentioned above can be found at the official competition web page [1].

3.4 Review of Relevant Research

Since the running of the competition, quite a few good papers have been published regarding this particular timetabling problem. Some of these describe modifications to algorithms that were official competition entries and claim excellent results. Some have gone on to look at other aspects of the problem. In this subsection we now review some of the most notable and relevant works in this problem area.

In [5], Arntzen and Løkketangen describe a two-stage tabu search algorithm for the problem. In the first stage, the algorithm uses a constructive procedure to build an initial feasible timetable, which operates by taking events one by one, and assigning them to feasible places in the timetable, according to some specialised heuristics that also take into account the potential number of soft constraint violations that such an assignment might cause. The order in which events are inserted is determined dynamically, and decisions are based upon the state of the current partial timetable. The authors report that these heuristics successfully build feasible timetables in over 90% of runs with the competition instances. Next, with feasibility having been found, Arntzen and Løkketangen opt to use tabu search in conjunction with simple neighbourhood operators in order to optimise the soft constraints. In the latter stage, feasibility is always maintained.

Cordeau, Jaumard, and Morales (available at [1]) also use tabu search to try and satisfy the soft constraints in their timetabling algorithm. However, this method is slightly different to Arntzen and Løkketangen above, because, when dealing with the soft constraints, the algorithm also allows a small number of hard constraints to be broken from time to time. The authors achieve this by introducing a partially stochastic parameter α that is then used in the following evaluation function:

$$f(tt) = \alpha h(tt) + s(tt) \tag{2}$$

where $h(tt)$ indicates the number of hard constraint violations in timetable tt, and $s(tt)$ the number of soft constraint violations. During the search, the parameter α helps to control the level of infeasibility in the timetable because if the number of hard constraint violations in tt increases, then α is also increased. Thus, as the number of infeasibilities rises, it also becomes increasingly unlikely that a search space move causing additional infeasibilities will be accepted. The authors claim that such a scheme allows freer movement about the search space.

Socha, Knowles, and Sampels have also suggested ways of applying the ant colony optimisation metaheuristic to this problem. In [47], the authors present two ant-based algorithms – an Ant Colony System and a MAX-MIN system – and provide a qualitative comparison between them. At each step of both algorithms, every ant first constructs a complete assignment of events to timeslots using heuristics and pheromone information, due to previous iterations of the algorithm. Timetables then undergo further improvements via a local search procedure, outlined in [42]. Indeed, the only major differences between the two approaches are in the way that heuristic and pheromone information is interpreted, and in the methodologies for updating the pheromone matrix. However, tests using a range of problem instances indicate that the MAX-MIN system generally achieves better results. A description of the latter algorithm – which was actually entered unofficially to the timetabling competition – can also be found at [1], where good results are reported.

Another good study looking at this problem is offered by Chiarandini *et al.* in [19]. In this research paper, which also outlines an unofficial competition entry, the authors present a broad study and comparison of various different heuristics and metaheuristics for the UCTP. After experimenting with a number of different approaches and also parameter settings (much of which was done automatically using the *F-Race* method of Birattari *et al.* [6]), their favoured method is a two-stage, hybrid algorithm that actually uses a variety of different search methods. In the first stage, constructive heuristics are initially employed in order to try and find a feasible timetable, although, as the authors note, these are usually unable to find complete feasibility unaided. Consequently, local search and tabu search schemes are also included to try and help eliminate any remaining hard constraint violations. Feasibility having been achieved, the algorithm then concentrates on satisfying the soft constraints and conducts its search only in feasible areas of the search space. It does this first by using variable neighbourhood search and then with simulated annealing. The annealing phase is reported to use more than 90% of the available run time of the total algorithm, and a simple reheat function for this phase is also implemented (this operates by resetting the temperature to its initial starting value when it is felt that the search is starting to stagnate). Extensive use of delta evaluation [39] is also made in an attempt to try and speed up the algorithm and, according to the authors, the final algorithm achieves results that are significantly better than the official competition winner.

Kostuch also uses simulated annealing as the main construct of his timetabling algorithm, described in [31]. Based upon his winning entry to the competition, this algorithm works by first gaining feasibility via simple graph colouring heuristics (plus a series of improvement steps if the heuristics prove inadequate) and then uses simulated annealing to try and satisfy the soft constraints by first ordering the timeslots, and then by swapping events between timeslots. One of the interesting aspects of Kostuch's approach is that when a feasible timetable is being constructed, efforts are made in order to try and schedule the events into just forty of the available forty-five timeslots. As the author notes, five of the available timeslots will automatically have penalties attached to them (due to the soft constraint S1) and so it could be a good idea to try and eliminate them from the search from the outset. Indeed, the author only allows the extra five timeslots to be opened if feasibility using forty timeslots cannot be achieved in reasonable time. (In reported experiments, the events in nine of the twenty instances were always scheduled into forty timeslots.) Of course, if an assignment to just forty timeslots is achieved, then it is possible to keep the five end-of-day timeslots closed and simply conduct the soft constraint satisfaction phase on the remaining forty timeslots. This is essentially what

Kostuch's algorithm does and, indeed, excellent results are claimed in [31].

Finally, in [35] Lewis and Paechter have proposed a "grouping genetic algorithm" (GGA) that is used exclusively for finding feasible timetables in this UCTP (i.e. the algorithm does not consider soft constraints). The rationale for this approach is that the objective of this (sub)problem may be viewed as the task of "grouping" events into an appropriate number of timeslots, such that all of the hard constraints are met. Furthermore, because, in this case, it is the timeslots that define the underlying building blocks of the problem (and not, say, the individual events themselves) the algorithm makes use of specialised genetic operators that try to allow these groups to be propagated during evolution[10]. Experiments in [35] show that performance of this algorithm can sometimes also be improved through the use of specialist fitness functions and additional heuristic search operators. One negative feature of this algorithm, however, is that whilst seeming to perform well with smaller instances (≈200 events), it seems less successful when dealing with larger instances (≈1000 events). This is mainly due to the fact that the larger groups encountered in the latter cases tend to present much more difficulty with regards to their propagation during evolution. Indeed, experiments in [35] show that in most cases, significantly better results can actually be gained when the evolutionary features of the algorithm (population, recombination etc.) are removed altogether, thus allowing the heuristic-search operator to work unaided. (This heuristic search-based algorithm forms a part of the algorithm that will be described in section 4.3 later.)

4 A Robust, Two-Stage Algorithm for the UCTP

Having reviewed a number of published works that have looked at this standardised version of the UCTP, in this section we will now describe an example two-stage algorithm that, in our experiences, has performed very well with many available benchmark instances for this problem. The feasibility-finding stage (sections 4.2 and 4.3) is particularly successful; with the twenty competition instances, for example, we will see that it is often able to achieve its goal in very small amounts of time. We will also see that it is able to cope very well with a large number of specially made "harder" instances of various sizes. Meanwhile, the second stage of our

[10] The resultant "grouping" genetic operators follow the methodologies used in similar algorithms for other "grouping problems" such as bin packing [29] and graph colouring [25, 27].

algorithm is concerned with the satisfaction of the soft constraints, which is attempted using two separate phases of simulated annealing that will be described in section 4.5.

4.1 Achieving Feasibility: Pre-compilation and Representational Issues

Before attempting to construct a feasible timetable, in our approach we first carry out some useful pre-compilation by constructing two matrices that are then used throughout the algorithm. We call these the *event-room* matrix and the *conflicts* matrix. Remembering that n represents the number of events and m the number of rooms, the Boolean $(n \times m)$ event-room matrix is used to indicate which rooms are suitable for which events. This can be easily calculated for an event i by identifying which rooms satisfy both the seating capacity and the features required by i. Thus if, room j is deemed suitable, then element (i, j) in the matrix is marked as true, otherwise it is marked as false.

The $(n \times n)$ conflicts matrix, meanwhile, can be considered very much like the standard adjacency matrix used for representing graphs. For our purposes, the matrix indicates which pairs of events can and cannot be scheduled into the same timeslot. Thus, if event i and event j have one of more common student, then elements (i, j) and (j, i) in the matrix are marked as true, otherwise false. As a final step, and following the suggestions of Carter [14], we are also able to add some further information to the matrix. Note that, in this problem, if we have two events, k and l, that do not conflict but can both only be placed into the same single room, then there can exist no feasible timetable in which k and l are assigned to the same timeslot. Thus, we may also mark elements (k, l) and (l, k) as true in the conflicts matrix.

With regards to the way in which an actual timetable will be represented in this algorithm, similarly to works such as [19, 32, 35, 47], we choose to use a two-dimensional matrix where rows represent rooms, and columns represent timeslots. We also choose to place the restriction that each cell in the matrix (i.e. each place[11] in the timetable) can be blank, or can contain *at most* one event. Note that this latter detail therefore actually encodes the third hard constraint into the representation, meaning that it is now impossible to double book a room.

[11] For the remainder of this chapter, when referring to a timetable, a *place* may be considered a timeslot/room pair. More formally, the set of all places $P = T \times R$.

4.2 Achieving Feasibility - The Construction Stage

An initial assignment of events to places (cells in the matrix) is achieved following the steps outlined in the procedure CONSTRUCT in fig. 2. This procedure is also used for completing partial timetables that can occur as a result of the heuristic search procedure, explained in the next subsection. Starting with an empty or partial timetable *tt* and a list of unplaced events U (in the first case $U = E$), this procedure first opens up a collection of timeslots, and then utilises the procedure INSERT-EVENTS that takes events one-by-one from U and inserts them into feasible places in the timetable *tt*. (The heuristics governing these choices are described in Table 1.) The entire construction procedure is completed when all events have been assigned to the timetable (and therefore $U = \emptyset$).

Of course, during this process, there is no guarantee that every event will have a feasible place into which it can be inserted. In order to deal with these, we therefore relax the requirement regarding the number of timeslots being used, and open up extra timeslots as and when necessary. Obviously, once all of the events have been assigned, if the number of timeslots being used $|T|$ is larger than the actual target amount, then the timetable may not actually be considered feasible (in the strict sense), and efforts will, of course, need to be made to try and rectify the situation. Methods for achieving this will be described in section 4.3.

With regards to the heuristics that are used in this construction process (Table 1), it is worth noting that those used for determining the order in which events are inserted are somewhat akin to the rules for selecting which node to colour next in the classical Dsatur algorithm for graph colouring [7]. However, in this case we observe that h_1 also takes the issue of room allocation into account. Heuristic h_1 therefore selects events based on the state of the current partial timetable, prioritising those with the least remaining feasible options. Ties are then broken by h_2, which chooses the event with the highest conflicts degree (which could well be the most problematic of these events). Once an event has been chosen, further heuristics are then employed for selecting a suitable place. Heuristic h_4 attempts to choose the place that will have the least effect on the future place options of the remaining unplaced events [5]. Heuristic rule h_5, meanwhile, is used to encourage putting events into the fuller timeslots, thereby hopefully packing the events into as few timeslots as possible. Finally, h_3 and h_6 add some randomisation to the process and, in our case, allow different runs to achieve different timetables.

```
CONSTRUCT (tt, U)
1. if (len(tt) < max_timeslots)
2.    Open (max_timeslots – len(tt)) new timeslots;
3. INSERT-EVENTS (tt, U, 1, max_timeslots);

INSERT-EVENTS (tt, U, l, r)
1. while (∃ e ∈ U with feasible places between timeslots l and r in tt)
2.    Choose an event e ∈ U with feasible places in tt using h1, breaking ties
      with h2, and further ties with h3;
3.    Pick a feasible place p for e using heuristic h4, breaking ties with h5 and
      further ties with h6;
4.    Move e to p;
5. if (U = ∅) end;
6. else
7.    Open ⌈|U| / m⌉ new timeslots;
8.    INSERT-EVENTS (tt, U, r, len(tt));
```

Fig. 2. The procedures CONSTRUCT and INSERT-EVENTS: Here, *tt* represents the current partial timetable and *U* is a set of unplaced events of cardinality $|U|$. Additionally, len(*tt*) represents a function that returns the number of timeslots currently being used by *tt*; max_timeslots represents the maximum number of timeslots that a timetable can use for it to be considered feasible (i.e. 45), and, as before, *m* represents the cardinality of the room set.

Table 1. Description of the various event and place selection heuristics used within the procedure INSERT-EVENTS.

Name	Description
h_1	Choose the event with the smallest number of feasible places to which it can be assigned in the current timetable
h_2	Choose the event which conflicts with the largest number of other events
h_3	Choose an event randomly
h_4	Choose the place that the least number of other unplaced events could be feasibly assigned to in the current timetable
h_5	Choose the place in the timeslot with the most events in
h_6	Choose a place randomly

4.3 Reducing the Number of Timeslots with a Heuristic Search Procedure

Although no hard constraints will be violated in any timetable produced by the construction procedure described above, it is, of course, still possible that more than the required number of timeslots will be used, thus rendering it infeasible. We therefore supplement the construction procedure with a heuristic search procedure (originally described in [35]) that operates as follows (see also fig. 3):

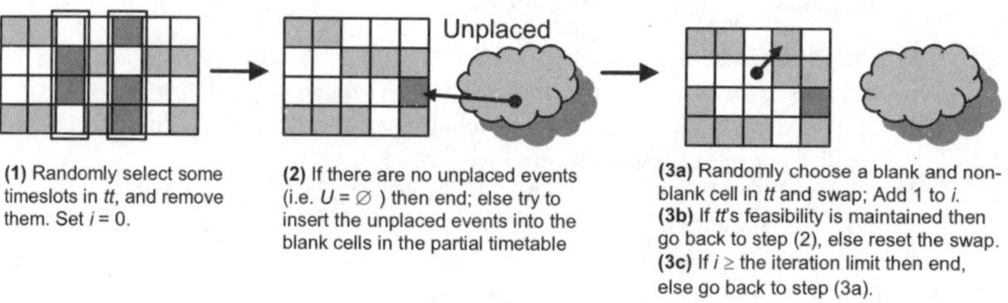

Fig. 3. Pictorial description of the heuristic search procedure used for attempting to reduce the number of timeslots used by a timetable.

Given a timetable *tt*, a small number of randomly selected timeslots are first removed (defined by a parameter *rm*, such that between one and *rm* timeslots are chosen randomly). The events contained within these are then put into the list of unplaced events *U*. Steps (2) and (3) of fig. 3 are then applied repeatedly until either *U* is empty, or an iteration limit is reached. If, as in the latter case, upon termination *U* still contains some events, then CONSTRUCT is used to create new places for these. Now, if the resultant timetable is using the required number of timeslots, then the process can be halted (a completely feasible timetable has been found), otherwise further timeslots are selected for removal, and the whole process is repeated.

4.4 Experimental Analysis

As it turned out, the construction procedure described in section 4.2 was actually able to cope quite easily with the twenty problem instances used for the International Timetabling Competition. Indeed, in our experiments feasible timetables using a maximum of 45 timeslots were found straight away in over 98% of trials without any need for opening up additional timeslots or invoking the heuristic search procedure. (Even in cases where the heuristic search procedure *was* needed, feasibility was still always achieved in less than 0.25 seconds of CPU time[12].) We also observed that the construction procedure was often actually able to pack the events into less than the available forty-five timeslots. For example, competition instance-15 required only 41.9 timeslots (averaged across twenty runs), and

[12] These trials, like all experiments described in this chapter, were conducted on a PC under Linux, using 1GB RAM, and a Pentium IV 2.66Ghz processor.

instance-3 required just 41.8. Others, such as competition instances 6 to 9, on the other hand, always required the full forty-five timeslots.

However, although these observations seem to highlight the strengths of our constructive heuristics in these cases, they do not really provide us with much information on the operational characteristics of the heuristic search procedure. For this reason, we therefore conducted a second analysis using an additional set of UCTP instances [2] that have been used in other studies [33-35] and which are deliberately intended to be "hard" with regards to achieving feasibility. These sixty instances are separated into three classes: small, medium, and large (containing approximately 200 events, 400 events and 1000 events respectively). Further details of the instances, including information on how we attempted to ensure their "hardness" can be found at [2] and [33]. Note, however, that each of these instances is known to have at least one feasible solution, and that for some of them there is also a known perfect solution. For the remaining instances, meanwhile, some are known to definitely *not* have perfect solutions[13], whilst, for others, this is still undetermined. See Table 2 below for further information.

In this second set of experiments, we conducted 20 trials per instance, using CPU time limits of 30, 200, and 800 seconds for the small, medium and large instances respectively (these match the time limits used in [35]). We also used parameters $rm = 1$, and an iteration limit of $10000n$. Note that our use of the number of events n in defining the latter parameter allows the procedure to scale with instance size.

Table 2 summarises the results of these experiments and entries that are highlighted indicate problem instances where feasibility was found in every individual trial. Here we can see that in many instances, particularly in the small and medium sets, when timetables using 45 timeslots were not achieved by the construction procedure, the heuristic search operator has successfully managed to remedy the situation within the imposed time limits. Additionally, even with problem instances where solutions using 45 timeslots were not always achieved, we see that the number of timeslots being used generally drops a noteworthy amount within the time limit.

Indeed, our use of the heuristic search procedure is further justified when we compare these results to those achieved by the GGA presented in [35]. From the above table we can see that, with this algorithm, we are

[13] We were able to determine that an instance had no perfect solution when the number of events was greater than $40m$ (where m represents the number of rooms). In these instances we know that at least $(n - 40m)$ events will always have to be assigned to the end-of-day timeslots, thus causing violations of soft constraint S1.

always able to achieve feasible timetables for fifteen, thirteen, and seven instances of the small, medium, and large instance sets respectively. This compares favourably with the work described in [35], where solutions to only eleven, six, and two problem instances were always found. It is also worth pointing out that the results in [35] were also gained after performing considerable parameter tuning with each instance set. Here, on the other hand, the results in Table 2 were gained with very little tuning (beyond our own intuitions), hinting that this algorithm might also be more robust with regard to what instances it is able to effectively deal with.

Table 2. Performance of the Heuristic Search Procedure with the Sixty "Harder" Instances. This table shows, for each instance, the mean and standard deviations of the number of timeslots being used (a) after the initial assignment by the construction procedure (Av. slots. init. ± σ), and (b) at the time limit (Av. Slots. end ± σ). Also shown is the number of timeslots used in the most successful runs (Best). All results are taken from 20 runs per instance and are rounded to one decimal place. Lastly, in column P we also provide some supplementary information about the instances: a "Y" indicates that we know there to be at least one perfect solution obtainable from the instance, an "N" indicates that we know that there definitely isn't a perfect solution, and "?" indicates neither.

	Small (30 seconds)				Medium (200 seconds)				Large (800 seconds)			
#	P	Av. slots. init. ± σ	Av. slots end ± σ	Best	P	Av. slots. init. ± σ	Av. slots end ± σ	Best	P	Av. slots. init. ± σ	Av. slots end ± σ	Best
1	Y	45.8 ± 0.9	44.7 ± 0.5	44	Y	45.8 ± 0.8	44.8 ± 0.4	44	Y	43.9 ± 0.7	43.9 ± 0.7	43
2	Y	45.0 ± 0.0	45.0 ± 0.0	45	Y	47.9 ± 1.4	44.6 ± 0.5	44	Y	49.2 ± 1.6	44.8 ± 0.4	44
3	?	50.0 ± 0.0	44.8 ± 0.4	44	?	47.1 ± 1.4	44.9 ± 0.3	44	Y	46.9 ± 0.8	44.9 ± 0.3	44
4	Y	50.5 ± 1.8	44.4 ± 0.5	44	N	50.4 ± 1.3	44.7 ± 0.5	44	N	52.1 ± 0.9	45.2 ± 0.4	45
5	?	57.6 ± 1.3	45.0 ± 0.0	45	N	51.0 ± 1.5	45.0 ± 0.2	44	N	54.1 ± 1.5	46.0 ± 0.0	46
6	Y	43.3 ± 1.6	43.3 ± 1.6	41	Y	56.8 ± 1.8	45.0 ± 0.2	44	N	59.8 ± 1.7	48.7 ± 0.5	48
7	?	53.0 ± 0.0	44.9 ± 0.3	44	?	62.2 ± 1.6	48.1 ± 0.6	47	N	67.1 ± 1.8	54.0 ± 0.6	52
8	N	55.7 ± 1.2	46.0 ± 0.4	45	Y	58.8 ± 1.4	44.9 ± 0.3	44	N	53.6 ± 1.7	45.0 ± 0.0	45
9	N	64.0 ± 0.0	45.5 ± 0.5	45	?	67.1 ± 1.8	47.8 ± 0.6	47	N	51.1 ± 1.0	45.1 ± 0.3	45
10	N	46.0 ± 0.0	45.0 ± 0.0	45	Y	46.0 ± 1.3	44.7 ± 0.5	44	N	51.7 ± 1.1	46.0 ± 0.0	46
11	Y	44.9 ± 0.4	44.9 ± 0.4	43	Y	60.3 ± 1.7	45.0 ± 0.2	44	N	53.3 ± 1.0	46.0 ± 0.0	46
12	N	45.0 ± 0.0	45.0 ± 0.0	45	?	51.2 ± 1.3	45.0 ± 0.2	44	Y	48.7 ± 1.0	45.0 ± 0.2	44
13	N	60.8 ± 0.9	45.1 ± 0.3	45	Y	63.7 ± 1.6	45.2 ± 0.5	44	Y	51.6 ± 0.7	45.0 ± 0.0	45
14	N	64.1 ± 0.8	46.7 ± 0.9	45	Y	55.4 ± 1.0	44.8 ± 0.4	44	Y	49.1 ± 0.9	45.0 ± 0.0	45
15	Y	45.0 ± 0.0	45.0 ± 0.0	45	N	59.8 ± 1.9	45.0 ± 0.0	45	Y	65.4 ± 1.2	45.6 ± 0.7	45
16	Y	60.9 ± 2.3	44.8 ± 0.4	44	?	75.1 ± 1.5	46.9 ± 0.7	46	Y	63.0 ± 1.3	45.9 ± 0.7	45
17	?	59.0 ± 0.0	45.0 ± 0.0	45	Y	65.6 ± 1.4	44.7 ± 0.5	44	?	88.9 ± 1.4	56.0 ± 1.2	54
18	N	53.2 ± 0.8	45.3 ± 0.5	45	?	89.6 ± 0.5	45.6 ± 0.6	45	?	77.0 ± 1.6	56.3 ± 1.1	54
19	N	73.6 ± 2.8	45.0 ± 0.0	45	N	92.4 ± 1.5	46.0 ± 0.5	45	?	81.7 ± 1.4	61.1 ± 0.8	60
20	N	46.0 ± 0.0	45.0 ± 0.0	45	N	77.7 ± 2.3	46.3 ± 0.6	45	?	76.1 ± 2.3	55.0 ± 0.7	54

4.5 Satisfying the Soft Constraints

Having now reviewed a seemingly effective and robust algorithm for achieving timetable feasibility, in this section we will now move on to the task of satisfying the soft constraints of the UCTP. Similarly to the ideas of White and Chan [51] and also Kostuch [31], our algorithm will attempt to do this in two phases: firstly, by seeking a suitable ordering of the timeslots (using neighbourhood operator N_1 – see fig. 4), and secondly by shuffling events around the timetable (using neighbourhood operator N_2). In both phases we will use simulated annealing (SA) for this task and, as we will see, the second SA phase will generally constitute the lengthiest part of this process. In this algorithm we also make extensive use of delta-evaluation [39], and the algorithm will halt when a perfect solution has been found or, failing this, when a predefined time limit is reached (in the latter case, the best solution found during the whole run will be returned).

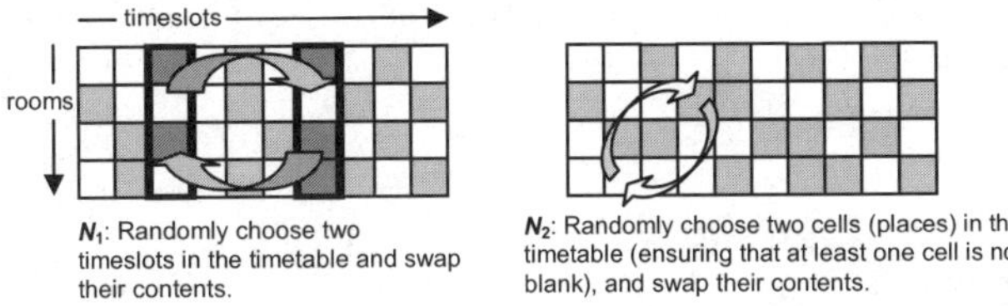

Fig. 4. The two neighbourhood operators used with the simulated annealing algorithm.

In both phases, SA will be used in the following way: starting at an initial temperature t_0, during the run the temperature t will be slowly reduced. At each value for t, a number of neighbourhood moves will then be attempted. Any move that increases the *cost* of the timetable (i.e. the number of soft constraint violations) will then be accepted with probability defined by the equation $\exp(-\delta/t)$, where δ represents the change in cost. Moves that reduce or leave unchanged the cost, meanwhile, will be accepted automatically.

In the next four subsections we will outline the particular characteristics of these two SA phases. We will then present an experimental analysis in section 4.5.5-6.

4.5.1 SA Phase-1 - Search Space Issues

This phase of SA is concerned with the exploration of the search space defined by neighbourhood operator N_1 (see fig. 4). Note that due to the structure of this timetabling problem (in particular, that there are no hard constraints that depend on the ordering of events), a movement in N_1 will always preserve feasibility.

It is also worth mentioning, however, that often there may be many feasible timetables that are not achievable through the use of N_1 alone. For example, the size of the search space offered by N_1 is $|T|!$ (i.e. the number of possible permutations of the timeslots). However, given that a feasible timetable must always have $|T| \leq 45$, this means that the number of possible solutions achievable with this operator will not actually grow with instance size. Also, if we were to start this optimisation phase with a timetable in which two events – say, i and j – were assigned to the same timeslot, then N_1 would never actually be able to change this fact. Indeed, if the optimal solution to this problem instance required that i and j were in *different* timeslots, then an exploration with N_1 would never actually be able to achieve the optimal solution in this case.

Given these issues, it was therefore decided that this phase of SA would only be used as a preliminary step for making quick-and-easy improvements to the timetable. Indeed, this also showed to be the most appropriate response in practice.

4.5.2 SA Phase-1 - Cooling Schedule

For this phase, an initial temperature t_0 is determined automatically by calculating the standard deviation in the cost for a small sample of neighbourhood moves. (We used sample size 100). This scheme of calculating t_0 is based upon the physical processes of annealing, which are beyond the scope of this chapter, but of which more details can be found in [50]. However, it is worth noting that in general SA practice, it is important that a correct value for t_0 is determined: a value that is too high will invariably waste run time, because it will mean that the vast majority of movements will be accepted, providing us with nothing more than a random walk about the search space. On the other hand, an initial temperature that is too low could also be detrimental, as it might cause the algorithm to be too greedy from the outset and make it more susceptible to getting stuck at local optima. In practice, our described method of calculating t_0 tended to allow approximately 75-85% of moves to be accepted, which is widely accepted as an appropriate amount in SA literature.

With regards to other features of the cooling schedule, because we only view this phase as a preliminary, during execution we choose to limit the number of temperatures that we will anneal at to a fixed value M. In order to have an effective cooling, this also implies a need for a cooling schedule that will decrement the temperature from t_0 to a value close to zero, in exactly M steps. We use the following cooling scheme:

$$\begin{aligned} \lambda_0 &= 1-\beta \\ \lambda_{i+1} &= \lambda_i + \tfrac{\beta+\beta}{M} \\ t_{i+1} &= t_i - \lambda_{i+1}\tfrac{t_0}{M} \end{aligned} \tag{3}$$

Here, β represents a parameter that, at each step, helps determine a value for λ. This λ-value is then used for influencing the amount of concavity or convexity present in the cooling schedule. Fig. 5 shows these effects in more detail.

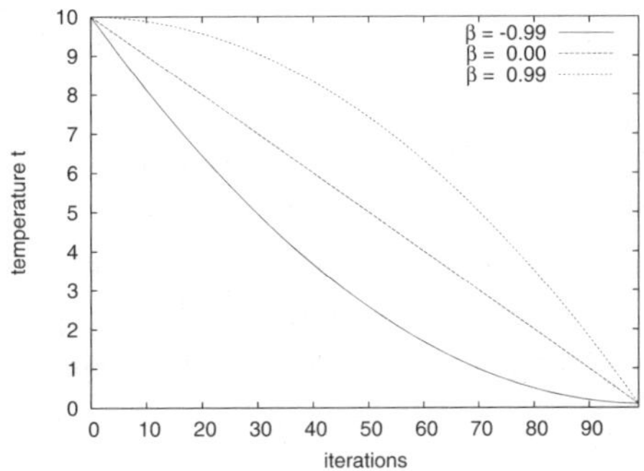

Fig. 5. The effects of the parameter β with the cooling scheme defined in eq. (3). For this example, $t_0 = 10.0$ and $M = 100$.

In our experiments, for this phase we set $M = 100$ and, in order to allow more of the run to operate at lower temperatures, we set $\beta = -0.99$. The number of neighbourhood moves to be attempted at each temperature was set at $|T|^2$, thus keeping it proportional to the total size of the neighbourhood (a strategy used in many SA implementations [3, 19, 31]).

4.5.3 SA Phase-2 - Search Space Issues

In this second and final round of simulated annealing, taking the best solution found in the previous SA phase, an exploration of the search space defined by neighbourhood operator N_2 is conducted (see fig. 4). However, note that, unlike neighbourhood operator N_1, moves in N_2 might cause a violation of one or more of the hard constraints. In our approach we deal with this fact by immediately rejecting and resetting any move that causes such an infeasibility to occur.

Before looking at how we will tie this operator in with the SA approach, it is first worth considering the large amount of flexibility that N_2 can offer the search. Suppose, for the sake of argument, that in a single application of the operator we elect to swap cells p and q:

- If p is blank and cell q contains an event e, then this will have the effect of moving e to a new place p in the timetable;
- If p contains an event e and cell q contains an event g, then this will have the effect of swapping the places of events e and g in the timetable.

Additionally,

- If p and q are in the same column, only the rooms of the affected events will change;
- If p and q are in the same row, only the timeslots of the affected events will change;
- If p and q are in different rows *and* different columns, then both the rooms and timeslots of the affected events will be changed.

As can be seen, N_2 therefore has the potential to alter a timetable in a variety of ways. In addition, we also note that the number of new solutions (feasible and infeasible) that are obtainable via any single application of N_2 is exactly:

$$\tfrac{1}{2}n(n-1)+nx-1 \tag{4}$$

(where x defines the number of blank cells in the timetable). Thus, unlike N_1, the size of the neighbourhood is directly related to the number of events n, and therefore the size of the problem. This suggests that for anything beyond very small instances, more time will generally be required for a thorough exploration of N_2's solution space.

4.5.4 SA Phase-2 - Cooling Schedule

For this phase, an initial temperature t_0 is calculated in a very similar fashion to SA phase-1. However, before starting this second SA phase we also choose to reduce the result of this calculation by a factor (c_2/c_1), where c_1 represents the cost of the timetable before SA phase-1, and c_2 the cost *after* SA phase-1. Our reason for doing this is that during our experiments, we observed that an unreduced value for t_0 was often so high, that the improvements achieved during the SA phase-1 were regularly undone at the beginning of the second. Reducing t_0 in this way, however, seemed to allow the second phase of SA to build upon the progress of SA phase-1, thus giving a more efficient run.

In order to determine when the temperature t should be decremented we choose to follow the methodologies used by Kirkpatrick *et al.* [30] and Abramson *et al.* [3] and define two values. The first of these specifies the maximum number of feasible moves that can be attempted at any value for t and, in our case, we calculate this with the formula: $\eta_{max}n$ (where η_{max} is a parameter that we will need to tune[14]). However, in this scheme t is also updated when a certain number of feasible moves have been accepted at the current temperature. This value is calculated with the formula $\eta_{min}(\eta_{max}n)$, where η_{min} is in the range (0, 1] and must also be tuned.

To decrease the temperature, we choose to use the traditional geometric scheme [30] where, at the end of each cycle, the current temperature t_i is modified to a new temperature t_{i+1} using the formula $t_{i+1} = \alpha t_i$, where α is a control parameter known as the cooling rate.

Finally, because this phase of SA will operate until a perfect solution has been found, or until we reach the imposed time limit, we also make use of a reheating function that is invoked when no improvement in cost is achieved for ρ successive values of t (and so the search has presumably become stuck at a local optimum). In order to calculate a suitable temperature to reheat to, we choose to use a method known as "reheating as a function of cost", which was originally proposed by Abramson, Krishnamoorthy, and Dang in [4]. In essence, this scheme determines a reheat temperature by considering the current state of the search; thus, if the best solution found so far has a high cost, then a relatively high reheat temperature will be calculated (as it is probably favourable to move the search to a new region of the search space). On the other hand, if the best solution found so far is low in cost, then a lower reheat temperature will be calculated, as it is probably the case that only small adjustments need to be made. In studies such as [4] and [26] (where further details can also be found) this has shown to be an effective method of reheating with this sort of problem.

4.5.5 Algorithm Analysis – 45 or 40 Timeslots?

For our experimental analysis of this SA algorithm, we performed two separate sets of trials on the 20 competition instances, using a time limit specified by the competition-benchmarking program[15]. For the first set, we simply used our construction and heuristic search procedures (section 4.2 and 4.3) to make any feasible timetable where a maximum of 45 timeslots was being used. The SA algorithm would then take this timetable and operate in the usual way. For our second set, however, we chose to make a slight modification to our algorithm and allowed the heuristic search pro-

[14] Note that our use of the number of events n in this formula keeps the result of this calculation proportional to instance size.

[15] This equated to 270 seconds of CPU time on our computers

cedure to run a little longer in order to try and schedule all of the events into a maximum of just 40 timeslots (we chose to allow a maximum of 5% of the total runtime in order to achieve this). Our reasons for making this modification were as follows:

When we were designing and testing our SA algorithm, one characteristic that we sometimes noticed was the difficulty that N_2 seemed to have when attempting to deal with violations of soft constraint S1: often, when trying to rid a timetable of a violation of S2 or S3, N_2 would do so by making use of an end-of-day timeslot. Or in other words, in trying to eliminate one constraint violation, the algorithm would often inadvertently cause another one. The reasons why such behaviour might occur start to become more evident if we look back at the descriptions of the three soft constraints in section 3.2. Note that S2 and S3 stand out as being slightly different to S1, because if an event e is involved in a violation of either S2 or S3, then this will not simply be down to the position of e in the timetable, it will also be due to the *relative positions* of the other events that have common students with e. By contrast, if e is causing a violation of S1, then this will be due to it being assigned to one of the five end-of-day timeslots, and has nothing to do with the relative positions of other events with common students to e. Thus, given that a satisfaction of S1 depends solely on not assigning events to the five end-of-day timeslots, a seemingly intuitive idea might therefore be to simply remove these five timeslots (and therefore constraint S1) from the search altogether. In turn, the SA algorithm will then only need to consider the remaining 40 (unpenalised) timeslots and only try to satisfy the two remaining soft constraints.

In our case, it turned out that our strategy of allowing the heuristic search procedure to run a little longer worked quite well: using the same experimental set-up as described in section 4.4, with the 20 competition instances the procedure was able to schedule all events into 40 timeslots in over 94% of cases. In the remaining cases (which, incidentally, never actually required more than 41 timeslots) the extra timeslots were labelled as end-of-day timeslots. However, in order to still lend special attention to the task of eliminating S1 violations, we used a slightly modified version of N_2 that would automatically reject any move that caused the number of events assigned to these end-of-day timeslots to increase, but would also eliminate these timeslots if they were ever to become empty during the SA process. In our case, this strategy would always eliminate the remaining end-of-day timeslots within the first minute-or-so of the run.

4.5.6 Results

Table 3 provides a comparison of these two sets of trials using 50 runs on each of the 20 instances. In both cases we used a cooling rate of $\alpha = 0.995$ and $\rho = 30$. Suitable values for η_{min} and η_{max} (the two parameters that we witnessed to be the most influential regarding algorithm performance), on the other hand, were determined empirically by running the algorithm at 11 different settings for η_{max} (between 1 and 41, incrementing in steps of 4) and 10 different values for η_{min} (0.1 to 1.0, in steps of 0.1). At each setting for η_{min} and η_{max} we then performed 20 separate runs on each of the 20 competition problem instances, thus giving a total of 400 runs per setting. The best performing values for η_{min} and η_{max} in both cases (i.e. the settings that gave the lowest average cost of the 400 runs when using 40 and 45 timeslots) were then used for our comparison.

It can be seen in Table 3 that when using just 40 timeslots the SA-algorithm is able to produce better average results in the majority of cases (17 out of the 20 instances). Additionally, the best results (from 50 runs) are also produced in 16 of the 20 instances, with ties occurring on a further 2. A Wilcoxon signed-rank test also reveals the differences in results produced in each set of trials to be significant (with a probability greater than 95%). The results gained when using 40 timeslots also compare well to other approaches. For example, had the best results in the Table 3 been submitted to the timetabling competition, then according to the judging criteria, this algorithm would have been placed second (although note that according to a Wilcoxon signed-rank test, there is actually no significant difference between these results and the competition winner, which, incidentally, also reported results that were the best found in 50 runs).

The reasons why we believe the use of just 40 timeslots to be advantageous have already been outlined in the previous subsection. However, it is also worth noting that although the entire search space will be smaller when we are using only 40 timeslots (because there will be $5m$ fewer places to which events can be assigned to) the removal of the end-of-day timeslots will also have the effect of reducing the number of blanks that are present in the timetable matrix. Indeed, considering that moves in N_2 that involve blanks (and therefore just one event) are, in general, more likely to retain feasibility than those involving two, this means that further restrictions will actually be added to a search space where movements are already somewhat inhibited. Considering that that one of the major requirements for the two-stage timetabling approach is for practical amounts of movements in feasible areas of the search space to be achievable (see section 2), there is thus a slight element of risk in reducing the number of timeslots in this way. For these instances, however, the strategy seems to be beneficial.

Table 3. Comparison of the two trial-sets using the 20 competition instances. In each case the average cost, standard deviation, and best cost (parenthesised) from 50 runs on each instance is reported.

Instance #	**1**	**2**	**3**	**4**	**5**	**6**	**7**	**8**	**9**	**10**
Using 45 slots with $\eta_{max} = 9$ and $\eta_{min} = 0.1$	85.9 ± 10.9. (68)	68.5 ± 8.2. (49)	86.9 ± 12.7. (63)	260.1 ± 23.4. (207)	190.1 ± 25.7. (133)	31.5 ± 8.8. (12)	42.3 ± 17.0. (19)	28.3 ± 7.2. (14)	52.2 ± 9.6. (31)	84.9 ± 8.0. (68)
Using 40 slots with $\eta_{max} = 5$ and $\eta_{min} = 0.9$	86.9 ± 17.6. (62)	53.5 ± 10.2. (39)	95.6 ± 18.8. (69)	231.8 ± 39.5. (176)	147.7 ± 29.5. (106)	22.8 ± 8.4. (11)	23.7 ± 13.3. (5)	22.2 ± 8.6. (10)	41.4 ± 13.7. (22)	91.7 ± 15.3. (70)

Instance #	**11**	**12**	**13**	**14**	**15**	**16**	**17**	**18**	**19**	**20**	**Av.**
Using 45 slots with $\eta_{max} = 9$ and $\eta_{min} = 0.1$	61.6 ± 9.7. (43)	147.5 ± 16.3. (109)	130 ± 14.1. (101)	107 ± 33.4. (55)	41.5 ± 8.5. (22)	47.2 ± 7.9. (29)	169.3 ± 26.5. (119)	45.9 ± 7.8. (27)	85.5 ± 14.7. (62)	9.5 ± 4.5. (1)	88.8 (61.6)
Using 40 slots with $\eta_{max} = 5$ and $\eta_{min} = 0.9$	60.6 ± 16.0. (38)	133.8 ± 28.1. (94)	128.2 ± 19.2. (101)	66.3 ± 20.7. (37)	33.2 ± 13.6. (14)	35.8 ± 12.6. (18)	129.4 ± 25.0. (94)	40.8 ± 9.7. (27)	84.9 ± 21.2. (55)	8.6 ± 6.1. (0)	77 (52.4)

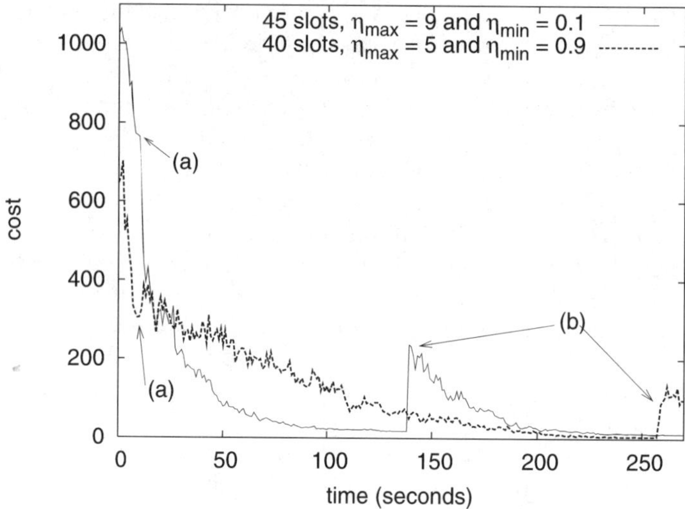

Fig. 6. Two example runs of the SA algorithm on competition instance-20. Points (a) indicate where the algorithm has switched from SA phase-1 to SA phase-2. Points (b) indicate where a reheating has occurred.

Finally, in fig. 6 we show two example runs of the SA algorithm using the parameters defined in Table 3. Here we can observe the contributions that both phases of SA lend to the overall search, and also the general effects of the reheating function (although in one case we can see that it is invoked too late to have a positive effect). Additionally, we can see that the second line (using 40 timeslots) actually starts at a markedly lower cost than the first, because the elimination of all S1 violations in this case, has actually resulted in a better quality initial solution. However, note that this line also indicates a slower progression through the search space during the first half of the run, which could well be due to the greater restrictions on movement within the search space that occur as a result of this condition.

5 Conclusions and Discussion

University timetabling in the real world is an important problem that can often be difficult to solve adequately, and sometimes impossible (without relaxing some of the imposed constraints). In this chapter we have mentioned that one of the most important issues for designers of timetabling algorithms is the question of how to deal effectively with both the hard constraints and the soft constraints, and have noted that when using metaheuristics, this is usually attempted in one of three ways: by using one-stage optimisation algorithms; by using two-stage optimisation algorithms; or by using algorithms that allow relaxations of some feature of the problem.

In this chapter we have given a detailed analysis of the so-called UCTP and have reviewed many of the existing works concerning it. In section 4 we have also provided a description and analysis of our own particular algorithm for this problem. As we have noted, the UCTP was used as the benchmark problem for the International Timetabling Competition in 2002-3. By formulating this problem and then encouraging researchers to write algorithms for it, we have attempted to avoid many of the difficulties that are often caused by the idiosyncratic nature of timetabling, and have provided a means by which researchers can test and compare their algorithms against each other in a meaningful and helpful way.

However, it should be noted that when conducting research in this way we must always be cautious about extrapolating strong scientific conclusions from the results. For example, whilst one timetabling algorithm may *appear* to be superior to another, these differences could be due to mundane reasons such as programming/compiler issues, or the parameters and/or seeds that are used. Superior performance might also simply occur because some algorithms are more suited to the particular constraints of this problem.

It is also worth bearing in mind that whilst the use of benchmark instances may facilitate analysis and comparison of algorithms, ultimately they do not necessarily allow insight into how these algorithms might fare with other kinds of problem instance. For example, in this chapter we have seen that many of the algorithms that have gained good results to the 20 competition instances – including our own – have done so using a two-stage optimisation approach. However, this apparent success could, in part, be due to the competition criteria for judging timetable quality (section 3.3), and also the fact that the instances are fairly easy to solve with regard to finding feasibility. This might therefore lend favour to the two-stage optimisation approach. Indeed, in cases where different judging criteria or different problem instances are used, perhaps some other sort of timetabling strategy would show more value.

In conclusion, when designing algorithms for timetabling, it is always worth remembering that in the real world many different sorts of constraints, problem instances, and even political factors might be encountered. The idiosyncratic nature of real-world timetabling indicates an advantage to those algorithms that are robust with respect to problem-class changes or to those that can easily be adapted to take account of the needs of particular institutions.

References

[1] http://www.idsia.ch/Files/ttcomp2002/

[2] http://www.emergentcomputing.org/timetabling/harderinstances

[3] D. Abramson, "Constructing School Timetables using Simulated Annealing: Sequential and Parallel Algorithms," *Management Science*, vol. 37, pp. 98-113, 1991.

[4] D. Abramson, H. Krishnamoorthy, and H. Dang, "Simulated Annealing Cooling Schedules for the School Timetabling Problem," *Asia-Pacific Journal of Operational Research*, vol. 16, pp. 1-22, 1996.

[5] H. Arntzen and A. Løkketangen, "A Tabu Search Heuristic for a University Timetabling Problem," in *Metaheuristics: Progress as Real Problem Solvers*, vol. 32, *Computer Science Interfaces Series*, T Ikabaki, K. Nonobe, and M. Yagiura, Eds. Berlin: Springer-Verlag, 2005, pp. 65-86.

[6] M. Birattari, T. Stützle, L. Paquete, and K. Varrentrapp, "A Racing Algorithm for Configuring Metaheuristics," presented at The Genetic and Evolutionary Computation Conference (GECCO) 2002, New York, 2002.

[7] D. Brelaz, "New methods to color the vertices of a graph," *Commun. ACM*, vol. 22, pp. 251-256, 1979.

[8] E. Burke, D. Elliman, and R. Weare, "Specialised Recombinative Operators for Timetabling Problems," in *The Artificial Intelligence and Simulated*

Behaviour Workshop on Evolutionary Computing, vol. 993, *Lecture Notes in Computer Science*. Berlin: Springer-Verlag, 1995, pp. 75-85.

[9] E. Burke, D. Elliman, and R. Weare, "The Automation of the Timetabling Process in Higher Education," *Journal of Education Technology Systems*, vol. 23, pp. 257-266, 1995.

[10] E. Burke, D. Elliman, and R. Weare, "A Hybrid Genetic Algorithm for Highly Constrained Timetabling Problems.," presented at Genetic Algorithms: Proceedings of the Sixth International Conference (ICGA95), 1995.

[11] E. Burke and M. Petrovic, "Recent Research Directions in Automated Timetabling," *European Journal of Operational Research*, vol. 140, pp. 266-280, 2002.

[12] E. K. Burke, D. G. Elliman, P. H. Ford, and R. Weare, "Examination Timetabling in British Universities: A Survey," in *Practice and Theory of Automated Timetabling (PATAT) I*, vol. 1153, *Lecture Notes in Computer Science*, E. Burke and P. Ross, Eds. Berlin: Springer-Verlag, 1996, pp. 76-92.

[13] E. K. Burke and J. P. Newall, "A Multi-Stage Evolutionary Algorithm for the Timetable Problem," *IEEE Transactions on Evolutionary Computation*, vol. 3, pp. 63-74, 1999.

[14] M. Carter, "A Survey of Practical Applications of Examination Timetabling Algorithms," *Operations Research*, vol. 34, pp. 193-202, 1986.

[15] M. Carter and G. Laporte, "Recent Developments in Practical Examination Timetabling," in *Practice and Theory of Automated Timetabling (PATAT) I*, vol. 1153, E. Burke and P. Ross, Eds. Berlin: Springer-Verlag, 1996, pp. 3-21.

[16] M. Carter, G. Laporte, and S. Y. Lee, "Examination Timetabling: Algorithmic Strategies and Applications," *Journal of the Operational Research Society*, vol. 47, pp. 373-383, 1996.

[17] M. Carter and G. Laporte, "Recent Developments in Practical Course Timetabling," in *Practice and Theory of Automated Timetabling (PATAT) II*, vol. 1408, *Lecture Notes in Computer Science*, E. Burke and M. Carter, Eds. Berlin: Springer-Verlag, 1998, pp. 3-19.

[18] S. Casey and J. Thompson, "GRASPing the Examination Scheduling Problem," in *Practice and Theory of Automated Timetabling (PATAT) IV*, vol. 2740, *Lecture Notes in Computer Science*, E. Burke and P. De Causmaecker, Eds. Berlin: Springer-Verlag, 2002, pp. 233-244.

[19] M. Chiarandini, K. Socha, M. Birattari, and O. Rossi-Doria, "An Effective Hybrid Approach for the University Course Timetabling Problem," *Technical Report AIDA-2003-05, FG Intellektik, FB Informatik, TU Darmstadt, Germany*, 2003.

[20] A. Colorni, M. Dorigo, and V. Maniezzo, "Metaheuristics for high-school timetabling," *Computational Optimization and Applications*, vol. 9, pp. 277-298, 1997.

[21] T. Cooper and J. Kingston, "The Complexity of Timetable Construction Problems," in *Practice and Theory of Automated Timetabling (PATAT) I*, vol. 1153, *Lecture Notes in Computer Science*, E. Burke and P. Ross, Eds. Berlin: Springer-Verlag, 1996, pp. 283-295.

[22] D. Corne, P. Ross, and H. Fang, "Evolving Timetables," in *The Practical Handbook of Genetic Algorithms*, vol. 1, L. C. Chambers, Ed.: CRC Press, 1995, pp. 219-276.

[23] P. Cote, T. Wong, and R. Sabourin, "Application of a Hybrid Multi-Objective Evolutionary Algorithm to the Uncapacitated Exam Proximity Problem," in *Practice and Theory of Automated Timetabling (PATAT) V*, vol. 3616, *Lecture Notes in Computer Science*, E. Burke and M. Trick, Eds. Berlin: Springer-Verlag, 2005, pp. 294-312.

[24] P. Cowling, S. Ahmadi, P. Cheng, and R. Barone, "Combining Human and Machine Intelligence to Produce Effective Examination Timetables," presented at The Forth Asia-Pacific Conference on Simulated Evolution and Learning (SEAL2002), Singapore, 2002.

[25] A. E. Eiben, J. K. van der Hauw, and J. I. van Hemert, "Graph Coloring with Adaptive Evolutionary Algorithms," *Journal of Heuristics*, vol. 4, pp. 25-46, 1998.

[26] S. Elmohamed, G. Fox, and P. Coddington, "A Comparison of Annealing Techniques for Academic Course Scheduling," in *Practice and Theory of Automated Timetabling (PATAT) II*, vol. 1408, *Lecture Notes in Computer Science*, E. Burke and M. Carter, Eds. Berlin: Springer-Verlag, 1998, pp. 146-166.

[27] E. Erben, "A Grouping Genetic Algorithm for Graph Colouring and Exam Timetabling," in *Practice and Theory of Automated Timetabling (PATAT) III*, vol. 2079, *Lecture Notes in Computer Science*, E. Burke and W. Erben, Eds. Berlin: Springer-Verlag, 2001, pp. 132-158.

[28] S. Even, A. Itai, and A. Shamir, "On the complexity of Timetable and Multicommodity Flow Problems," *SIAM Journal of Computing*, vol. 5, pp. 691-703, 1976.

[29] E. Falkenauer, *Genetic Algorithms and Grouping Problems*: John Wiley and Sons, 1998.

[30] S. Kirkpatrick, C. Gelatt, and M. Vecchi, "Optimization by Simulated Annealing," *Science*, pp. 671-680, 1983.

[31] P. Kostuch, "The University Course Timetabling Problem with a 3-Phase Approach," in *Practice and Theory of Automated Timetabling (PATAT) V*, vol. 3616, *Lecture Notes in Computer Science*, E. Burke and M. Trick, Eds. Berlin: Springer-Verlag, 2005, pp. 109-125.

[32] R. Lewis and B. Paechter, "New Crossover Operators for Timetabling with Evolutionary Algorithms," presented at The Fifth International Conference on Recent Advances in Soft Computing RASC2004, Nottingham, England, 2004.

[33] R. Lewis and B. Paechter, "Application of the Grouping Genetic Algorithm to University Course Timetabling," in *Evolutionary Computation in Combinatorial Optimization (EvoCop)*, vol. 3448, *Lecture Notes in Computer Science*, G. Raidl and J. Gottlieb, Eds. Berlin: Springer-Verlag, 2005, pp. 144-153.

[34] R. Lewis and B. Paechter, "An Empirical Analysis of the Grouping Genetic Algorithm: The Timetabling Case," presented at the IEEE Congress on Evolutionary Computation (IEEE CEC) 2005, Edinburgh, Scotland, 2005.

[35] R. Lewis and B. Paechter, "Finding Feasible Timetables using Group Based Operators," *(Forthcoming) Accepted for publication in the IEEE Trans. Evolutionary Computation*, 2006.

[36] Z. Michalewicz, "The Significance of the Evaluation Function in Evolutionary Algorithms," presented at The Workshop on Evolutionary Algorithms, Institute for Mathematics and Its Applications, University of Minnesota, Minneapolis, Minnesota, 1998.

[37] B. Paechter, R. Rankin, A. Cumming, and T. Fogarty, "Timetabling the Classes of an Entire University with an Evolutionary Algorithm," in *Parallel Problem Solving from Nature (PPSN) V*, vol. 1498, *Lecture Notes in Computer Science*, T. Baeck, A. Eiben, M. Schoenauer, and H. Schwefel, Eds. Berlin: Springer-Verlag, 1998, pp. 865-874.

[38] J. T. Richardson, M. R. Palmer, G. Liepins, and M. Hilliard, "Some Guidelines for Genetic Algorithms with Penalty Functions.," in *the Third International Conference on Genetic Algorithms*, J. D. Schaffer, Ed. San Francisco, CA, USA: Morgan Kaufmann Publishers Inc, 1989, pp. 191-197.

[39] P. Ross, D. Corne, and H.-L. Fang, "Improving Evolutionary Timetabling with Delta Evaluation and Directed Mutation," in *Parallel Problem Solving from Nature (PPSN) III*, vol. 866, *Lecture Notes in Computer Science*, Y. Davidor, H. Schwefel, and M. Reinhard, Eds. Berlin: Springer-Verlag, 1994, pp. 556-565.

[40] P. Ross, D. Corne, and H. Terashima-Marin, "The Phase-Transition Niche for Evolutionary Algorithms in Timetabling," in *Practice and Theory of Automated Timetabling (PATAT) I*, vol. 1153, *Lecture Notes in Computer Science*, E. Burke and P. Ross, Eds. Berlin: Springer-Verlag, 1996, pp. 309-325.

[41] P. Ross, E. Hart, and D. Corne, "Genetic Algorithms and Timetabling," in *Advances in Evolutionary Computing: Theory and Applications*, A. Ghosh and K. Tsutsui, Eds.: Springer-Verlag, New York., 2003, pp. 755- 771.

[42] O. Rossi-Doria, J. Knowles, M. Sampels, K. Socha, and B. Paechter, "A Local Search for the Timetabling Problem," presented at Practice And Theory of Automated Timetabling (PATAT) IV, Gent, Belgium, 2002.

[43] O. Rossi-Doria, M. Samples, M. Birattari, M. Chiarandini, J. Knowles, M. Manfrin, M. Mastrolilli, L. Paquete, B. Paechter, and T. Stützle, "A Comparison of the Performance of Different Metaheuristics on the Timetabling Problem," in *Practice and Theory of Automated Timetabling (PATAT) IV*, vol. 2740, *Lecture Notes in Computer Science*, E. Burke and P. De Causmaecker, Eds. Berlin: Springer-Verlag, 2002, pp. 329-351.

[44] W. Salwach, "Genetic Algorithms in Solving Constraint Satisfaction Problems: The Timetabling Case," *Badania Operacyjne i Decyzje*, 1997.

[45] A. Schaerf, "Tabu Search Techniques for Large High-School Timetabling Problems," in *Proceedings of the Thirteenth National Conference on Artificial Intelligence*. Portland (OR): AAAI Press/ MIT Press, 1996, pp. 363-368.

[46] A. Schaerf, "A Survey of Automated Timetabling," *Artificial Intelligence Review*, vol. 13, pp. 87-127, 1999.

[47] K. Socha and M. Samples, "Ant Algorithms for the University Course Timetabling Problem with Regard to the State-of-the-Art," in *Evolutionary Com-*

putation in Combinatorial Optimization (EvoCOP 2003), vol. 2611, *Lecture Notes in Computer Science*. Berlin: Springer-Verlag, 2003, pp. 334-345.

[48] H. Terashima-Marin, P. Ross, and M. Valenzuela-Rendon, "Evolution of Constraint Satisfaction Strategies in Examination Timetabling," presented at The Genetic and Evolutionary Computation Conference (GECCO), 2000.

[49] J. M. Thompson and K. A. Dowsland, "A Robust Simulated Annealing based Examination Timetabling System," *Computers and Operations Research*, vol. 25, pp. 637-648, 1998.

[50] P. van Laarhoven and E. Aarts, *Simulated Annealing: Theory and Applications*. Reidel, The Netherlands: Kluwer Academic Publishers, 1987.

[51] G. White and W. Chan, "Towards the Construction of Optimal Examination Schedules," *INFOR*, vol. 17, pp. 219-229, 1979.

Optimum Oil Production Planning using an Evolutionary Approach

Tapabrata Ray[1] and Ruhul Sarker[2]

[1]School of Aerospace, Civil and Mechanical Engineering
[2]School of Information Technology and Electrical Engineering
University of New South Wales at the Australian Defence Force Academy, Northcott Drive, Canberra 2600, Australia
Email: {t.ray, r.sarker}@adfa.edu.au

Summary. In this chapter, we discuss a practical oil production planning problem from a petroleum field. A field typically consists of a number of oil wells and to extract oil from these wells, gas is usually injected which is referred as gas-lift. The total gas used for oil extraction is constrained by daily availability limits. The oil extracted from each well is known to be a nonlinear function of the gas injected into the well and varies between wells. The problem is to identify the optimal amount of gas that needs to be injected into each well to maximize the amount of oil extracted subject to the constraint posed by the daily gas availability. The problem has long been of practical interest to all major oil exploration companies as it has a potential of deriving large financial benefits. Considering the complexity of the problem, we have used an evolutionary algorithm to solve various forms of the production planning problem. The multiobjective formulation is attractive as it eliminates the need to solve such problems on a daily basis while maintaining the quality of solutions. Our results show significant improvement over existing practices. We have also introduced a methodology to deliver robust solutions to the above problem and illustrated it using the six-well problem. Furthermore, we have also proposed a methodology to create and use a surrogate model within the framework of evolutionary optimization to realistically deal with such problems where oil extracted from a well is a nonlinear function of gas injection and a piecewise linear model may not be appropriate.

T. Ray and R. Sarker: *Optimum Oil Production Planning using an Evolutionary Approach*, Studies in Computational Intelligence (SCI) **49**, 273–292 (2007)

1 Introduction

Petroleum, either oil or gas, is a finite and scare resource upon which modern society is heavily dependent on. Hence, mankind is forced to rationalize and optimize its production and consumption. In this chapter, we consider a crude oil production system. In the system, there is a underground oil reservoir and the reservoir has a number of wells. There are two basic methods of extracting oil from such reservoirs (Kosmidis et al., 2005): (i) naturally flowing and (ii) gas lift. In the first one, the oil is able to flow naturally to surface, while the second requires injection of high pressure gas to facilitate oil extraction. The gas lift is considered as the most economic method for artificial lifting of oil (Aaytollahi et al, 2004 and Camponogara and Nakashima, 2005).

In this study, we consider gas lift extraction method. As it will be discussed later, for a given well, the oil production per day can be expressed as a nonlinear function of gas injected into the well in that day. The oil production per day increases with the increase of gas used to a certain level and then decreases. That means an excessive use of gas may increase the gas cost, as well as production cost, without providing any benefit in terms of oil production volume. For a given amount of gas used, the amount of oil extraction significantly varies from well to well. That means the nonlinear function of gas usage versus oil extracted varies from well to well. As a result, an inappropriate gas allocation to different wells, under limited gas availability, will reduce the overall production and hence profitability from the entire reservoir. So the single objective gas lift optimization problem is to allocate a limited amount of gas to a number wells in a reservoir while maximizing the total oil production in a day. However, the amount gas may vary from day to day. That means, the management has to re-solve the problem if the amount of gas is different. In such situations, it is appropriate to solve the problem as bi-objective problem where the objectives would be:

- Maximize oil production and
- Minimize gas used.

Prior research in gas-lift optimization only devoted to single objective optimization problem using either a single well model (Fang and Lo, 1996) or multiple wells model (Dutta-Roy and Kattapuram, 1997). A range of methodologies was used in solving this problem such as equal-slope method (Nishikiori et al., 1989), linear programming (Fang and Lo, 1996), mixed integer linear programming (Kosmidis et al., 2005) quadratic programming (Dutta-Roy and Kattapuram, 1997), dynamic programming (Camponogara and Nakashima, 2005) and others. In this research, we consider a six well and a fifty six well problem. We define the problem as a single and a multiobjective problem and use evolutionary algorithm to solve the mathematical models. Evolutionary algorithms have been used to solve a number of multiobjective optimization problems from the domain of operations research in recent years (Sarker et al., 2002). Excellent comprehensive review of evolutionary multiobjective optimization appears in Coello (1999).

This chapter is organized as follows. Following introduction, we present a mathematical model of the problem. The following section presents the algorithm used for solving the problem. The last two sections discuss about results and conclusions.

2 Mathematical Model

As mentioned earlier, the oil production per day from a given well is a function of the gas injected into int. However, there is no standard function which can be used to determine the production level for all the wells. The practice is to collect production data i.e. amount of gas injected versus amount of oil extracted at a number of discrete points for each well and then generate an approximate function. Using these discrete data points, the researchers construct a function either as piece-wise linear [1] or as a quadratic function [3]. Although both functions have drawbacks in estimating the production level accurately, we would use piece-wise linear function as this method is widely used and easy to model.

The mathematical model of the problem is formulated as follows:

Parameters:

N the number of wells

I_n the number of line segments (gas used axis – x) in the function in well n

G_{ni} slope of the function (oil produced per unit of gas used) at the line segment i in well n

GL limit of gas usage in all wells per day

U_{ni} upper limit of gas usage rate at the individual line segment i in well n

Variables:

X_n is the gas used in well n

x_{ni} is the gas usage in segment i in well n

$$S_{ni} = \begin{cases} 0 & if\ x_{ni} < U_{ni} \\ 1 & if\ x_{ni} = U_{ni} \end{cases}$$

The relationships between X_n and x_{ni} are as follows:

$$X_n = \sum_i x_{ni} \qquad \forall n$$

Where

$$0 \le x_{ni} \le U_{ni} \quad \forall n, i$$

In addition, for a given value of X_n (which is in segment i in well n), all $x_{n(i+1)}$ will be equal to zero and all $x_{n(i-1)}$ will be at the upper bound. However, the value of x_{ni} will be greater than zero and less than or equal to the upper bound.

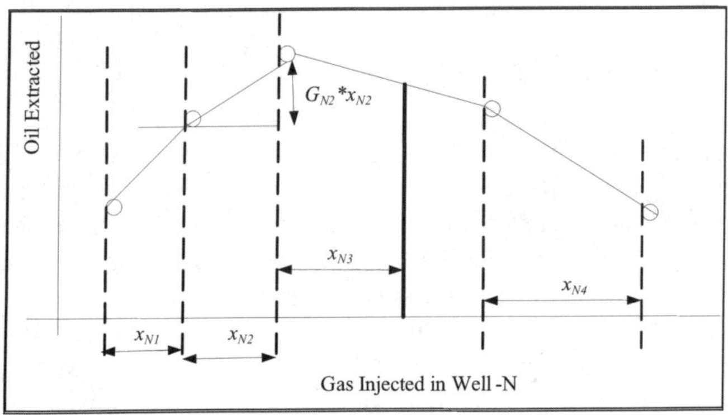

Fig. 1. Schematic Diagram of Solution Representation

2.1 Single Objective Model

We first present a single objective model for the above problem. The objective of the problem is to maximize the total daily production.

$$\text{Maximize } Z = \sum_n \sum_i G_{ni} x_{ni}$$

<u>Constraints</u>:

(i) Calculating the gas usage level in each well

$$X_n = \sum_i x_{ni} \qquad \forall n$$

(ii) To make the constraint (i) meaningful for a piece wise linear curve, the following condition must hold.

$$U_{ni}S_{ni} \le x_{ni} \le U_{ni}S_{n(i-1)} \qquad \forall n,i$$

Here S_{n0} = 1 and S_{nI_n} = 0. This constraint will basically set x_{n4} and all other $x_{ni\ (i>=4)}$ =0 as in Fig. 1.

(iii) Gas limitation: The total use of gas in all wells in a given day must be less than or equal to the available gas.

$$\sum_n X_n \le GL$$

(iv) Nonnegativity constraints

$$x_{ni} \ge 0,\ X_n \ge 0,\ S_{ni} \in (0,1),\ \forall n,i$$

In a later section, we will use the model to solve the 30 day planning problem where the daily oil production and the gas use is represented by $\sum_n \sum_i G_{nit} x_{nit}$ and $\sum_n X_{nt}$ (t = 1 to 30) respectively.

2.2 Multiobjective Model

In reality we want to maximize the oil production using minimum possible gas. So the problem can be defined as a bi-objective problem where the objectives are:

- Maximization of oil production and
- Minimization of gas use

The corresponding objective functions are as follows:

Maximize $Z_1 = \sum_n \sum_i G_{ni} x_{ni}$

Minimize $Z_2 = \sum_n X_n$

Subject to the above constraints

2.3 Solution Approach

We have developed an evolutionary algorithm to solve the above problem. This algorithm is a variant of NSGA-II [8 and 9] and has a major difference in the process of population reduction. In the process of population reduction from a size of 2M to M, the method insists on maintaining not only the end points of the objective space but also maximum and minimum values of the variables. The process is certainly more computationally expensive than NSGA-II and can be thought as a diversity maintaining mechanism which might be useful for problems where the diversity in the variable space is important. Before introducing the algorithm, it is necessary to introduce the notion of Pareto optimal design and nondominated design. A design $\mathbf{x}^* \in F$ is termed Pareto optimal if there does not exist another $\mathbf{x} \in F$, such that $f_i(\mathbf{x}) \leq f_i(\mathbf{x}^*)$ for all $i = 1,...,k$ objectives and $f_j(\mathbf{x}) < f_j(\mathbf{x}^*)$ for at least one j. Here, F denotes the feasible space (i.e., regions where the constraints are satisfied) and $f_j(\mathbf{x})$ denotes the jth objective corresponding to the design $\mathbf{x}$. If the design space is limited to M solutions instead of the entire F, the set of solutions are termed as nondominated solutions. Since in practice, all the solutions in F cannot be evaluated exhaustively, the goal of multiobjective optimization is to arrive at the set nondominated solutions with a hope that it is sufficiently close to the set of Pareto solutions. Diversity among these set of solutions is also a desirable feature as it means making a selection from a wider set of design alternatives.

The pseudo-code of the Multi-objective Constrained Algorithm (MCA) is provided below.

(a) $t \leftarrow 0$
(b) Generate M individuals representing a population: $Pop(t) = \{I_1, I_2, \ldots, I_M\}$ uniformly in the parametric space.
(c) Evaluate each individual: Compute their objectives and constraints i.e., $f_k(I_i)$ *and* $c_j(I_i)$; for $i = 1,2,\ldots,M$ individuals, $k = 1,\ldots,O$ objectives and $j = 1,\ldots,S$ constraints.
(d) **Select** two parents P1 and P2. (The procedure for selection is described below).
(e) **Create** two children C1 and C2 via crossover and mutation of P1 and P2.
(f) Repeat steps (d) and (e) until M children are created.
(g) Evaluate M children.
(h) Merge M parents and M Children to form a population of size 2M.
(i) **Retain** better performing M solutions from the above 2M solutions.
(j) $t \leftarrow t+1$
(k) If $t < T_{max}$ then repeat steps (d) through (j) Else Stop. $where T_{max}$ denotes the maximum number of generations.

The procedure for selecting a parent P1 is described below and the same applies to selecting P2.

(a) Select two individuals (S1 and S2) from the population of M solutions using a uniform random selection.
(b) If S1 is feasible and S2 is infeasible: S1 is selected as the parent and vice versa.
(c) If both S1 and S2 are infeasible: One which has the minimum value of the maximum violated constraint is selected as the parent. Compute $max(g_i(x); i = 1, S)$ for S1 and S2 and choose the one which has the minimum value. ($g_i(x)$ denotes the constraints).
(d) If both S1 and S2 are feasible and S1 dominates S2: S1 is selected as parent and vice versa.
(e) If both S1 and S2 are feasible and none dominates each other: The parent is a random choice between S1 and S2.

We have used simulated binary crossover (SBX) and the polynomial mutation for the real variables as adapted in NSGA-II to create two children from a pair of parents.

The procedure to retain M solutions from a set of 2M solutions is presented below:
(a) Rank the set of 2M solutions. (The procedure for ranking is described below).
(b) If the number of rank=1 solutions (i.e. non-dominated solutions) is less than M, select top M solutions based on their rank and copy them to the new population.
(c) If the number of rank=1 solutions is more than M, follow the following steps:
   i. Select the solutions which have a minimum value (assuming minimization in both objectives) in any of the objectives and copy them to the new population.
   ii. For every variable, copy two solutions to the new population which has its minimum and the maximum value if they have not been copied yet (including step i). If the number of solutions copied is more than M/2, than M/2 randomly selected solutions are allowed to stay in the population.
   iii. For the remaining rank=1 solutions, the sequence of who goes in first to the new population is decided as follows:
      1. Compute the score by inserting the solution into the new population one at a time. The score is the minimum Euclidean distance computed between the solution attempting to enter with all other existing solutions in the new population based on the objective function space. Scaled values are used for the score

computation i.e. the objective space is scaled using the maximum and minimum values in each dimension based on the set of rank=1 solutions.

2. The solution with the highest score is allowed to go into the new population unless it has been copied earlier in which case the solution with the next score goes in.
3. The steps (1) and (2) are repeated until the new population has a size of M.

The procedure for rank computation is as follows:

(a) Separate the set of 2M solutions to a set of feasible and a set of infeasible solutions.
(b) Perform a non-dominated sorting to assign ranks to the solutions in the feasible set.
(c) Rank the solutions in the infeasible set based on their maximum value of the violated constraint.
(d) Update the ranks of the solutions in the infeasible set by adding the rank of the worst feasible solution to each.

The assumptions behind the procedures in the algorithm are:

(a) A feasible solution is always preferred over an infeasible solution. This is a commonly adopted practice, although one might argue that it's better to retain a marginally infeasible solution rather than a bad feasible solution.
(b) Step (i) in the above procedure ensures that the endpoints in the objective space are inserted into the new population and the extent of the non-dominated front is preserved.
(c) Step (ii) is a means to maintain variable diversity i.e. to include a possibility of retaining variable values which might be useful.

3 Results and Discussion

In this research, we have used representative data for the six wells problem and the fifty six well problem from Buitrago et al (1996). The data for each well represent a relationship between oil extraction and gas usage for a number of discrete points. For modelling purposes, we consider a linear interpolation between any two consecutive data points. That means we use a piece wise linear function. The functions for sample data for a six-well problem are shown in Appendix-A. The gas-lift optimization model formulated above was solved using the algorithm presented in an earlier section. The runs were conducted on a Desktop (IBM ThinkCentre: Intel Pentium 4, 2.8GHz, 1GB RAM). Source Codes were complied using Visual C++ Version 6.0, Enterprise Edition.

3.1 Six Well Problem: Single Objective Formulation

The problem is to maximize the total oil that can be extracted from six wells subject to a constraint which limits the maximum gas injection volume of 6000 thousand standard cubic feet(MSCF) into each well. The problem was solved with the following set of parameters (a) population size of 100; (b) generations of 100 and 200 (c) probability of crossover of 0.7 and 0.9 (d) probability of mutation of 0.1 and 0.2 (e) distribution index for crossover as 10 and 20 (f) distribution index of mutation as 10 and 20 and finally (g) with random seeds of 0.2, 0.4 and 0.6. The above set of parameters relate to 96 independent runs. The results are summarized in Table 1.

The best result obtained from the above set of trials corresponds to an oil production of 3663.99 barrels per day(BPD) whereas Buitrago et al (1996) reported a value of 3629.0 BPD. This corresponds to a benefit of 35 BPD which is significant. The gas (MSCF) to be injected in each of the six wells are (475.524996, 743.445098, 1350.921920, 827.749079, 1199.584681, 0.212049) respectively. The gas injection into each well was allowed to vary between 0 and 6000 and treated as a real variable. The average CPU time for an optimization run with a population size of 100 and a generation of 100 is around 3.04 seconds.

Table 1. Summary of Results for the Six Well Problem

Best	3663.99
Worst	3653.90
Average	3660.20
Median	3660.77

3.2 Six Well Problem: Multiobjective Formulation

We have also solved the multiobjective formulation of the six well problem. The first objective is to maximize the total oil extracted from the wells and second objective minimizes the total volume of gas injected into the wells. The nondominated set of solutions obtained using a population size of 100, generation of 200, probability of crossover as 0.9, probability of mutation as 0.1, distribution index of crossover as 10 and distribution index of mutation of 20 and random seed of 0.2 is presented in Figure 2.

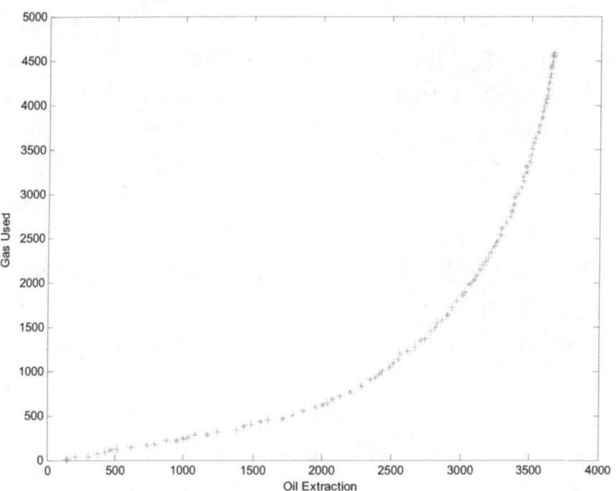

Fig. 2. Nondominated Set of Solutions for the Multiobjective Problem

One can observe a well spread set of solutions with different oil extraction values and corresponding gas usage. If there are requirements that state a minimum oil extraction volume has to be achieved, a constraint can easily be added into the model to take that into consideration.

3.3 Six Well Problem: Robust Formulation

Often in reality, there is a need to ensure that the gas allocation plan is robust i.e. the performance of the solution (oil extracted) should not largely vary upon marginal variations in the amount of gas injected to the wells. Such a solution is termed as a *robust solution* and it should also ensure that the solution does not violate constraints during the course of marginal variation. We define a 1% of the variable range i.e. 60 units of variation in the amount of gas injected to each well around the operating point and a randomly sampled neighborhood of 50 points to assess the robustness of a solution. The problem is then solved as a multiobjective optimization problem where the first objective is the maximization of the oil extracted while the second relates to minimizing the standard deviation of the oil extracted based on the neighborhood samples as defined above. The first constraint remains the same, i.e the amount of gas usage in any well should be less than 6000 MSCF and the second additional constraint ensures no violations among the neighborhood samples.

The results of the problem are presented in Figure 3.

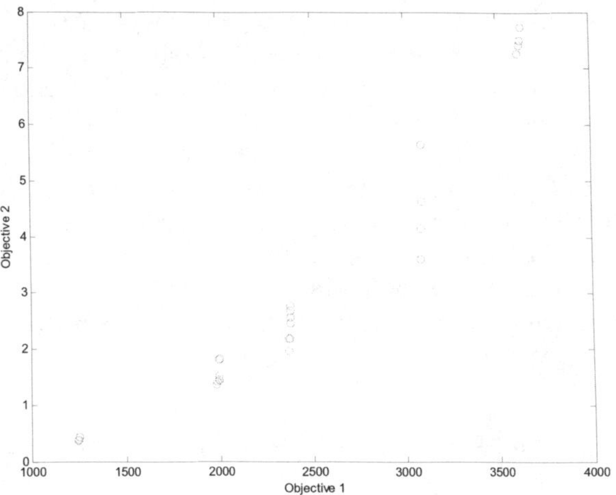

Fig. 3. Results of the Robust Formulation

It is interesting to observe that there are distinct bands of solutions. A typical solution from the rightmost band with objective function values of (3628.231005, 7.499015) has the gas injection volumes (MSCF) as (517.191202, 832.566250, 1032.058910, 722.134616, 1184.509518, 129.363873). The solution corresponding to the performance maximized design in the earlier section has objective values of (3663.99, 8.33384) and the gas injection volumes are (475.524996, 743.445098, 1350.921920, 827.749079, 1199.584681, 0.212049). It is also interesting to observe that the performance maximized solution (result of the earlier section) violates constraint 2 of the robust formulation by 1807 units. This indicates that the solution of the performance maximized plan is likely to violate the neighborhood constraint. It is expected that the robust solution with an objective value of 3628.231005 will have a performance less than the performance maximized solution which is 3663.99 as the variance of the robust solution is 7.499015 which is lower than 8.33384.

3.4 Six Well Problem: Single Objective Formulation with Surrogate Assistance

One of the important issues that we have not addressed so far relates to the process of generating approximating functions i.e. oil extracted as a function of the gas injected into each well. In this study, we have used a piecewise linear function simply because that was used by an earlier study. In order to deal effectively with such classes of problems, we have incorporated a function approximation module that is able to generate approximating functions based on piecewise linear, quadratic functions, multilayer perceptrons, radial basis function networks and kriging.

Using a quadratic response surface model for each well, we could obtain optimal solutions to the single objective optimization problem with an objective value of 3659.139691 and the gas injection volumes as (489.646790, 715.501427, 1069.667052, 1088.063162, 1194.282731 and 17.523259). The results are fairly close to that of the piecewise linear model as the functions can be reasonably well approximated using a quadratic function.

3.5 Fifty-Six Well Problem: Single Objective Formulation

In order to demonstrate that our methodology is suitable even for larger problems, we took up the fifty six well problem. The problem is to maximize the total oil that can be extracted from fifty six wells subject to a constraint which limits the maximum injection volume of 6000 (MSCF) into each well. The problem was solved with the following set of parameters (a) population size of 100; (b) generations of 100 and 200 (c) probability of crossover of 0.7 and 0.9 (d) probability of mutation of 0.1 and 0.2 (e) distribution index for crossover as 10 and 20 (f) distribution index of mutation as 10 and 20 and finally (g) with random seeds of 0.2, 0.4 and 0.6. The above set of parameters relate to 96 independent runs. The results are summarized in Table 2.

Table 2. Summary of Results for the Fifty Six Well Problem

Best	22033.4
Worst	21222.4
Average	21622.3
Median	21651.2

The best result obtained from the above set of trials corresponds to an oil production of 22033.4 barrels per day(BPD) whereas Buitrago et al (1996) reported a value of 21789.9 BPD. This corresponds to an increase of 243 BPD. The gas injected (MSCF) in each well is (812.344738, 447.222418, 150.800916, 25.725667, 2.681695, 428.400480, 443.113410, 550.422412, 1431.407125, 9.943476, 1186.702945, 1797.354002, 0.292891, 381.387919, 855.172364, 966.752560, 572.051953, 835.340977, 53.061313, 1417.768635, 674.159487, 260.742450,

0.062342, 738.513865, 142.199457, 3.961657, 1.579943, 180.758969, 25.616667, 2.859420, 2.491896, 175.831471, 500.263909, 1.078196, 211.798944, 0.031785, 1.727461, 242.665409, 313.904987, 13.693065, 329.857509, 0.099835, 1199.252135, 34.839834, 59.724609, 2.538431,1.812014, 2572.130336, 0.075651, 0.044171, 0.371710, 5.910134, 11.686541, 16.888570, 5.103219, 2274.166799) respectively. The average CPU time for a run with a population size of 100 and a generation of 100 is around 4.03 seconds. A typical progress plot from a run for the fifty six well problem is presented in Figure 4.

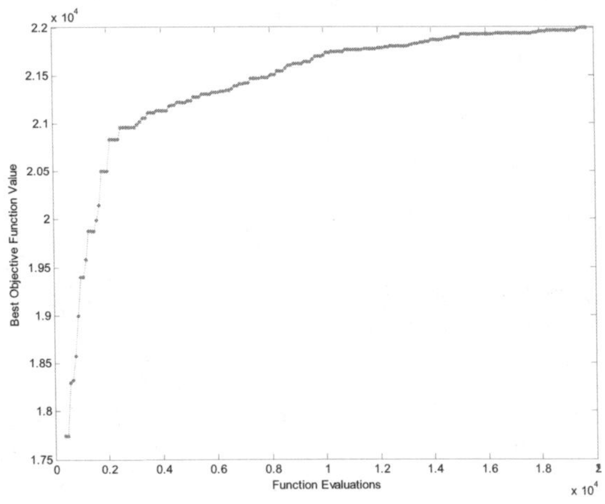

Fig. 4. Progress plot for a typical run of the fifty six well problem

3.6 Fifty-Six Well Problem: Multiobjective Formulation

We have also solved the multiobjective formulation of the fifty-six well problem. The first objective is to maximize the total oil extracted from the wells and second objective minimizes the total volume of gas injected into the wells. The nondominated set of solutions obtained using a population size of 100, generation of 200, probability of crossover as 0.9, probability of mutation as 0.1, distribution index of crossover as 10 and distribution index of mutation of 20 and random seed of 0.2 is presented in Figure 5.

Once again, one can notice the spread of solutions along the front. For both the single objective optimization problems, our algorithm reported better results than what was reported in literature (Buitrago et al (1996). We have also solved the multiojective versions of the problem as such an approach gives an overview of how many BPD can be produced given an amount of gas on a daily basis instead of resolving the problem everyday.

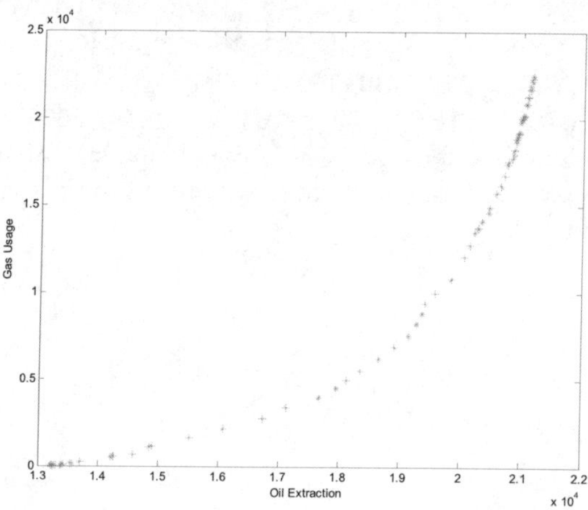

Fig. 5. Nondominated Set of Solutions for the Multiobjective Problem

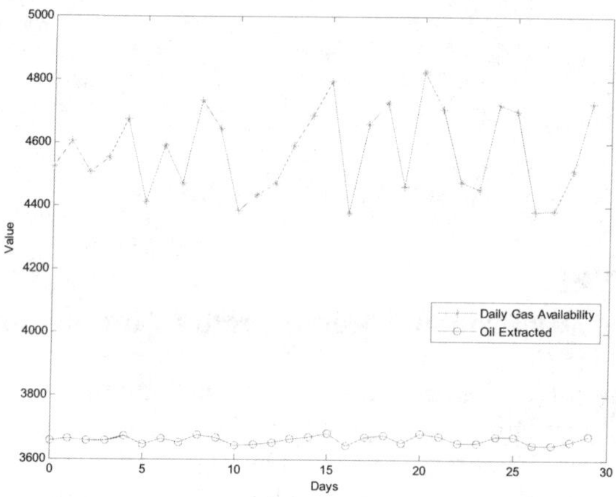

Fig. 6. Daily Gas Availability and Oil Extracted over a 30 day period

3.7 Thirty-Day Reservoir Optimization Problem:

In reality, the problems listed needs to be solved on a regular basis. We have assumed a daily gas availability variation of 4600 +/- 460 for a period of 30 days. The problem was solved daily and the results are presented in Figure 6 and Figure 7.

From Figure 6, one can observe that an increase in gas availability, results in an increase in oil extraction. This behavior is consistent i.e. ups and downs in the gas availability curve correspond to ups and downs in the oil extracted curve when solving optimally.

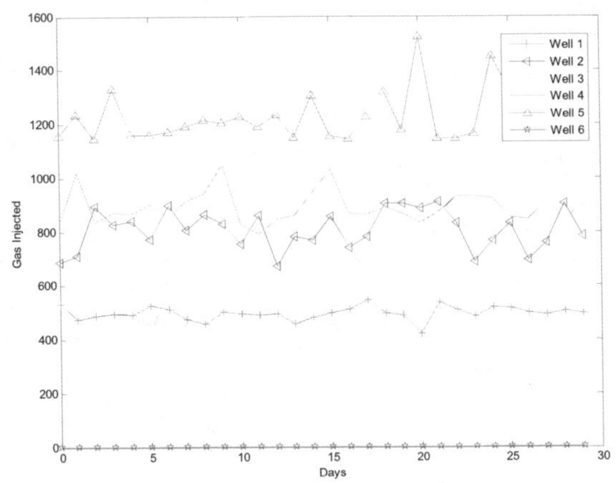

Fig. 7. Daily Gas Injection to the Wells over a 30 day period

The part of the curve presented in Figure 8 is actually a segment of the nondominated curve presented in Figure 2. To further analyze and validate our findings, we solved a MO problem with additional constraints of gas availability lying between 4600 /- 460 so that we obtain the segment of the nondominated front that is of interest to us. The plot of the nondominated solitions of the above MO problem and the solutions listed in Figure 8 are plotted on the same scale in Figure 9. It is interesting to observe that the solutions to the MO problem are effectively of the same quality as the solutions obtained by solving single objective formulations on a daily basis. This means it is possible to generate the whole range of solutions using a MO formulation and use it as required on a daily basis instead of resolving the problem on a daily basis.

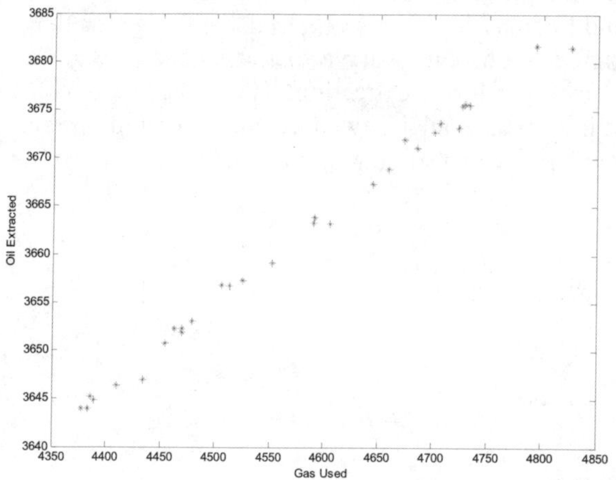

Fig. 8. Variation of Volume of Oil Extracted with Gas Availability

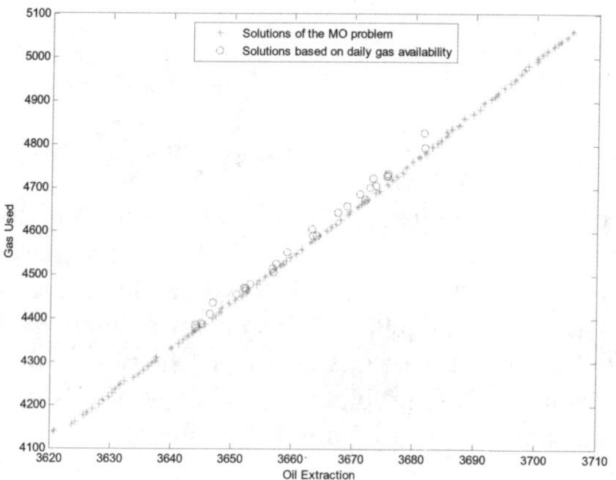

Fig. 9. Comparison of the Quality of Solutions between a MO approach and an approach that relies on SO optimization on a daily basis.

If there are constraints on daily gas availability and also on the amount of oil extraction, the same set of results of the MO problem can be directly used without the need to resolve the problem. As an example, if we assume that the forecasted daily gas availability and amount of oil extracted for a period if 30 days is provided to us(Fig. 10), a plot of them over the solutions of the MO problem will appear as Fig. 11. It is clear from Fig. 11, that the points below the MO line is not feasible to achieve and the additional amount of gas required or the reduced oil extraction limits can be obtained from the solutions of the MO problem.

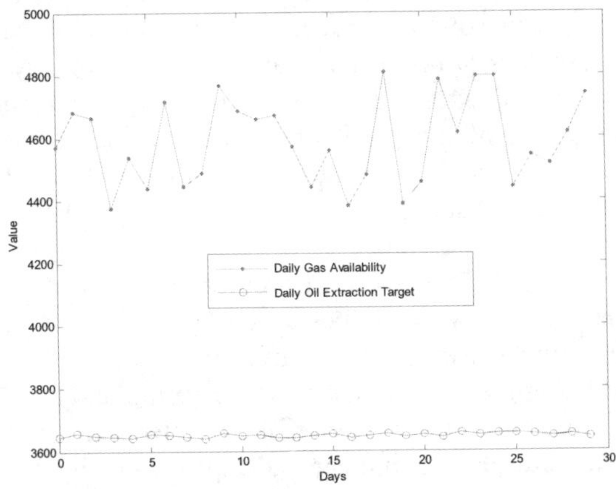

Fig. 10. Projected Oil Extraction and Gas Availability for 30 days

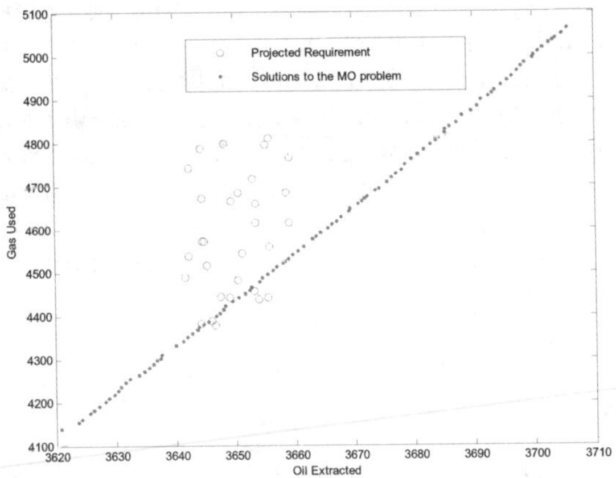

Fig. 11. Solutions to the MO problem and the projected 30day requirements

4 Conclusions

In this chapter, we have introduced a practical gas-lift optimization problem and solved the single and multiobjective versions of the problem using our multiobjective evolutionary algorithm. For both the six well and the fifty six well problem, our method reported better results than that of previous reports. For the six well problem, we reported a solution with 35 BPD (12775 barrels improvement on a yearly basis) more and for the fifty six well problem our solution provides 243 BPD (88695 barrels improvement on a yearly basis) more than the results reported by Buitrago et al (1996). We have also extended our study to include the results of multiobjective formulations of the problem to highlight some of the additional benefits that can be derived using our multiobjective evolutionary algorithm. The consistency and the efficiency of the algorithm have been demonstrated through multiple runs.

We have also solved the six well problem to yield robust solution and compared the solution with the performance maximized design. Furthermore, to deal with realistic problems from the field, we have introduced a surrogate modeling framework that allows creation and use of approximating functions to capture the behavior of gas injection versus oil extracted functions which are often nonlinear and a piecewise linear approach may not be sufficiently accurate.

Our MO approach is attractive as it eliminates the need to solve gas lift optimization problems on a daily basis. Since, our method does not rely on functional or slope continuity of the objective or constraint functions, the method can be easily coupled with functional approximation models/ surrogate models of oil extraction versus gas usage for each well. Since the oil extraction versus gas usage is known to be a nonlinear function that varies across wells, we are currently exploring the possibility of automatically creating surrogate models based on data and then using them within the optimization framework to derive optimal gas allocation in each well for the maximum oil extraction from the field.

References

1. Coello Coello, C. A.: A Comprehensive Survey of Evolutionary-Based Multiobjective Optimization Techniques, Knowledge and Information Systems: An International Journal, (1999) 269-308
2. Camponogara, E., Nakashima, P.: Solving a Gas Lift Optimization Problem using Dynamic Programming, European Journal of Operational Research, (2005) Article in Press (Available on-line)
3. Deb, K., Pratap, A., Agarwal, S., Meyarivan, T.: A Fast and Elitist Multiobjective Genetic Algorithm: NSGAII. IEEE Transaction on Evolutionary Computation, (2002) 181-197

4. Deb, K., Pratap, A., Agarwal, S., Meyarivan, T.: A Fast Elitist Non-dominated Sorting Genetic Algorithm for Multi-objective Optimization: NSGAII, Proceedings of the Parallel Problem Solving from Nature VI Conference, Paris, France, (2000) 849-858
5. Dutta-Roy, K., Kattapuram, J.: A New Approach to Gas Lift Allocation Optimization, SPE Western Regional Meeting, SPE 38333 (1997)
6. Nishikiori, N., Render, R. A., Doty, D. R., Schmidt, Z.: An Improved Method for Gas Lift Allocation Optimization, 64th Annual Technical Conference and Exhibition of SPE, SPE18711 (1989)
7. Sarker, R., Liang, K., Newton, C.: A New Evolutionary Algorithm for Multiobjective Optimization, European Journal of Operational Research, (2002) 12-23
8. Aaytollahi, S., Narimani, M., Moshfeghian, M.: Intermittent Gas Lift in Aghajari Oil Field, a Mathematical Study, Journal of Petroleum Science and Engineering, (2004) 245-255
9. S. Buitrago, S., Rodriguez, E., Espin, D.: Global Optimization Techniques in Gas Allocation for Continuous Flow Gas Lift Systems, SPE Gas Technology Conference, Calgary, Canada, April 28 – May 1, 1996, SPE35616, (1996) 375-383.
10. Kosmidis, V., Perkins, J., Pistikopoulos, E.: A Mixed Integer Optimization Formulation for the Well Scheduling Problem on Petroleum Fields, Computers and Chemical Engineering, (2005) 1523-1541
11. Fang, W. Y., Lo, K. K.: A Generalized Well-Management Scheme for Reservoir Simulation, SPE Reservoir Engineering, (1996) 116-120

Appendix A

Production vs gas usage for six wells

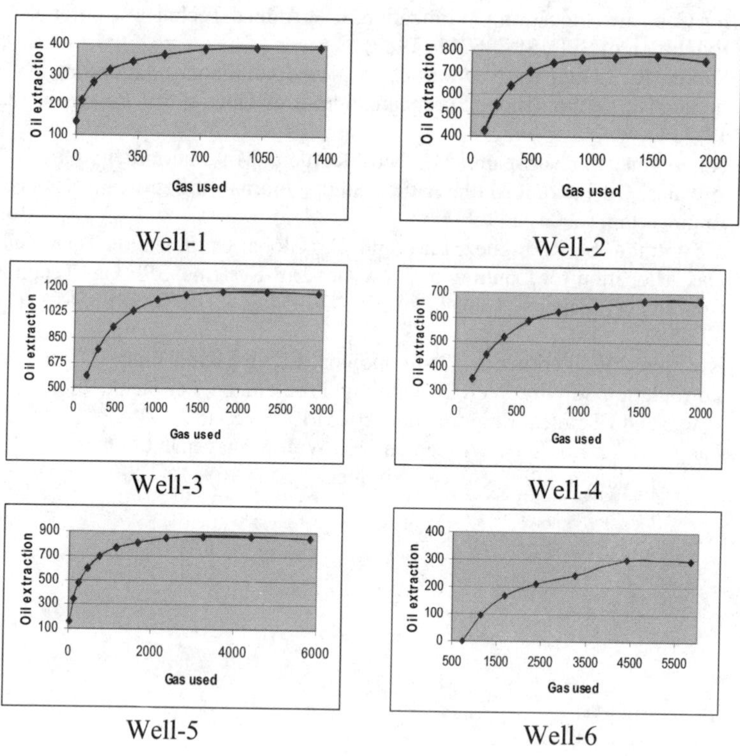

Well-1 Well-2

Well-3 Well-4

Well-5 Well-6

A Hybrid Evolutionary Algorithm for Service Restoration in Power Distribution Systems

Isamu Watanabe[1], Ikuo Kurihara[1], and Yoshiki Nakachi[2]

[1] System Engineering Research Laboratory,
Central Research Institute of Electric Power Industry (CRIEPI),
2-11-1, Iwado Kita, Komae-shi, Tokyo 201-8511, Japan
{isamu,kurihara}@criepi.denken.or.jp,
[2] Chubu Electric Power Co., Inc.,
20-1, Kitasekiyama, Ohdaka-cho, Midori-ku, Nagoya 459-8522, Japan
Nakachi.Yoshiki@chuden.co.jp

Abstract. This paper proposes a hybrid evolutionary algorithm for addressing service restoration problems in power distribution systems. The authors have already proposed an optimization algorithm based on a genetic algorithm (GA), called the two-stage GA, and this algorithm has been shown to perform well for small systems. However, it is difficult to apply to large-scale systems from the viewpoint of computation time. To improve the time performance of the algorithm, the authors introduce three kinds of speed-up strategy: a local search procedure, greedy algorithm, and efficient maximum flow algorithm. Computational results with several test systems show that the proposed hybrid algorithm can dramatically reduce the computation time compared with the two-stage GA, and can be applied to real-scale systems.

1 Introduction

Power distribution systems are usually operated in a radial configuration, but possess a kind of meshed structure that allows for several operating configurations. This means that in large distribution systems, normally open (disconnecting) lines between neighboring parts of the network are generally provided. When a fault occurs, in order to restore as many loads as possible, the areas isolated by the fault should be supplied by transferring loads in the out-of-service areas to other distribution feeders via network reconfigurations. This procedure is called *service restoration.* Figure 1 shows a simple example of service restoration. To restore the isolated area by the fault on switch S12, the network is reconfigured by altering the status of switches S11, S13, and S16. By reconfiguring the network, the isolated area can be supplied from other power sources. Finding a new network configuration is not the only concern in service restoration. It is also important to find the optimal sequence of switching operations. In order not to interrupt loads while changing configurations, the switching operations must be performed sequentially. Hence, the task of reaching a particular configuration can be regarded as scheduling of switching operations, that is, finding a sequence of connections and disconnections of line sections (closing and opening switches).

I. Watanabe et al.: *A Hybrid Evolutionary Algorithm for Service Restoration in Power Distribution Systems,* Studies in Computational Intelligence (SCI) **49**, 293–311 (2007)

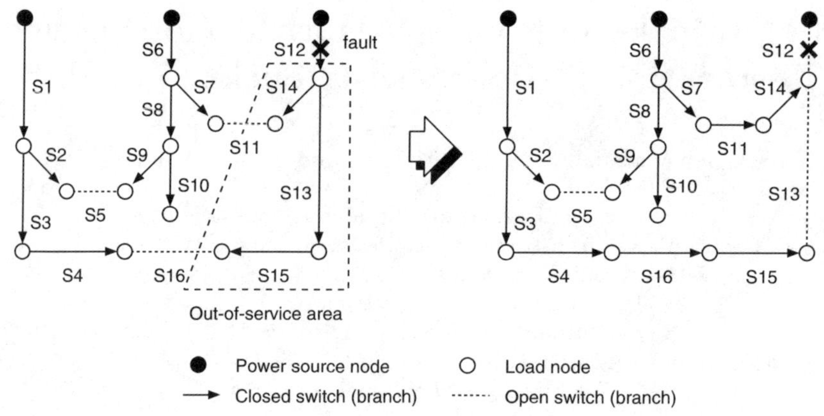

Fig. 1. A simple example of service restoration

Because an effective service restoration strategy plays a key role in improving system reliability, there has been considerable research focused on this problem. The problem has been addressed with methods such as integer programming [1, 2], knowledge-based expert systems [3, 4], artificial neural networks [5], fuzzy reasoning [6–8], and heuristic search [8–10]. Although these approaches can solve the problem with rather less computational burden, the results are only approximations and local optima. In addition, it is difficult to find a global optimum in a real-scale system that would have a large number of switches. In recent years, some meta-heuristic approaches have been used for service restoration in power distribution systems: simulated annealing [11–13], tabu search [14, 15], ant colony optimization [16], and genetic algorithm [17–19]. Moreover, a comparative study of meta-heuristic approaches to service restoration has also been reported [20]. However, as far as the authors know, many studies with these meta-heuristic techniques have been conducted on the network reconfiguration and loss minimization problems, but the application of these techniques to the scheduling of switching operations is limited. Although a multiobjective service restoration problem including the minimization of the number of necessary switching operations has also been proposed (e.g., [15, 21]), it is insufficient from the standpoint of optimizing the whole restoration process. To ensure minimal reduction in system reliability, not only should the number of switching operations be minimized, but also energy not supplied (ENS).

To tackle the ENS minimization problem, the authors have proposed an optimization algorithm based on a genetic algorithm (GA), called two-stage GA [22]. This algorithm uses a GA as an optimization procedure in each stage. The first stage creates radial network configurations, and the second stage searches for an optimal sequence of switching operations that minimizes ENS for each configuration. The second stage is embedded in the first stage and calculates the fitness of the whole. Although this algorithm performs well for small systems, it is difficult to apply to large systems from the viewpoint of computation time. In this paper, a hybrid evolutionary algorithm is proposed for the ENS minimization problem. To improve the time performance of the algorithm, the authors introduce three kinds of speed-up strategy: a local search procedure, greedy algorithm for scheduling of switching operations, and efficient maximum flow algorithm.

The remainder of this paper is organized as follows. The following section describes a formulation of the service restoration problem. Section 3 reviews the existing GA-based algorithm [22]. In Section 4, the hybrid evolutionary algorithm is proposed and Section 5 presents several numerical results to demonstrate the effectiveness of the proposed algorithm. Finally conclusions are drawn.

2 Problem Formulation

Service restoration in power distribution systems involves operating the switches to restore as many loads as possible for the out-of-service area following a fault. As mentioned above, not only the final network configuration but also the sequence of switching operations is important in service restoration, because the sequence of operations has a considerable influence on system reliability. Moreover, the switching operations must be performed sequentially, due to technical limitations.

In this section, we formulate the service restoration problem to minimize ENS under network operating constraints.

2.1 Constraints

Not every configuration is a feasible solution to the service restoration problem. Thus, it is necessary to specify which configurations are feasible, and which are not. The constraints that should be considered in this problem are as follows:

(a) *Radial network constraint*
Network configuration must retain radial topology even during the restoration process.
(b) *Power source constraint*
The total loads of each sub-system must not exceed the capacity of the corresponding power source.
(c) *Line capacity constraint*
The line currents (power flows) must not exceed the maximum values related to the line sections.

In addition to the constraints listed above, the following additional requirements should be satisfied in the problem. If these requirements cannot be met, a certain value will be added to the objective function as a penalty.

(d) Power not supplied (PNS) should decrease monotonically.
(e) All de-energized loads should be restored after completing restoration.

The relationship between PNS and ENS is shown in Fig. 2.

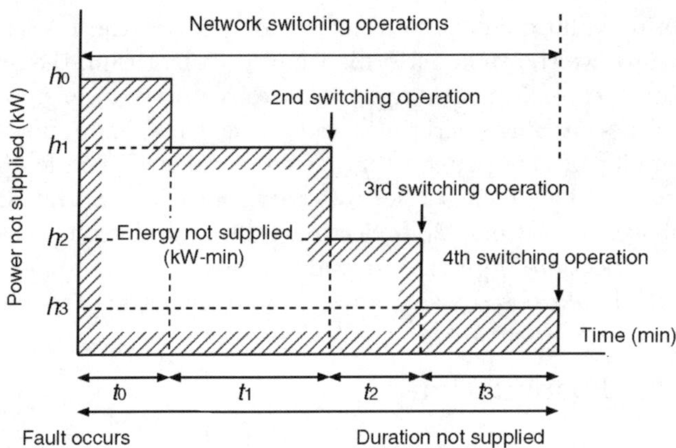

Fig. 2. An example of the service restoration process

2.2 Objective Function

Let h_k be PNS after the k-th switching operation, and t_k be the time interval needed for the $(k + 1)$-th operation. ENS to be minimized in this problem is defined as

$$\mathrm{ENS} = \sum_{k=0}^{K-1} (h_k \times t_k) \ , \tag{1}$$

where K is the number of switching operations. For convenience, the fault occurrence is defined as the 0-th switching operation. PNS just after a fault is h_0 and the time needed for the first switching operation is t_0. PNS after the first operation is h_1 and the time needed for the second operation is t_1, and so on. The penalty to be added to the objective function is also defined as

$$P = w_1 \sum_{k=1}^{K} \max\{h_k - h_{k-1}, 0\} + w_2 h_K \ , \tag{2}$$

where w_1 and w_2 are the weighting factors related to the additional requirements (d) and (e), respectively. Therefore, the objective function z to be minimized is as follows;

$$z = \mathrm{ENS} + P \ . \tag{3}$$

3 Two-stage Genetic Algorithm

For the ENS minimization problem defined in the previous section, the authors have proposed an optimization algorithm based on a genetic algorithm, called the two-stage GA [22]. A GA is a method for search and optimization that imitates the process of natural selection and evolution. Due to their ability to find global optimal solutions for large-scale combinatorial optimization problems, GAs have

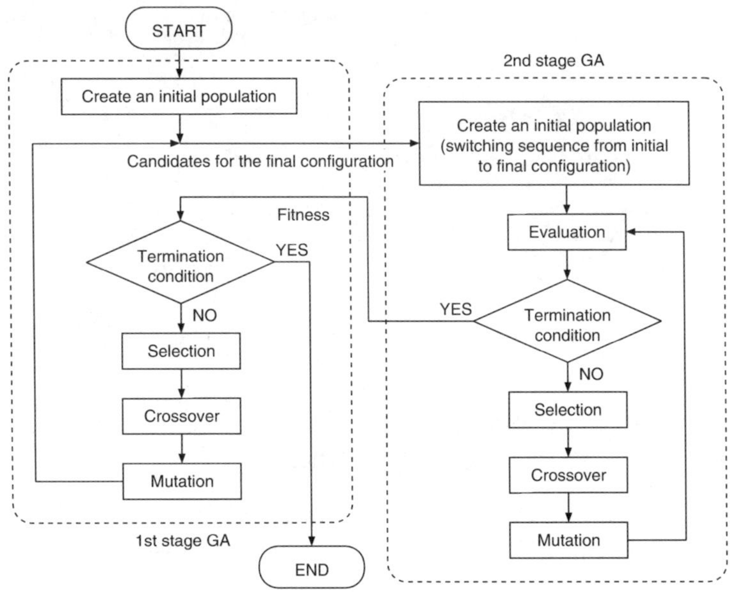

Fig. 3. Two-stage genetic algorithm for service restoration

been found to be an efficient method for solving power system problems, including the service restoration problem [18, 19]. This algorithm uses a GA as an optimization procedure in each stage. The candidates for the final configuration are created in the first stage and passed to the second stage. Then the second stage evaluates them by finding an optimal sequence of switching operations, and returns their fitness to the first stage. The flowchart of the two-stage GA is shown in Fig. 3.

One of the most important factors which affect a GA's performance is the interaction of its coding of candidate solutions with the operators it applies to them. In particular, all chromosomes should represent feasible solutions. To improve the performance of the two-stage GA, we adopt the *edge-set representation* [23] for a radial network and the *random keys representation* [24] for a sequence of switching operations. By using these representations, all chromosomes generated by initialization, crossover and mutation can represent feasible solutions. The details of chromosome representations adopted in each stage are described in the followings.

3.1 Optimization of Network Configuration

In the first stage, radial network configurations are created as the candidates for the final configuration. As the fitness of each configuration is evaluated in the second stage, the first stage GA only searches for optimal radial network configurations. Hence, the problem dealt with in the first stage is regarded as a conventional network reconfiguration problem.

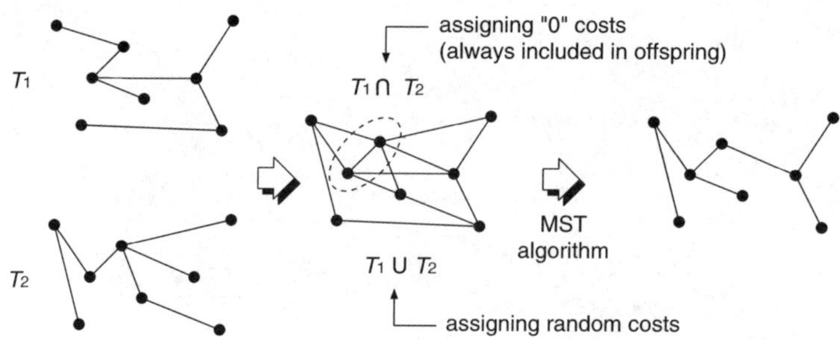

Fig. 4. Crossover operation in the first stage GA

Representing Radial Network. The radial network configuration can be roughly modeled as an undirected graph, where

- the edges (branches) represent line switches, and
- the nodes represent collections of power system equipment containing feeder segments and power sources.

The network reconfiguration problem is therefore regarded as finding an optimal spanning tree that satisfies the constraints on the corresponding graph. Much work has been done on representing spanning trees for evolutionary search (e.g., [25, 26]). Recent studies have indicated the general usefulness of representing spanning trees directly as lists of their edges and applying operators that always yield feasible trees [23, 27]. The first stage GA uses the edge-set representation and unique operators based on minimum spanning tree (MST) algorithms.

Initialization. Kruskal's algorithm builds an MST on a weighted graph G by examining G's edges in order of increasing weight [28]. If the algorithm instead examines G's edges in random order, it returns a random spanning tree on G. This algorithm is called KruskalRST [23]. In the two-stage GA, the initial population is produced by the random spanning tree algorithm, KruskalRST.

Crossover. In GA, offspring should represent solutions that combine substructures of their parental solutions. To provide this *heritability*, a crossover operator must build a spanning tree that consists mostly of edges found in the parents. It is also beneficial to favor edges that are common to both parents [23]. This can be done by applying the random spanning tree algorithm to the graph $G' = (V, T_1 \cup T_2)$, where T_1 and T_2 are the edge sets of the parental trees. Figure 4 illustrates an example of this crossover operation.

Mutation. A mutated chromosome should usually represent a solution similar to that of its parent. To provide this *locality*, a mutation operator must make a small change in a parent solution. This means that a mutated chromosome

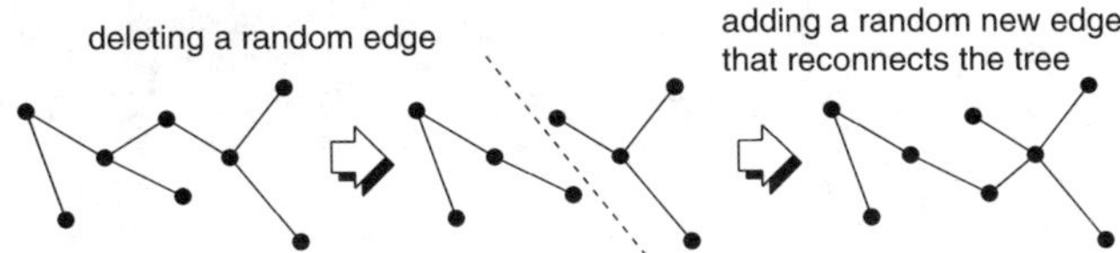

Fig. 5. Mutation operation in the first stage GA

should represent a tree that consists mostly of edges also found in its parent. This can be done by deleting a random edge from T and replacing it with a random new edge that reconnects the tree. Figure 5 shows this operation.

3.2 Scheduling of Switching Operations

The second stage GA searches for an optimal sequence of switching operations that minimizes ENS for each configuration created in the first stage. The best values obtained in this stage are regarded as the fitness values of configurations.

Determination of Operation Switches. In this problem, each switch is assumed to be operated only once at most. Under this assumption, the switches to be operated can be determined as the ones whose state changes from open to closed, and vice versa. Let T_0 and T_K be the edge set of the initial and the candidate for the final configuration, respectively. The edge set E^d corresponding to the switches to be operated is determined as the symmetric difference of the two sets T_0 and T_K,

$$E^d = (T_0 \cup T_K) \setminus (T_0 \cap T_K) \ . \tag{4}$$

Representing Sequence of Switching Operations. The solution for service restoration is a sequence of operations that specify which switches need to change their state. Such a sequence is represented as a string of the switch number, where the first element of the string is interpreted as the switch that needs to be closed, the next element is taken as representing the switch to be open, and so on. Therefore, a sequence of switching operations is represented as a permutation of only the edges whose state changes.

The second stage GA uses random keys [24] to represent a sequence of switching operations. This representation encodes a permutation with $(0, 1]$ random numbers. These values are used as the sort keys to decode the permutation. An example of the random keys encoding for permutation is shown below.

$$(0.46, 0.91, 0.33, 0.75, 0.51) \ .$$

As shown in Fig. 6, by sorting in ascending order, we get the sequence,

$$3 \rightarrow 1 \rightarrow 5 \rightarrow 4 \rightarrow 2 \ .$$

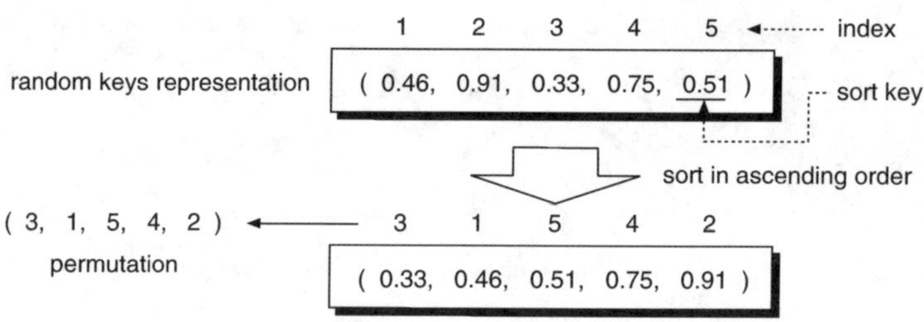

Fig. 6. A simple example of random keys representation

Random keys representation can use every traditional crossover operator. The two-stage GA uses a one-point crossover in the second stage. The mutation operator replaces the value of a gene by a uniform $(0, 1]$ random number.

3.3 Fitness Evaluation

PNS can be calculated as the difference between the sum of loads and the maximum flow. The maximum flow for a distribution system is defined as the sum of power flows from each power source. Let ℓ_i be the capacity (demand) of the load node i, L be the number of load nodes, and $f_k^{\max}$ maximum flow after the k-th switching operation. PNS after the k-th operation is calculated using the following equation:

$$h_k = \sum_{i=1}^{L} \ell_i - f_k^{\max}, \quad \forall k = 1, 2, \ldots, K \ . \tag{5}$$

From the equations (1) and (5), we can obtain ENS for a candidate for the final configuration. In this ENS minimization problem, the fitness function is defined as the inverse of the objective function.

4 Hybrid Evolutionary Algorithm

As described in the previous section, the authors have already proposed the two-stage GA as an optimization method for the service restoration problem [22]. The two-stage GA performs well for small-scale systems, but it is difficult to apply the algorithm to large-scale systems from the viewpoint of computation time. To improve the time performance of the two-stage GA, the authors propose here a hybrid evolutionary algorithm combined with the following three speed-up strategies:

- Local search procedure
- Greedy algorithm for scheduling of switching operations
- Efficient maximum flow algorithm

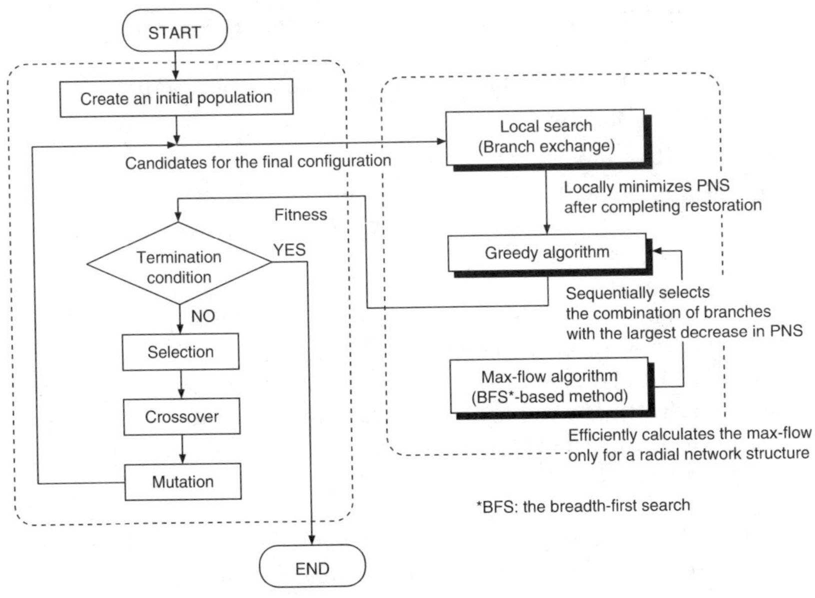

Fig. 7. A hybrid evolutionary algorithm for service restoration

The flowchart of the proposed algorithm is shown in Fig. 7, and the details of each speed-up strategy are described below. The test system used in this section includes three power source (transformer) nodes, 241 load nodes, and 347 branches. All the following experiments were performed on an AMD Athlon64 processor 3200+ 2.2 GHz. The algorithm was coded in C and the code was compiled with gcc 3.3.2 -O3.

4.1 Local Search Procedure

To find promising final radial configurations earlier in the first stage of the two-stage GA, all the candidates for final configuration obtained by GA are applied with a local search procedure to minimize PNS after completing restoration. In this study, the branch exchange method [29] is used as a local search procedure.

The branch exchange method is known as one of the most effective local search procedures for addressing the service restoration problem, and the basic concept of the method can be explained as follows. From the initial radial configuration (in this study, the candidate for final configuration obtained by GA), a loop is constructed by adding one branch, and then by removing another branch, the network configuration is brought back to radial. Here, it is not realistic to consider all possible combinations of adding and removing branches; the branch for removal is therefore selected from among branches adjacent to the added branch. This procedure is iteratively repeated until the objective function cannot be reduced by any branch exchange operation.

Table 1. Effectiveness of Branch Exchange Method

Branch Exchange Method	Not Applied	Applied
Average PNS† (kW)	1201.7	13.8
# of Full Restoration (100 trials)	0	66

†Power Not Supplied

To verify the effectiveness of the branch exchange method, the result of comparing the network configuration before and after applying the branch exchange is presented. The results obtained by applying only the branch exchange method to randomly generated 100 radial networks are shown in Table 1. In this preliminary experiment, the greedy algorithm was not applied because we are interested only in the final network configurations, not in the switching operations. Table 1 shows that PNS after completing restoration can be reduced dramatically by using the branch exchange method. In addition, the final radial configuration completely restored can be obtained 66 times in 100 trials.

4.2 Greedy Algorithm for Scheduling of Switching Operations

In the second stage of the two-stage GA, it is necessary to calculate the maximum flow repeatedly to search for the optimal switching sequence. However, this is a very time-consuming task. To improve the time performance of the proposed algorithm, not only should the maximum flow be calculated efficiently, the calculation frequency of the maximum flow should also be decreased. In this study, instead of the second stage GA, a greedy algorithm that sequentially selects the combination of branches with the largest decrease in PNS is used to reduce the calculation frequency of the maximum flow as much as possible.

Figure 8 shows how the proposed greedy algorithm works. Let us consider the set of six branches from A to F that should be operated. First we select the combination of branches with the largest decrease in PNS (i.e. branches A and B) from among all nine feasible combinations, and schedule this combination as the first switching operation. Next, after recalculating PNS for all four feasible combinations without branches A and B, select and schedule the combination with the largest decrease in PNS (i.e. branches C and D) as the second switching operation. Finally, schedule the combination of branches E and F as the final switching operation.

The switching sequence obtained by the greedy algorithm might not satisfy the first additional requirement. Should this be the case, a penalty defined in (2) will be added to the objective function.

4.3 Efficient Maximum Flow Algorithm

As mentioned above, PNS after the k-th operation can be calculated by subtracting the maximum flow from the total load in the system. Because the total

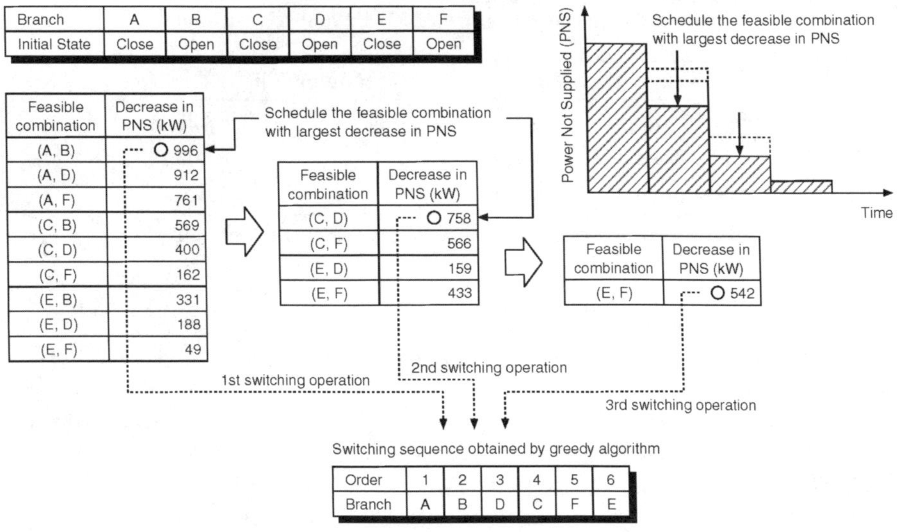

Fig. 8. Greedy algorithm for scheduling of switching operations

load is assumed to be constant in this problem, the PNS calculation time might be able to be reduced by efficiently calculating the maximum flow, and to speed up the algorithm. Many maximum flow algorithms have been proposed so far, but the two-stage GA uses the Ford-Fulkerson algorithm [28] that is one of the most traditional algorithms. The effect when the Ford-Fulkerson algorithm was replaced with a more efficient one was investigated here.

The results of a comparative study of the following three kinds of algorithms are presented.

- Ford-Fulkerson algorithm [28]
- Maximum flow algorithm using maximum adjacency (MA) ordering [30]
- Proposed algorithm based on the breadth-first search

The maximum flow algorithm using MA ordering[3] is known as one of the most efficient maximum flow algorithms today. The first two algorithms are designed for a general network structure, but what this problem needs is an algorithm only for a radial network configuration. Speed-up of the maximum flow calculation can be expected by assuming the network configuration to be radial. The proposed algorithm based on the breadth-first search (BFS) is an efficient algorithm intended for a radial network configuration. The proposed algorithm begins at the power source node and explores all the neighboring nodes. It sends as much flow as possible from the source node to the nearest

[3] Let an undirected graph be given, which has n vertices. An ordering $v_1, v_2, \ldots, v_n$ of vertices is called an *MA ordering* if an arbitrary vertex is chosen as v_1, and after choosing the first i vertices $v_1, \ldots, v_i$, the $(i+1)$-th vertex v_{i+1} is chosen from the vertices u that have the largest number of edges between $\{v_1, \ldots, v_i\}$ and u [31].

Table 2. Time Performance of Max-Flow Algorithms

Max-Flow Algorithm	Total Running Time (s)
Ford-Fulkerson	62.26
MA Ordering [30]	1.05
BFS-based Method†	0.10

†Proposed algorithm based on the breadth-first search

nodes. Then for each of those nodes, it explores their unexplored neighboring nodes, and so on.

The total running time of applying the three algorithms listed above to randomly generated 1000 radial networks are shown in Table 2. By replacing the Ford-Fulkerson algorithm with the algorithm using MA ordering, the maximum flow calculation can be sped up by about 60 times. Moreover, using the proposed algorithm may be expected to make the calculation about ten times faster than using MA ordering.

5 Numerical Experiments

In order to verify the effectiveness of the proposed algorithm, computational results with several test distribution systems are reported here.

5.1 Computational Results with Small Test System

To perform a comparison between the two-stage GA and the proposed algorithm, experiments were conducted using a small test system. The small test system used in the experiments includes three power source nodes, 27 load nodes and 40 branches. Figure 9 illustrates the network structure of the system. The number in a circle shows the load capacity (demand). The number next to a branch and the number in parentheses show the branch number and the capacity of the branch, respectively. The sum of the load capacity is 137 kW.

The following two simulation cases were investigated.

- CASE-1: A fault occurs on branch #12.
- CASE-2: A fault occurs on branch #0.

CASE-2 is a more serious fault case than CASE-1 because branch #0 is an important one that directly connects the power source to the network. If branch #0 has a fault, the whole load needs to be supplied with only the remaining two power sources.

In the experiments, the two-stage GA and the proposed algorithm used a kind of steady state genetic algorithm (SSGA) for creating candidates for final network configurations. Initially, a certain number (*pop*) of parent individuals

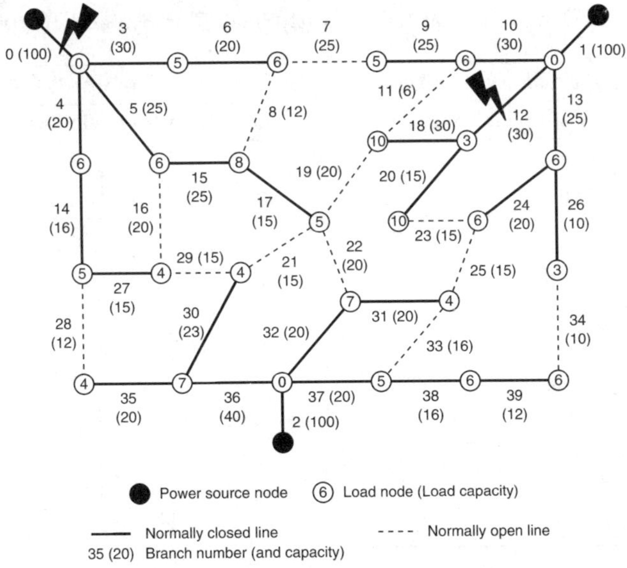

Fig. 9. Small test system (3 power source nodes, 27 load nodes and 40 branches)

Table 3. Parameter Settings for Each Algorithm

	Two-stage GA		Proposed
	1st stage	2nd stage	algorithm
population size (*pop*)	20	8	10
crossover rate (p_c)	0.8	0.8	0.8
mutation rate (p_m)	0.06	0.10	0.06
max. generation ($g_{\max}$)	50	5	10

are set up in the population, and the same number of offspring are created from the parents. Then, the *pop* best individuals are selected as the new population from the union of parents and offspring. Roulette and tournament selections were used in the two-stage GA and the proposed algorithm, respectively. A set of GA-related parameters used is listed in Table 3. The time interval t_k was set to 1.0 for all k, and the weighting factors w_1, w_2 were set to 10.

The computational results are presented in Table 4. All the experiments were performed using 100 different random seeds. In this table, each column of each algorithm shows the average values of the objective function, average computation time and success rate, respectively. The average computation time is defined as the average time required to find an optimal solution or perform the algorithm until the maximum generation. The second column of the table shows the optimal values obtained by an exact method based on dynamic programming [32].

Table 4. Computational Results with Small Test System (100 trials)

	Optimal (kW-min)	Two-stage GA			Proposed Algorithm		
		Average	Time (s)	%Success	Average	Time (s)	%Success
CASE-1	46	46.2	17.30	82%	46.0	<0.01	100%
CASE-2	86	90.1	28.73	40%	86.0	0.10	100%

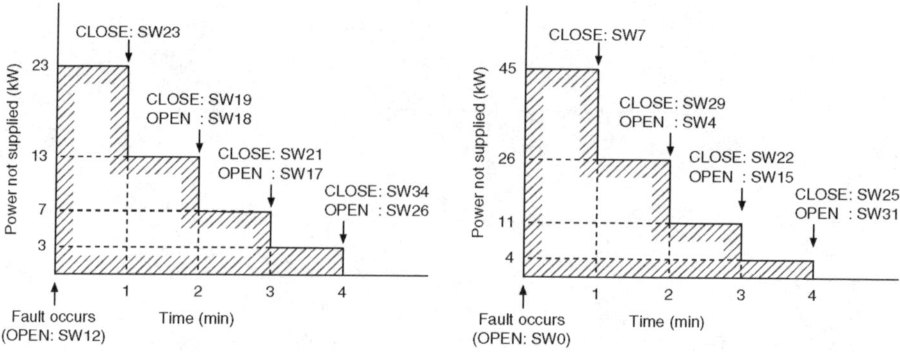

Fig. 10. Optimal restoration process (Left: CASE-1, Right: CASE-2)

The results show that the proposed algorithm can find the optimal solution within 0.1 seconds for each fault case. Compared with the two-stage GA, the proposed algorithm can dramatically reduce the computation time. Figure 10 shows the optimal restoration process, and the details of switching operations for each fault case are presented in Fig. 11 and 12.

5.2 Computational Results with Large Test System

In order to assess the scalability of the proposed algorithm, the algorithm was applied to a large test system. The large test system used in the experiments includes three power source nodes, 241 load nodes and 347 branches. Two fault cases (CASE-3 and CASE-4) were investigated. The parameter settings used in the algorithm are the same ones as shown in Table 3.

The computational results are presented in Table 5. All the experiments were performed using 100 different random seeds. The optimal values obtained by an exact method based on dynamic programming [32] are also presented. Due to computation time, the two-stage GA is inapplicable to this problem. Table 5 shows that the proposed algorithm can find good solutions in a considerably shorter time for each case. Figure 13 depicts the best sequence of switching operations obtained. For each case, PNS decreases monotonically and all de-energized loads are finally restored.

The above results indicate that the proposed algorithm is a promising one for addressing the service restoration problem.

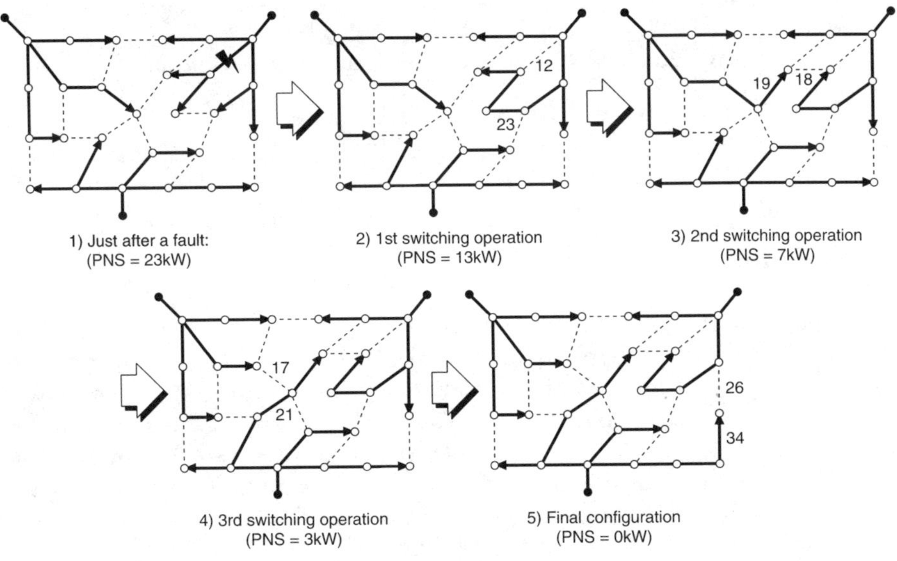

Fig. 11. Optimal sequence of switching operations (CASE-1)

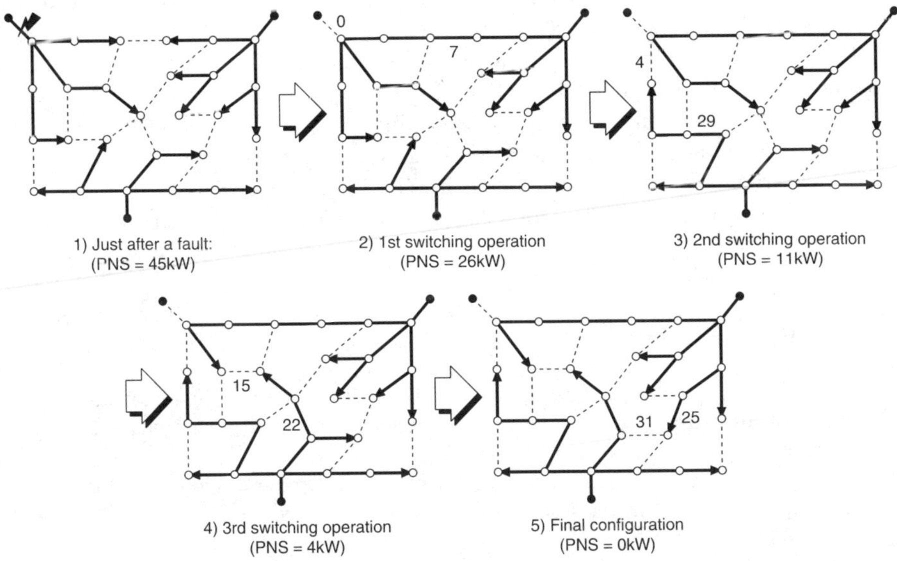

Fig. 12. Optimal sequence of switching operations (CASE-2)

Table 5. Computational Results with Large Test System (100 trials)

	Optimal	Proposed Algorithm			
	(kW-min)	Best (kW-min)	Avg. (kW-min)	Stdev. (kW-min)	Time (s)
CASE-3	4214.6	4219.7	4232.4	2.53	4.32
CASE-4	5113.2	5137.3	5142.9	18.56	7.85

6 Conclusions

In this paper, a hybrid evolutionary algorithm is proposed for service restoration in power distribution systems. To improve the time performance of the two-stage GA, the authors introduced three kinds of speed-up strategy: a local search procedure, greedy algorithm, and efficient maximum flow algorithm on radial networks. The computational results with the small test system show that the proposed algorithm can dramatically reduce the computation time compared with the two-stage GA. Moreover, the results with the large test system show that the algorithm can be applied to real-scale systems.

From the viewpoint of minimizing ENS during the restoration process, it might not be the best strategy to minimize PNS after completing restoration by the local search procedure. However, to obtain the optimal solution within a limited computation time, it is very important to limit the search area only to promising radial configurations. In that sense, a local search by the branch exchange method may be effective.

The search by greedy algorithm essentially means that only one switching sequence is evaluated for each final radial configuration. That is, if the final radial configuration is fixed, then the switching sequence is indirectly fixed, too. In this study, a deterministic method by greedy algorithm was used to speed up the algorithm at the expense of optimality. The development of a fast and flexible optimization algorithm for scheduling of switching operations remains a task for the future.

References

1. Aoki, K., Kuwabara, H., Satoh, T., Kanezashi, M.: Outage state optimal load allocation by automatic sectionalizing switches operation in distribution systems. IEEE Trans. on Power Delivery **2**(4) (1987) 1177–1185
2. Aoki, K., Nara, K., Itoh, M., Satoh, T., Kuwabara, H.: A new algorithm for service restoration in distribution systems. IEEE Trans. on Power Delivery **4**(3) (1989) 1832–1839
3. Liu, C.C., Lee, S.J., Venkata, S.S.: An expert system operational aid for restoration and loss reduction of distribution systems. IEEE Trans. on Power Systems **3**(2) (1988) 619–626
4. Lee, H.J., Park, Y.M.: A restoration aid expert system for distribution substations. IEEE Trans. on Power Delivery **11**(4) (1996) 1765–1770

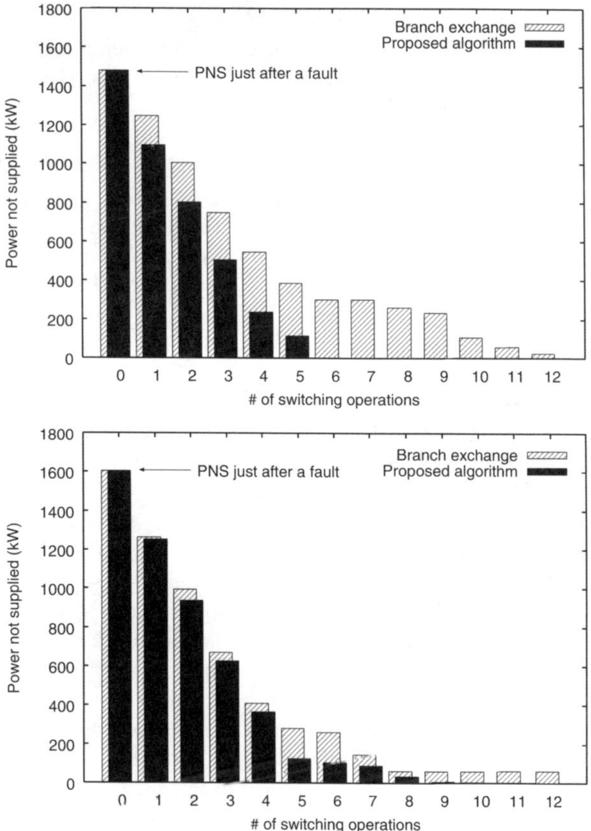

Fig. 13. Best switching sequence obtained by the proposed algorithm (Upper: CASE-3, Lower: CASE-4)

5. Bretas, A.S., Phadke, A.G.: Artificial neural networks in power system restoration. IEEE Trans. on Power Delivery **18**(4) (2003) 1181–1186
6. Hsu, Y.Y., Kuo, H.C.: A heuristic based fuzzy reasoning approach for distribution system service restoration. IEEE Trans. on Power Delivery **9**(2) (1994) 948–953
7. Lin, W.M., Chin, H.C.: Preventive and corrective switching for feeder contingencies in distribution systems with fuzzy set algorithm. IEEE Trans. on Power Delivery **12**(4) (1997) 1711–1716
8. Zhou, Q., Shirmohammadi, D., Liu, W.H.E.: Distribution feeder reconfiguration for service restoration and load balancing. IEEE Trans. on Power Systems **12**(2) (1997) 724–729
9. Morelato, A.L., Monticelli, A.: Heuristic search approach to distribution system restoration. IEEE Trans. on Power Delivery **4**(4) (1989) 2235–2241
10. Shirmohammadi, D.: Service restoration in distribution networks via network reconfiguration. IEEE Trans. on Power Delivery **7**(2) (1992) 952–958
11. Chiang, H.D., Jean-Jumeau, R.: Optimal network reconfigurations in distribution systems: Part 1: A new formulation and a solution methodology. IEEE Trans. on

Power Delivery **5**(4) (1990) 1902–1909
12. Chiang, H.D., Jean-Jumeau, R.: Optimal network reconfigurations in distribution systems: Part 2: Solution algorithms and numerical results. IEEE Trans. on Power Delivery **5**(3) (1990) 1568–1574
13. Jeon, Y.J., Kim, J.C., Kim, J.O., Shin, J.R., Lee, K.Y.: An efficient simulated annealing algorithm for network reconfiguration in large-scale distribution systems. IEEE Trans. on Power Delivery **17**(4) (2002) 1070–1078
14. Toune, S., Fudo, H., Genji, T., Fukuyama, Y., Nakanishi, Y.: A reactive tabu search for service restoration in electric power distribution systems. In: Procs. of the 1998 IEEE International Conference on Evolutionary Computation, Anchorage, AK (1998) 763–768
15. Genji, T., Oomori, T., Miyazato, K., Hayashi, N., Fukuyama, Y.: Service restoration in distribution systems aiming higher utilization rate of feeders. In: The Fifth Metaheuristcis International Conference (MIC2003), Kyoto, Japan (2003)
16. Watanabe, I.: An ACO algorithm for service restoration in power distribution systems. In: Procs. of the 2005 IEEE Congress on Evolutionary Computation, Edinburgh, UK, IEEE Press (2005) 2864–2871
17. Fukuyama, Y., Chiang, H.D.: A parallel genetic algorithm for service restoration in electric power distribution systems. In: Procs. of the 1995 IEEE International Conference on Fuzzy Systems. Volume 1., Yokohama, Japan (1995) 275–282
18. Chavali, S., Pahwa, A., Das, S.: A genetic algorithm approach for optimal distribution feeder restoration during cold load pickup. In: Procs. of the 2002 IEEE Congress on Evolutionary Computation, Honolulu, HI (2002) 1816–1819
19. Luan, W.P., Irving, M.R., Daniel, J.S.: Genetic algorithm for supply restoration and optimal load shedding in power system distribution networks. Procs. of IEE Generation Transmission and Distribution **149**(2) (2002) 145–151
20. Toune, S., Fudo, H., Genji, T., Fukuyama, Y., Nakanishi, Y.: Comparative study of modern heuristic algorithms to service restoration in distribution systems. IEEE Trans. on Power Delivery **17**(1) (2002) 173–181
21. Matos, M.A., Melo, P.: Multiobjective reconfiguration for loss reduction and service restoration using simulated annealing. In: Procs. of IEEE Budapest Power Tech'99, Budapest, Hungary (1999)
22. Watanabe, I., Nodu, M.: A genetic algorithm for optimizing switching sequence of service restoration in distribution systems. In: Procs. of the 2004 IEEE Congress on Evolutionary Computation, Portland, OR, IEEE Press (2004) 1683–1690
23. Raidl, G.R., Julstrom, B.A.: Edge sets: An effective evolutionary coding of spanning trees. IEEE Trans. on Evolutionary Computation **7**(3) (2003) 225–239
24. Bean, J.C.: Genetic algorithms and random keys for sequencing and optimization. ORSA Journal on Computing **6**(2) (1994) 154–160
25. Palmer, C.C., Kershenbaum, A.: Representing trees in genetic algorithms. In: Procs. of the First IEEE Conference on Evolutionary Computation, Piscataway, NJ, IEEE Press (1994) 379–384
26. Rothlauf, F.: Representations for Genetic and Evolutionary Algorithms. Springer-Verlag (2006)
27. Julstrom, B.A., Raidl, G.R.: Initialization is robust in evolutionary algorithms that encode spanning trees as sets of edges. In: Procs. of the 2002 ACM Symposium on Applied Computing, ACM Press (2002) 547–552
28. Ahuja, R.K., Magnanti, T.L., Orlin, J.B.: Network Flows: Theory, Algorithms, and Applications. Prentice-Hall (1993)

29. Aoki, K., Nara, K., Satoh, T., Kitagawa, M., Yamanaka, K.: New approximate optimization method for distribution system planning. IEEE Trans. on Power Systems **5**(1) (1990) 126–132
30. Fujishige, S.: A maximum flow algorithm using MA ordering. Operations Research Letters **31**(3) (2003) 176–178
31. Nagamochi, H., Ibaraki, T.: Graph connectivity and its augmentation: applications of MA orderings. Discrete Applied Mathematics **123** (2002) 447–472
32. Carvalho, P., Ferreira, L., Rojao, T.: Dynamic programming for optimal sequencing of operations in distribution networks. In: Proc. of the 15th Power Systems for Computation Conference. (2005)

Particle Swarm Optimisation for Operational Planning: Unit Commitment and Economic Dispatch

P. Sriyanyong, Y.H. Song and P.J. Turner

School of Engineering and Design,
Brunel University, Uxbridge, UB8 3PH, UK.

Summary. This chapter proposes the application of a Particle Swarm Optimisation (PSO) algorithm to Unit Commitment (UC) and Economic Dispatch (ED) problems, which occur in the operational planning of a power system. To solve the UC problem, PSO is applied to update the Lagrange multipliers and is also incorporated into the Lagrange Relaxation method to improve its performance. For the ED problem, PSO is integrated into a modified heuristic search to enhance the searching efficiency. The research shows that the proposed methods can provide solutions with good quality and stable convergence characteristics whilst their implementation is simple and their computation time is reasonable.

1 Introduction

Unit Commitment (UC) is a problem in power system operation that determines the schedule of generating units to meet electricity demand and operating constraints over a time horizon. Basically, Economic Dispatch (ED), as a sub-problem of UC, determines the optimal scheduling of generation for a particular time that minimises the total production cost and satisfies equality and inequality constraints. Recently, a number of computation techniques such as Simulated Annealing (SA), Genetic Algorithm (GA), Evolutionary Programming (EP), Tabu Search (TS) and Particle Swarm Optimisation (PSO) have been applied to solve these problems. Compared to other methods, PSO can solve the problems quickly with

P. Sriyanyong et al.: *Particle Swarm Optimisation for Operational Planning: Unit Commitment and Economic Dispatch*, Studies in Computational Intelligence (SCI) **49**, 313–347 (2007)

high quality solutions and stable convergence characteristics and it is easily implemented.

In this chapter, the methodology to apply PSO to solve the UC problem is proposed, where PSO is used to update Lagrange multipliers and is also incorporated into the Lagrange Relaxation method (LR) to improve its performance. Moreover, PSO is integrated into a modified heuristic search to solve the ED problem. The simulation results from these applications show that the proposed methods can solve the UC and ED problems efficiently and effectively with high quality solutions. The organization of this chapter is as follows: section 2 introduces evolutionary computation techniques in power systems, in section 3, the overview of PSO is presented. Section 4 shows the applications of PSO to the UC and ED problems and finally, conclusions are drawn in section 5.

2 Evolutionary Computation Techniques in Power Systems

Security of supply is a critical issue in the operational planning of a modern power system as electricity is a basic need to every sector in the economy. While electricity demand changes instantaneously during the course of a day, each generating unit itself has operating limits and at the same time is uneconomic to store electricity. Thus, it is a crucial role of electric utilities to maintain the reliability and continuity of electricity supply whilst providing least cost of operation. To meet these contradictory objectives, the operation of a power system must deal with a number of dynamic issues. In the case of hourly generation planning, Economic Dispatch (ED) schedules the outputs of all committed generating units, which are previously identified by the Unit Commitment (UC) problem.

Nowadays the modern power system is more dynamic and contains a number of constraints. Thus, the accurate solutions to the ED and UC problems are essential in order to operate the power system in an economic and efficient manner. Therefore a number of computation techniques have progressively been proposed to solve these critical issues. Overviews of some of these computation methodologies are presented below.

In the Iron Age, blacksmiths discovered that the formation of crystals in a solid is dependent on its cooling time; the slower the cooling, the more perfect the crystals form [1]. Simulated Annealing (SA) applies this idea in its computational algorithm. The basic principle of SA is that the parameter of the objective function is analogous to the "Temperature" of the metal in an annealing process. A change in the parameter of the objective function or the "Temperature" is then basically measured throughout an iterative computation. The transition at each iteration will automatically be accepted if the change in "Temperature" is negative. The transition at iteration when the change in "Temperature" is positive will also be accepted however if the rate of change of such "Temperature" remains within the Boltzmann based probability distribution. In this case, the additional procedure called "Cooling Schedule" is required to lower the "Temperature" and the computation will continue iteratively until it reaches the "Freezing Temperature", a condition where no further change in the "Temperature" occurs [1],[2]. Generally, the SA algorithm is able to deal with arbitrary systems. It is based on a local search technique and is regarded as a powerful method in terms of its ability to find a near global optimal solution. When combined with a probabilistic approach, SA is also able to find a solution outside a local optimum [3],[4]. Setting the parameters for SA is difficult however and the computation speed will be slow when the method is applied to complicated power systems [4].

The Genetic Algorithm (GA) technique is based on a stochastic global search method which mimics some of the processes of natural evolution and selection [3]. The principal idea of this algorithm comes from the natural world where each species is required to adapt to a complicated changing environment so that it can maximise the likelihood of its survival. The knowledge that each species gains is encoded in its chromosomes, which continually transform when reproduction occurs. Over a period of time, the changes in these chromosomes give rise to species that are more likely to survive, and so have a greater chance of passing their improved characteristic onto future generations [1] . The GA is analogous to the idea of chromosomes in nature where the computation method identifies candidate solutions which are encoded by a finite bit string [3]. Each chromosome exchanges information through a naturally random process so that solutions can evolve to be close to the optimum. The sequence of calculations will continue and repeat until termination conditions are satisfied. The strength of a GA is that it only requires information of the objective function. Thus, a GA can deal with a non-smoothing, discontinuous and non-differentiable function [3]. Since the computation of GA requires encoding and decoding schemes however, it takes a longer time to reach an optimal solution. Sometimes, it is found that a GA has a problem with its computation efficiency and convergence [1].

The fundamental concept of Evolutionary Programming (EP) is on a similar basis to GAs in that it maintains populations of potential solutions and uses a mechanism to select the optimum from a set of those populations [5]. Rather than using generic specific operators as observed in nature as a GA does however, EP sets its control parameter from real values of the problem that will be investigated. In addition, EP primarily bases its algorithm on mutation and selection while GAs traditionally use crossover [6].

Particle Swarm Optimisation (PSO) is developed from a similar basis as the previous methods. It applies a simplified social model, which for instance zoologists might use to explain the movement of individuals within a group [7], to solve an optimisation problem. PSO is one of the modern algorithms used to solve global optimisation problems [8]. In its algorithm, PSO initialises a population of random solutions each of which is defined as a “particle”. Initially, every particle flies into a problem hyperspace at a random velocity. Thereafter, each particle adjusts its travelling speed dynamically corresponding to the flying experiences of itself and its colleagues [4],[5]. The PSO computation will repeat until it finds a global optimal solution. The PSO is simple to implement compared to other methods and it can quickly solve a problem to give high quality solutions and stable convergence characteristics [5],[9]. In addition, PSO is robust in solving continuous non-linear optimisation problems and unlike other techniques, it has a flexible and well-balanced mechanism to enhance and adapt the global and local exploration abilities [10]. PSO does have a drawback however in that the algorithm seems sensitive to the tuning of some of its weights or parameters. Therefore there is much research into the potential of PSO to solve complex power system problems [9].

This chapter proposes the application of Particle Swarm Optimisation (PSO) to solve the Unit Commitment (UC) and Economic Dispatch (ED) problems and the simulation results from both show that the proposed methods can efficiently and effectively provide quality solutions and stable convergence characteristic with a simple implementation process.

3 Overview of Particle Swarm Optimisation

In 1995, Kennedy and Eberhart introduced a new evolutionary computation technique called Particle Swarm Optimisation (PSO) [11]. Similar to other evolutionary computation techniques, PSO employs the principle of a random initialised population and the concept of evaluation and modification of a population to find the optimal solution. In contrast however, PSO does not utilise the operators during the modification step (e.g. mutation and crossover) as a GA does since it can update itself [12],[13]. Mathematically, the fundamental model of PSO can be expressed by the following [14].

Let a swarm have n particles in a d-dimensional search space. At the tth iteration, $x_i^t = (x_{i1}^t, x_{i2}^t, \ldots, x_{id}^t)$ expresses the position of the ith particle and $pbest_i^t = (pbest_{i1}^t, pbest_{i2}^t, \ldots, pbest_{id}^t)$ shows the best previous position of the ith particle. In addition, the best position among all the particles is represented by $gbest_d^t = (gbest_1^t, gbest_2^t, \ldots, gbest_d^t)$. The velocity of the ith particle can be represented by $v_i^t = (v_{i1}^t, v_{i2}^t, \ldots, v_{id}^t)$. Each of the population, called a particle or agent, can be updated or changed to the new position according to the current velocity, the difference between the current position and the best value itself (*pbest*) and its group (*gbest*) [15]. To update the velocity of the ith particle, there are a number of algorithms as discussed below [13]:

1) Original PSO algorithm (OPSO)
The modified velocity can be calculated from:

$$v_{id}^{t+1} = v_{id}^t + c_1 \times rand_1 \times (pbest_{id} - x_{id}^t) + c_2 \times rand_2 \times (gbest_d - x_{id}^t) \tag{1}$$

where the values of both c_1 and c_2 are just 2 and both $rand_1$ and $rand_2$ are random numbers between 0 and 1.

2) Basic PSO algorithm (BPSO)
As the OPSO does not adapt the velocity step size, it may lead to a poor searching ability. Consequently, BPSO utilises an inertia weight (w) in order to balance the global and the local searches. The updated velocity in the BPSO is calculated by:

$$v_{id}^{t+1} = w \cdot v_{id}^t + c_1 \times rand_1 \times (pbest_{id} - x_{id}^t) + c_2 \times rand_2 \times (gbest_d - x_{id}^t) \tag{2}$$

where w is 0.9 at the first iteration and linearly decreases to 0.4 at the final iteration [14].

3) Constriction factor PSO algorithm (CPSO)
CPSO has been proposed by Clerc [16-18] so as to ensure convergence of the PSO algorithm. The updated velocity in the CPSO can be expressed by:

$$v_{id}^{t+1} = k \times [v_{id}^t + c_1 \times rand_1 \times (pbest_{id} - x_{id}^t) + c_2 \times rand_2 \times (gbest_d - x_{id}^t)] \tag{3}$$

$$k = \frac{2}{\left|2 - \varphi - \sqrt{\varphi^2 - 4\varphi}\right|}, \quad \varphi = c_1 + c_2, \quad \varphi > 4 \tag{4}$$

where φ is generally set to 4.1, both c_1 and c_2 are set to 2.05 and k is 0.729 as presented in [18].

4) Original PSO including inertia weight and constriction factor (CBPSO)
In this algorithm, both the inertia weight and constriction factor are incorporated into the Original PSO, as presented in [19]. The modified velocity of each particle can be calculated as follows:

$$v_{id}^{t+1} = k \times [w \cdot v_{id}^{t} + c_1 \times rand_1 \times (pbest_{id} - x_{id}^{t}) + c_2 \times rand_2 \times (gbest_d - x_{id}^{t})]. \tag{5}$$

Subsequently, the modified position of each particle can be calculated as shown in the following equations:

$$x_{id}^{t+1} = x_{id}^{t} + v_{id}^{t+1} \tag{6}$$

where

v_{id}^{t} : velocity of particle i at iteration t in d-dimensional space;

$V_{d,\min} \le v_{id}^{t} \le V_{d,\max}$; $i = 1, 2, \ldots, n$, $d = 1, 2, \ldots, m$,

x_{id}^{t} : current position of particle i at iteration t,

w : inertia weight factor,

t : number of iterations,

n : number of particles in a group,

m : number of members in a particle,

k : constriction factor,

c_1, c_2 : acceleration constant,

$rand_1, rand_2$: uniformly distributed random number between 0 and 1.

4 Particle Swarm Optimisation with Applications in Power Systems

4.1 Application of PSO in Unit Commitment

4.1.1 Introduction

Unit Commitment (UC) is a problem in power system operation that determines the schedule of generating units to meet electricity demand and operating constraints over a time horizon [20],[21].

4.1.2 Problem formulation

The objective of the UC problem is to minimise the sum of generation cost and start-up cost over a short term period where the objective function can be mathematically formulated by the following equation [20],[21]:

$$Minimise: TC = \sum_{t=1}^{T}\sum_{i=1}^{N}\left[F_i(P_{it}) + ST_i(1-U_{i(t-1)})\right]U_{it} \tag{7}$$

Subject to the following constraints:

a) Power balance

$$\sum_{i=1}^{N} P_{it}U_{it} = P_{Dt} \tag{8}$$

b) Spinning reserve

$$\sum_{i=1}^{N} P_{i,\max}U_{it} - P_{Dt} - SR_t \geq 0 \tag{9}$$

c) Operating limit

$$P_{i,\min}U_{it} \leq P_{it} \leq P_{i,\max}U_{it} \tag{10}$$

d) Minimum up/down time

$$U_{it} = 1 \quad for \quad \sum_{h=t-T_{i,up}}^{t-1} U_{ih} < T_{i,up}$$
$$U_{it} = 0 \quad for \quad \sum_{h=t-T_{i,down}}^{t-1} (1 - U_{ih}) < T_{i,down} \tag{11}$$

In this study, the start-up cost is calculated as follows [22]:

$$ST_i = \begin{cases} HSC_i, & if \quad T_{i,off} \le T_{i,down} + CSH_i, \\ CSC_i, & \text{otherwise.} \end{cases} \tag{12}$$

List of symbols

TC : total production cost ,

$F_i(P_{it})$: fuel cost of generator i given by a quadratic function

$F_i(P_{it}) = a_i P_{it}^2 + b_i P_{it} + c_i$,

ST_i : start-up cost of unit i,

U_{it} : the on/off status of unit i at hour t,

P_{it} : the generation output of unit i at hour t,

P_{Dt} : load demand at hour t,

SR_t : spinning reserve at hour t,

$P_{i,min}$: minimum power output of unit i,

$P_{i,max}$: maximum power output of unit i,

$T_{i,up}$: minimum up time of unit i,

$T_{i,down}$: minimum down time of unit i,

N : total number of generators,

T : total number of hours ,

$T_{i,off}$: the unit's off time,

HSC_i : the unit's hot start-up cost,

CSC_i : the unit's cold start-up cost, and

CSH_i : the cold start hour.

4.1.3 Methodology

Lagrange Relaxation (LR) is an optimization approach to solve the UC problem. Although the computation of LR is fast, it has the problems with numerical convergence and poor quality of solution [21],[23]. To overcome these, it is proposed to incorporate the PSO algorithm into the LR method so as to improve its performance. The implementation processes of the proposed method are explained below:

A. Lagrange Relaxation (LR)

The basic concept of the LR procedure is to relax or ignore the coupling constraints of the UC problem (e.g. power balance and spinning reserve constraints). In addition, it decomposes the main problem into sub-problems which are easier to solve. In the LR method, Lagrange multipliers (λ_t and μ_t) are integrated into the main problem in order to create the penalty terms [20],[24],[25]. Therefore, the Lagrange function can be formed as follows [20],[24]:

$$L(P,U,\lambda,\mu) = TC(P_{it},U_{it}) + \sum_{t=1}^{T}\lambda_t(P_{Dt} - \sum_{i=1}^{N}P_{it}U_{it}) + \sum_{t=1}^{T}\mu_t(P_{Dt} + SR_t - \sum_{i=1}^{N}P_{i,\max}U_{it}). \tag{13}$$

From the concept of dual optimisation, we can obtain the values of the Lagrange multipliers by maximising the Lagrange function (L) with respect to the Lagrange multipliers λ_t and μ_t, whilst minimising with respect to P_{it} and U_{it}, that is

$$q^*(\lambda,\mu) = \max_{\lambda_t,\mu_t} q(\lambda,\mu) \tag{14}$$

where

$$q(\lambda,\mu) = \min_{P_{it},U_{it}} L(P,U,\lambda,\mu) \tag{15}$$

and

$$\min q(\lambda,\mu) = \sum_{i=1}^{N}\min\sum_{t=1}^{T}[F_i(P_{it}) + ST_i(1 - U_{i(t-1)}) - \lambda_t P_{it} - \mu_t P_{i,\max}]U_{it} \tag{16}$$

subject to the operating limit constraints and minimum up/down time constraints. Generally, two-state dynamic programming is applied to solve this problem. The main idea of applying two-state dynamic programming is to find the path that offers minimum total cost, which consists of the summation of the fuel cost and start-up cost, up to the current hour. So in each hour, there are two possible states for a generator (i.e. "1" = on and "0"= off), and their total costs are compared to make a decision for choosing the minimum cost in that path. Moreover, minimum up/down time constraints will also be taken into account as illustrated in Fig. 3. Thereafter, the minimum cost and its path from the previous hour will be stored and the same processes carried out as in the earlier step until the final hour [23].

B. Incorporation of PSO into LR for UC

Conventionally, LR uses a gradient method to update the Lagrange multipliers. It has a limitation however in that sometimes the solution is trapped in a local optimum causing a convergence problem [26]. In this section, a new hybrid method (LR-PSO) is proposed to solve the UC problem. PSO is applied to update the Lagrange multipliers and is also incorporated into the LR method to improve its performance. The flow chart describing the procedures of the proposed method is shown in Fig. 1.

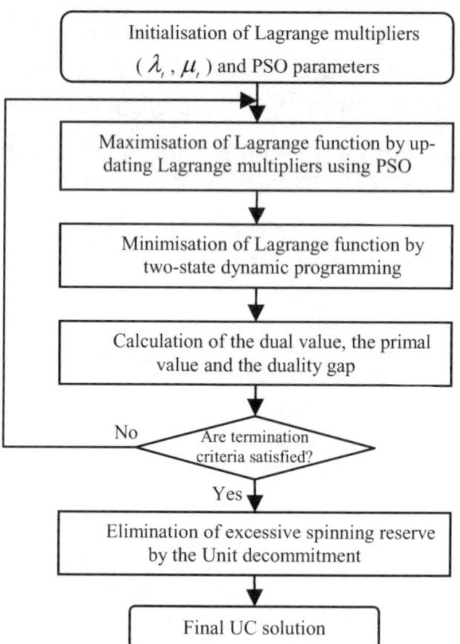

Fig. 1. The basic flow chart of the proposed method

The steps of the computation method as presented in Fig. 1 are discussed below.

Step 1: Initialisation of Lagrange multipliers and PSO parameters

- Generate an initial population of particles (λ_t and μ_t). Normally, each particle is generated randomly within an allowable range. The members of the population are stored in a matrix form which defines the Lagrange multipliers as shown in Fig. 2.
- Subsequently, initialise the parameters of the PSO algorithm (e.g. population size, initial/final inertia weight, velocity of particle, acceleration constant, constriction factor, the maximum generation and the duality gap).

$$
\begin{array}{c|cccccccc}
 & \multicolumn{8}{l}{\textit{Hour}} \\
 & 1 & 2 & \cdot\ \cdot & T & 1 & 2 & \cdot\ \cdot & T \\
\hline
1 & \lambda_1^1 & \lambda_2^1 & \cdots & \lambda_{T-1}^1 \ \ \lambda_T^1 & \mu_1^1 & \mu_2^1 & \cdots & \mu_{T-1}^1 \ \ \mu_T^1 \\
2 & \lambda_1^2 & \lambda_2^2 & \cdots & \lambda_{T-1}^2 \ \ \lambda_T^2 & \mu_1^2 & \mu_2^2 & \cdots & \mu_{T-1}^2 \ \ \mu_T^2 \\
\cdot & \vdots & \vdots & \cdots & \vdots \ \ \vdots & \vdots & \vdots & \cdots & \vdots \ \ \vdots \\
\cdot & \lambda_1^{N-1} & \lambda_2^{N-1} & \cdots & \lambda_{T-1}^{N-1} \ \ \lambda_T^{N-1} & \mu_1^{N-1} & \mu_2^{N-1} & \cdots & \mu_{T-1}^{N-1} \ \ \mu_T^{N-1} \\
N & \lambda_1^N & \lambda_2^N & \cdots & \lambda_{T-1}^N \ \ \lambda_T^N & \mu_1^N & \mu_2^N & \cdots & \mu_{T-1}^N \ \ \mu_T^N \\
\end{array}
$$

(Rows: *Individual No.*; left block: λ; right block: μ)

Fig. 2. Population in the form of a matrix

- Define each particle as *pbest* and the best position of all particles as *gbest*.

Step 2: Maximisation of Lagrange function by updating Lagrange multipliers using PSO

- Calculate the evaluation value or dual value (q) of each individual (λ_t, μ_t) as follows.

$$
\begin{aligned}
q(\lambda,\mu) = & \sum_{i=1}^{N}\sum_{t=1}^{T}[F_i(P_{it}) + ST_i(1-U_{i(t-1)})]U_{it} \\
& -\sum_{i=1}^{N}\sum_{t=1}^{T}(\lambda_t P_{it} + \mu_t P_{i,\max})U_{it} + \sum_{t=1}^{T}\left(\lambda_t P_{Dt} + \mu_t (P_{Dt} + SR_t)\right)
\end{aligned}
\tag{17}
$$

- Compare each evaluation value with the previous *pbest*. If the current value is more, let it be *pbest*. Similarly, if the best value in *pbest*'s group is more than *gbest*, let the value be *gbest*.

- Update the member velocity (v) of each individual (λ_i, μ_i) by Eq. (5).

- If $v_{id}^{(t+1)} > V_{d,\max}$, then $v_{id}^{(t+1)} = V_{d,\max}$ or if $v_{id}^{(t+1)} < -V_{d,\max}$, then $v_{id}^{(t+1)} = -V_{d,\max}$. The maximum velocity can be calculated as follows [10]:

$$V_{d,\max} = \frac{(x_{id,\max} - x_{id,\min})}{N} \tag{18}$$

  where N is a chosen number of intervals.

- Update the member position of each individual (λ_i, μ_i) from Eq. (6).

Step 3: Minimisation of Lagrange function by two-state dynamic programming

- Minimise the Lagrange function using two-state dynamic programming for P_{it} and U_{it}, where $i = 1...N$, and $t = 1...T$.

- Use the forward dynamic programming (FDP) to solve the dual problem. The objective is to minimise q. Fig. 3 illustrates the concept of two-state dynamic programming [21].

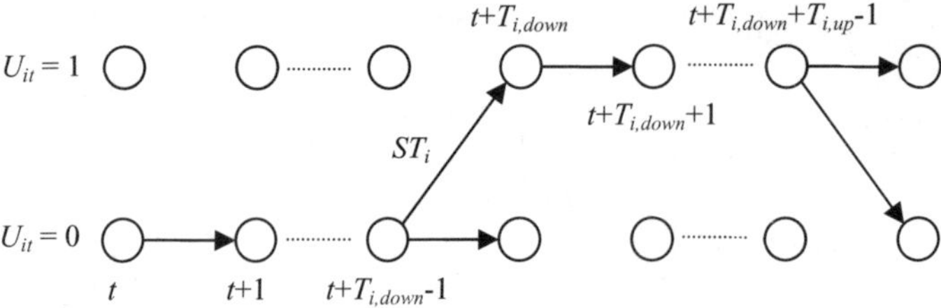

Fig. 3. Two-state dynamic programming

Step 4: Calculation of the dual value, the primal value, and the duality gap

- Determine the dual value from Eq. (17) using P_{it} and U_{it} obtained from step 3.

- To solve the economic dispatch problem, use U_{it} from step 3 to obtain P_{it}^{*}, and then calculate primal value (J).

$$J = \sum_{t=1}^{T}\sum_{i=1}^{N}[F_i(P_{it}^{*}) + ST_i(1 - U_{i(t-1)})]U_{it} \tag{19}$$

- The difference between the primal and dual problem, called the duality gap (ε), is used as a terminating criterion. The duality gap can be calculated from

$$\varepsilon = \frac{J - q}{q}. \tag{20}$$

Step 5: If either the predefined maximum number of generations is reached, or the duality gap is less than a setting threshold, then stop. The latest P_{it}^{*} is the optimal solution. Otherwise, return to Step 2.

Step 6: Elimination of excessive spinning reserve by the Unit decommitment

According to [23], over committed units in some hours may lead to an excessive spinning reserve requirement that will result in high total production cost. Accordingly, the elimination of excessive spinning reserve is necessary to apply for this case. To deal with this problem, a heuristic search method called the Unit decommitment is adopted, as proposed by [23]. The procedures of the Unit decommitment are shown in Fig. 4.

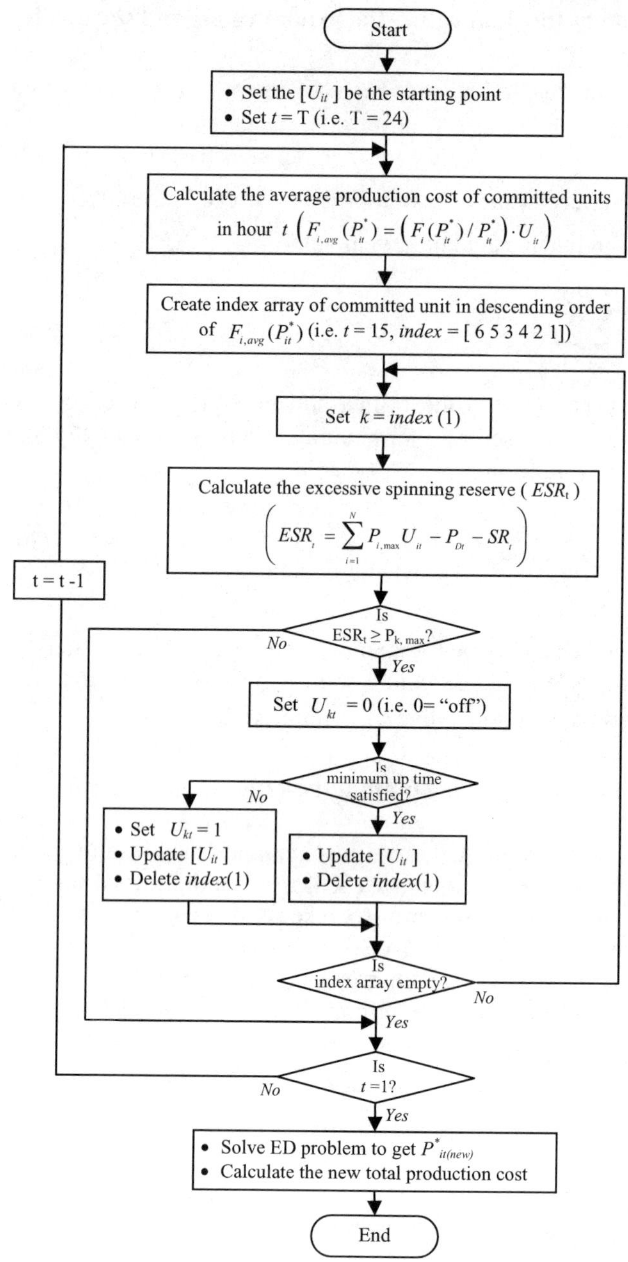

Fig. 4. Flow chart of the Unit decommitment for eliminating excessive spinning reserve

4.1.4 Experimental Results

In this section, the proposed LR-PSO method is applied to solve the UC problem. In order to illustrate its effectiveness, the method is applied to two different systems, namely a 3-unit 4-hour system and a 10-unit 24-hour system. The data used in both cases are adopted from [20] and [27] and the details are in Appendix A. The simulations are carried out using Matlab. The parameters of the PSO used in all simulations are: initial inertia weight (w_{max}) = 0.9; final inertia weight (w_{min}) = 0.4; acceleration constants (c_1, c_2) = 2.05 and constriction factor (k) = 0.729.

Case A: 3-unit, 4-hour system

In this case, the set parameters of the proposed method are population size = 40, number of runs = 30, maximum number of generations = 18, and duality gap = 0.02. For the LR method [20], the parameters for simulation are $\alpha_1 = 0.01$ and $\alpha_2 = 0.002$. Furthermore, spinning reserve is not considered; therefore the elimination of excessive spinning reserve section will be neglected. To examine the effectiveness of the proposed method, the simulation results are compared with the results obtained from the LR method, which is re-implemented. The optimal solution is $20162.75 as reported in [21]. From the simulation results, the proposed method reaches the optimal solution ($20162.75) in every run. Table 1 presents the optimal solution obtained from the proposed method. Furthermore, the comparison of the average convergence curves between the proposed method and the LR method are demonstrated in Fig. 5. It can be observed that the LR method has a numerical convergence problem. In the 10th iteration, the duality gap of the proposed method is 0.02 which satisfies the stopping criterion while for the LR method, the stopping criterion is satisfied in the 18th iteration. From the comparison of the two methods, it is clearly shown that the proposed method is more effective than the LR method in terms of overcoming the convergence problem and reaching the duality gap.

Table 1. The optimal solution obtained from the proposed method

Hour	Load (MW)	Unit Number			Fuel cost ($)
		1	2	3	
1	170	0	0	170	1264.50
2	520	0	320	200	4616.00
3	1100	500	400	200	11400.00
4	330	0	130	200	2882.25
Total					20162.75

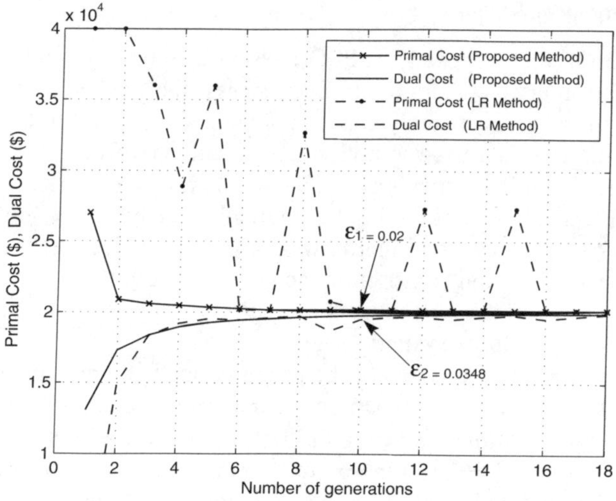

Fig. 5. Comparison of convergence curves between the proposed method and the LR method (ε_1 - Duality gap of the proposed method and ε_2 - Duality gap of the LR method)

Case B: 10-unit, 24-hour system

For this case, the set parameters of the proposed method is population size = 100. Since the spinning reserve will be taken into account, it is assumed to be 10% of the load demand. To investigate the effect of excessive spinning reserve, two versions of the LR-PSO method, one including and one without the Unit decommitment are simulated. Tables 2, 3 and 4 illustrate the best solution obtained from the LR method, the proposed LR-PSO method without the Unit decommitment, and the proposed LR-PSO method, respectively. The total cost of the LR method is $565823 while the total cost of the proposed method without applying the Unit decommitment is $565275 which is $548 less than the LR method. In addition, the total cost obtained from the proposed method is $455 cheaper than the total cost obtained from the proposed LR-PSO method without applying the Unit decommitment. The saving in total cost is a consequence of elimination of excessive spinning reserve in the 15th hour of the LR-PSO method without the Unit de-commitment, the detail of which is presented in Table 3. Table 5 compares the result of the proposed method with those of other research studies. From the simulation results, it can therefore be concluded that the performance of the proposed LR-PSO method is better than other methods in terms of total production cost. Since the simulations were carried out on different types of computers, the computation time will not be compared here.

Table 2. The best solution obtained from the LR method

Hour	Load (MW)	Generation schedule (MW) U1	U2	U3	U4	U5	U6	U7	U8	U9	U10	Fuel cost ($)	Startup cost ($)	Total cost ($)
1	700	455	245	0	0	0	0	0	0	0	0	13683.13	0	13683.13
2	750	455	295	0	0	0	0	0	0	0	0	14554.50	0	14554.50
3	850	455	370	0	0	25	0	0	0	0	0	16809.45	900	17709.45
4	950	455	340	0	130	25	0	0	0	0	0	19145.70	560	19705.70
5	1000	455	390	0	130	25	0	0	0	0	0	20020.02	0	20020.02
6	1100	455	360	130	130	25	0	0	0	0	0	22387.04	1100	23487.04
7	1150	455	410	130	130	25	0	0	0	0	0	23261.98	0	23261.98
8	1200	455	455	130	130	30	0	0	0	0	0	24150.34	0	24150.34
9	1300	455	455	130	130	85	20	25	0	0	0	27251.06	860	28111.06
10	1400	455	455	130	130	162	33	25	10	0	0	30057.55	60	30117.55
11	1450	455	455	130	130	162	73	25	10	10	0	31916.06	60	31976.06
12	1500	455	455	130	130	162	80	25	43	10	10	33890.16	60	33950.16
13	1400	455	455	130	130	162	33	25	10	0	0	30057.55	0	30057.55
14	1300	455	455	130	130	85	20	25	0	0	0	27251.06	0	27251.06
15	1200	455	440	130	130	25	20	0	0	0	0	24605.73	0	24605.73
16	1050	455	310	130	130	25	0	0	0	0	0	21513.66	0	21513.66
17	1000	455	260	130	130	25	0	0	0	0	0	20641.82	0	20641.82
18	1100	455	360	130	130	25	0	0	0	0	0	22387.04	0	22387.04
19	1200	455	415	130	130	25	20	25	0	0	0	25341.60	430	25771.60
20	1400	455	455	130	130	162	33	25	10	0	0	30057.55	60	30117.55
21	1300	455	455	130	130	85	20	25	0	0	0	27251.06	0	27251.06
22	1100	455	360	130	130	25	0	0	0	0	0	22387.04	0	22387.04
23	900	455	420	0	0	25	0	0	0	0	0	17684.69	0	17684.69
24	800	455	345	0	0	0	0	0	0	0	0	15427.42	0	15427.42
Total												561733.23	4090	565823.23

Table 3. The best solution obtained from the proposed LR-PSO method without the Unit decommitment

Hour	Load (MW)	Generation schedule (MW) U1	U2	U3	U4	U5	U6	U7	U8	U9	U10	Fuel cost ($)	Startup cost ($)	Total cost ($)
1	700	455	245	0	0	0	0	0	0	0	0	13683.13	0	13683.13
2	750	455	295	0	0	0	0	0	0	0	0	14554.50	0	14554.5
3	850	455	370	0	0	25	0	0	0	0	0	16809.45	900	17709.45
4	950	455	455	0	0	40	0	0	0	0	0	18597.67	0	18597.67
5	1000	455	390	0	130	25	0	0	0	0	0	20020.02	560	20580.02
6	1100	455	360	130	130	25	0	0	0	0	0	22387.04	1100	23487.04
7	1150	455	410	130	130	25	0	0	0	0	0	23261.98	0	23261.98
8	1200	455	455	130	130	30	0	0	0	0	0	24150.34	0	24150.34
9	1300	455	455	130	130	85	20	25	0	0	0	27251.06	860	28111.06
10	1400	455	455	130	130	162	33	25	10	0	0	30057.55	60	30117.55
11	1450	455	455	130	130	162	73	25	10	10	0	31916.06	60	31976.06
12	1500	455	455	130	130	162	80	25	43	10	10	33890.16	60	33950.16
13	1400	455	455	130	130	162	33	25	10	0	0	30057.55	0	30057.55
14	1300	455	455	130	130	85	20	25	0	0	0	27251.06	0	27251.06
15	1200	455	440	130	130	25	20	0	0	0	0	24605.73	0	24605.73
16	1050	455	310	130	130	25	0	0	0	0	0	21513.66	0	21513.66
17	1000	455	260	130	130	25	0	0	0	0	0	20641.82	0	20641.82
18	1100	455	360	130	130	25	0	0	0	0	0	22387.04	0	22387.04
19	1200	455	415	130	130	25	20	25	0	0	0	25341.60	430	25771.6
20	1400	455	455	130	130	162	33	25	10	0	0	30057.55	60	30117.55
21	1300	455	455	130	130	85	20	25	0	0	0	27251.06	0	27251.06
22	1100	455	360	130	130	25	0	0	0	0	0	22387.04	0	22387.04
23	900	455	420	0	0	25	0	0	0	0	0	17684.69	0	17684.69
24	800	455	345	0	0	0	0	0	0	0	0	15427.42	0	15427.42
Total												561185.19	4090	565275.2

Table 4. The best solution obtained from the proposed LR-PSO method

Hour	Load (MW)	Generation schedule (MW)										Fuel cost ($)	Startup cost ($)	Total cost ($)
		U1	U2	U3	U4	U5	U6	U7	U8	U9	U10			
1	700	455	245	0	0	0	0	0	0	0	0	13683.13	0	13683.13
2	750	455	295	0	0	0	0	0	0	0	0	14554.50	0	14554.50
3	850	455	370	0	0	25	0	0	0	0	0	16809.45	900	17709.45
4	950	455	455	0	0	40	0	0	0	0	0	18597.67	0	18597.67
5	1000	455	390	0	130	25	0	0	0	0	0	20020.02	560	20580.02
6	1100	455	360	130	130	25	0	0	0	0	0	22387.04	1100	23487.04
7	1150	455	410	130	130	25	0	0	0	0	0	23261.98	0	23261.98
8	1200	455	455	130	130	30	0	0	0	0	0	24150.34	0	24150.34
9	1300	455	455	130	130	85	20	25	0	0	0	27251.06	860	28111.06
10	1400	455	455	130	130	162	33	25	10	0	0	30057.55	60	30117.55
11	1450	455	455	130	130	162	73	25	10	10	0	31916.06	60	31976.06
12	1500	455	455	130	130	162	80	25	43	10	10	33890.16	60	33950.16
13	1400	455	455	130	130	162	33	25	10	0	0	30057.55	0	30057.55
14	1300	455	455	130	130	85	20	25	0	0	0	27251.06	0	27251.06
15	1200	455	455	130	130	30	0	0	0	0	0	24150.34	0	24150.34
16	1050	455	310	130	130	25	0	0	0	0	0	21513.66	0	21513.66
17	1000	455	260	130	130	25	0	0	0	0	0	20641.82	0	20641.82
18	1100	455	360	130	130	25	0	0	0	0	0	22387.04	0	22387.04
19	1200	455	415	130	130	25	20	25	0	0	0	25341.60	430	25771.60
20	1400	455	455	130	130	162	33	25	10	0	0	30057.55	60	30117.55
21	1300	455	455	130	130	85	20	25	0	0	0	27251.06	0	27251.06
22	1100	455	360	130	130	25	0	0	0	0	0	22387.04	0	22387.04
23	900	455	420	0	0	25	0	0	0	0	0	17684.69	0	17684.69
24	800	455	345	0	0	0	0	0	0	0	0	15427.42	0	15427.42
Total												560729.80	4090	564819.80

Table 5. Comparison of simulation results

Method	Total production costs ($)
LR [27]	565,825
GA [27]	565,825
HPSO [28]	574,153
LR*	565,823
LR-PSO**	565,275
LR-PSO	564,820

* Re-implemented the LR method

** The proposed LR-PSO method without unit decommitment

4.1.5 Summary of application of PSO in Unit Commitment

This section presents a new methodology, called LR-PSO or Particle Swarm Optimisation (PSO) combined with Lagrange Relaxation method (LR), to solve the UC problem. Applying the LR-PSO method improves the performance of the LR method since PSO is used to update the Lagrange multipliers. To illustrate its performance, the proposed method is tested on 3-Unit 4-Hr system and 10-Unit 24-Hr system. Compared with the LR, Genetic Algorithm (GA) and Hybrid Particle Swarm Optimisation (HPSO) methods, the proposed LR-PSO method has provided a satisfactory performance in terms of solution quality. Furthermore, it could be extended to solve a large-scale system and a profit-based unit commitment problem under the competitive environment of power systems.

4.2 Application of PSO in Economic Dispatch

4.2.1 Introduction

Economic Dispatch (ED) is a sub-problem of the general UC problem. In essence, the ED problem is to determine the optimum scheduling of generation at a particular time that minimises the total production cost while satisfying an equality constraint and inequality constraints i.e. power balance constraint and operating limits [29]. For the sake of simplicity, the cost function of the standard ED problem is generally a single quadratic function; however, in practical ED problems, valve-point loadings have been included. Taking the valve-point loadings into account will increase multiple local minimum points in the cost function and make the problem more difficult [6].

4.2.2 Problem formulation

In general, the mathematical model of the ED problem is as follows [20]:

$$Minimise : TC = \sum_{i=1}^{N} F_i(P_i) \tag{21}$$

Subject to:

a) Power balance constraint

$$\sum_{i=1}^{N} P_i = P_D \tag{22}$$

b) Operating limit constraints

$$P_{i,\min} \leq P_i \leq P_{i,\max} \tag{23}$$

For the standard ED problem (smooth cost functions), the generator's fuel cost function can be represented by:

$$F_i(P_i) = a_i P_i^2 + b_i P_i + c_i \tag{24}$$

whilst in the ED problem with non-smooth cost functions, the fuel cost function with valve-point loadings can be expressed as [30] :

$$F_i(P_i) = a_i P_i^2 + b_i P_i + c_i + \left| e_i \times \sin(f_i \times (P_{i,\min} - P_i)) \right|. \tag{25}$$

Examples of smooth and non-smooth cost functions are shown in Fig. 6(a) and 6(b), respectively.

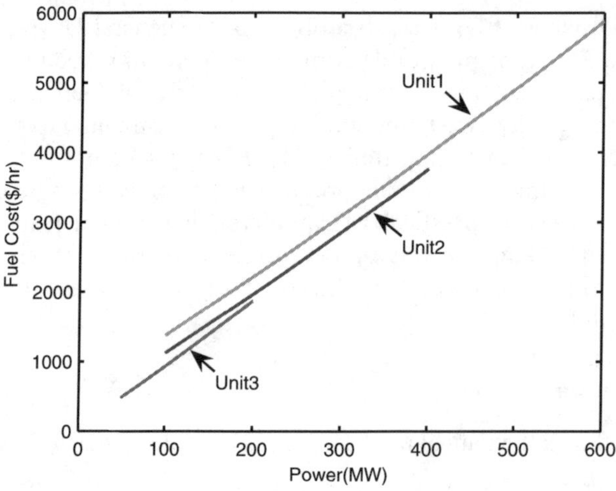

Fig. 6(a). An example of input-output curve with smooth cost function

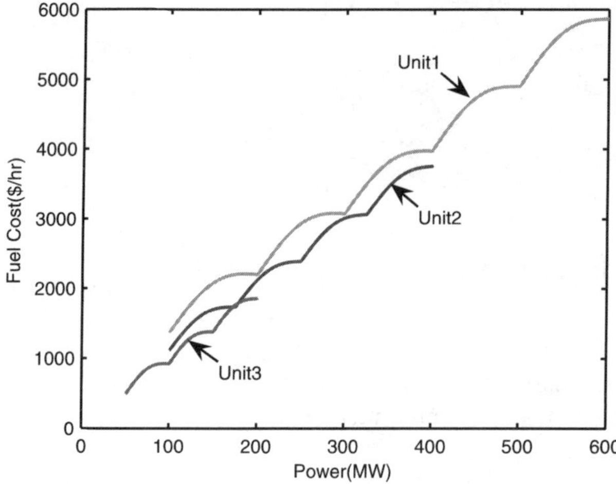

Fig. 6(b). An example of input-output curve with valve point loadings

List of symbols

TC : total production cost,

$F_i(P_i)$: fuel cost of ith generator; generator's fuel cost can be calculated from Eq. (24) or Eq. (25),
where a_i, b_i and c_i are coefficients of the fuel cost function e_i and f_i are coefficients from the valve-point loading of the ith generator,

P_i : power output of ith generator,

P_D : power demand,

$P_{i,min}$: minimum power output of ith generator,

$P_{i,max}$: maximum power out put of ith generator,

N : number of generators.

4.2.3 Implementation of PSO approach in ED problem

The major steps of the proposed PSO approach in the ED problem are summarised below:

1) Initialisation of experimental parameters: the positions of the particles, $x_i = [P_{i1}, P_{i2}, \ldots, P_{id}]$ and the velocities of the particles, $v_i - [v_{i1}, v_{i2}, \ldots, v_{id}]$.
2) Updating the velocities and positions.
3) Modification of the particles using a heuristic method.
4) Evaluation of the particles in the population.
5) Updating of *pbest* (best position found by the ith particle) and *gbest* (best position found by the group)
6) If the termination criteria are satisfied, then stop. Otherwise, return to step 2.

The process can be expressed in details as follow:

Step 1: Initialisation of experimental parameters, the positions, and velocities of particles

- Initialise the experimental parameters, which are population size, initial inertia weight (w_{max}), final inertia weight (w_{min}), acceleration constants (c_1,c_2), the maximum generation, generation limit, fuel cost coefficients and power demand.
- Generate initial population randomly by using the modified heuristic search the main concept of which is to initialise the particles,

subject to power balance and operating limit constraints. The heuristic search is modified from [7] in order to enhance its performance. The modified search procedures are therefore given in Fig. 7.

- Initialise the velocities of particles.
- Calculate the fitness values (*TC*) of each particle using Eq. (21).
- Let each fitness value be *pbest*.
- Search for the best position (the least cost) of all particles (the *pbest* values) and let this value be *gbest*.

Step 2: Updating of the velocities and positions

- Update the velocity and position of each particle by Eq. (2) and Eq. (6), respectively.
- If $x_{id}^{(t+1)} > P_{i,max}$, then $x_{id}^{(t+1)} = P_{i,max}$ or if $x_{id}^{(t+1)} < P_{i,min}$, then $x_{id}^{(t+1)} = P_{i,min}$. Otherwise, set $x_{id}^{(t+1)} = x_{id}^{(t+1)}$.

Step 3: Modification of the particles

Computation continues on the basic idea described in step 1, except the procedure for finding the final value (P_{i1}), which is illustrated in sections 1 and 2 of Fig. 8. The final value in this step (P_{iL}) can be randomly calculated instead of calculating only the first element of each particle as shown in Fig. 7 because of a lack of diversity in the values. To carry this out, the updated swarm in step 2 is used as the starting point. Fig. 8 shows the procedures for this heuristic modification.

Step 4: Evaluation of the particles

- Calculate each fitness value using the fitness function as shown in Eq. (21).

Step 5: Updating of *pbest* and *gbest*

- If the current fitness value is less than the previous *pbest*, let it be *pbest*. Correspondingly, let the best value *pbest* be *gbest* if it is less than *gbest*.

Step 6: If the current iteration is more than the maximum number of iterations, then stop and take the latest *gbest* to be the final solution. Otherwise, return to Step 2.

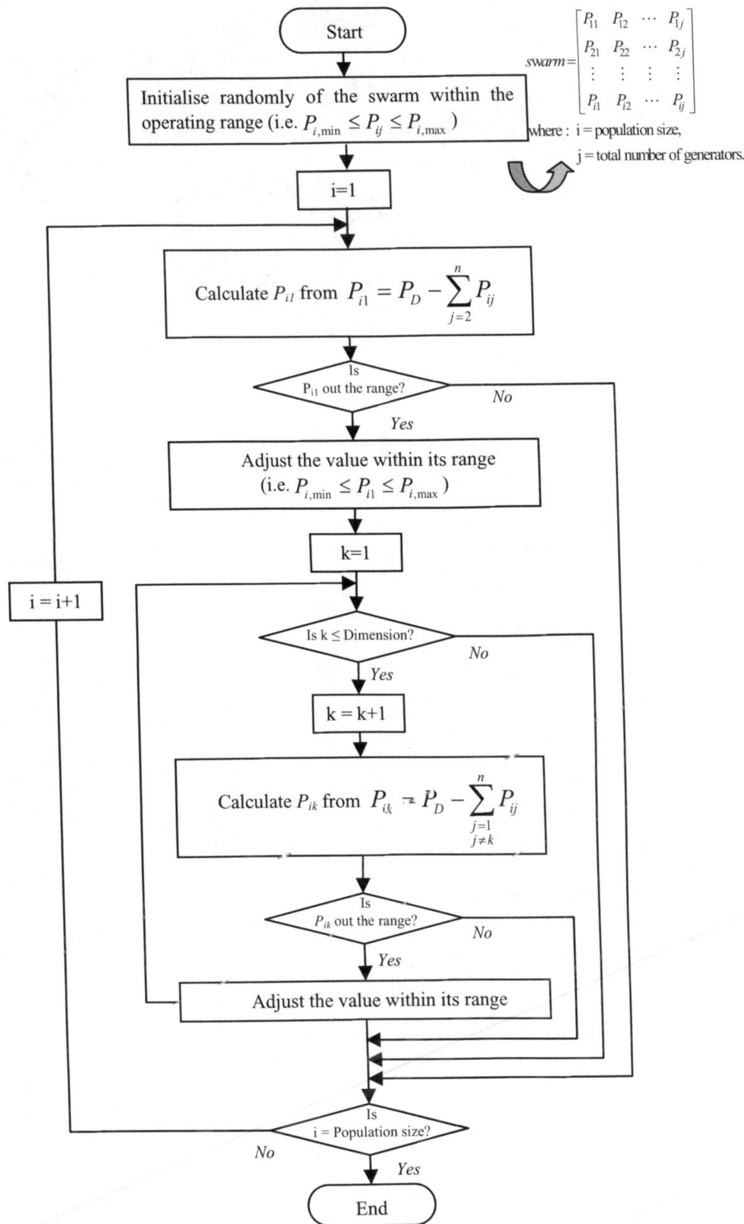

Fig. 7. Flow chart of the modified heuristic search for initialisation

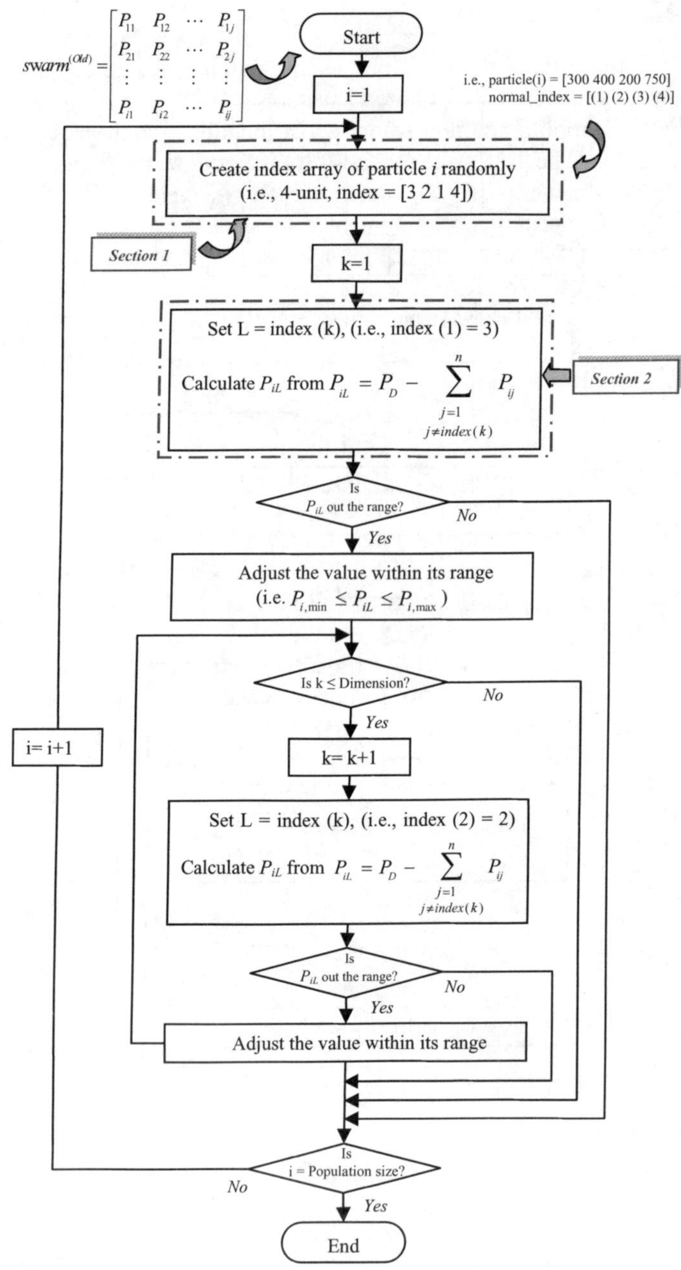

Fig. 8. Flow chart of the modified heuristic search for particles' modification

4.2.4 Experimental Results

To investigate the efficiency of the proposed method, three different systems have been considered. The first two systems consist of the standard 3-unit system with smooth and non-smooth cost functions given in [20] and [30], respectively. The last system is a 40-unit system with non-smooth cost functions as shown in [6]. The data for each system are shown in Appendix B. The simulations are implemented in Matlab and executed on a Pentium IV, 3GHz personal computer with 512 MB RAM, where parameters of PSO in all case studies are initial inertia weight (w_{max}) = 0.9, final inertia weight (w_{min}) = 0.4, and acceleration constants (c_1, c_2) = 2.

Case A: 3-generator system

In this case, the population size and the maximum number of generations are set to 10 and 300 respectively, and the power demand is set to 850 MW. From the previous work, the global solution is $8194.35612 as presented in [20]. Table 6 compares the simulation results of the proposed method with the modified Hopfield neural network (MHNN) [29], the improved evolutionary programming (IEP) [31], the numerical method (NM) [20] and the modified PSO (MPSO) [7]. From the simulation of this case, the computation time is 0.42 seconds and the results show that the method can achieve a global solution. Fig. 9 illustrates not only the convergence characteristic to the optimal solution but also the effect of the different initial positions.

Table 6. Comparison of total costs with different methods for case A

Method	Generating Unit			Total Power (MW)	Total cost ($)
	P_1	P_2	P_3		
MHNN [29]	393.800	333.100	122.300	849.2	8187.00000
IEP [31]	393.170	334.603	122.227	850	8194.35614
NM [20]	393.170	334.604	122.226	850	8194.35612
MPSO [7]	393.170	334.604	122.226	850	8194.35612
Proposed Method	393.170	334.604	122.226	850	8194.35612

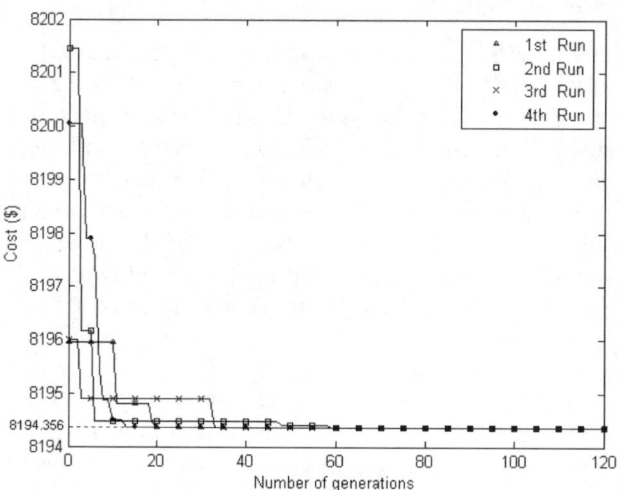

Fig. 9. Convergence characteristics of the proposed method with different initial conditions in case A

Case B: 3-generator system with valve-point loadings

In this case, the population number is 20 while the maximum number of generations and the power demand remain the same as Case A. As reported in [32], the optimal solution is $8234.07. Table 7 illustrates the comparison of simulation results with various methods (i.e. the genetic algorithm (GA) [30], the IEP [31], the evolutionary programming (EP) [33], the Taguchi method (TM) [34] and the MPSO [7]). In this case study, it is observed that the proposed method also provides the optimal solution ($8234.07) and its computation time is 0.5 seconds. Similarly, the convergence curves with different initial conditions are illustrated in Fig. 10.

Table 7. Comparison of total costs with different methods for case B

Method	Generating Unit			Total Power (MW)	Total cost ($)
	P1	P2	P3		
GA [30]	300.00	400	150	850	8237.60
IEP [31]	300.23	400.00	149.77	850	8234.09
EP [33]	300.26	400	149.74	850	8234.07
TM[34]	300.27	400	149.73	850	8234.07
MPSO [7]	300.27	400	149.73	850	8234.07
Proposed method	300.27	400	149.73	850	8234.07

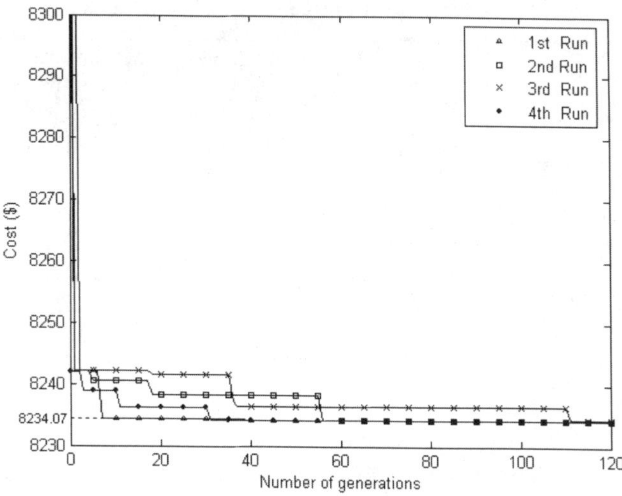

Fig. 10. Convergence characteristics of the proposed method with different initial conditions in case B

Case C: 40- generator system with valve-point loadings

The setting parameters of Case C are population size = 60, maximum number of generations = 1500, and power demand = 10500MW. Table 8 summarises the mean cost, the minimum cost and the mean computation time of the proposed method compared with the classical evolutionary programming (CEP) [6], the fast EP (FEP) [6], the modified fast EP (MFEP) [6], the improved FEP(IFEP) [6], the TM [34], and the MPSO [7] methods respectively.

In this case, the mean cost of the proposed method represents an average over 20 different initial runs. The results show that the mean cost and the minimum cost of the proposed method are the cheapest. The proposed method is better than the others in terms of the quality of its solution. The best solution obtained from the proposed method is shown in Table 9 and the plot of the best cost is illustrated in Fig. 11. However, it is difficult to compare the computation time of this proposed method to any other research as basically the simulations are carried out by different types of computers.

Table 8. Comparison of simulation results among various methods for case C

Method	Mean cost ($)	Min. cost ($)	Mean Computation time (s)
CEP*[6]	124,793.48	123,488.29	1956.93
FEP*[6]	124,119.37	122,679.71	1039.16
MFEP*[6]	123,489.74	122,647.57	2196.10
IFEP*[6]	123,382.00	122,624.35	1167.35
TM [34]	123,078.21	122,477.78	94.28
MPSO [7]	-	122,252.26	-
Proposed method**	122,304.70	122,190.63	14.56

* Simulations were executed on Pentium II, 350 MHz, 128-MB RAM

** Simulations were executed on Pentium IV, 3 GHz, 512-MB RAM

Table 9. The best simulation result obtained from the proposed method for case C

Unit	Power (MW)	Cost ($)	Unit	Power (MW)	Cost ($)
1	114.0000	978.1563	21	550.0000	5575.3293
2	113.9992	978.1432	22	550.0000	5575.3293
3	120.0000	1544.6534	23	550.0000	5558.0493
4	180.2295	2153.9843	24	524.3592	5078.9631
5	97.0000	853.1776	25	523.8558	5286.9053
6	140.0000	1596.4643	26	550.0000	5785.6643
7	300.0000	3216.4240	27	10.0104	1140.7632
8	299.9700	3051.8101	28	10.0107	1140.7701
9	300.0000	3071.9895	29	10.0000	1140.5240
10	130.0035	2502.1451	30	96.9997	853.1732
11	94.0000	1893.3054	31	190.0000	1643.9913
12	94.0000	1908.1668	32	190.0000	1643.9913
13	125.0000	2541.6813	33	190.0000	1643.9913
14	394.2939	6415.2276	34	200.0000	2101.0170
15	304.5691	5172.3846	35	200.0000	2043.7270
16	304.5581	5172.1210	36	199.9984	2043.7119
17	489.3338	5297.8828	37	110.0000	1220.1661
18	489.2954	5289.1100	38	110.0000	1220.1661
19	511.4847	5545.3790	39	110.0000	1220.1661
20	511.4307	5544.1887	40	511.5979	5547.8322
Total Power(MW) and Total Cost($)				10,500.000	122,190.626

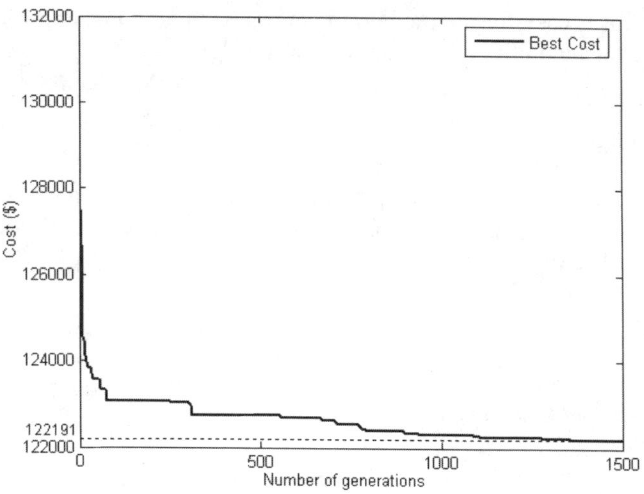

Fig. 11. Convergence characteristic of the proposed method for case C

4.2.5 Summary of application of PSO in Economic Dispatch

In this section, PSO is integrated into a modified heuristic search to solve the ED problem with both smooth and non-smooth cost functions. In a computation process, PSO searches for a global solution while a modified heuristic handles equality and inequality constraints. To illustrate its efficiency and effectiveness, the proposed method is applied to 3 different case studies. For the first two cases, a 3-unit system with smooth and non-smooth cost functions, the simulation results show that the proposed method succeeds in reaching the global solution and has a reasonable computation time. The result from the last case, which is a 40-unit system with non-smooth cost functions, clearly confirms that the proposed method is more powerful than other methods.

5 Conclusions

This chapter presents the applications of PSO to both the UC and ED problem. To solve the UC problem, PSO is applied to update the Lagrange multipliers and is also incorporated into the LR method to improve its performance. For the ED problem, PSO is integrated into a modified heuristic search to enhance its searching efficiency. Even though PSO is very simple to implement with high quality solutions and stable convergence characteristics, it has a shortcoming in that there can be a lack of diversity.

Therefore, it is easy for the method to get trapped into a local optimum. To overcome this, the PSO would require further study.

Acknowledgement

The first author of this chapter gratefully acknowledges financial support from the Royal Thai Government and King Mongkut's Institute of Technology North Bangkok, Thailand.

Appendix

Appendix A: Unit Commitment

Table 10. Unit data for the 3-unit system [20]

	Unit 1	Unit 2	Unit 3
P_{max} (MW)	600	400	200
P_{min} (MW)	100	100	50
a ($/MW^2 h)	0.002	0.0025	0.005
b ($/MW h)	10	8	6
c ($/h)	500	300	100

Table 11. Load demand for 4-hour

Hour	1	2	3	4
Load (MW)	170	520	1100	330

Table 12. Unit data for the 10-unit system [27]

	Unit 1	Unit 2	Unit 3	Unit 4	Unit 5
P_{max} (MW)	455	455	130	130	162
P_{min} (MW)	150	150	20	20	25
a ($/MW^2 h)	0.00048	0.00031	0.002	0.00211	0.00398
b ($/MW h)	16.19	17.26	16.60	16.50	19.70
c ($/h)	1000	970	700	680	450
min up (h)	8	8	5	5	6
min down (h)	8	8	5	5	6
hot start cost ($)	4500	5000	550	560	900
Cold start cost ($)	9000	10000	1100	1120	1800
Cold start hrs (h)	5	5	4	4	4
Initial status (h)	8	8	-5	-5	-6

	Unit 6	Unit 7	Unit 8	Unit 9	Unit 10
P_{max} (MW)	80	85	55	55	55
P_{min} (MW)	20	25	10	10	10
a ($/MW^2 h)	0.00712	0.00079	0.00413	0.00222	0.00173
b ($/MW h)	22.26	27.74	25.92	27.27	27.79
c ($/h)	370	480	660	665	670
min up (h)	3	3	1	1	1
min down (h)	3	3	1	1	1
hot start cost ($)	170	260	30	30	30
Cold start cost ($)	340	520	60	60	60
Cold start hrs (h)	2	2	0	0	0
Initial status (h)	-3	-3	-1	-1	-1

Table 13. Load demand for 24-hour

Hour	Load (MW)	Hour	Load (MW)
1	700	13	1400
2	750	14	1300
3	850	15	1200
4	950	16	1050
5	1000	17	1000
6	1100	18	1100
7	1150	19	1200
8	1200	20	1400
9	1300	21	1300
10	1400	22	1100
11	1450	23	900
12	1500	24	800

Appendix B: Economic Dispatch

Table 14. Unit data for test case A (3-generator system with smooth cost functions) where *a, b, c* are cost coefficients in the production cost function [20]

Generator	P_{min}(MW)	P_{max}(MW)	*a*	*b*	*c*
1	150	600	0.001562	7.92	561
2	100	400	0.001940	7.85	310
3	50	200	0.004820	7.97	78

Table 15. Unit data for test case B (3-generator system with non-smooth cost functions) where *a, b, c, e* and *f* are cost coefficients in the production cost function [30]

Generator	P_{min}(MW)	P_{max}(MW)	*a*	*b*	*c*	*e*	*f*
1	100	600	0.001562	7.92	561	300	0.0315
2	100	400	0.001940	7.85	310	200	0.042
3	50	200	0.004820	7.97	78	150	0.063

Table 16. Units data for test case C (40-generator system with non-smooth cost functions) [6]

Generator	P_{min}(MW)	P_{max}(MW)	*a*	*b*	*c*	*e*	*f*
1	36	114	0.00690	6.73	94.705	100	0.084
2	36	114	0.00690	6.73	94.705	100	0.084
3	60	120	0.02028	7.07	309.54	100	0.084
4	80	190	0.00942	8.18	369.03	150	0.063
5	47	97	0.0114	5.35	148.89	120	0.077
6	68	140	0.01142	8.05	222.33	100	0.084
7	110	300	0.00357	8.03	287.71	200	0.042
8	135	300	0.00492	6.99	391.98	200	0.042
9	135	300	0.00573	6.60	455.76	200	0.042
10	130	300	0.00605	12.9	722.82	200	0.042
11	94	375	0.00515	12.9	635.20	200	0.042
12	94	375	0.00569	12.8	654.69	200	0.042
13	125	500	0.00421	12.5	913.40	300	0.035
14	125	500	0.00752	8.84	1760.4	300	0.035
15	125	500	0.00708	9.15	1728.3	300	0.035
16	125	500	0.00708	9.15	1728.3	300	0.035
17	220	500	0.00313	7.97	647.85	300	0.035
18	220	500	0.00313	7.95	649.69	300	0.035
19	242	550	0.00313	7.97	647.83	300	0.035
20	242	550	0.00313	7.97	647.81	300	0.035
21	254	550	0.00298	6.63	785.96	300	0.035
22	254	550	0.00298	6.63	785.96	300	0.035
23	254	550	0.00284	6.66	794.53	300	0.035
24	254	550	0.00284	6.66	794.53	300	0.035
25	254	550	0.00277	7.10	801.32	300	0.035
26	254	550	0.00277	7.10	801.32	300	0.035
27	10	150	0.52124	3.33	1055.1	120	0.077
28	10	150	0.52124	3.33	1055.1	120	0.077
29	10	150	0.52124	3.33	1055.1	120	0.077
30	47	97	0.01140	5.35	148.89	120	0.077
31	60	190	0.00160	6.43	222.92	150	0.063
32	60	190	0.00160	6.43	222.92	150	0.063
33	60	190	0.00160	6.43	222.92	150	0.063
34	90	200	0.0001	8.95	107.87	200	0.042
35	90	200	0.0001	8.62	116.58	200	0.042
36	90	200	0.0001	8.62	116.58	200	0.042
37	25	110	0.0161	5.88	307.45	80	0.098
38	25	110	0.0161	5.88	307.45	80	0.098
39	25	110	0.0161	5.88	307.45	80	0.098
40	242	550	0.00313	7.97	647.83	300	0.035

References

[1] Y.-H. Song, "Introduction" in Modern Optimisation Techniques in Power Ststems, Y.-H. Song, Ed.: Kluwer Academic Publishers, 1999, pp. 1-13.

[2] E. Aarts and J. Korst, *Simulated Annealing and Boltzmann Machines*: John Wiley & Sons, 1989.

[3] P. Attaviriyanupap, H. Kita, E. Tanaka, and J. Hasegawa, "A hybrid EP and SQP for dynamic economic dispatch with nonsmooth fuel cost function," *IEEE Trans. Power Syst.*, vol. 17, pp. 411-416, May 2002.

[4] T. A. A. Victoire and A. E. Jeyakumar, "Hybrid PSO-SQP for economic dispatch with valve-point effect," *Electric Power Systems Research*, vol. 71, pp. 51-59, 2004.

[5] K. Y. Lee and M. A. El-Sharkawa, *A Tutorial Course on Evolutionary Computation Techniques for Power System Optimization*. Seoul, Korea: IFAC Symposium on Power Plants and Power, Sep. 2003.

[6] N. Sinha, R. Chakrabarti, and P. K. Chattopadhyay, "Evolutionary programming techniques for economic load dispatch," *IEEE Trans. Evol. Comput.*, vol. 7, pp. 83-94, Feb. 2003.

[7] J.-B. Park, K.-S. Lee, J.-R. Shin, and Kwang Y. Lee, "A particle swarm optimization for economic dispatch with nonsmooth cost functions," *IEEE Trans. Power Syst*, vol. 20, pp. 34-42, Feb. 2005.

[8] B. Zhao, C. X. Guo, and Y. J. Cao, "A multiagent-based particle swarm optimization approach for optimal reactive power dispatch," *IEEE Trans. Power Syst*, vol. 20, pp. 1070-1078, May 2005.

[9] Z.-L. Gaing, "Discrete particle swarm optimization algorithm for unit commitment," *IEEE Power Eng. Soc. General Meeting*, vol. 1, pp. 418-424, Jul. 2003.

[10] M. A. Abido, "Optimal power flow using particle swarm optimization," *International Journal of Electrical Power & Energy Systems*, vol. 24, pp. 563-571, 2002.

[11] J. Kennedy and R. Eberhart, "Particle swarm optimization," *in Proc. IEEE Int. Conf. Neural Networks*, vol. 4, pp. 1942 - 1948, Nov. 1995.

[12] X. Hu, R. C. Eberhart, and Y. Shi, "Engineering optimization with particle swarm," *in Proc. IEEE Swarm Intelligence Symposium(SIS'03)*, pp. 53-57, Apr. 2003.

[13] R. C. Eberhart and Y. Shi, "Particle swarm optimization: developments, applications and resources," *in Proc. Congr. Evol. Comput.*, vol. 1, pp. 81-86, May 2001.

[14] Y. Shi and R. C. Eberhart, "Empirical study of particle swarm optimization," *in Proc. Congr. Evol. Compt.*, vol. 3, pp. 1945-1950, Jul. 1999.

[15] Y. Fukuyama, "Particle Swarm Optimization Techniques with applications in Power System," in *Evolutionary Computation Techniques for Power System Optimization*, K. Y. Lee and M. A. El-Sharkawa, Eds. Seoul, Korea: Tutorial given at The IFAC Symposium on Power Plants and Power, Sep. 2003, pp. 45-62.

[16] M. Clerc and J. Kennedy, "The particle swarm-explosion, stability, and convergence in a multidimensional complex space," *IEEE Trans. on Evolutionary Computation*, vol. 6, no. 1, pp. 58-73, February 2002.

[17] M. Clerc, "The swarm and the queen: towards a deterministic and adaptive particle swarm optimization," *in Proc. Congr. Evol. Compt.*, vol. 3, pp. 1951-1957, Jul. 1999.

[18] R. C. Eberhart and Y. Shi, "Comparing inertia weights and constriction factors in particle swarm optimization," *in Proc. Congr. Evol. Compt.*, vol. 1, pp. 84 - 88, Jul. 2000.

[19] N. Higashi and H. Iba, "Particle swarm optimization with Gaussian mutation," *in Proc. IEEE Swarm Intelligence Symposium (SIS'03)*, pp. 72-79 Apr, 2003.

[20] A. J.Wood and B. F. Wollenberq, *Power Generation, Operation & Control*, 2 ed. New York: John Wiley, 1984.

[21] C.-P. Cheng, C.-W. Liu, and C.-C. Liu, "Unit commitment by Lagrangian relaxation and genetic algorithms," vol. 15, pp. 707-714, May 2000.

[22] K. A. Juste, H. Kita, E. Tunaka, and J. Hasegawa, "An evolutionary programming solution to the unit commitment problem," *IEEE Trans. Power Syst*, vol. 14, pp. 1452-1459 Nov. 1999.

[23] W. Ongsakul and N. Petcharaks, "Unit commitment by enhanced adaptive Lagrangian relaxation," *IEEE Trans. Power Syst*, vol. 19, pp. 620-628, Feb. 2004.

[24] S. O. Orero and M. R. Irving, "A combination of the genetic algorithm and lagrangian relaxation decomposition techniques for the generation unit commitment problem," *Electric Power Systems Research*, vol. 43, pp. 149-156, 1997.

[25] S. Sen and D. P. Kothari, "Optimal thermal generating unit commitment: a review," *International Journal of Electrical Power & Energy Systems*, vol. 20, pp. 443-451, 1998.

[26] P. Attaviriyanupap, H. Kita, E. Tanaka, and J. Hasegawa, "A hybrid LR-EP for solving new profit-based UC problem under competitive environment," *IEEE Trans. Power Syst*, vol. 18, pp. 229-237 Feb. 2003.

[27] S. A. Kazarlis, A. G. Bakirtzis, and V. Petridis, "A genetic algorithm solution to the unit commitment problem," *IEEE Trans. Power Syst*, vol. 11, pp. 83-92, Feb. 1996.

[28] T.-O. Ting, M. V. C. Rao, C. K. Loo, and S. S. Ngu, "Solving Unitcommitment Problem Using Hybrid Particle Swarm Optimization," *Journal of Heuristics*, vol. 9, pp. 507-520, 2003.

[29] J.H.Park, Y.S.Kim, I.K.Eom, and K.Y.Lee, "Economic load dispatch for piecewise quadratic cost function using Hopfield neural network," *IEEE Trans. Power Syst*, vol. 8, pp. 1030-1038, Aug. 1993.

[30] D. C. Walters and G. B. Sheble, "Genetic algorithm solution of economic dispatch with valve point loading," *IEEE Trans. Power Syst*, vol. 8, pp. 1325 - 1332, Aug. 1993.

[31] Y.-M. Park, J. R. Won, and J. B. Park, "A new approach to economic load dispatch based on improved evolutionary programming," *Eng. Intell. Syst. Elect. Eng. Commu.*, vol. 6, pp. 103-110, Jun. 1998.

[32] W.-M. Lin, F.-S. Cheng, and M.-T. Tsay, "An improved tabu search for economic dispatch with multiple minima," *IEEE Trans. Power Syst*, vol. 17, pp. 108 - 112, Feb. 2002

[33] H.-T. Yang, P.-C. Yang, and C.-L. Huang, "Evolutionary programming based economic dispatch for units with non-smooth fuel cost functions," *IEEE Trans. Power Syst*, vol. 11, pp. 112-118, Feb. 1996

[34] D. Liu and Y. Cai, "Taguchi method for solving the economic dispatch problem with nonsmooth cost functions," *IEEE Trans. Power Syst*, vol. 20, pp. 2006-2014, Nov. 2005.

Evolutionary Generator Maintenance Scheduling in Power Systems

Keshav P. Dahal[1] and Stuart J. Galloway[2]

[1]School of Informatics, University of Bradford, Bradford, UK.
[2]Institute of Energy and Environment, University of Strathclyde, Glasgow, UK.

Summary. This chapter considers the development of metaheuristic and evolutionary-based solution methodologies to solve the generator maintenance scheduling (GMS) problem of a centralized electrical power system. The effective maintenance scheduling of power system generators is very important to power utilities for the economical and reliable operation of the power system. To demonstrate the application and capabilities of the proposed algorithms a GMS test problem is formulated as an integer programming problem using a reliability based objective function and typical problem constraints. The implementation of a genetic algorithm (GA) and a simulated annealing (SA) heuristic and the effect of varying the GA and SA parameters on the performance of these approaches are presented. The application of a GA/SA hybrid approach is also investigated. This approach uses the probabilistic acceptance criterion of SA within the GA framework. The implementation and performance of the proposed solution techniques are discussed. The application of an inoculated initial population with some heuristically developed solutions are also demonstrated. Results contained in this chapter demonstrate that the GA/SA hybrid technique is more effective than approaches based solely on GA or solely on SA, offering an effective alternative for solving the GMS problems within a realistic timeframe.

1. Introduction

A centralized electrical power system can comprise of a large number of different power generators with different running costs. This system operates under fluctuations of consumer demand. In order to maintain power

K.P. Dahal and S.J. Galloway: *Evolutionary Generator Maintenance Scheduling in Power Systems*, Studies in Computational Intelligence (SCI) **49**, 349–382 (2007)

quality and reliability as well as to maximize profits a power system operator requires scheduling for the operation and maintenance of power generators. The overall generation cost to the system can be reduced by providing better scheduling of the generators for operation and maintenance. This problem is fundamental to existing system operation and there have been a number of management strategies proposed. Indeed, the increase in complexity of these strategies can be seen as a measure of the important role that the scheduling plays in reducing the overall cost and maximizing the reliability [3, 5, 11, 14]. Generator maintenance scheduling (GMS) is one of the main decision-making problems in power generation management.

In modern power systems the demand for electricity has greatly increased with the related expansion in power system size. This has resulted in higher numbers of generators being connected to the power system and lower generation reserve margins, making the GMS problems more complicated. The reserve margin is the amount of generation capacity that must be maintained on standby in case of shortages at times of high demand. The peculiarities of these problems are as a consequence of the following power system features: infeasibility of storing the generated electrical energy, the limited capacity of the transmission network, the need for an adequate amount of reserve capacity, and the impact of parallel operation of inter-connected power systems upon the necessary level of reserve for each member of a power pool [26, 35].

The GMS is a sub-problem of an integral long-term operational planning problem. It is vital for a power generating utility to determine when its generators should be taken off-line for preventive maintenance. This is primarily because other short-term and long-term planning activities such as unit commitment, generation dispatch, import/export of power and generation expansion planning are directly affected by such decisions [3, 5, 32, 33, 35].

GMS plays a very important role in the economical and reliable operation of a power system. The traditional approaches for dealing with the maintenance problem are that the preventive maintenance is periodically scheduled and repair work is carried out after a fault has occurred [33, 38]. The purpose of these activities is to keep a healthy operating condition of equipment, to reduce the frequency of equipment faults, to extend life time and to increase reliability and economic benefits. The annual costs for electric utilities to carry out this work are significant. Along with the large direct costs for maintenance there are a lot of hidden costs, such as the cost of undelivered energy during maintenance time, the increase in cost due to the replacement of efficient generating equipment with less efficient plant, the maintenance cost of equipment for added reserve capacity to maintain reliability etc. Therefore, even a small improvement on maintenance scheduling can lead to savings of significant value [9].

The high order of dimensionality, the non-linearity and the stochastic nature of a power system forces the development of new optimization methods for GMS problems. Researchers have focused much attention to new theoretical approaches to these scheduling problems from a power system planning, design and operational management point of view [11, 19]. Recently there has been an upsurge in the use of methods that mimic natural processes to solve complex problems [17, 27, 33]. This chapter investigates the formulation of the generator scheduling problem in a centralized power system and the development of metaheuristic-based solution methods to solve it.

The chapter is organized as follows. The following section gives a background to the formulation of a general GMS problem and reviews solution techniques. Section 3 provides the mathematical formulation for the test problem developed for this work. This formulation uses a reliability based objective with general unit and system constraints. Section 4 introduces the metaheuristic approaches and their implementation for the test GMS problem. The applications of GA, SA and GA-based hybrid approaches for the test problem and the effects of technique parameters on the performance are detailed in Section 5. This section also compares the performances of all proposed solution methods. The final section summarizes the chapter.

2. GMS problem formulations and solution methods

2.1 Problem formulations

The goal of GMS is to calculate a maintenance timetable for generators in order to maintain a high level of system reliability, reduce total operating cost, extend generator life time and relax any pressure for new installation. This is to be achieved while satisfying constraints on the individual generators and the power system. The maintenance schedule of each generating unit should be optimized in terms of the objective function under a series of constraints. The selection of objectives and constraints depends on the particular needs of a given GMS problem, the data available, the accuracy to be sought, and the chosen methodology for solving the problem. [19, 38].

Objective functions

In [11], [19], [29] and [38] the essential features of the GMS problem are analyzed, including various objectives and imposed constraints for solving the GMS problem. There are generally two categories of objectives in GMS, based on economic cost [2, 4, 9, 18, 37, 38] and reliability [1, 2, 3, 14, 37, 38].

The most common objective based on economic cost is to minimize the total operating cost over the operational planning period [4, 38]. The operating cost includes two components: the production cost and the maintenance cost. The production cost is mainly the cost of fuel needed for units to produce a certain amount of electrical energy [19]. The maintenance cost of units may include costs associated with the maintenance activities and also with the down time of units. The production cost alone could also be chosen as the objective function [18, 38]. Production cost minimization however, often requires many approximations or computationally intensive simulation to yield a solution. It was reported in the literature that minimizing production cost (which is the main part of the operating cost for thermal plants) is an insensitive objective for GMS [34, 37].

A number of reliability definitions such as expected lack of reserve, expected energy not supplied (EENS) and loss of load probability (LOLP), which are based on power system measures can be used as reliability criteria for the formulation of objective functions for GMS [3, 38].

The levelizing of the reserve is the most common reliability criterion [10, 14]. The reserve of the system during any period t is the sum of capacity of all installed generators, minus the peak load forecast for that period and the sum of capacity of the generators that are in pre-scheduled maintenance. The levelizing of the reserve generation over the entire operational planning period can be realized by maximizing the minimum reserve of the system during any time period [10]. This usually leads to a levelizing of reserves unless there is one interval with an extremely low reserve, either due to a large predicted load or a lot of pre-scheduled maintenance. In the case of a large variation of reserve, minimizing the sum of squares of the reserves can be an effective approach [14]. Minimizing the sum of the individual loss of load probabilities (LOLP) for each interval can also form an objective based on a reliability criterion under the condition of load uncertainty and the random forced outage of units [38].

In view of the relatively small difference in costs corresponding to the most expensive and least expensive feasible maintenance schedule, it is often better to use reliability indices rather than the production costs in formulating an objective function [14, 19, 37, 38].

Constraints of the GMS problem

Any maintenance timetable should satisfy a given set of constraints. The general constraints for GMS problems are related to generating units, the power system itself and available resources [11, 18, 19, 26]. Regarding the requirement of satisfying constraints, it is possible to make a distinction between constraints which must not be violated, known as hard constraints, and constraints which could be relaxed, known as soft constraints.

The following constraints are related to generators:

- Maintenance window constraint - defines the possible times and the duration of maintenance, and the limitation on the earliest and latest times for the maintenance of each generator.
- Exclusion constraint - prevents the simultaneous maintenance of a set of generators.
- Sequence constraint - restricts the initiation of maintenance on some generators after a period of maintenance of some other generators.

The following constraints result from power system operation:

- Load constraints - considers the demand and minimum reserve on the power system for each time interval.
- Reliability constraint - considers the risk level associated with a given maintenance schedule, which can be expressed via a number of reliability indices, such as minimum reserve margin and acceptable level of EENS or LOLP.
- Transmission capacity constraint - specify the limit of transmission capacity in an interconnected power system.
- Geographical constraint - limits the number of generators under maintenance in a region.

The following constraints are related to the resource availability for maintenance:

- Crew constraints - considers the manpower availability for maintenance work for the given period.
- Material resource constraint - specifies the limits on the resources needed for maintenance during each period, including spare parts, special tools etc.

These constraints can be divided into hard and soft constraints. The first and second groups of the constraints described above are usually the hard constraints, which represent physical limits and strict operational rules of the power system. The third group of constraints are not rigid.

2.2 Review of GMS solution techniques

The GMS problem has been tackled using a range of discrete optimization methods with varying degrees of success. The effective solution techniques perform systematic searches through the solution space so that the great number of unfeasible or inferior solutions are eliminated. Kralj and Petrovic [19] review the development of different mathematical programming, heuristic and expert system approaches for solving GMS problems. Conventionally these approaches have been used for solving GMS as a discrete optimization problem over a number of time steps

Mathematical programming methods for GMS are mainly based on integer programming (IP), branch and bound (BAB) and dynamic programming (DP). The IP formulation of the GMS problem has been presented in a number of works, from the relatively simple linear model of Dopazo and Merrill [9] to the complex model of Mukerji *et al.* [29]. Egan *et al.* [10] treat GMS as an IP problem, successively applying the BAB technique to solve a small sized problem. DP has been applied for solving GMS problems [37, 35]. In this case the method is a compromise between sequential and simultaneous maintenance scheduling. Together with its numerical efficiency, the important features of DP include the ability to incorporate maintenance carried out in several phases during the scheduling period. The dimension barrier (computational requirement) problem, as recognized by [35, 37], limits the practical application of this method. Many power utilities are using heuristic-based techniques and knowledge based expert systems to overcome the limitations of mathematical programming methods for their maintenance scheduling [19].

The goal programming approach for GMS with multiple objectives has been proposed in [20, 28]. The approach uses two sequential optimization stages, first minimizing system operating cost, and then levelizing reserve margin by introducing an upper bound on the system operating cost. This approach has been applied effectively to a medium sized thermal system [20] and to a hydro-thermal system in [28]. Decomposition and hierarchical approaches have been proposed in [24, 25] to tackle the complexity of the problem iteratively. These methods decompose the complex multi-area problem into one master and several sub-problems that are related to the original. The sub-problems being more straight forward to solve.

Fustar and Hsieh [16] have developed an expert system for monitoring the generator maintenance schedule and assisting with necessary revisions of the schedules. Knowledge of the GMS domain is represented as a number of heuristic rules to find a solution with minimum deviations from an existing yearly schedule. The method is problem specific and approximate. Lin *et al.* [23] presented an expert system for GMS. As the objective function varies for different operating conditions, an operation index is identified to determine an appropriate strategy for the decision making process. The rules are obtained from domain experts and embedded in the knowledge base. The approach is applied to a simplified model of a realistic power system and the authors find improvements in the results compared with the existing operational practice. Heuristic approaches are usually problem specific and require significant information from operator, which can be tedious [37].

In order to overcome some of these limitations a number of metaheuristic and soft computing based approaches for maintenance have been studied. These include genetic algorithm (GA) [1, 2, 35], simulated annealing (SA) [31], ant colony optimization [15], evolutionary programming [13], fuzzy logic [12], agent technology [36] and evolutionary-based hybrids [4, 12, 21, 22]. The applications of evolutionary-based techniques to power systems are reviewed in [27]. Our previously published work [6-8] presented GA and SA-based approaches for the GMS problem. This chapter details the development of a variety of improved GA-based approaches in order to more quickly obtain the optimum or near optimum solution for a GMS case study. Three solution encodings are investigated, and the effect of GA and SA parameters in the performance is analyzed. Five different population pools are created from heuristic schedule for the initialization of the GA-based approaches to achieve better performances.

3. GMS Test Problem and Implementation

3.1 Test problem description

Here a test problem of scheduling the maintenance of 21 units over a planning period of 52 weeks is considered. This test problem is derived from the example presented in [37] with some simplifications and additional constraints, and has been previously studied in [6-8, 15].

Table 1 gives the capacities, allowed periods (maintenance windows) and required duration of maintenance outages for each unit. Thirteen units are to be maintained in the first half of the year, i.e. week 1-26, and the remaining eight in the second half of the year. The last column of Table 1 indicates the manpower required for each week of maintenance outage. The power system peak load is 4739 MW, and there are 20 technical staff available for maintenance work in each week.

Table 1. Data for the Test GMS problem.

Unit	Capacity (MW)	Allowed period	Outage (weeks)	Manpower required for each week
1	555	1-26	7	10+10+5+5+5+5+3
2	555	27-52	5	10+10+10+5+5
3	180	1-26	2	15+15
4	180	1-26	1	20
5	640	27-52	5	10+10+10+10+10
6	640	1-26	3	15+15+15
7	640	1-26	3	15+15+15
8	555	27-52	6	10+10+10+5+5+5
9	276	1-26	10	3+2+2+2+2+2+2+2+2+3
10	140	1-26	4	10+10+5+5
11	90	1-26	1	20
12	76	27-52	3	10+15+15
13	76	1-26	2	15+15
14	94	1-26	4	10+10+10+10
15	39	1-26	2	15+15
16	188	1-26	2	15+15
17	58	27-52	1	20
18	48	27-52	2	15+15
19	137	27-52	1	15
20	469	27-52	4	10+10+10+10
21	52	1-26	3	10+10+10

The problem formulation uses the reliability criteria of minimising the sum of squares of the reserves in each week as an objective. Each unit must be maintained (without interruption) for a given duration within an allowed period, and the available manpower is limited. The allowed period of each generator is the time intervals within which its maintenance activity is to be scheduled. This information is the result of technical assessment and the experience of the maintenance personnel in ensuring adequate maintenance frequency and thereby avoiding outages during the peak load season. In the test problem the system peak load is taken to be constant throughout the year - although this is not the norm for genuine GMS problems, it will not limit the validity of the presented approaches. Due to its complexity the exact optimum solution for this problem is unknown.

3.2 Mathematical model

Mathematically, the GMS test problem can be formulated as an integer programming problem by using integer variables associated with answers to "When does maintenance start?" or alternatively by using conventional binary variables associated with answers to "When does maintenance occur?" [9]. However, a formulation using the variables associated with the first question takes care of the constraints on the periods and duration of maintenance and hence, the number of unknowns is reduced. The answer to the first question also automatically provides the answer to the second. The variables are bounded by the maintenance window constraints. However, for clarity the problem is first formulated using binary variables that indicate the start of maintenance of each unit at each time. The following notation is introduced:

i index of generating units
I set of generating unit indices
N total number of generating units
t index of periods
T set of indices of periods in planning horizon
e_i earliest period for maintenance of unit i to begin
l_i latest period for maintenance of unit i to end
d_i duration of maintenance for unit i
P_{it} generating capacity of unit i in period t
L_t anticipated load demand for period t
M_{it} manpower needed by unit i at period t
AM_t available manpower at period t

Suppose $T_i \subset T$ is the set of periods when maintenance of unit i may start, so $T_i=\{t \in T: e_i \leq t \leq l_i-d_i+1\}$ for each i. The maintenance start indicator is defined as

$$X_{it} = \begin{cases} 1 \text{ if unit } i \text{ starts maintenanc e in period } t \\ 0 \text{ otherwise} \end{cases},$$

for unit $i \in I$ in period $t \in T_i$. It is convenient to introduce two further sets. Firstly let S_{it} be the set of start time periods such that if the maintenance of unit i starts at period k that unit will be in maintenance at period t, so $S_{it}=\{k \in T_i: t-d_i+1 \leq k \leq t\}$. Secondly, let I_t be the set of units which are allowed to be in maintenance in period t, so $I_t=\{i: t \in T_i\}$. Then the problem can be formulated as a quadratic 0-1 programming problem as below. A reliability criterion will be used as the objective function in the formulation of the test GMS problem. The leveling of the reserve generation over the planning period can be used as a reliability criterion. The net reserve of the system during any period t is the total installed capacity minus the peak load forecast for that period (L_t) and the reserve loss due to the pre-scheduled outages. The reserve can be levelized by maximizing the minimum net reserve of the system during any time period [38]. In the case of a large variation of reserve, minimizing the sum of squares of the reserves can be an effective approach. In this application this is used as an objective function. Hence the objective is to minimize the sum of squares of the reserve generation,

$$\underset{X_{it}}{Min} \left\{ \sum_{t \in T} \left(\sum_{i \in I} P_{it} - \sum_{i \in I_t} \left(\sum_{k \in S_{it}} X_{ik} P_{ik} \right) - L_t \right)^2 \right\} \tag{1}$$

subject to the maintenance window constraint,

$$\sum_{t \in T_i} X_{it} = 1 \quad \text{for all } i \in I, \tag{2}$$

the manpower constraint,

$$\sum_{i \in I_t} \sum_{k \in S_{it}} X_{ik} M_{ik} \leq AM_t \quad \text{for all } t \in T, \tag{3}$$

and the load constraint,

$$\sum_{i \in I} P_{it} - \sum_{i \in I_t} \sum_{k \in S_{it}} X_{ik} P_{ik} \geq L_t \quad \text{for all } t \in T. \tag{4}$$

Equations (1)-(4) define a general mathematical model for a general GMS problem formulated as a quadratic 0-1 programming problem. Further constraints may be posed involving the reliability, transmission capacity and maintenance in local areas of the power system.

The classical way to tackle such problems is by using Integer programming or Dynamic programming and heuristic methods. However, although effective there are often problems concerning the applicability of these techniques to problem of this size as discussed in section 2.2. Instead, the work reported here considers the use of metaheuristic techniques for the solution of the general mathematical model of the GMS problem.

The sum of squares of the reserves, i.e. the objective value given by (1), measures the reliability of the power system. The lower the values of the objective, the more uniformly the reserve margin is distributed, and the higher the reliability. The average reserve level (471.4 MW) gives the lower bound of the objective value (115.56×10^5) for the test GMS problem, which is a theoretical value that gives an uniform reserve margin over the scheduling period (not considering the discrete value of units capacity, the maintenance window or crew constraints).

3.3 Heuristic solutions

As described above the optimum schedule for the test GMS problem is unknown due to its complexity. In order to compare the schedules obtained from the metaheuristic based approaches, two heuristic schedules have been developed. These heuristic schedules will also be used to initialise the metaheuristic-based approaches.

- *Heuristic schedule 1*: This schedule has been developed by ranking the generating units in order of decreasing capacity to maintain a level reserve generation margin while considering maintenance window constraints and load constraints. The resulting solution respects all constraints imposed in the problem except the manpower constraint, which is violated in three periods. The objective value of the solution, which is the sum of the squares of the reserves (*SSR*) throughout the planning period given by (1) is 134.61×10^5. The amount of manpower constraints violation (*TMV*) given by (3) is 22.
- *Heuristic schedule 2*: This schedule has been developed by ranking the generating units in order of decreasing weekly manpower requirements for maintenance work to maintain a level manpower utilization throughout the scheduling period while considering maintenance window constraints. The resulting solution is infeasible as the load constraint is violated in four time periods and the manpower constraint is violated in one time period. The objective value of the solution (*SSR*) is 149.7×10^5. The amount of manpower constraint violation (*TMV*) given by (3) is 3 and the amount of load violation (*TLV*) given by (4) is 161.

4. Metaheuristic techniques

4.1 Introduction

Computer-based metaheuristic techniques such as genetic algorithms (GAs) and simulated annealing (SA) are completely distinct from classical mathematical programming and trial-and-error heuristic methods [17]. These two search techniques originate from different metaphors of natural behaviour.

GAs replicate the principles of the theory of evolution, i.e. selection and inheritance [17, 27, 33]. They are based on natural genetic and evolutionary mechanisms that operate on populations of solutions. A population of candidate solutions is maintained by the GA throughout the solution process. Initially this population of solutions is generated randomly or by other means. During each iteration step, a selection operator is used to choose two solutions from the current population. The selection is based upon the measured goodness of the solutions in the population - this is being quantified by an evaluation function. The selected solutions are then subjected to crossover. The crossover operator exchanges sections between these two selected solutions with a defined crossover probability (CP). One of the resulting solutions is then chosen for application of a mutation operator, whereby the value at each position in the solution is changed with a defined mutation probability (MP). The algorithm is terminated, when a defined stopping criterion is reached. The reader is referred to [17] for a more theoretical and mathematical description of the GA technique.

SA is drawn from an analogy between thermodynamics and combinatorial problems. In recent years these techniques have attracted the attention of many researchers attempting to solve complex scheduling problems. They have been applied widely to solve different types of scheduling and optimisation problems [17, 31, 33]. The SA method is based on the analogy between the physical annealing process of a solid and the problem of finding the minimum or maximum of a given function depending on many parameters, as encountered in combinatorial optimisation problems [17, 33]. SA is an iterative search process that maintains a single point in the search space and repeatedly makes trial moves from the current point. First an initial solution and an initial temperature (T_0) are selected. As the algorithm progresses a new trial solution is generated by introducing some changes to the current solution and the temperature is reduced according to a specified cooling schedule. The goodness of the new solution is measured using an evaluation function. If the new solution is an improvement, it is accepted unconditionally, otherwise it is accepted with a probability given by

$$P(\Delta E)=exp(-\Delta E)/T_s), \quad (5)$$

where ΔE is the improvement in the solution and T_s is the temperature at stage s. The process takes place in a series of stages. Progression through successive stages leads to a gradual reduction in the probability of accepting non-improved trial solutions. The algorithm is terminated, when a defined stopping criterion is reached. Again, the reader is referred to [17] for a more theoretical and mathematical description of the SA technique.

A number of decisions must be made in order to implement the SA and GA methods for solving an optimisation problem. Firstly, there are problem specific decisions which are concerned with the search space (and thus the representation) of feasible solutions, the form of the evaluation function, and the operators used to generate new trial solutions. The second class of decisions are generic and involve the parameters of the techniques themselves. The following two sub-sections describe the representation and evaluation of candidate solutions for the test GMS problem. These two issues are common for the implementation of both of these techniques.

4.2 Problem encoding

The encoding of the problem using an appropriate representation is a crucial aspect of the implementation of a metaheuristic approach for solving an optimization problem. Different types of candidate solutions may be used to encode the set of parameters for the evaluation function. GMS problems can be solved using three types of representations:

- binary representation
- binary for integer representation
- integer representation

In the binary representation the GMS problem (1) - (4) is encoded by using a one-dimensional binary array as follows.

$$[X_{1,e1}, X_{1,(e1+1)}, \dots, X_{1,(l1-d1+1)}, X_{2,e2}, X_{2,(e2+1)}, \dots$$
$$\dots, X_{2,(l2-d2+1)}, \dots, X_{N,eN}, X_{N,(eN+1)}, \dots X_{N,(lN-dN+1)}]$$

This binary string (chromosome) consists of sub-strings which each contain the variables over the whole scheduling period for a particular unit. The size of the GA search space for this type of representation is $2^{\sum_{i=1}^{N}(l_i-d_i-e_i+2)}$.

For each unit $i=1,2,\dots,N$, the maintenance window constraint (2) forces exactly one variable in $\{X_{it}: t \in T_i\}$ to be one and the rest to be zero. The solution of this problem thus amounts to finding the correct choice of positive variable from each variable set $\{X_{it}: t \in T_i\}$, for $i=1,2,\dots,N$. The index t of this positive variable indicates the period when maintenance for unit i starts. In order to reduce the number of variables the indices of the positive variables from $\{X_{it}: t \in T_i\}$, for $i=1,2,\dots,N$, can be taken as new variables. The advantage of this approach is the possibility of using an integer encoding for these new variables in a genetic structure consisting of a string of integers, each one of which represents the maintenance start period of a unit. For this representation the string length is equal to the number of units (N) and the string is $t_1,t_2,...,t_i,...,t_N$, where t_i is an integer, $e_i \le t_i \le l_i - d_i + 1$, for each $i=1,2,...,N$, which indicates the maintenance start period for unit i. This type of representation automatically considers the maintenance window constraint (2) and greatly reduces the size of the GA search space to $\prod_{i=1}^{N} (l_i - d_i - e_i + 2)$.

The integer formulation of the problem can also be encoded by using binary (or Gray) code to represent the integer variables in the GA structures. For example, with t_i defined as above, suppose the number of possible values of t_i is $l_i - d_i - e_i + 2 = 32$, then a 5 bit binary pattern may be used to represent the possible variable values. This representation is called 'binary for integer'. In this case if the number of variable values is not a power of 2, some of the binary values will be redundant. To overcome this problem, some integer values are represented by two or more bit patterns. The string length in this situation is $b_1 + b_2 + ... + b_N$ and the GA search space is $2^{\sum_{i=1}^{N} b_i}$, where b_i is the number of the binary bits used to represent the integer variable values for unit i and equals the least positive integer greater than or equal to $log_2(l_i - d_i - e_i + 2)$. This redundancy increases the size of the search space since $2^{\sum_{i=1}^{N} b_i} \ge 2^{\sum_{i=1}^{N} \log_2 (l_i - d_i - e_i + 2)} = \prod_{i=1}^{N} (l_i - d_i - e_i + 2)$.

The GMS test problem has been encoded using each of the three representations described above and the performance of the GA investigated in section 5.1.

4.3 Evaluation function

An evaluation function is used to calculate the merit of a candidate. The evaluation function, E, formulated for the GMS test problem considered in this work is a weighted sum of the objective and penalty functions for violations of the constraints. The penalty value for each constraint violation is proportional to the amount by which the constraint is violated, hence

$$E = \omega_O \, SSR + \omega_W WCV + \omega_M \, TMV + \omega_L \, TLV, \tag{6}$$

where SSR is the sum of squares of reserves as in equation (1), WCV is the window constraints violation of equation (2), TMV is the total manpower violation of equation (3) and TLV is the total load violation of equation (4).

The weighting coefficients ω_O, ω_W, ω_M and ω_L are chosen such that the penalty values for the constraint violations dominate over the objective function, and the violation of the relatively hard window and load constraints (equations (2) and (4)) give a greater penalty value than for the relatively soft crew constraint. This balance is set because a maintenance window feasible solution with a high reliability but requiring more manpower may well be considered acceptable for a power utility, as the unavailable manpower may be hired. In general the penalty value for the constraint violations dominates over the objective function. Feasible solutions with low evaluation measures have high fitness values while unfeasible solutions with high evaluation measures take low fitness values. After a series of experimentations the weighting coefficients ω_O, ω_W, ω_M and ω_L are fixed to 10^{-5}, 1000, 4 and 2 respectively for the test problem considered.

One thing to note here is that the 'binary for integer' and 'integer' representations mentioned previously automatically satisfy the maintenance window constraints (2) and therefore, the *WCV* component in evaluation function (6) is always zero. Therefore for the 'binary for integer' and 'integer' the evaluation function is given by,

$$E = \omega_O\, SSR + \omega_M\, TMV + \omega_L\, TLV. \tag{7}$$

It is equation (7), the reduced form of equation (6), that will be used as part of the metaheuristic solution techniques considered in this work while using the 'binary for integer' and 'integer' representations.

5. Implementation and experimental results

This section presents the implementation of the GA, SA and their hybrid approaches to the GMS test problem. The performance of the different algorithms will be analysed in this section. The design of these approaches to give the best performance in terms of finding good solutions to the test GMS problem has been established after extensive experimentation. The adopted experimentation approach involved conducting 10 runs with particular selections of parameters and identifying the best solution (lowest evaluation value). The average of the best solutions from each of the 10 experiments is also determined. These methods have been implemented using the Reproductive Plan Language, RPL2 [30] and run on a Sun Sparcstation 1000.

5.1 GA application

Scope and GA experimentation

The test GMS problem has been encoded using each of the three representations described previously. GA performance is generally dependent on the GA operators and parameters used. A detailed analysis of the performance of the algorithm with the different approaches, operators and parameters for this test problem is presented here. Comparisons were made with the performance of GAs using steady state (SS) and generational (GN) population updating approaches [17], varying key GA parameters and employing different GA operators. The GA parameters which were varied include the crossover and mutation probabilities.

The total number of trials (iterations) for each run was fixed to 30,000 and the population size (PS) was fixed at 100. These parameters were defined by analysis of the convergence of the GA technique after a number of experiments. The elitism operator has been applied in all cases. As suggested in [30], the stochastic universal sampling (SUS) selection method has been used for the GN GA and the tournament selection method has been applied for SS GA to choose parents from the population pool for genetic manipulation. The two-point crossover and random mutation operators were used for all experiments. The crossover operator chooses two points at which to break each of the two selected parent strings, and then swaps the middle parts of solution strings. The mutation operator randomly alters a numeric value within the allowed range at each position in the solution string with a given mutation probability. The crossover probability (CP) and mutation probability (MP) were varied over the range of [0.6, 1.0] and [0.001, 0.1] respectively.

Results for the test GMS problem for a total of 900 runs of the genetic algorithm using these different scenarios are summarized in Table 2. The table shows the average evaluation measure of the best solutions of 10 GA runs, each using a different random initial population.

Effect of GA representations and operators

Table 2 presents the results obtained from the generational GA (left block labeled as ‘GN GA’) and the steady state GA (right block labeled as ‘SS GA’). The results show that both the GN and SS GAs are sensitive to variation of the crossover and mutation probabilities, particularly for the ‘binary’ representation. It can be seen that both of the GAs are generally more sensitive to variations of MP than that of CP. For most of the cases with the 'binary' representation both GAs gave the best performance at smaller mutation probabilities. With the 'binary for integer' representation the GN GA generally performed better with MP in the range of [0.001, 0.01] whereas the SS GA gave better performance with MP in the range of [0.005, 0.05]. With the 'integer' representation the SS GA gave the best performance at higher crossover and mutation probabilities than the GN GA.

For the 'binary' and 'binary for integer' representations, the crossover points may be within a gene (a sub-string of a genetic structure which represents one particular unit). Hence the crossover operator may split genes and introduce changes within them. Theoretically, the splitting of genes by the crossover operator seems undesirable. In the case of integer representation, this sub-string splitting does not occur and the individual variable values are preserved in crossover. In this case only the mutation operator is responsible for creating a new integer value for a gene.

Table 2. Summary of GA performance results.

			GN GA					SS GA				
			MP					MP				
Type			.001	.005	0.01	0.05	0.1	.001	.005	0.01	0.05	0.1
Binary	CP	0.6	588	616	1.2e5	7.6e6	6.1e6	840	809	2.1e4	4.6e6	6.5e6
		0.8	368	488	6420	7.0e6	6.6e6	831	818	826	5.1e6	7.6e6
		1.0	663	583	8.9e4	3.1e6	7.2e6	3937	2138	1.3e5	6.3e6	9.4e6
Binary for integer		0.6	229	355	278	495	853	400	397	406	354	389
		0.8	209	474	194	392	502	384	382	371	363	362
		1.0	303	197	181	403	746	211	172	185	174	264
Integer		0.6	174	161	155	176	229	171	156	153	150	161
		0.8	197	158	156	178	253	175	152	152	151	170
		1.0	184	159	164	180	260	172	160	157	147	162

Using a binary representation for the test problem, only 21 out of 496 bits in the string are '1' and the rests are '0'. A high mutation probability increases the chance of changing these '0's into '1's, dragging the solution into the maintenance window unfeasible region. Hence the search space is very large and most of it represents unfeasible solutions. Therefore, a high mutation probability has the potential to disrupt and degrade the search process using the binary representation of the GMS problem. With higher mutation probabilities the GA could not find a maintenance window feasible solution even after 30,000 trials. However, with lower mutation probabilities the GA found maintenance feasible solutions but converged prematurely.

The 'binary for integer' and 'integer' encodings of the GMS problem result in every candidate solution being maintenance window feasible, which causes a great reduction in the size of the search space. The GA search is thus limited within the maintenance feasible region. In this case a higher mutation probability increases the exploration for the global minima within this limited region, reducing the chance of premature convergence. However, a very high mutation probability causes more randomness in the GA search, reducing the exploitation of the solutions previously found.

One point to be noted for the 'binary for integer' representation is that the actual mutation probability for changing integer values is much greater than the prescribed mutation probability. As explained previously, in the 'binary for integer' representation the variable states (integers) are denoted by a binary (or Gray) code with a number of binary bits in a string, for example 5 bit strings are used for each unit for our test problem. The mutation operator takes each bit and decides whether or not to change that bit with the given mutation probability. In particular, the given mutation probability is the probability of mutating each binary bit. However, a change in at least one of these 5 bits by the mutation operator results in a change in the corresponding integer value. The actual mutation probability m_a of changing the integer value for a given binary mutation probability m can be calculated as $m_a=1-(1-m)^{b_i}$, where b_i is the number of bits used to represent the integer variable for unit i. For example, if the given mutation rate $m=0.05$ and a 5 bit representation is used, the actual mutation probability is $m_a=0.23$. In order to have the actual mutation probability $m_a=0.05$, the mutation probability m should be taken as about 0.01. However, the distribution of the new integer values following mutation is not uniform in this case.

In the integer encoding of a GMS problem each gene is an integer, which represents the number of the time period in which maintenance work begins for a unit. The mutation operator takes each integer and with the given mutation probability changes the value within the allowed integer interval. The distribution of the new integer value within the interval is approximately uniform during mutation.

During a run of a GA the optimum value of the mutation probability may vary. An adaptive mutation operator has also been considered for the GMS problem, which dynamically varies the mutation probability depending on the Hamming distance between parents selected for crossover. The actual mutation probability is always less than or equal to the initially prescribed value. GA runs were carried out for the integer representation with three prescribed mutation values of 0.005, 0.01 and 0.05. The results show little difference between the performance with the traditional mutation operator and adaptive mutation operator for this representation.

Performance comparison

Table 3. Comparison of the best performing GAs for different problem encodings.

	Binary	Binary for integer	Integer
GA approach to give best results	GN	SS	SS
CP to give best results	0.8	1.0	1.0
MP to give best results	0.001	0.005	0.05
average evaluation value (over 10 runs)	368	172	146.71
best evaluation value (over 10 runs)	209	145	137.91
CPU time (one run)	45s	32s	34s
Size of GA search space	2^{496}=2.05x10^{149}	2^{105} = 4.06x10^{31}	6.23 x 10^{28}

Table 3 presents a performance comparison for the different representations. The values of the GA parameters which resulted in the best average evaluation value for each representation from Table 2, were taken for this comparison. It is apparent from the Table 2 and Table 3 that the integer representation is the best choice for GMS problems. The GA with the integer representation found a solution with an evaluation value of 137.91 for the test GMS problem, which is better than the solution found by the GA with the binary for integer representation (145), and significantly better still than that of the GA with binary representation (209). Moreover, the computational time taken for 30,000 trials by the GA with the integer representation is significantly shorter than that for the binary representation and slightly greater than that for the 'binary for integer' representation. The computational time is very much dependent on the string length as the mutation operator checks each locus of the string. The higher the MP used the greater the computational time. The size of GA search space for each representation is also presented in Table 3. The great advantage of the 'integer' representation for a GMS problem is that it greatly reduces the search space for the GA. As the SS GA with the integer representation is found to be the most effective for solving GMS problems in terms quality of the solution, further analysis will be concentrated on this GA design alone.

5.2 GA with Inoculation

A simple GA generates candidate solutions to fill its initial population pool by sampling the search space at random. This is by no means the only approach available. The GA search process may be started from an initial population that contains one or more meaningful individuals, instead of randomly generated individuals. A simple way of producing a hybrid method is to generate the initial population by using some form of pre-processing heuristic before a genetic algorithm is invoked. In this case attempts are made to improve solutions encountered by the GA. If a stochastic method for generating reasonable solutions is known, this can be an effective way of accelerating, and sometimes improving the absolute performance of the genetic algorithm [17]. The practice of including known good solutions in the initial population is often known as inoculation [30]. Here a GA approach with this type of augmentation is referred to as an inoculated GA.

The performance of a simple steady state GA is investigated by seeding one or more heuristically developed solutions in the initial population pool for the test GMS problem. The following five different ways of creating an initial population have been considered.

Case 1: All individuals in the initial population pool are generated randomly (same as simple steady state GA).

Case 2: The initial population pool is seeded with *Heuristic schedule 1.* The rest of the individuals in the pool are generated randomly.

Case 3: The initial population pool is seeded with *Heuristic schedule 2.* The rest of the individuals in the pool are generated randomly.

Case 4: The initial population pool is seeded with both *Heuristic schedule 1* and *Heuristic schedule 2.* The rest of the individuals in the pool are generated randomly.

Case 5: The initial population contains *Heuristic schedule 1, Heuristic schedule 2,* 33 individuals derived (through mutation) from *Heuristic schedule 1* and 33 individuals derived from *Heuristic schedule 2.* The rest of the population are filled with randomly generated solutions.

The design of the simple steady state GA to give the best performance in terms of finding good solutions to the test GMS problem has been established in Table 3. This identified SS GA with the integer representation, a crossover probability (CP) =1.0, mutation probability (MP) =0.05 and population size (PS) =100 as the best performing, this form of the SSGA is used for the inoculated GA studies.

For each case 10 GA runs were carried out with different initial populations but generated in the same way. The average evaluation value of initial solutions, the average evaluation values of the best solutions and the evaluation value of the best solution over 10 GA runs for each case are shown in Table 4.

Table 4. Average results of 10 GA runs seeded with different initial populations.

Initial pool	Average evaluation value of initial individuals	Average evaluation value of best individuals	Evaluation value of the best individual
Case 1	5782.63	146.71	137.91
Case 2	5732.15	155.44	139.53
Case 3	5734.76	142.67	139.96
Case 4	5668.15	148.38	139.77
Case 5	3326.20	146.56	140.11

It can be observed from Table 4 that the cases where *Heuristic schedule 1* was seeded (*case 2, case 4* and *case 5*), the average evaluation values of the best solutions over 10 GA runs were worse than that obtained with *Heuristic schedule 2* seeded (*case 3*). The GA with the random initial population pool (*case 1*) even performed better than those used in *case 2* and *case 4*. Further investigation showed that out of 10 GA runs, five runs in *case 2*, two runs in *case 4* and one run in *case 5* were trapped into a solution whose evaluation value was 163.62. This point in the search space may be a local minimum in the neighbourhood of the seeded *Heuristic schedule 1*, to which the GA converged. However, seeding with one individual representing *Heuristic schedule 2* (*case 3*), whose evaluation value is much greater than that of *Heuristic schedule 1,* significantly improved the average performance of the GA. It would appear that *Heuristic schedule 2* has some building blocks of good solutions which are spread to other individuals during the GA search process through the action of the crossover operator. The results obtained show that the inclusion of a superfit solution in initial population pool may not always enhance the performance of these methods. Some care must be taken to ensure sufficient diversity is present in the population during the search process for the methods to proceed effectively.

5.3 SA application

Like in the GA implementation, a number of problem specific and technique specific decisions must be made in order to implement the SA method. The problem specific decisions include the encoding of feasible solutions, the form of the evaluation function, and the move operator used to generate new trial solutions in neighborhood structures. The technique specific decisions involve the parameters of the SA technique, such as the initial temperature, the cooling schemes and stopping conditions.

The integer representation, which was found to be the most effective for GMS problems, is used in this work. The same evaluation function represented by equation (7) is used to assign a figure of merit, which represents the quality of that solution.

The move operator specifies the algorithm for generating a new trial solution from the current solution. The move operator employed in this work randomly selects one variable (i.e. one unit's maintenance start time) from the integer strings to be changed. The selected variable is then changed to a random value in the allowed range.

The criterion for stopping the algorithm can be expressed either in terms of a minimum value of the temperature, or in terms of the 'freezing' of the system at the current solution. 'Freezing' may be identified by the number of iterations (or 'temperatures') that have passed without a move being accepted or by the number of accepted moves in a stage dropping below a given value. However, the simplest rule of all is to fix the total number of iterations and this is done in the reported work. The number of iterations needs to be carefully tuned with other parameters to ensure that it corresponds to a sufficiently low temperature to ensure convergence.

Initialisation

The initialisation of the SA method involves the selection of an initial temperature (T_0) and an initial solution in the search. The initial solution may be generated at random or by using a simple heuristic method. Here the SA method has been tested with both random and heuristically developed initial solutions.

If the final solution is to be independent of the starting solution, the initial temperature T_0 must be high enough to allow an almost free exchange of neighbouring solutions. Generally the value of T_0 is chosen in such way that it is greater than the maximum possible difference between the evaluation values of two solutions [17].

Here a number of SA runs were made to observe the effect of initialisation. Fig. 1 illustrates the final solutions after 30,000 iterations obtained from 10 SA runs, each with a different random initial solution, for five different initial temperatures (T_0) values in the range 10^{-2} to 10^6.

When T_0 is very low, the SA algorithm performs as a downhill method, accepting few uphill moves, and hence tends to find a local optimal solution in the neighbourhood of the initial solution. In this case the evaluation value of the final solution is dependent on the starting point in the search space - this is demonstrated by the wide spread of data points in Fig. 1.

When T_0 is very high, almost all of the generated solutions are accepted and therefore, the solutions move more randomly through the domain in the early stages of the cooling process. Hence with a high T_0 value the final solution is less dependent on the initial solution.

Two further tests were conducted using a consistent T_0 of 10,000 with *Heuristic schedule1* and *Heuristic schedule 2* as initial trial solutions respectively. The final solutions averaged over 10 SA runs for these cases do not differ much from the average solution obtained with a random initial solution for T_0=10,000.

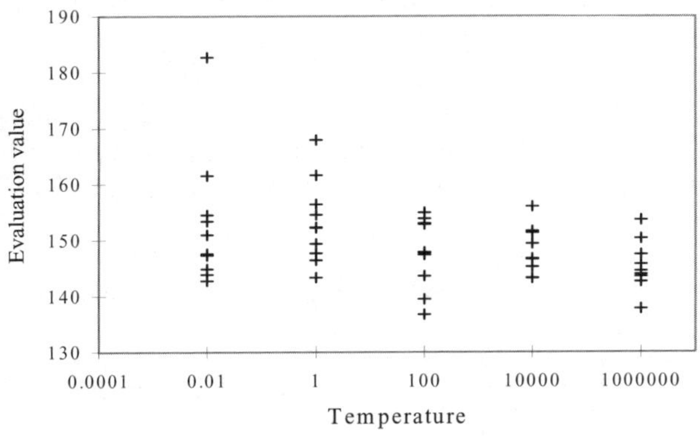

Fig. 1. Effect of initialisation on the performance of the SA method.

Cooling schedule

In SA the temperature is reduced as the algorithm progresses according to a cooling schedule. The cooling schedule may be adapted by using a large number of iterations at a small number of temperatures or a small number of iterations (or just one iteration) at a large number of temperatures. The number of iterations at each temperature and the rate at which the temperature is reduced are important factors in controlling the performance of the SA method. The following two cooling schedules, which occur most widely in practice, have been employed for the test GMS problem.

(a) Cooling Schedule A - executes just one iteration at each temperature, but reduces the temperature geometrically according to the relation, $T_i=\alpha T_{i-1}$, where T_i is the temperature at iteration i and α is the cooling parameter, $0<\alpha<1$.

(b) Cooling Schedule B - executes a number of iterations at each temperature before reducing the temperature according to the relation,

$T_s = \alpha T_{s-1}$ and $i_s = i_{s-1} + 1$, where T_s is the temperature at stage s, and i_s is the number of iterations at stage s.

In SA the number of iterations increases with successive temperatures since it is important to spend more time searching at lower temperatures to ensure that the neighbourhood of a local optimum has been fully explored.

A number of experiments were conducted to investigate the best value of the cooling parameter α for cooling schedules (a) and (b). Cooling schedule (a) has been tested for three values of α: 0.9990, 0.9995 and 0.9999. Likewise, cooling schedule (b) has been tested with α: 0.835, 0.910 and 0.968. The temperature reduction profile of cooling schedule (a) corresponds to that of cooling schedule (b) with their respective values of α.

Table 5 compares the average evaluation value of the final solutions obtained from 10 SA runs, the number of iterations until convergence (recognised by no subsequent improvement in the solution) and the associated computational time for these cases.

Table 5. Average results of 10 SA runs for cooling schedules (a) and (b) with different values of α.

	Cooling schedule (a)			Cooling schedule (b)		
α	0.9990	0.9995	0.9999	0.835	0.910	0.968
Average evaluation of final solutions	147.16	146.06	143.88	148.99	146.02	142.98
No. of iterations for convergence	29,000	30,000	141,000	18,000	28,000	105,000
Computational time for convergence (s)	25	26	122	23	26	102

The results in Table 5 show that if the value of α is sufficiently close to unity, it is reasonable to expect that a good solution can be found. However, the convergence of the method with a very high value of α is very slow, since it takes longer for the temperature to decrease sufficiently as the cooling parameter increases towards unity. Therefore there is a trade-off between the quality of the solution, and the time taken for the method to converge, this depends on the value of α. The available computational time (defining the total number of SA iterations) must be considered during the selection of a value of α in order to make the SA method efficient in finding a good solution. For example, with 30,000 iterations cooling schedule (a) with α=0.9995 (cooling schedule (b) with α=0.91) gives the best performance of the method for the test GMS problem. Of course, if a large computational time can be allowed, then α should be set to a larger value in order to obtain a better solution as seen in Table 5. Experimental results suggest that the type of cooling schedule does not have a great effect on the performance of the method if the overall temperature profile is approximately the same over the total number of trials.

The experimental results show that the SA with the initial temperature T_0 = 10,000, and the stage-wise cooling schedule (b), with number of stages $s = 300$ and α=0.91, gives the best performance for 30,000 fitness evaluations per run. With this SA design 10 SA runs were made. The average evaluation value of the best solutions obtained over 10 SA runs was 146.06 and the best solution had evaluation value of 140.49.

5.4 GA/SA hybrid technique

GA/SA Architecture

It has been demonstrated that the performance of a GA approach can be improved by combining it with other techniques. A GMS problem is considered using a GA combined with SA in [21]. In [22] a Tabu Search technique was coupled with a GA/SA hybrid method. Local search methods were combined with the evolutionary approach to tackle the GMS problem in [4]. In each case hybridization improved the convergence of the algorithms. Most of these applications formulated the problems using the economic objective with typical problem constraints, and a binary string representation to encode a candidate solution.

The work reported here proposes a GA/SA hybrid method using an integer representation to encode candidate solutions combining the GA and SA approaches. The mechanism of this hybrid GA/SA approach for a minimization problem is shown in Fig. 2. The hybrid approach maintains a population of candidate solutions throughout the solution process using a steady state population updating approach, which directly inserts a new solution into the population pool replacing a less fit solution. First an initial population of candidate solutions is generated randomly (or by other means) and an initial temperature is selected. The initial temperature should be large enough to allow the free movement of a trial solution in the search space during the early stages of the search process.

In each iteration the GA/SA hybrid approach selects two solutions from the population pool and applies a crossover operator. One of the "crossovered" solutions is randomly selected to undergo mutation. The resulting solution replaces an existing member of the population pool. This solution is inserted in a controlled manner, by taking account of its evaluation value and the stage reached in the search process. To implement this, the probabilistic acceptance approach of the simple SA, as expressed by equation (1), is incorporated into the GA algorithm to decide whether the new solution should be included in the population. The evaluation value of a newly created solution (E_{new}) is compared with the evaluation value of the best amongst its parents ($E_{current}$) to calculate the increase in evaluation value ($\Delta E = E_{new} - E_{current}$). The new solution is then accepted with probability given by $P(\Delta E)=exp(-\Delta E/T)$, where T is the temperature which defines the stage in the process. An acceptance of a new solution replaces the worst solution of the population pool.

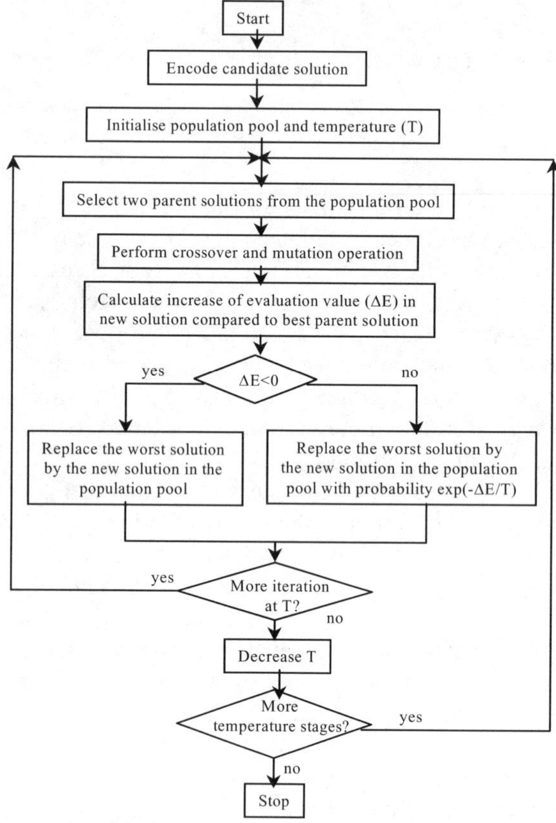

Fig. 2. The algorithm of the proposed GA/SA hybrid method.

The features for the GA and SA adapted in the hybrid approach have been borrowed from the results described in the previous section. In summary these are: integer representation, evaluation function expressed by equation (7), a steady state approach, tournament selection, two-point crossover, random mutation, a population size=100, a tournament pool size=10, an initial temperature=10,000 and a stage-wise cooling schedule. A specific temperature value defines a stage of the GA/SA process. The stage-wise cooling schedule executes a number of genetic operations (iterations) at a temperature (i.e. at one stage) before reducing the temperature according to equation, $T_s=\alpha T_{s-1}$, where T_s is the temperature at stage s and α is the cooling parameter. In the reported experimentation the genetic operations have been performed 100 times for each temperature. The number of temperature alterations (or stages) was fixed to 300, giving 30,000 fitness evaluations per run of the algorithm.

GA/SA sensitivity analysis

Table 6. GA/SA performance with different values of cooling parameter.

α	Average evaluation value
0.92	148.03
0.95	145.78
0.98	352.24

In order to determine the best value of the cooling parameter (α) for the proposed hybrid method, a number of experiments have been performed and the results obtained are summarized in Table 6. Three values of α covering a relatively wide range have been selected for the experiments based on previous experience with the SA method. With α=0.92, the temperature decrease is very rapid and the algorithm lacks in exploration, concentrating more on exploitation in the neighborhood of a solution in the population pool. With α=0.98, the temperature does not drop sufficiently far within 30,000 iterations and the method works as a simple GA technique. The cooling schedule with α=0.95 provides a good compromise between the exploitation and exploration during the search process and this is supported by observing the best performance of the algorithm for 30,000 iterations. This cooling parameter value is used for further investigation of the GA/SA method.

The sensitivities of the method to variation of crossover probability (CP) and mutation probability (MP) have also been established. Results were obtained for varying CP in the range [0.6, 1.0] and MP in the range [0.001, 0.1]. For each value of CP and MP 10 independent experiments were performed using the same collection of 10 random initial populations. The sensitivity of the GA/SA approach to variation of CP and MP is depicted in Fig. 3. For comparison, the sensitivity of the simple GA to variation of CP and MP is also depicted in Fig. 3. It can be observed that the performance of the GA/SA is generally less sensitive than that of the simple GA for the given range of crossover probability. The performance of the GA/SA and the simple GA method does not differ much for MP=0.001. However, for higher MP values the GA/SA method is more robust than the simple GA method alone in terms of consistently finding better results. Although mutation can introduce new information to solutions, it can also destroy useful information. In the simple GA the mutation operator becomes disruptive as MP increases as seen in the climbing evaluation value of the graph. In the GA/SA hybrid method, the SA probabilistic acceptance test tends to preserve the positive effects and counter the adverse effects of the mutation operator. That is, new solutions, even those whose evaluation function values are lower than those of current solutions, are fully accepted at the beginning of the search, thus introducing more diversity amongst the candidate solutions. However, at later stages of the search process, the chance of mutated solutions of lesser fitness being accepted will be low.

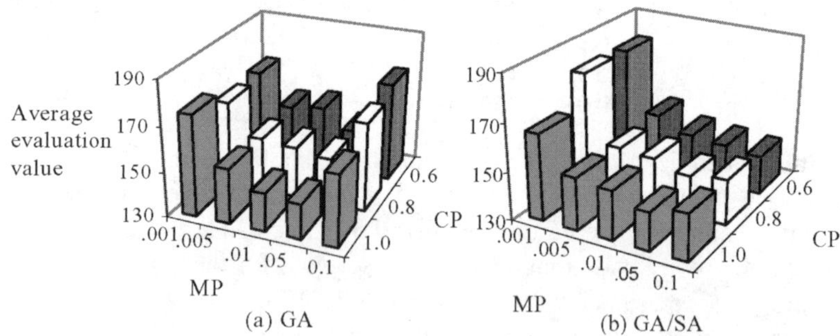

Fig. 3. Sensitivity of the steady state GA and GA/SA to variation of CP and MP.

The best average performance of the GA/SA hybrid is found with CP=1.0 and MP=0.05. These CP and MP values match the optimum values identified for the simple steady state GA. The average evaluation value of the best solutions obtained over 10 GA/SA runs with these parameters is 145.78 and the best solution has evaluation value of 138.12.

5.5 Inoculated GA/SA

Previously it was found that seeding a heuristic solution to the initial population pool improved the average GA performance. To extend this work the GA/SA hybrid approach also had its initial population seeded, this being referred to as the inoculated GA/SA approach. The initial population pool is seeded using *Heuristic solution 2* with the remainder of the candidate solutions in the pool generated randomly (initial population *case 3* as described in section 5.2). This case was found to give the best performance for the inoculated GA. The steady state population updating structure with population size=100, CP=1.0, MP=0.05 and cooling parameter α=0.95 has been used, these having been identified to give the best performance for the GA/SA approach. The average evaluation value of the best solutions found by the inoculated GA/SA is 141.71 and the best solution found has an evaluation measure of 139.10.

5.6 Performance comparison of different methods

Table 7. Comparison of results obtained using different methods.

	Average of best solutions	Best solutions			
	Evaluation value	Evaluation	Objective ($\times 10^5$)	TMV	TLV
Heuristic schedule 1	-	222.61	134.61	22	0
Heuristic schedule 2	-	483.70	149.7	3	161
GA	146.71	137.91	137.91	0	0
SA	146.06	140.49	140.49	0	0
Inoculated GA	142.67	139.95	139.96	0	0
GA/SA	145.78	138.12	138.12	0	0
Inoculated GA/SA	141.71	139.10	139.10	0	0

From Table 7 the average performance of the inoculated GA is demonstrably better than the average performance of the SA, which is slightly better than that of the simple steady state GA.

The average performance of the GA/SA hybrid is slightly better than that of the GA and SA methods individually. Although the performance of the hybrid approach with the best parameters does not differ significantly from the performance of the simple GA, it is an important point to note that the approach is much more robust in the sense that it consistently gives good results over a wide range of CP and MP values. The GA/SA performance does not vary much as long as the parameters are within the reasonable ranges.

It can be observed from Table 7 that the average evaluation value of the best solutions found by the inoculated GA/SA is 141.71, which is an improvement over that found by other approaches. The best solution found by the inoculated GA/SA approach has an evaluation measure of 139.10, which is lower (better) than that of the best solution found by the SA, heuristic and inoculated GA. This evaluation value however, is slightly greater than the evaluation measure of the best solutions found by the GA (137.91) and GA/SA (138.12). On the basis of by comparing the average of best solutions over multiple experiments, the expectation is that the inoculated GA/SA method will be more consistent at finding a better solution. Furthermore, the best parameters of the SA and GA are generally decided upon after a number of experiments. From the inoculated GA/SA performance analysis it can be seen that the approach can produce a good solution for a wide range of technique parameters. Hence, the parameter selection process in the GA/SA method involves considerably less experiments than that of the GA and SA.

6. Chapter summary

The GMS is a long-term operation planning problem. The solution to this problem is a timetable to take generators off-line for preventive maintenance. The decision must take into account power system reliability, maintenance crew limitations and constraints on the individual generators and the power system. This chapter has presented the development and applications of GA, SA and GA-based hybrid approaches for GMS problem. The design, implementation, performance, sensitivity and results obtained of these approaches have been discussed.

A detailed analysis of the use of different GA designs, problem representations, genetic operators and GA parameters has been presented for a test GMS problem. Different types of solution representations (binary, binary for integer and integer) have been employed and discussed. As the GMS problem variables are integer, representing them directly as integers in a genetic structure has many advantages. The most significant of these is the great reduction in the GA search space. Furthermore, this type of representation is obvious and easy for decoding and a meaningful crossover and mutation operator can be applied. The sensitivity of these GA approaches to generational and steady state population updating approaches as well as to the variations of the crossover probability and the mutation probability have also been established. The results obtained show that the GA is generally sensitive to these GA features, and can find good solutions of the GMS problem if an appropriate problem encoding, GA approach, evaluation function and GA parameters are selected.

The applications of SA approach using an integer representation have been demonstrated for a test problem of GMS. The effects of the initial temperature, initial solution, and cooling parameters for two types of cooling schedules were studied for the SA method. The test results show that the selection of an initial solution and a cooling schedule does not greatly affect the performance of the SA method provided the initial temperature is high enough to allow the free movement of a trial solution in the search space in the early stages of the search process. The initial temperature should be selected considering the problem domain and the setting of the cooling parameter for a chosen cooling schedule. This is performed by making a trade-off between the computational time and the quality of solutions.

The study of the inoculated GA using heuristically derived solutions in the initial population shows that the inoculation can enhance the performance of the GA. However, the inclusion of a superfit solution in the initial population pool may lead to only a local optimum being identified.

A GA/SA approach has been designed by incorporating the SA probabilistic acceptance test for every newly created solution within a GA framework. The application of a GA/SA approach using an integer representation has been demonstrated for the test GMS problem. The sensitivity of the approach to the variation of cooling parameter, crossover probability and mutation probability has been studied. An inoculated GA/SA using a seeded initial population pool has also been employed to the test GMS. The performance and results obtained from these GA/SA approaches have been compared with those of other techniques. The test results show that the GA/SA approach is sensitive to the cooling parameter; this should be selected to make a good compromise between exploration and exploitation of the search space for the given number of iterations (computational time).

The best parameters of the SA and GA are generally decided upon after a number of experiments. The results presented in this chapter show that the GA/SA approach is more robust and stable for solving GMS problems in a wide range of technical controlling factors, such as cooling parameter, crossover probability and mutation probability than the simple SA or simple GA. Hence, the parameter selection process in the GA/SA method involves fewer experiments than that in the GA method. Furthermore, the hybrid method also improved the convergence of the simple GA. The study of the inoculated GA/SA using a heuristically derived solution in the initial population shows that inoculation can enhance the performance of the GA/SA approach. Comparing the individual average results of different approaches considered, the inoculated GA/SA approach gives the best average performance.

The GA/SA evolutionary-based approaches obtained feasible schedules whereas developed heuristic schedules 1 and 2 were infeasible (due to non-satisfaction of constraints). Mathematical programming approaches cannot be readily applied due to the complexity and combinatorial explosion of the problem. Although the evolutionary-based approaches do not guarantee the global optimal solution, these techniques have successfully found feasible, near optimal solutions. It is a significant achievement to obtain a good solution to a complex problem like GMS.

References

[1] A. Abdulwhab, R. Billinton, A.A. Eldamaty, S.O. Faried, "Maintenance Scheduling Optimization Using a Genetic Algorithm (GA) with a Probabilistic Fitness Function", Electric Power Components and Systems, vol. 32, no. 12, pp. 1239-1254, 2004

[2] S. Baskar , P. Subbaraj , M.V.C. Rao and S. Tamilselvi, "Genetic algorithms solution to generator maintenance scheduling with modified genetic operators", IEE Proceedings - Generation, Transmission and Distribution, vol. 150, no. 01, pp. 56-66, 2003.

[3] L. Bertling, R. Allan, R. Eriksson, "A Reliability-Centered Asset Maintenance Methodfor Assessing the Impact of Maintenance in Power Distribution Systems", IEEE Transactions on Power Systems, vol. 20, no. 1, February 2005.

[4] E.K. Burke, A.J. Smith, "Hybrid Evolutionary Techniques for Maintenance Scheduling Problem", IEEE Trans. on Power Systems, vol. 15, no. 1, pp. 122-128, 2000.

[5] D. Chattopadhyay, "Life-Cycle Maintenance Management of Generating Units in a Competitive Environment", IEEE Trans. on Power Systems, vol. 19, no. 2, pp. 1181-1189, 2004.

[6] K.P. Dahal, C.J. Aldridge, J.R. McDonald, "Generator maintenance scheduling using a genetic algorithm with a fuzzy evaluation function" Fuzzy Sets and Systems, vol. 102, pp. 21-29, 1999.

[7] K.P. Dahal, J.R. McDonald and G.M. Burt, "Modern heuristic techniques for scheduling generator maintenance in power systems", Transactions of Institute of Measurement and Control, vol. 22, pp. 179-194, 2000.

[8] K.P. Dahal, G.M. Burt, J.R. McDonald, S.J. Galloway, "GA/SA-based hybrid techniques for the scheduling of generator maintenance in power systems", Proceedings of IEEE Congress of Evolutionary Computation (CEC2000), 567-574, San Diego, 2000.

[9] J.F. Dopazo, H.M. Merrill, "Optimal generator maintenance scheduling using integer programming", IEEE Trans. PAS-94(5):1537-1545, 1975.

[10] Egan, G.T., Dillon, T.S. and Morsztyn, K. 1976. 'An experimental method of determination of optimal maintenance schedules in power systems using

branch-and-bound technique'. IEEE Transactions on Systems, Man and Cybernetics. SMC-6, 538-547.

[11] M.Y. El-Sharkh, R. Yasser, and A.A. El-Keib, "Optimal maintenance scheduling for power generation systems—A literature review," in *Proc. Maintenance and Rel. Conf.*, vol. 1, May 01–20.10, 1998, pp. 20.01–20.10.

[12] M.Y. El-Sharkh, A.A. El-Keib, "Maintenance Scheduling of Generation and Transmission Systems Using Fuzzy Evolutionary Programming", IEEE Trans. on Power Systems, vol. 18, no. 2, pp. 862-866, 2003.

[13] M.Y. El-Sharkh., A.A. El-Keib, "An evolutionary programming-based solution methodology for power system generation and transmission maintenance scheduling", Electric Power Systems Research, vol. 65, no. 1, pp. 35-40, 2003.

[14] J. Endrenyi et al, "The Present Status of Maintenance Strategies and the Impact of Maintenance on Reliability", IEEE Trans. on Power Systems, vol. 16, no. 4, pp. 638-646, 2001.

[15] W.K. Foong, H.R. Maier, A.R. Simpson, "Ant colony optimization for power plant maintenance scheduling optimization", Proceedings of the Genetic and Evolutionary Computation conference, pp. 249-256, 2005.

[16] Fustar, S. and Hsieh, J. (1988) A knowledge based method for revision of yearly generator maintenance scheduling, *IEEE Symposium on Expert Systems Application to Power Systems*, Stockholm-Helsinki 9-23.

[17] Glover F. and Kochenberger G., (eds.) Handbook of Meta-Heuristics, Kluwer, 2003.

[18] K.-Y. Huang and H.-T. Yang, "Effective algorithm for handling constraints in generator maintenance scheduling", IEE Proceedings - Generation, Transmission and Distribution, vol. 149, no. 03, pp. 274-282, 2002.

[19] B.L. Kralj, and R. Petrovic, (1988) Optimal preventive maintenance scheduling of thermal generating units in power systems - A survey of problem formulations and solution methods, *European Journal of Operational Research* **35,** 1-15.

[20] B. Kralj, and N. Rajakovic, "Multiobjective programming in power system optimization: new approach to generator maintenance scheduling", *Electrical power and Energy Systems* **16,** 211-220, 1994.

[21] H. Kim, K. Nara and M. Gen, "A method for maintenance scheduling using GA combined with SA", Computers and Industrial Engineering 27, 477-480, 1994.

[22] H. Kim, Y. Hayashi and K. Nara, "An algorithm for thermal unit maintenance scheduling through combined use of GA, SA and TS", IEEE Transactions on Power Systems 12, 329-335, 1997.

[23] Lin, C.E., Huang, C.J., Huang, C.L., Liang, C.C. and Lee, S.Y. (1992) An expert system for generator maintenance scheduling using operation index, IEEE *Transactions on Power Systems* **7,** 1141-1148.

[24] M. Marwali and M. Shahidehpour, "A Probabilistic Approach to Generation Maintenance Scheduler with Network Constraints," Electric Power and Energy Systems, Vol. 21, pp. 533-545, 1999.

[25] M.K.C. Marwali and S.M. Shahidehpour, "Coordination Between Long-Term and Short-Term Generation Scheduling with Network Constraints", IEEE Trans. on Power Systems, vol. 15, no. 3, pp. 1161-1167, 2000.
[26] M. Marwali and M. Shahidehpour, "Integrated Generation and Transmission Maintenance Scheduling with Network Constraints," IEEE Transactions on Power Systems, Vol. 13, No. 3, pp. 1063-1068, 1998.
[27] V. Miranda, D. Srinivasan and L.M. Proença, "Evolutionary computation in power systems", Electrical Power & Energy Systems. 20, 89-98, 1998.
[28] Moro, L.M. and Ramos, A. (1999) Goal programming approach to maintenance scheduling of generating units in large scale power systems, *IEEE Transactions on Power Systems* 14, 1021-1028.
[29] R. Mukerji, H.M. Merrill, B.W. Erickson, J.H. Parker, R.E. Friedman, "Power plant maintenance scheduling: optimising economics and reliability", IEEE Transactions on Power Systems, Vol. 6, No. 2, May 1991, pp. 476-483.
[30] "Reproductive Plan Language (RPL2) - User manual", Quadstone Ltd, 1997.
[31] T. Satoh and K. Nara, "Maintenance scheduling by using simulated annealing method", IEEE Transactions on Power Systems. 6, 850-857, 1991.
[32] M. Shahidehpour and M. Marwali, Maintenance Scheduling in a Restructured Power System. Norwell, MA: Kluwer, May 2000.
[33] Y.-H. Song (ed) "Modern Optimisation Techniques in Power Systems", Kluwer Academic Publishers, 1999.
[34] Y. Wang, E. Handschin "A new genetic algorithm for preventive unit maintenance scheduling of power systems", International Journal of Electrical Power & Energy Systems, vol. 22, no. 5, pp. 343-348, 2000.
[35] X. Wang, and J.R. McDonald, (Eds.) Modern Power System Planning, McGraw-Hill, London, 247-307, 1994.
[36] X. Xu M. Kezunovic, "Mobile Agent Software Applied in Maintenance Scheduling", North American Power Symposium (NAPS 2001), Texas, 2001.
[37] Z. Yamayee and K. Sidenblad, "A computationally efficient optimal maintenance scheduling method", IEEE Transactions on Power Apparatus and Systems PAS-102, 330-338, 1983.
[38] H.H. Zurn, V.H. Quintana, "Several objective criteria for optimal generator preventive maintenance", IEEE Transactions on Power Apparatus and Systems, Vol. PAS-96, No. 3, May/June 1977, pp. 984-992.

Evolvable Fuzzy Scheduling Scheme for Multiple-Channel Packet Switching Network

Ju Hui Li[1], Meng Hiot Lim[2], Yew Soon Ong[3], and Qi Cao[4]

[1] School of EEE, Block S1, Nanyang Technological University, Singapore 639798
pg01896341@ntu.edu.sg
[2] School of EEE, Block S1, Nanyang Technological University, Singapore 639798
emhlim@ntu.edu.sg
[3] School of SCE, Block N4, Nanyang Technological University, Singapore 639798
asysOng@ntu.edu.sg
[4] School of EEE, Block S1, Nanyang Technological University, Singapore 639798
pg04780942@ntu.edu.sg

Evolvable fuzzy system (EFS) relies on dynamic adaptation of the heuristics which are coded as fuzzy rules. If the mechanisms for fuzzy inferencing are realized in hardware within the framework of the EFS in order to satisfy the time-critical requirement of the application, the system can be regarded as a form of evolvable fuzzy hardware. This chapter considers the application of evolutionary fuzzy system for cell scheduling in ATM network. Besides being able to adapt to the changing environment, the EFS also enjoys a special property whereby the cell delay can be conveniently adjusted by explicitly tuning a parameter in the fitness function. In this chapter, we explore the application of EFS to solve multiple channel scheduling. We show that any number of input channels can be accommodated while still preserving the special advantages of the EFS.

1 Introduction

For modern networks, effective management of bandwidth resource is critical because of the diversity in the services required. Bandwidth sharing is thus inevitable for efficient bandwidth utilization. Multiplexer is a network component to administer the sharing of high speed link for different input channels through *time division*. The output capacity is divided into different time slots assigned to various input traffic flows according to a specific control scheme. The effectiveness of the control scheme can be gauged by the

J.H. Li et al.: *Evolvable Fuzzy Scheduling Scheme for Multiple-Channel Packet Switching Network*, Studies in Computational Intelligence (SCI) **49**, 383–403 (2007)

level that the desired *quality of service* (QoS) requirements of the input flows are met. QoS includes factors such as *cell delay* and *cell loss*. Sometimes the balance of cell losses among different input flows and the isolation of contract following flows from the effects of mal-functioning flows are also considered. The *cell delay* factor mainly refers to the overall waiting time of the cells in the buffers before being serviced by the multiplexer. The *cell loss* factor indicates the number of cells being dropped when the input buffer overflows. This is inevitable when the sum of the actual input bit rates is larger than the link capacity that the output channel can offer. The balance of cell losses refers to the requirement that the system should not be biased towards certain input flows while always dropping packets from other input channels. This is a form of fairness to be achieved by the scheduling scheme, as specified in [1]. Another form of fairness can be considered during the occurrence of congestion. For such a situation, only the flows which override their reserved bit rates should suffer significant QoS degradation.

Each input flow should negotiate with the network system on its bit rate requirement before the admission of its connection. Regarding the required bit rate, the connection would be set up only if there is enough link capacity to accommodate the input flow. After the success of the admission process, the bit rate of the traffic flow cannot exceed its reserved bit rate. Based on a QoS consideration, significant QoS degradation should only be suffered by the misbehaving flow channels. This approach can effectively restrain the aggressive source domination, which could be a symptom of systematic vicious attacks on a network. The delay suffered by a cell unit in a network system includes propagation delay, queuing delay and service time at every queuing point. In this chapter, we focus on a single node switching system, thus only queuing delay is of major concern in the following parts.

The common services supported by modern networks include *constant bit rate* (CBR), *variable bit rate* (VBR), *available bit rate* (ABR), and *unspecified bit rate* (UBR). CBR always reserves a static bandwidth, and it can transmit cells at the *peak cell rate* (PCR) for any duration of time. CBR is normally used to support telephone, video conferencing and entertainment video. VBR is for the services that require variable bandwidth resource. It is characterized by the PCR, *sustainable cell rate* (SCR), and *maximum burst size* (MBS). ABR and UBR are complementary services to fully take advantage of the bandwidth resource. The cell flow characteristic of CBR can be modeled by Fig. 1. VBR can be modeled by a two-state *Markov chain model* in which the

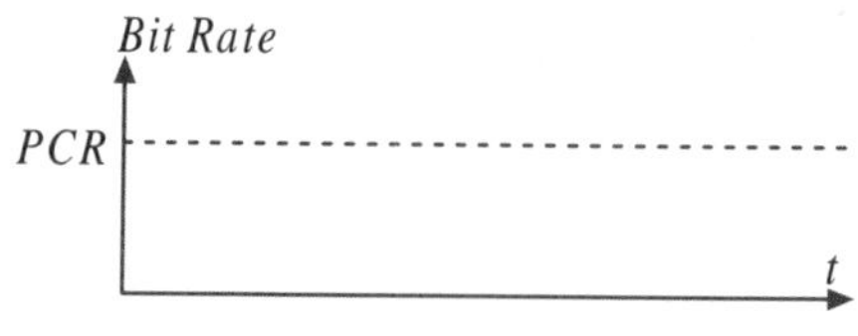

Fig. 1. Model of CBR

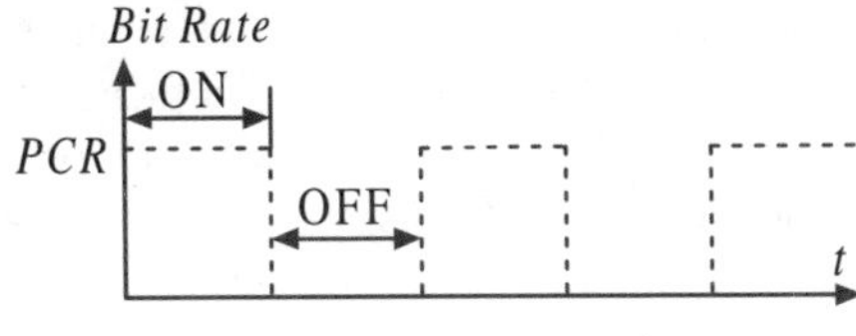

Fig. 2. Model of VBR

cell flow has ON (*burst*) period and OFF (*silence*) period [2-7]. The source transmits cells equidistantly during the ON period while keeping silent during the OFF period. VBR's cell flow characteristic is depicted by Fig. 2

There are many schemes which have been proposed to address the cell-scheduling problem. Some of the schemes will be discussed in Section 2. EHW(*evolvable hardware*), which has been attracting greater interests in the research community was also applied on the cell-scheduling problem domain. EHW is a kind of hardware that can evolve its architecture or behaviour to adapt to its working environment. Due to the complexity of the evolutionary algorithms and the huge solution space, the advantage of EHW to solve cell-scheduling was not very obvious. For the purpose of promoting EHW's application, we proposed the EFS(*evolvable fuzzy system*) to address the cell-scheduling problem. The proposed EFS is capable of adapting to the dynamic environment instantly and is thus viable to be applied in the real-time problem domain. The efficacy of the EFS in cell-scheduling involving 2 input channels has been demonstrated in earlier work [19, 20, 22]. In order to further demonstrate the applicablity of EFS, we extend the work to multi-channel application, a more general problem on cell-scheduling.

The remaining of this chapter is organized as follows. In Section 2, some conventional schemes for cell-scheduling is described. In Section 3, the evolution scheme of the EFS for three input channels is described. In Section 4, the encoding scheme of the fuzzy system for *genetic algorithm* (GA) in the evolutionary system is introduced. This is followed by the introduction of the fitness function adopted in the evolutionary system for multi-channel applications in Section 5. In Section 6, the simulation results of the proposed evolutionary system are compared with other schemes such as GPS (*general processor sharing*), MCRR(*minimized cycle round robin*), and FCFS (*first-come first-serve*). We conclude in Section 7 with discussion on the properties of EFS and some comments on possible extension of our work.

2 Related Research

Various algorithms or systems have been designed for cell-scheduling. For example, the *generalized processor sharing* (GPS) is an idealized scheduling scheme using a *fluid model* [8]. It is assumed that every input channel is serviced by an independent server simultaneously with a reserved link capacity. GPS has two significant properties, appropriate delay bound and fair bandwidth allocation. Although GPS has significant properties, it is not realizable due to the ideal fluid model it adopts. Nevertheless, GPS is used by many algorithms as a benchmark or to generate time stamps. The *weighted fair queuing* (WFQ) is a scheme which selects cells based on the simulation of the GPS system [9]. The cell that finishes its transmission first in the GPS system will be selected for transmission in the WFQ system. This system can guarantee a service for each cell no later than the GPS system, but sometimes it may service the cell much earlier than the GPS system. In order to overcome the disadvantages of WFQ, *worst-case fair weighted fair queuing*(WF2Q) has been proposed [10]. WF2Q can approximate to GPS system very effectively. Other schemes such as *virtual clock* [11], *start-time fair queuing* (SFQ) [12] and *self-clocked fair queuing* (SFCQ) [13] use time stamps to control the sequence of cells transmission. There is significant computation complexity involved during the scheduling process or in the queuing process for every cell in the above systems.

Another approach of cell scheduling is *round robin* (RR) scheme whereby within one round, each input channel is serviced according to a predetermined quota. *Weighted round robin* (WRR) is a modified round robin scheme [14]. Each input channel is assigned a weighted value based on its reserved bandwidth resource. The weighted factor is the service quota each channel can acquire during a service round. After its quota runs out, the channel will not be serviced any more until the next service round. WRR can strictly guarantee the bandwidth allocation ratio regardless of the misbehaviour of other input flows. But its major drawback is its queuing delay and delay jitter. To overcome this, the *minimized cycle round robin* (MCRR) is proposed, which is an improvement of WRR in two aspects. First, the quota of each cell flow is determined in such a way that there is no common factors among them. The second aspect of improvement is that one cell from each cell flow is serviced for each visit [16]. This modification is similar to the WRR1 [15]. In order to maintain good isolation of the ill effects from misbehaving cell flows and to achieve good delay guarantees without incurring high computational complexity during the scheduling process, multi-class WRR has been proposed [15]. In multi-class WRR, the cell flows are categorized as $group_1$, $group_2$, $group_3$ and so on, in the order of decreasing bandwidth ratio. For the first round robin cycle, all the sessions in $group_1$ are serviced, followed by the first few sessions of $group_2$. In the next round robin cycle, the sessions from $group_1$ are serviced again, and the sessions in $group_2$ starting from where it left off before, followed by the sessions of $group_3$, and so on. This is called a *minicycle*. The maximum length of the *minicycle* is the smallest value of D_i, which corresponds to the maximum length of the round robin slot whereby all the sessions of $group_i$ must be visited. The bandwidth ratio denoted by the parameter $\phi_i = \frac{1}{D_i}$, is allocated to every session of $group_i$. One major drawback of multiclass WRR is that ϕ_i cannot be an arbitrary value. For example, suppose ϕ_1 is 0.7, then D_1 is $\frac{1}{0.7} \approx 1.4$. Since D_1 should be an integer, D_1 is either 1 or 2. If it is assigned the value 1, $group_1$ will monopolize the complete bandwidth. If it is 2, $group_1$ will only utilize half of the bandwidth resource.

Both the *fluid model* approach and the *round robin* approach have their own advantages and disadvantages. GPS time stamp based approach incurs high computational overhead during the queuing or de-queuing process. This can significantly affect the efficiency of the cell scheduling. Round robin approach has the lowest computational complexity but poor delay guarantee. The traditional schemes are very inflexible and cannot adapt to the system's environment in an efficient way. In order to address this shortcoming, *evolvable hardware* (EHW) system was proposed [7, 17, 18]. Liu. *et. al.* reported the applications of *functional evolvable hardware* in ATM cell scheduling. In the reported system, a circuit to compute the desired weighting factors for the WRR scheme was successfully evolved. Due to the large search space and hence the large number of generations required for the evolutionary search, the reported *functional evolvable hardware* is not applicable for real-time applications. The time cost for one circuit evolution process is very high. The second reason why it is not applicable is that there is no reasonable scheme to trace and adapt to the dramatic changes of the input cell flows. In order to promote the application of EHW for real-time applications, Li and Lim [17] proposed *evolvable fuzzy system* (EFS) for the cell scheduling application. This scheme is suitable for implementation as *evolvable fuzzy hardware* (EFH) [19, 20]. EFH can effectively address some existing open issues outlined in [21]. In the proposed system, two cell flows were mainly considered, one being delay sensitive and another being loss sensitive. The system adopted the principle of "locality" to justify the training scheme and fixed the number of generations for the evolution process. The "locality" principle is based on the assumption that the cell flow pattern of subsequent time period tends to be similar to the current cell flow pattern. The validity of such an assumption has been shown for CBR or for the bursty period of VBR using the flow models in Fig. 1 and 2. Based on the simulation presented in the published paper, the prediction error during the changes from OFF to ON period or from ON to OFF period is not very significant when the time window for evolution is small enough. Thus the evolvable system is trained using the current cell flow to search for a good fuzzy rule set to control the subsequent cell flows. The number of generations for the evolution process can be determined experimentally. It should be sufficient for EFS to find good chromosomes. These two schemes endowed the EFS with real-time applicability. In order to further demonstrate the capability of EFS, we employ it on a more general cell-scheduling problem domain. Through the analysis and simulations presented in the following Sections of this chapter, we will demonstrate how the EFS model can be extended to handle multi-channel applications.

3 Evolution Scheme

The block architecture of the EFS can be described by Fig. 3. $class_{\#}$ represents the different classes of input cell flow. $BUF_{\#}$ is the queuing buffer for each input channel. It is used to temporarily store data packets when congestion occurs. The size of $BUF_{\#}$ corresponds to the size of the time window, consistent with the principle of "locality". $TB_{\#}$ is the buffer for the storage of training patterns. It mainly stores the information of each cell, in particular the arrival time and the arriving bit rate. When one of the $TB_{\#}$ is full, the evolution process is triggered. Thus the fuzzy rule set is optimized for every small period. MP is a multiplexer to send the cells in the buffers through the OUT channel under the control of an RFIC. RFIC refers to *reconfigurable fuzzy inference chip*, a fuzzy processing unit whereby the context can be re-configured on-line without any setup overhead [22]. The value of SEL indicates the buffer from which the cells are to be sent. The *evolution module* is a component that realizes the GA search. Every chromosome generated by the *evolution module* is evaluated by the *scheduling model* block. The block simulates the scheduling behaviour of the MP unit. $TB_{\#}$ supplies the cell flows to the *scheduling model*, which carries out scheduling according to the control of the chromosome being evaluated. After all the cell units stored in $TB_{\#}$ are transmitted by the *scheduling model*, the fitness value is calculated for each chromosome. In order to prevernt the evolution system from taking a too long time to evolve and, to make the evolvable system suitable for the

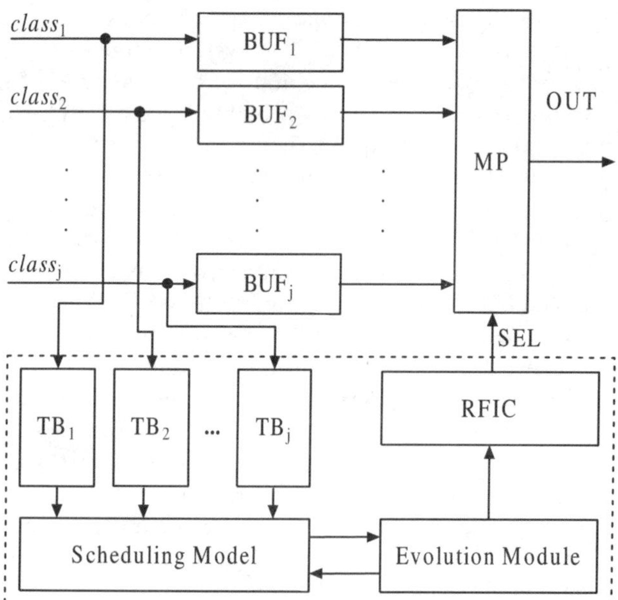

Fig. 3. The System Architecture For Multiple Input Channels Application

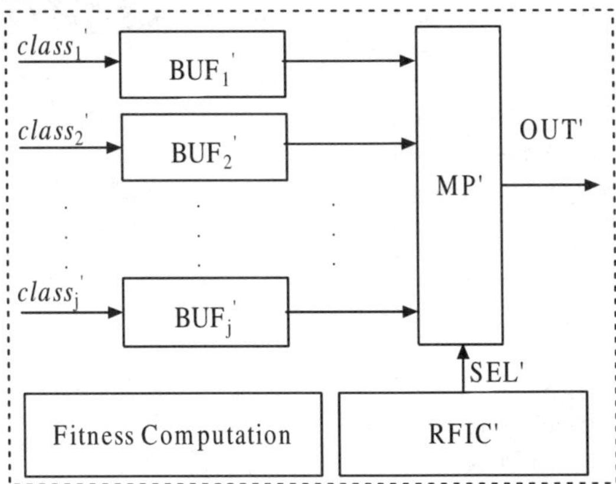

Fig. 4. The Architecture of The *Scheduling Model*

application of real-time problem domain, number of generations adopted in GA can be fixed. The best chromosome after a fixed number of generations will be downloaded into the RFIC. At the beginning of the evolution, the content in $TB_\#$ is modified based on the arrival time of the last cell unit in $BUF_\#$ to prepare the training data. The architecture of the *scheduling model* is as shown in Fig. 4. It includes RFIC′, MP′, $BUF'_\#$ and the *fitness computation* block. RFIC′ is also a *reconfigurable fuzzy inference chip*. MP′ is an emulation of the MP. It simulates the functionality of the MP block by calculating the queuing delay for every cell stored in $BUF'_\#$. $class'_\#$ is the cell flow from $TB_\#$. The *fitness computation* block in Fig. 4 computes the fitness value for each evaluated chromosome.

The evolution process of the EFS can be described by the procedural codes listing in Algorithm 1. The function of $TB_\#\ Modification$ routine is to prepare the training data for the evaluation process. This process generates packet information such as arrival time and bit rate based on the last few cells in $BUF_\#$. The *Initialization of* $BUF'_\#$ routine copies the cell information from $BUF_\#$ in Fig. 3 to $BUF_\#'$ in Fig. 4. *Evolution and Evaluation* routine carries out the evolution and evaluation. The routine for the *Evaluation* module in Algorithm 1 is described further by Algorithm 2. The *Enqueuing* process adds the information of the arriving cells to $BUF'_\#$. *Evaluation process* implements the fuzzy inference and cell scheduling. The *Fitness Computation Process* carries out fitness computation.

4 Fuzzy System

In this section, the fuzzy terms involved in the proposed EFS are first introduced. For the purpose of GA search in the solution space of the fuzzy

Algorithm 1 Evolutionary Process

Notations:

m: The size of $\mathrm{TB}_{\#}$
v: The size of $\mathrm{BUF}'_{\#}$ and $\mathrm{BUF}_{\#}$
g: The number of the generation
p: Population size
$q_{\#}$: The number of cells in $\mathrm{BUF}'_{\#}$
c: The number of evolution cycle
l: Chromosome length in bits
j: Channel number

$\mathrm{TB}_{\#}$ Modification:

```
/*Construct TB# based on the bit rate
when evolution is triggered*/
    for i < m do
        modify TB1(i);
        modify TB2(i);
              ⋮
        modify TBj(i);
    end for;
```

Initialization of $\mathrm{BUF}'_{\#}$:

```
/*Copy cell information from BUF# to BUF#'*/
    for i < j do
        copy BUF1(i) to BUF1(i)';
        copy BUF2(i) to BUF2(i)';
              ⋮
        copy BUFj(i) to BUFj(i)';
    end for;
```

Evolution and Evaluation:

```
    for i < c do
        for i1 < p do
            Initial population generation;
            Call Evaluation();
        end for;
        for i2 < g do
            for i3 < p do
                Toss for crossover;
                Select chromosomes for crossover;
                Do crossover operation;
                Toss for mutation;
                Do mutation operation;
                Call Evaluation();
            end for;
        end for;
    end for;
```

Algorithm 2 Evaluation

Enqueuing: /*Upon the arrival of cells*/
  modify $\mathrm{BUF}_1(i)'$;
  $q_1 = q_1 + 1$;
or / and
  modify $\mathrm{BUF}_2(i)'$;
  $q_2 = q_2 + 1$;
  ⋮
or / and
  modify $\mathrm{BUF}_j(i)'$;
  $q_j = q_j + 1$;

Evaluation: /*Simulate the cell transmission*/
 while $((\exists i, 1 \leq i \geq j) q_i > 0)$ and (no cell is sending) do
  fuzzy inference;
   /*according to the fuzzy inference result*/Calculate the queuing delay of the oldest unit in BUF_i;
  Delete the oldest unit from BUF_i;
 end while;
Fitness Computation:
/*Compute the fitness value*/
 Fitness Computation;

rule set, genetic coding of the fuzzy rule set is necessary. In the second part of this section, the encoding scheme of the fuzzy rule set is introduced. In the following parts, three input channels ($j = 3$) are assumed for simplicity of discussion.

4.1 Fuzzy Terms

For the fuzzy system, we define three fuzzy variables c_1, c_2 and c_3. $c_\#$ can be defined as $\frac{L_\#}{L_{max}}$. $L_\#$ represents the number of empty units in $\mathrm{BUF}_\#$. L_{max} is the length of the buffers. $c_\#$ can be characterized by the fuzzy terms $\{S, M, L\}$. The membership functions for these fuzzy terms are chosen to be triangular as in Fig. 5.

The output of the fuzzy system can be described by the terms set $\{0, 1, 2, 3\}$. "0" indicates the absence of fuzzy rule for the corresoponding input condition. "1", "2" and "3" mean that the cell in the corresponding buffer is to be transmitted. For the fuzzy system, the fuzzy rules are in the form "if $< antecedent_1 >$ and $< antecedent_2 >$ and $< antecedent_3 >$ then $< conclusion >$". Two examples of rules are "if $< c_1$ is $S >$ and $< c_2$ is $S >$ and $< c_3$ is $S >$ then 1" and "if $< c_1$ is $M >$ and $< c_2$ is $M >$ and $< c_3$ is $M >$ then 2". The first rule means that the cell in BUF_1 is to be transmitted while the second rule indicates that the cell in BUF_2 is to be sent. The Mamdani

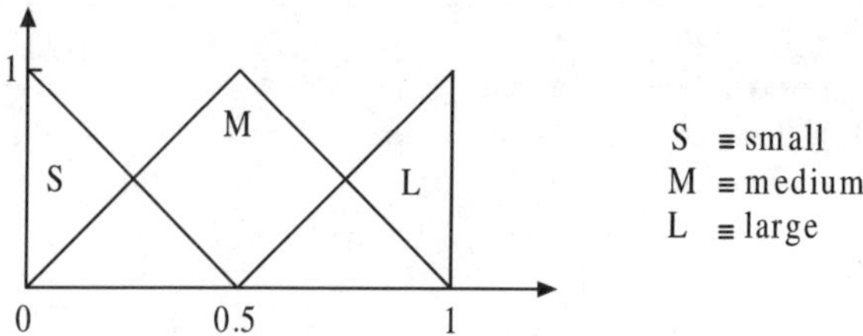

Fig. 5. The membership functions for $c_{\#}$

Table 1. A Startup Rule Set

$c_3 = S$		c_1 S	c_1 M	c_1 L
c_2	S	1	2	2
	M	1	3	3
	L	1	3	3

$c_3 = M$		c_1 S	c_1 M	c_1 L
c_2	S	1	2	2
	M	1	2	3
	L	1	2	3

$c_3 = L$		c_1 S	c_1 M	c_1 L
c_2	S	1	2	3
	M	1	2	3
	L	1	2	3

implication is adopted for fuzzy inference. When some rules have the same conclusion, the final conclusion for these rules will be aggregated. The output of the fuzzy system is derived by comparing the degree of the different conclusions. The one with biggest degree will be used as the final output. An example of rule set for the startup of the system is shown in Table 1. All these rules can work together to fulfill the control scheme. The inference results related to "1", "2" or "3" can be aggregated respectively and the final value of the aggregated results determines whether $class_1$, $class_2$ or $class_3$ is to be serviced.

4.2 GA Encoding

According to the rule set presented in Table 1, the core rule set can be coded as "122133133, 122123123, 123123123". The first 9 genes correspond to the case when $c_3 = S$. The second 9 genes correspond to $c_3 = M$, and the last 9 genes represent the rules for $c_3 = L$. Each sub-table grouping in Table 1 can be interpreted in a row-wise manner. The one-point mutation scheme and one-point crossover operation are adopted in the GA implementation. For selecting the chromosomes to be mutated or crossed, the roulette wheel selection scheme is adopted.

5 Fitness Function

Fitness function is an important component which directly affects the behavior of the evolvable system. In [17] and [19], where only two cell flows were considered, the QoS of $class_1$ is positively correlated to that of $class_2$. Hence, only the cell delay of $class_1$ needs to be considered in the fitness function. For a multi-channel application, it is far too complicated to distinguish between cell flows that are delay sensitive and cell flows which are loss sensitive. In this chapter, the main objective is to benchmark the performance of EFS with that of GPS. Hence the average cell delay and bandwidth allocation of each cell flow are included in the fitness function.

The fitness function can be described by Eqs. 1-8 as follow.

$$\beta_i = 10 \times (\frac{n_i}{BitRate_i} - 1) \quad (1)$$

$$\epsilon_i\prime = \begin{cases} \lambda \times 10^{\beta_i} \; if \; \beta_i < 0 \\ \lambda \qquad\quad else \end{cases} \quad (2)$$

$$\epsilon_i = \begin{cases} \epsilon_i\prime \; if \; \epsilon_i\prime < \epsilon_i \\ \epsilon_i \;\; else \end{cases} \quad (3)$$

$$Ratio_i = \gamma \times \left| \frac{n_i}{n_1 + n_2 + n_3} - \frac{m_i}{m_1 + m_2 + m_3} \right| \quad (4)$$

$$ReqDelay_i = \alpha_i \times \rho \times \upsilon \quad (5)$$

$$RealDelay_i = \frac{\sum_{h=1}^{m_i} Delay(h)}{m_i} \quad (6)$$

$$Delay_i = |RealDelay_i - ReqDelay_i| \quad (7)$$

$$F = \kappa - \sum{}_{i=1}^{3} \epsilon_i \times (Delay_i + Ratio_i) \quad (8)$$

The meaning of the variables in the above equations are outlined in Table 2.

If $BitRate_i$ is larger than n_i, indicating that $class_i$ is misbehaving, β_i will be negative and hence ϵ_i in Eq. 3 will be much smaller than λ. This will put less emphasis on the significance of the QoS of $class_i$ in Eq. 8. In Eq. 2, λ is a constant value. If β_i is larger than 0, $\epsilon_i\prime$ in Eq. 2 is assigned the value of λ, otherwise it is left unchanged. Eq.3 captures the smallest value of ϵ_i during the control process. This equation indicates that if one cell flow overrides its contract, its QoS is subsequently deemphasized even if it recovers. This approach elimiates the bad effects on future flow patterns by past misbehaviour. m_i in Eq.4 is the number of cells from $class_i$ transmitted during the evaluation process. $\frac{m_i}{m_1+m_2+m_3}$ indicates the real bandwidth allocation during the evaluation process by the evaluated chromosome for $class_i$. $\frac{n_i}{n_1+n_2+n_3}$ represents the desired bandwidth ratio. Thus, $\left|\frac{n_i}{n_1+n_2+n_3} - \frac{m_i}{m_1+m_2+m_3}\right|$ is the error of the bandwidth ratio allocation for $class_i$. $ReqDelay_i$ indicates the desired average delay for cells from $class_i$. $RealDelay_i$ is the average delay suffered by cells from $class_i$ during evaluation. Eq.8 describes the calculation of the

Table 2. Definition of Variables

Variable	Definition
β_i	Status of contract following of $class_i$
$BitRate_i$	Bit rate of the flow patterns stored in TB_i
n_i	Subscribed bandwidth of $class_i$
λ	A constant for the fitness weight factor
ϵ_i	Fitness weight factor for $class_i$
m_i	Number of cells transmitted from $class_i$
γ	Weight factor for bandwidth error
$Ratio_i$	Error of bandwidth ratio for $class_i$
α_i	Tuning factor of the average cell delay
ρ	Transmission time of one cell unit
υ	Size of $\text{BUF}_{\#}$
$ReqDelay_i$	Required average delay for $class_i$
$RealDelay_i$	Average delay suffered by cells from TB_i
$Delay_i$	Error of average delay for $class_i$
$\text{Delay}(h)$	Delay suffered by every cells
κ	A constant
F	Fitness value

fitness value of each evaluated chromosome. This fitness model expresses the objective of small error in bandwidth allocation ratio and small difference in average delay between EFS and the desired average delay.

6 Simulations

In this section, the computation complexity of the evolution and evaluation process is first analyzed. Based on the analysis of the computation complexity, a paremeter will be set in the following simulations to consider the time taken by the evolution and evaluation processes. EFS has a special property whereby the QoS performance can be tuned through parameteric adjustment of the fitness function. In this section, the good performance achieved in terms of bandwidth allocation and average delay is demonstrated by comparisons with GPS, MCRR and FCFS. Then, the property of tunability is demonstrated by tuning α_i. For the simulations, it is supposed that the capacity of the output channel is 155.52 MHz. For the CBR flows, the bit rate is constant and the gap between each packet is one bit time period. For VBR flow, the bit rate in every ON period is constant which is a normally distributed random variable. The gap between two adjacent ON periods is also a normally distributed random variable. The time gap between two adjacent packets is fixed as in the CBR case. All the simulations are carried out for 2 seconds of cell flows using a C++ program.

6.1 Computation Complexity Analysis

In order to assess the performance of the whole EFS for real-time application, it is useful to determine the computation cost taken by the evolution process. The faster the evolution, the better the system reacts to the changing patterns. It is assumed that multiplication and addition are basic operations requiring the same computation time. Exponential operation can be realized by means of a lookup table. Thus the time taken by an exponential operation is only the memory access time. The number of input channels considered is 3 (j=3). All the notations are the same as in Algorithms 1 and 2. The variables involved in the complexity analysis are described in Table 3.

The computation complexity of Algorithm 1 and Algorithm 2 are summarized in Table 4. For Algorithm 2, during the whole evaluation, there is maximally $m \times j$ cells arriving from $TB_{\#}$. Each arriving cell takes two operations for modifying $BUF_{\#'}$ and incrementing the cell counter $q_{\#}$ (# represents numbers between 1 and j). Thus the computation cost for the *Enqueuing* part in Algorithm 2 is at most $m \times j \times 2$ basic operations. The fuzzy inference can be realized by hardware implementation and it takes one basic operation timing cycle. Hence, the number of computations required by the *Evaluation* in Algorithm 2 is about $j \times (m+v) \times 3$. The value 3 is the number of operations within the *while* loop. Each operation is a fundamental operation. For the whole evaluation process, there are at most $j \times (m+v)$ cells to be processed. So $j \times (m+v)$ times of *while* loop will be executed. The computation time taken by the *Fitness Computation* in Algorithm 2 is about 22, since each

Table 3. Variables Description

Variable	Description
j	Number of input channels
m	Size of $TB_{\#}$
v	Size of $BUF'_{\#}$ and $BUF_{\#}$
g	Number of generations
p	Population size
c	Number of evolution cycle
l	Chromosome length in bits

Table 4. Computation Complexity of Algorithm1 & Algorithm 2

Algorithm	Operations	Computation Complexity
Algorithm 1	TB# Modification	$m \times j$
	Initialization of BUF#′	$\rho \times j$
	Evolution & Evaluation	$g \times \rho \times E$
Algorithm 2	Enqueuing	$2 \times j \times m$
	Evaluation	$3 \times j \times (m+v)$
	Fitness Computation	22

calculation in Eqs.1-8 in Section 5 takes about one operation timing cycle and there are about 22 operations in total. Thus the total number of operations for Algorithm 2 is $(m \times j \times 2) + [j \times (m + v) \times 3] + 22$ which can be represented by E.

For Algorithm 1, *$TB_{\#}$ modification*, filling out $TB_{\#}$ based on the corresponding bit rate when evolution is triggered, takes $m \times j$ operations. During *Initialization of $BUF_{\#'}$*, the copying of cell information from $BUF_{\#}$ to $BUF_{\#'}$, requires a computation cost of $j \times v$. In the *Evolution and Evaluation*, the first internal *for* loop requires a total of $p \times (l + E)$ operations in which l is for the initialization cost of every chromosome. "E" is the computation cost of Algorithm 2. This is followed by two embedded *for* loops, which requires approximately $g \times p \times (5 + E)$ operations in total. The value "5" is for the 5 instructions before the *Evaluation* operation. In Algorithm 1, the *Evaluation* subroutine is the most computationally intensive. In order to solve the bottleneck of the evolution speed, the p *Evaluation* processes can be realized through parallel hardware implementation. Hence, the *Evolution and Evaluation* routine in Algorithm 1 can be modified to Algorithm 3.

Adopting Algorithm 3, the number of operations required by Algorithm 1 is approximately $c \times [(p \times l + E) + g \times (5 + E)]$ operations. The parameters adopted in the following simulations are $m = 100$, $v = 100$, $g = 10$, $p = 10$, $j = 3$, $c = 2$ and $l = 9$. Here the generation number is fixed to 10. The bigger the number of generations, the better the solution derived. The best chromosome searched for the specified number of generations is deemed to be good enough for the scheduling control purpose.

Algorithm 3 Modified Evolution

Evolution and Evaluation:

```
for i < c do
    for i1 < p do
        Initial Population generation;
    end for;
    Call Evaluation();
    for i2 < g do
        for i3 < p do
            Toss for crossover;
            Select chromosomes for crossover;
            Do crossover operation;
            Toss for mutation;
            Do mutation operation;
        end for;
        Call Evaluation();
    end for;
end for;
```

Based on the above parameters, the total number of operations required by Algorithm 1 is approximately 50000 operations. Given a state-of-the-art microprocessor of 3000MIPS (*Million Instructions Per Second*), the evolution process requires less than 20 μs of computation time. If hardware pipeline is adopted in the implementation, the computation cost can be further reduced. In the following simulations, a computation time of 20 μs is considered.

6.2 Normal QoS Performance

Scenario$_1$

In this scenario, $class_1$ is CBR with a reserved bit rate of 65.52 MHz. $class_2$ and $class_3$ are also CBRs with reserved bit rate of 50 MHz and 40 MHz respectively. Assume that for unknown reasons, $class_1$'s actual bit rate is 100 MHz while that of $class_2$ and $class_3$ adhere to their contracts. For this case, the parameters in the fitness function can be set as follows, in which $\alpha_{\#}$ can be calculated by equation $\frac{t_i}{\rho \times \upsilon}$. Here, t_i refers to the possible transmission delay suffered by cells from $class_i$ in the GPS system based on its reserved bit rate. The parameters setting for the simulation is $\alpha_1 = 0.0237$, $\alpha_2 = 0.0312$, $\alpha_3 = 0.0387$, $\lambda = 1000$, $\gamma = 1000$, $\rho = 2.73$, $\upsilon = 100$ and $\kappa = 10^8$. λ should be a value which is much larger than 1. According to Eq.4 and 8, $\gamma = 1000$ emphasize the significance of the error of bandwidth allocation ratio to be 1000 times that of the delay error. The physical meaning of λ is to emphasize the QoS of contract following and de-emphasize the QoS of contract breaking flows. If this value is too big, the average cell delay will have much less effect on the final fitness value. If it is very small, the value of the second part in Eq.8 for contract breaking flows will have similiar effect on the final fitness value as the contract following flows. The average delay for the four schemes after simulation is presented in Table 5(a).

In the table, it can be seen that FCFS dealt with all the flows equally, thus the three flows show similar average delay. MCRR achieved much bigger delays for the contract following flows than that achieved by GPS. EFS achieved good average delay for the input flows of $class_2$ and $class_3$, which observe their contracts. In this scenario, the delay of $class_2$ and $class_3$ achieved by EFS is very close to that of GPS.

The bandwidth utilization ratio for this scenario is as shown in Table 6(a). In this scenario, the bandwidth ratio is sampled for every 50 μS. The means and standard deviations for the three input flows for the various schemes are recorded. Based on this table, MCRR's performance is closer to that of GPS than EFS and FCFS in terms of mean and standard deviation. But the difference in performance of these four schemes is very slight. By considering both the QoS of delay and the bandwidth allocation, EFS is much better than MCRR and FCFS.

Scenario$_2$

In this scenario, $class_1$ is VBR with the reserved bit rate of 65.52 MHz while $class_2$ and $class_3$ are CBRs with the reserved bit rates of 50 MHz and 40 MHz respectively. Assume that for some unknown reasons, $class_1$'s actual bit rate varies between 55.52 MHz and 105.52 MHz randomly during its ON period and $class_2$ and $class_3$ adhere to their contracts well. For this scenario, the parameters setting can be the same as that of $scenario_1$. From simulation, the average delay for the four schemes, is presented in Table 5(b). From the table, it can be seen that FCFS achieves similar average delays for all the three cell flows. MCRR achieved better delay performance for the contract following flows, but the delay for $class_2$ and $class_3$ are still not good. On the other hand, EFS achieved good average delay for the input flows that adhere to their contracts. In the table, the delay of $class_2$ and $class_3$ achieved by EFS is very close to that of GPS.

The corresponding bandwidth utilization ratio for this scenario is as shown in Table 6(b). In this scenario, $class_1$ is a bursty flow. The congestion may only occur during the ON periods of $class_1$, the bandwidth utilization ratio is averaged over each bursty period of $class_1$. Based on the mean and standard deviation values in Table 6(b), EFS performs better than MCRR and FCFS. The mean values for the bandwidth allocation ratio of the three flows achieved by EFS are closer to that of GPS than using MCRR or FCFS. Furthermore, the bandwidth allocation deviations by EFS are also much smaller than that of MCRR and FCFS.

Table 5. Average Delay for Scenario$_1$, Scenario$_2$ and Scenario$_3$

(a) SCENARIO$_1$

	GPS(μs)	EFS(μs)	MCRR(μs)	FCFS(μs)
$class_1$	650.2	647.9	620.5	647.4
$class_2$	8.5	11.9	155.9	648.7
$class_3$	10.6	15.5	86.5	650.3

(b) SCENARIO$_2$

	GPS(μs)	EFS(μs)	MCRR(μs)	FCFS(μs)
$class_1$	558.8	555.2	557.2	589.2
$class_2$	8.5	10.1	59.3	588.9
$class_3$	10.6	13.0	39.0	589.5

(c) SCENARIO$_3$

	GPS(μs)	EFS(μs)	MCRR(μs)	FCFS(μs)
$class_1$	1421.6	147.2	1767.8	63.0
$class_2$	4.9	6.2	68.0	82.9
$class_3$	10.6	12.2	859.9	63.6

Table 6. Bandwidth utilization ratio in $scenario_1$, $scenario_2$ and $scenario_3$

(a) $SCENARIO_1$

	GPS		EFS		MCRR		FCFS	
	mean	dev	mean	dev	mean	dev	mean	dev
$class_1$	0.4214	$1.15{\times}e^{-4}$	0.4205	0.0012	0.4208	$3.66{\times}e^{-4}$	0.4218	0.0042
$class_2$	0.3215	$8.56{\times}e^{-5}$	0.3219	$7.41{\times}e^{-4}$	0.3217	$2.17{\times}e^{-4}$	0.3213	0.0023
$class_3$	0.2572	$5.46{\times}e^{-5}$	0.2575	$4.61{\times}e^{-4}$	0.2575	$1.65{\times}e^{-4}$	0.2570	0.0019

(b) $SCENARIO_2$

	GPS		EFS		MCRR		FCFS	
	mean	dev	mean	dev	mean	dev	mean	dev
$class_1$	0.4208	0.0036	0.4200	0.0112	0.4266	0.0914	0.4219	0.0150
$class_2$	0.3219	0.0021	0.3221	0.0053	0.3183	0.0501	0.3212	0.0084
$class_3$	0.2574	0.0071	0.2579	0.0076	0.2551	0.0455	0.2569	0.0067

(c) $SCENARIO_3$

	GPS		EFS		MCRR		FCFS	
	mean	dev	mean	dev	mean	dev	mean	dev
$class_1$	0.2160	0.0134	0.2890	0.0447	0.1209	0.1501	0.3150	0.0174
$class_2$	0.4961	0.0312	0.4515	0.0490	0.6976	0.2497	0.4331	0.0311
$class_3$	0.2879	0.0178	0.2595	0.0082	0.1816	0.1456	0.2520	0.0138

Scenario$_3$

In this scenario, $class_1$ is CBR with reserved bit rate of 30 MHz while $class_3$ is CBR with reserved bit rate of 40 MHz. $class_2$ is VBR with reserved bit rate of 85.52 MHz, but its actual bit rate may varies between 55.52 MHz and 85.52 MHz randomly during its ON period. In this scenario, $class_2$ and $class_3$ follow their contracts well but the actual bit rate of $class_1$ is 50 MHz. The related parameters in the fitness function can be set as follows according to the GPS model: $\alpha_1 = 0.0519$, $\alpha_2 = 0.0183$ and $\alpha_3 = 0.0387$. The average delays for the four schemes based on simulations are presented in Table 5(c). From the table, it can be seen that FCFS achieved average delays within a tight range for the three flows as in $scenario_1$ and $scenario_2$. MCRR achieved poor delay performance for the three flows. EFS achieved good average delay in this scenario. The delay of $class_2$ and $class_3$ in this scenario achieved by EFS is very close to that of GPS. The delay for $class_1$ achieved by EFS is quite different from that achieved by GPS. This is mainly due to the fact that EFS is a bandwidth conserving scheduling system, which means that there is always cell unit being transmitted as long as there are cell units waitting in the $BUF_{\#}$. It attempts to achieve full bandwidth utilization. $class_2$ in this scenario sometimes transmitted below its subscribed bandwidth, the residual bandwidth in this case goes to $class_1$.

Table 7. The tuned average delay for $scenario_1$

	GPS(μs)	EFS(μs)	MCRR(μs)	FCFS(μs)
$class_1$	650.2	645.2	620.5	647.4
$class_2$	8.5	50.0	155.9	648.7
$class_3$	10.6	14.6	86.5	650.3

The bandwidth utilization ratio for this scenario is as shown in Table 6(c). For this scenario, $class_2$ is a bursty flow. The congestion may only occur during the ON period of $class_2$, the bandwidth utilization ratio is averaged over each bursty period of $class_2$. MCRR shows a very good overall bandwidth utilization ratios of 0.1931, 0.5494 and 0.2575. They are close to the corresponding reserved bandwidth ratios of $\frac{30}{155.52} = 0.1929$, $\frac{85.52}{155.52} = 0.5499$ and $\frac{40}{155.52} = 0.2572$. However, its bandwidth utilization ratios during the bursty periods are poor. As shown in the table, its performance is far from that of GPS while EFS achieved good performance, very close to that of GPS.

6.3 QoS Tunability

A special property of EFS is its QoS tunability. Suppose the average delay of $class_2$ in $scenario_1$ needs to be 60 μs. The value of α_2 can be calculated by $\frac{60}{\rho \times v} = \frac{60}{2.73 \times 100} = 0.2198$. The values of α_1 and α_3 are the same as that in the previous simulation of $scenario_1$. The average delay of EFS for all the flows with $\alpha_2 = 0.2198$ is listed in Table 7. Comparing Table 7 with Table 5, it can be deduced that after tuning the value of α_2, only the average delay of $class_2$ for EFS shows a significant change. The average delay of $class_2$ increases significantly. The achieved average delay of $class_2$ is less than the desired value of 60 μs. This can be attributed to the EFS being an open loop system rather than a closed-loop feedback system. If the average delay needs to be equal to the desired value, an adaptive feedback mechanism can be adopted to tune the value of α_2 automatically. This property of *tunability* is very useful in the case where one or more good performing flows need bigger queuing delay. By tuning the parameters of $\alpha_{\#}$, these requirements can be satisfied without affecting the other good performing flows.

7 Conclusions and future directions

In this chapter, we have presented EFS as a new traffic control scheme for packet-switching networks. EFS is presented in this chapter to demonstrate its capability to address the more general cell-scheduling problem. It is based on evolutionary algorithm and fuzzy control, thus it can be regarded as a third approach which is independent of the GPS approach and RR approach. After the analysis of the system computation complexity, EFS was simulated on three different input scenarios. In the first scenario, all the three schemes had similar performance as that of GPS in terms of bandwidth allocation. However, when the cell delay QoS was taken into consideration, the performance of EFS was much better than the other two. In $scenario_2$ and $scenario_3$, EFS showed very good average delay for the good performing flows while MCRR and FCFS achieved much bigger average delay. The bandwidth allocation ratio achieved by EFS was close to that of GPS in terms of mean and deviation. These simulations demonstrated that without adopting the clock schemes as in [11, 13] and so on, the QoS performance for cell-scheduling very close to that of GPS can also be achieved by EFS. Besides effective management of delay and bandwidth allocation, another desirable property of EFS is the QoS tunability. This was also demonstrated through simulations on $scenario_1$. If different average delay is desired for the good performing flows, the corresponding parameter in the fitness function can be tuned accordingly. In general, larger average delay can be achieved by increasing the corresponding parametric value.

Based on our work reported in this chapter, EFS can stand on its own as a useful technique to handle dynamic switching applications. EFS demonstrated strong performance based on simulations, with potential deployment in real-time applications. Further study on its performance in network switching application needs to be carried out. In particular, the evolution time can be further improved. For the purpose of reducing the computation time and advancing the system's performance, further study on the hardware architecture and hardware implementation of the evaluation process, a crucial part of the EFS, is required. And further, EFS can be employed to address the even more general scheduling problem such as the packet-switching problem in which the packet has variable length. Another promising application domain is the multi-protocol label switching system where similiar scheduling philosophy can also be employed.

References

1. R. Zhang, Y.A. Phillis, and J. Man, "A Fuzzy Approach to the Balance of Drop and Delay Priorities in Differentiated Services Networks", *IEEE Transactions on Fuzzy Systems*, vol. 11, no. 6, December 2003.
2. E.P. Rathgeb, "Modeling and Performance Comparison of Policing Mechanisms for ATM Networks", *IEEE J. Select. Areas Commun.*, vol. 9, no. 3, April 1991.
3. L. Zhang, "Virtual clock: A new traffic control algorithm for packet switched network", *ACM Trans. Comput. Syst.* vol. 9. no. 2. pp. 101-124, May 1991.
4. T. Lizambri, F. Duran and S. Wakid, "Priority Scheduling and Buffer Management for ATM Traffic Shaping", in *Proc. of 7th IEEE Workshop on Future Trends of Distributed Computing Systems, FTDCS'99*, pp. 36-43, Dec. 20-22, 1999, Cape Town, South Africa.

5. B. Maglaris, D. Anastassiou, P. Sen, G. Karlsson and J.D. Robbins, "Performance models of statistical multiplexing in packet video communications", *IEEE Trans. Commun.*, vol. 36, no. 7, pp. 834-844, July 1988.
6. R. Guerin, H. Ahmadi, M. Naghshineh, "Equivalent Capacity and Its Application to Bandwidth Allocation in High-Speed Networks", *IEEE J. Select. Areas Commun.*, vol. 9, no. 7, pp. 968-981, Sept. 1991.
7. W.X. Liu, M. Murakawa and T. Higuchi, "ATM Cell Scheduling by Function Level Evolvable Hardware", *Evolvable Systems: From Biology to Hardware, First International Conference, ICES 1996 (LNCS 1259)*: pp. 180-192.
8. A.K. Parekh, R.G. Gallager, "A Generalized Processor Sharing Approach to Flow Control in Integrated Services Networks: The Single-Node Case", *IEEE/ACM Trans. Networking.*, vol. 1, no. 3, pp. 344-357, June 1993.
9. A. Demers, S. Keshav, and S. Shenker, "Analysis and simulation of a fair queuing algorithm", *Journal of Internetworking Research and Experience*, pp. 3-26, Oct. 1990.
10. J.C.R. Bennett and H. Zhang, "WF2Q: Worst-case fair weighted fair queuing", in *Proc. IEEE INFOCOM, 1996*, pp. 120-128.
11. L. Zhang, "VirtualClock: A new traffic control algorithm for packet switching networks", in *Proc. of the ACM Symposium on Communications Architectures & Protocols*, pp. 19-29, September 24-27, 1990, Philadelpia, PA, USA.
12. P. Goyal, H.M. Vin, and H. Cheng, "Start-Time Fair Queuing: A Scheduling Algorithm for Integrated Services Packet Switching Networks", *IEEE/ACM Trans. Networking*, vol. 5, no. 5, pp. 690-704, Oct. 1997.
13. J. Davin and A. Heybey, "A simulation study of fair queuing and policy enforcement", *Computer commun. Rev.*, vol. 20, no. 5, pp. 23-29, Oct. 1990.
14. M. Shreedhar and G. Varghese, "Efficient Fair Queuing Using Deficit Round-Robin", *IEEE/ACM Trans. on Networking*, vol. 4, no. 3, pp. 375-385, June 1996.
15. H.M. Chaskar, and U. Madhow, "Fair Scheduling With Tunable Latency: A Round-Robin Approach", *IEEE/ACM Trans. Networking*, vol. 11, no. 4, pp. 592-601, Aug. 2003.
16. Y. Liang, "A Simple and Effective Scheduling Mechanism Using Minimized Cycle Round Robin", in *Proc. of IEEE International Conference on Communications*, vol. 4, pp. 2384-2388, New York, NY, April, 2002.
17. J.H. Li and M.H. Lim, "Evolvable fuzzy system for ATM cell scheduling", in *Proc. of 5th Int. Conf. Evolvable Syst.: From Biology to Hardware, ICES 2003 (LNCS 2606)*, pp. 208-217, Springer-Verlag, 2003.
18. W.X. Liu, M. Murakawa and T. Higuchi, "Evolvable Hardware for On-line Adaptive Traffic Control in ATM Networks", in *Proc. of the Second Annual Conference on Genetic Programming, 1997*, pp. 504-509, July 13-16, 1997, Stanford, California.
19. J.H. Li, M.H. Lim and Q. Cao, "An Intrinsic Evolvable and Online Adaptive Evolvable Fuzzy Hardware Scheme for Packet Switching Network", in *Proc. NASA/DoD Conference on Evolvable Hardware, 2004*, pp. 109-112, June 24-26, 2004, Seattle, Washington, USA.
20. J.H. Li, M.H. Lim and Q. Cao, "Evolvable Fuzzy Hardware for Real-Time Embedded Control for Packet-Switching", *Evolvable Machines: Theory & Practics*, N. Nedjah, L. de Macedo Mourelle (Eds), vol. 161, pp. 205-227, Springer-Verlag, Heidelberg, 2004.

21. X. Yao and T. Higuchi. "Promises and Challenges of Evolvable Hardware", *IEEE Trans. on Systems, Man and Cybernetics, Part C, Applications and Reviews*, vol. 29, no. 1, pp. 87-97, Feb. 1999.
22. M.H. Lim, Q. Cao, J.H. Li and W.L. Ng, "Evolvable Hardware Using Context Switchable Fuzzy Inference Processor", *IEE Proc. - Comput. Digit. Tech.*, vol. 151, no. 4, pp. 301-311, July 2004.

A Multi-Objective Evolutionary Algorithm for Channel Routing Problems

Chi Keong Goh, Wei Ling Lim, Yong Han Chew and Kay Chen Tan

Department of Electrical and Computer Engineering, National University of Singapore, 4, Engineering Drive 3, Singapore 117576

Summary. The channel routing problem (CRP) is derived from detailed routing model in VLSI design. The objectives of the problem can vary from reducing the number of horizontal tracks to minimizing the number of vias, length of wires used etc. It is not known how these objectives interact with one another, although it is believed that they are conflicting in nature. Unlike traditional single-objective optimization approaches, this paper presents a multiobjective evolutionary algorithm (MOEA) for CRP. Specialized genetic operators for solving the CRP are devised. In addition, a new method of random routing is introduced for better routing performance. Some standard benchmark problems are solved in this paper using the proposed algorithm to validate its performance. It is shown that the proposed algorithm is consistent and is able to obtain very competitive results as compared to well-known approaches.

1 Introduction

In the physical design process for VLSI circuits, the logical structure of the circuit is transformed into its physical layout through the processes of partitioning, placement, routing and finally compaction. This research focuses on the task of detailed routing, specifically the channel routing problem (CRP) which connects pins of signal nets in a rectangular region called a channel, in accordance to certain routing constraints and objectives. The result of this detailed routing has a strong influence on the fabrication yield and production costs of the given circuit.

It is not known to what extent the various objectives of the CRP are conflicting in nature, although it is highly likely that it is not possible to find a

C. Keong Goh et al.: *A Multi-Objective Evolutionary Algorithm for Channel Routing Problems*, Studies in Computational Intelligence (SCI) **49**, 405–436 (2007)

single optimal solution for most such problems. Despite this, past research [1, 2, 4, 7, 19, 30] converted the objective vector into a scalar function, thus optimizing the problem as though it is having single objective. Unfortunately, such conversions involve the determination of weighting coefficients for each of the objectives and this poses impracticalities due to difficulties in ascertaining such parameters.

Evolutionary algorithms (EAs) are stochastic search methods that simulate the process of evolution. They incorporate the concepts of reproduction, mutation, and crossover of chromosomes and the Darwinian principle of "survival of the fittest". As channel routing problems belong to the class of NP-hard problems [29], it is not possible to solve it in polynomial time. The stochastic nature and ability to handle large complex problems of EAs make them well suited for combinatorial optimization problems like the CRP. However, it is likely that the objectives of the CRP are conflicting in nature and should be best tackled by means of multiobjective (MO) optimization.

In addition, many existing routing algorithms [6, 21, 26] allow violations in the routing process and results in infeasible solutions. Such solutions would require further difficult manual routing to eradicate the infeasibilities and so defeat the whole purpose of designing an algorithm to solve the CRP automatically. On the other hand, those methods that do not permit violations [9, 20] may be too random in nature and often conduct the search in limited areas of the feasible region that yields poor results. As such, more computational resources had to be spent to obtain optimal solutions by using larger population and generation sizes. In some cases, the algorithms remain locked in the non-optimal region of solution space.

In contrast to conventional approaches, a multiobjective evolutionary algorithm (MOEA) [31] is proposed and applied to find the possible tradeoff solutions of the CRP. In order to deal with problem effectively, different operators that exploit problem knowledge such as channel update and problem-specific mutation operators are developed. Furthermore, a novel method of random routing is designed to utilize the effort of searching efficiently without compromising on the diversity and quality of the solutions. Analysis on the statistical performance of the proposed method is performed and the results obtained are compared to the best solutions found in literature according to the authors' best knowledge.

This paper is divided into 5 sections. Section 2 gives an overview of channel routing problem while Section 3 illustrates the proposed MOEA. The application and evaluation of the proposed method using benchmark problems and relevant quantitative performance metrics are given in Section 4. It also presents a comparative study of the proposed algorithm to other published results. Conclusions are drawn in Section 5.

2 Channel Routing Problems

In the physical design process of Very Large Scale Integration (VLSI) circuits, the logical structure of the circuit is transformed into its physical layout. Detailed routing is one of the tasks in this process performed at the final stage. In channel routing problem, a detailed router connects pins of signal nets in a rectangular region, called a channel which is shown in Fig. 1. The pins are located exclusively on the upper and lower boundaries, along the length of the channel. Pins belonging to the same net are connected together within the channel region.

The process of routing is highly constrained and there exist numerous routing constraints, such as number of layers, minimal space between wires, minimum channel width and minimum wire width [19]. The quality of this routing process has a strong influence on the performance and production costs of the circuit. With such complexity, the channel routing problem is NP-complete [29] and there is no known deterministic algorithm that can solve them in polynomial time. Some detailed routing models are presented in Section 2.1 while Section 2.2 gives a very brief note on common objective functions and constraints imposed in channel routing problems. The discussion continues with a review on evolutionary approaches to CRP in Section 2.3.

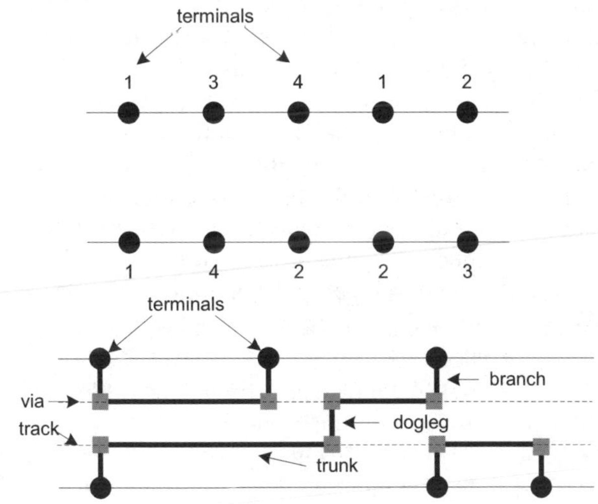

Fig. 1. A channel

2.1 Detail Routing Models

Many different ways of modeling the channel routing problem exist in literature [15]. The models used serve different purposes and are designed to resolve different issues that arise in the process of routing. Five models are discussed here. The first three are major graph-based detail-routing models, which are popular with most researchers. The fourth model, the gridless model, is a generalization of the previous three models. The fifth model, the multi-layer model, is an extension of models that allow only two layers to multiple layers.

In planar routing [15] different routes are required to be vertex-disjoint and it is often called the "river routing" model, because rivers cannot cross without their waters merging. This model is very restrictive, as many instances of detailed routing problems cannot be routed without crossing wires. This model is often used for routing chip inputs and outputs to the pads on chip boundary, or routing wires on single layer in routing schemes where the layer assignment is determined by technological constraints, such as power supply routing.

A Knock-Knee routing [15] is more general than planar routing as it allows wires to cross. Fig. 2 shows a knock-knee where two nets bend at a grid-vertex. Thus, at least two layers and sometimes even four layers are needed. This feature is not allowed in the Manhattan model. Unfortunately, the assignment of layers is found to be non-trivial and NP-hard. Furthermore, technological constraints like preferred wiring directions for different wiring layers do not fit well with this model.

In a Manhattan model, the detail-routing graph is assumed to be a partial grid graph and different routes are required to be edge-disjoint. There exist two variants in Manhattan routing: Restrictive routing and Unrestricted Overlap routing [15]. In the restrictive routing model, two layers are used typically and all vertical wire segments are routed on one layer, while horizontal wire segments are routed on the other layer. For the unrestrictive overlap model, both vertical and horizontal segments are routed on both layers, as long as the routes remain edge-disjoint. However, long stretches of overlapping wires in unrestricted overlap may introduce crosstalk problems in chip technologies. In both models, contact cuts (vias) can be placed at grid vertices to join wire segments on both layers. The model with unrestricted overlap is chosen to be used in this work.

The preceding three routing models are based on planar graphs and as a result of that, they can only model real technological constraints imprecisely. The gridless routing model is an attempt to move away from the detailed-routing graph and use the continuous plane as the basic routing domain.

It incorporates design rules of the fabrication process, so as to model real geometrics more precisely and achieve tighter optimization [15].

In PCB wiring, there are typically many more than two routing layers. The multilayer routing model [15] addresses this by representing each wire segment explicitly. As the determination of layer assignment for wire segments is done at detailed routing, the grid-based routing models are ideal for multilayer routing.

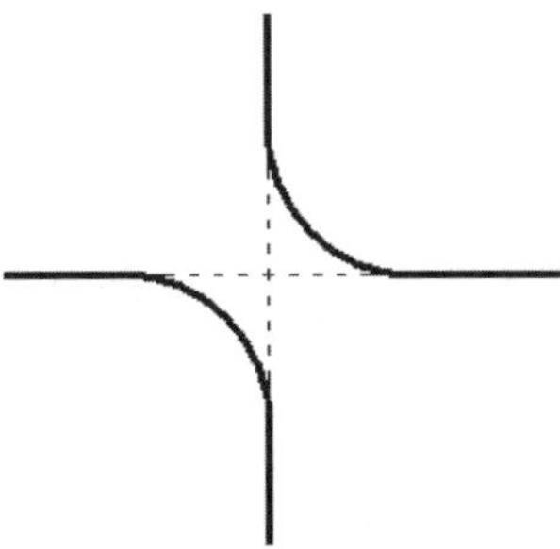

Fig. 2. A Knock-Knee in routing

2.2 Properties of Manhattan Routing

As the focus of this research is about solving channel routing in the Manhattan model, this section is dedicated to present relevant concepts in this model. Lower bounds on channel width, as well as vertical and horizontal constraints in the Manhattan routing model had been studied in the literature. However, its mathematical structure has not been known very well. The following are some concepts employed by others in the optimization of the channel routing problem.

Lower bounds and routability

The lower bound (Close density) of channel width for a restrictive Manhattan model is defined by [15] as follows. Let I be an instance of the channel routing problem. The closed density,

$$d_c(I) := \max_x |N_x| \tag{1}$$

where x is an arbitrary real number and N_x is the set of (nontrivial) nets that have one terminal on a column $\leq x$ and another in a column $\geq x$. From there, it can be deduced that in a Manhattan model with unrestricted overlap, the closed density can be defined as follows [9].

$$\tilde{d}_c(I) := \max_x |N_x| / 2 \tag{2}$$

The routability [8] of restricted CRP is determined by the Vertical Constraint Graph (VCG) [13]. A VCG is simply a directed graph with its nodes representing the nets in the channel and its branches representing the relative position of the horizontal parts of the net from the top to the bottom of the channel. It is formed by scanning each column in the channel and adding an edge pointing from the node representing the net that has a pin on the top of the channel, to the node representing a net that has a pin at the bottom of the channel.

The minimum number of rows in a restricted CRP is determined by the Horizontal Constraint Graph (HCG). Like the VCG, the HCG is a directed graph with nodes that represents the nets in the channel. However, its branches represent overlapping nets. The HCG is constructed by checking the rows and adding an edge between the nodes with overlapping intervals. The interpretation of the HCG is that the vertical segment of the node pointing to a second node should be routed on the right hand side of the vertical part of the net being pointed to. The concepts of the vertical and horizontal graphs had been used extensively in work found in literature [4, 16, 26, 32] to solve the channel routing problem.

Constraints and objectives

As mentioned in Section 2.1, routing of pins is highly constrained and it is performed in accordance to certain constraints and quality factors, so as to obtain good feasible routing solutions. Four constraints for interconnections were defined in [20]. A net is to be routed using Manhattan geometry with unrestricted overlap. Two layers are available for routing. A net may change from one layer to another using a contact window called a 'via'. Different nets cannot cross each other on the same layer and must respect a minimum distance rule.

The list below includes four characteristics that are used frequently to measure quality of channel routing in previous research.

- 100% routing: Manual routing of a few unrouted nets is time consuming and often involves the rerouting of most, if not all of the routed nets. Hence, it is important that the router completes routing the channel.
- Minimum area: The channel length is fixed, but the channel width (number of rows of the channel) is allowed to vary during the process of routing. It is desirable that the routed channel occupies the minimum area possible.

- Netlength: The shorter the length of the inter-connection wires, the smaller the propagation delay and smaller the routing area. The signal quality also improves.
- Number of vias: The introduction of via between 2 interconnection layers may result in more routing area, longer propagation delay and lower fabrication yield. The fewer the number of vias, the better the routing quality.

2.3 Evolutionary Algorithms for CRP

The use of evolutionary algorithms, particularly genetic algorithms, to solve CRP is found in [19], [26], and several other publications. As the available works are many, it is impossible to discuss each of them. Instead, an overview on the various methods of implementation used by these genetic algorithms is presented in this section.

Representation

Davidenko *et al.* [4] used a novel approach to reduce the channel with unrestricted overlaps to a binary string representation. In their work, only nets with horizontal constraints are represented in the chromosome and they are arranged according to the top and bottom terminals on the channel. A bit '1' for pin *i* with respect to *j* implied that track for pin *i* should appear above that of pin j in the channel.

Other structures used are vectors and matrices. Lienig and Thulasiraman [20], represented the chromosome using a three-dimensional lattice structure. Finally, the work by Lin *et al.* [21] represented the chromosome as a four-dimensional matrix that stored information of the net, as well as the layer, row and column occupied by the net.

Initial Population

The initial populations for the different evolutionary algorithm are created through random routing techniques. In some cases [6, 21, 26], violations are allowed in the initial population and the individuals with infeasibilities are penalized in their fitness.

In HGA [9], a pin is chosen from the set of all pins of the channel one at a time and is connected to pins of the same net. There are five modes of connection, namely left, right, forward, backward, and layer change. The choice of the mode chosen is dependent on probability, with the highest probability assigned to the mode that will give the shortest distance.

In GAP [20], a completely randomized routing technique is proposed. A set $S = \{s_1, \ldots s_i, \ldots s_k\}$ of all pins of the channel that are not connected and a set $T = \{t_u, \ldots t_j, \ldots t_v\}$ of pins having at least one connection to another pin are kept. The random routing of (s_i, t_j) is done by extending a vertical line from both s_i and t_j until they reached obstacles. A position between the start and end points of each of the lines is chosen and from this position, horizontal lines are extended in a similar way. Each of the channel layers have a preferred direction and each extension line has a $^2/_3$ chance of being routed on the layer associated with the extension line. Additional rows are inserted randomly if a connection is not found within i iterations. The random routing algorithm continued in this manner until either a connection is found or the number of rows inserted exceeded the maximum threshold. If the maximum threshold is exceeded, the entire channel is destroyed.

Fitness Evaluation and Selection

To our knowledge, all existing works used a weighted sum of evaluation criteria as the fitness function. The criteria used are specific to the implementations and includes penalty, constraints and quality factors. Typical criteria used are the number of violations, vias, tracks, netlength, vertical constraints, horizontal constraints and other cost factors. Most works are not clear on how selection was performed. In [20], roulette wheel selection is used, where each individual $p_i \in P_c$ is selected with a probability $F(p_i) / \sum_{p_i \in P_c} F(p)$. Göckel *et al.* [9] also used the roulette wheel selection.

Genetic Operators

Most literature found use either a 1-point or 2-point cut along the channel length as the crossover operator. In most cases, two parents give off two offsprings and each of the offspring inherits different portions of the cut channel from both parents.

Many different types of mutation have been proposed in literature. For typical binary representation, normal binary mutation is used [4]. Besides that, mutation could involve individuals being created [6], vertical and horizontal violations solved and most algorithms [9, 21] implemented a 'rip up and re-route' mutation operator.

In [26], an "Inter-Cluster Mutation" (ICM) is proposed. In the algorithm, a branch and bound method is used to partition the VCG into subproblems. The use of the ICM is to avoid disruption of optimal partition information gained by the algorithm over the previous generations. If a normal mutation operation is used, such optimal information can not be retained, as the operator will randomly cause the destruction of such information. The ICM consists of five different functions and they are: MoveNet, ExchangeNets, MergeCluster, BreakCluster and Vertical Swap. In [20], four different types of mutation operators are used to effectively explore the solution space of the highly constrained CRP.

Parallel Genetic Algorithms

Various works [17, 18, 24, 25, 27] have also attempted to use Parallel GA to solve the CRP. There are two main approaches, namely the Island model [18] and the Master-Slave model [24].

In the Island-model, instead of using parallel GA for the sole purpose of speeding up the algorithm, the theory of punctuated equilibria is employed to achieve better results. Using this theory, independent subpopulations of individuals have their own fitness functions and they evolved in isolation. There is an exchange of individuals, called migration, when a state of equilibrium has been reached throughout all the subpopulations. Nine processors with 50 individuals each are used, thus creating a total population of 450 individuals. The processors are connected by an interconnected network with torus topology and each processor had exactly four neighbors.

In the Master-slave model, both features of global and local selection are used. The root processor executed the conventional GA with global selection on total population and the remaining processors performed the conventional GA with local selection on subpopulations. After twenty-four generations, the total population on the root processor and the subpopulations on the remaining populations are interchanged and the process is executed for four generations. This algorithm is designed to achieve a faster runtime and to obtain good global optimal solution, with full utilization of the parallel system.

3 MOEA for Channel Routing

In this section, the proposed MOEA for solving channel routing problem is presented. Section 3.1 gives the actual Manhattan model of channel routing problem to be solved. Section 3.2 summarizes relevant principles for multiobjective optimization using evolutionary algorithms. All details involved in the proposed algorithm, such as genetic operators, representation and fitness evaluation are included.

3.1 Problem Definition

A problem definition of the Manhattan model is presented as below:

- The two-layer Manhattan model with unrestricted overlap was chosen for the following reasons:
- This two-layer model can be easily altered to allow for multiple layers.
- This model does not pose restrictions that result in impracticalities, unlike the Knock-Knee and Planar model.
- This model has fewer complexities than the gridless model.
- There already exist many research results to compare performance of benchmark test cases on this model.

This model is able to solve problems with multiple cyclical constraints, unlike the restrictive Manhattan model. Furthermore, the following assumptions were made in the problem definition:

- All routing is to be done within the channel.
- Doglegs are allowed.
- Multiple objectives to be optimized are conflicting in nature.

3.2 A Multiobjective Evolutionary Optimization Algorithm for CRP

Multi-objective Evolutionary Algorithm has the capability of searching for a set of Pareto optimal solutions for MO optimizations. The MOEA maintains a population of individuals and each of which is assigned a fitness value for different objectives to be satisfied. These solutions undergo a simulated evolution process of selection, crossover and mutation. In the selection process, a mating population is formed when individuals are selected for reproduction. The selected individuals are then manipulated by genetic operators to produce offspring.

Variable Length Chromosome Representation

A $m \times n$ vector (Fig. 3) is used as the chromosome representation, where m is the number of terminals and n is the number of rows. Each element of the vector contains information such as the integer value of pins occupying the two different layers and a binary value specifying the presence of a via. Since information such as the desired number and location of the terminals are not known *a prior*, the variable length chromosome implemented here allows for dynamic adaptation of routing design as opposed to a fixed matrix structure.

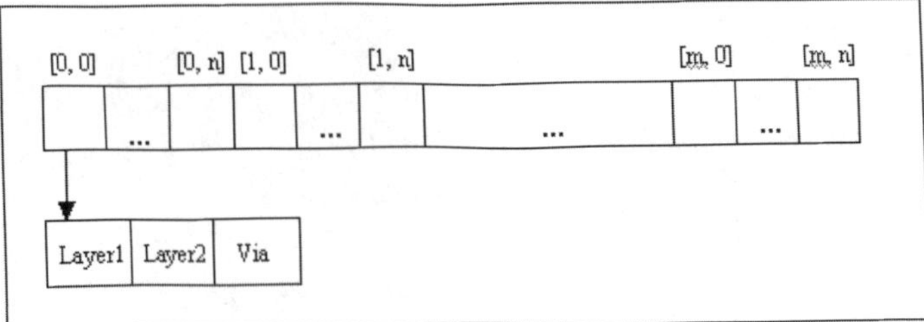

Fig. 3. Chromosome representation

Random Routing

Through an implementation and testing of the random routing algorithm proposed in [20], it is found that the method tends to result in channels with a large number of vias and thus the solution space is 'expensive' in terms of the number of vias. As such, a new method is proposed to avoid such problems, as well as to maintain a balance between different objectives within the search space. Two functions in the algorithm required the use of random routing algorithm and they are the initial population router and the re-router, which is used when the channel went through genetic operations.

Initial Router: The purpose of the initial router is to randomly route the given channel, so as to create a population of solutions that is diverse, while complying all the constraints present in CRP. Violations are not allowed in solutions produced by this routing process.

The channel is first assigned a fixed starting number of tracks, τ with $d_c(I) \leq \tau \leq 2 * d_c(I)$, where $d_c(I)$ is defined in Section 2.2.1. The algorithm then initialized an empty channel with τ number of tracks. Next, let $\boldsymbol{U} = \{u_1, u_2, \ldots, u_n\}$ be the set of all terminal pins of the channel that is not routed yet and $\boldsymbol{R} = \{r_1, r_2, \ldots, r_m\}$ be the set of terminal pins of the channel that are routed. Initially $\boldsymbol{R} = \phi$.

Choose a pin, u_i, of net N_j randomly from set $\boldsymbol{U}$. This pin is set as the *current* pin. Search through set $\boldsymbol{R}$ to look for routed pins of N_j. If one or more such pins are found, randomly choose one. Otherwise, search set $\boldsymbol{U}$ to obtain a pin of net N_j through a random selection. Once a pin is found, it is set as the *target* pin.

Next, using the relative x (horizontal) and y (vertical) positions of the *current* pin and the *target* pin, determine the *Preferred Horizontal* and *Preferred Vertical* directions from the following functions. Using the *Pre-*

ferred Horizontal and *Preferred Vertical* direction, one of the following routing states is determined.

$$Preferred\ \ Vertical\ \ direction = \begin{cases} -1 & if\ \ target_y - current_y < 0 \\ 1 & if\ \ target_y - current_y > 0 \\ 0 & otherwise \end{cases} \tag{3}$$

$$Preferred\ \ Horizontal\ \ direction = \begin{cases} -1 & if\ \ target_x - current_x < 0 \\ 1 & if\ \ target_x - current_x > 0 \\ 0 & otherwise \end{cases} \tag{4}$$

Next, both the *current* and *target* pins are routed in accordance to the routing state chosen. Each routing state dictated the preferred horizontal and vertical routing motion. For instance, if state 8 is chosen, the *current* pin appeared along the top terminal, while the *target* pin along the bottom terminal, on the right of *current* pin. A random choice of either moving down or right is made and an extension line is drawn along the chosen direction until an obstacle is met or the edge of the channel is reached. If either the preferred horizontal or vertical direction is '0', a random choice of moving left-right or up-down is made in the respective cases. If both preferred directions are '0', the pin is routed to the target position. Each time an obstacle is met in the current routing state, an attempt is made to move the extension line to the next layer. If it failed, a new routing state is assigned based on the new position of the current pin after extension.

During the routing process, the algorithm constantly checked if the current routing extension is connected to routed connections of the same net. If it is so, routing for the current pin is stopped, instead of forcing it to be routed to the target. This is done to improve the quality of solutions and to save computational resources spent on performing the process of random routing. A new pin is then randomly selected from set ***U*** and the entire process of random routing is repeated until all the pins are routed. After the channel has been successfully routed, an update is done on the channel to remove redundancies, like redundant tracks or vias. If an extension line is completely 'stuck' and impossible to route anymore, a new row is added and all routed connections are adjusted. After a maximum number of rows, Row_{max}, has been added and if the channel is still not completely routed, the channel is discarded. The entire routing process is repeated to create a new individual.

A simple illustration of the routing algorithm can be seen in the following example (Fig. 4). Details on performing the process of random routing on this channel are as shown in Fig. 5. The current pin chosen to be routed is highlighted by a circle; the target pin is highlighted by a square, while extension lines are highlighted in a shaded box. The routing algorithm is designed to allow for some degree of guidance to route each given pin towards its destination, while at the same time randomness is introduced in the decision to route either vertically or horizontally. By doing so, redundant cyclical connections cannot occur during the routing process and computation time is saved on checking through the channel for such redundancies.

Re-router: The re-route algorithm is used to re-connect disconnected pins during the genetic operations of crossover and mutation. The re-route algorithm is essentially the same as that of the initial router, though some features have been included to make the process of re-routing more robust. In addition to the random routing algorithm, if the current pin cannot be routed anymore, instead of adding one more row into the channel, all extension lines are deleted and the status of the current pin and target pin are exchanged, i.e. current pin became target pin and vice versa. The routing process proceeds on and another attempt is made to connect the two points. If this failed, the pins' statuses are reset back and all extension lines, together with all connections belonging to the net blocking the extension lines are deleted. Re-routing is then repeated. The maximum number of deletions allowed for any channel is limited to $DelNet_{max}$. The value of $DelNet_{max}$ is set to three in this research. This is to ensure that schema from the parent channels are not totally destroyed due to the deletions. Otherwise, the re-routing process will cause the MOEA to be merely a random search process.

1	3	4	1	2
0, 0, 0	0, 0, 0	0, 0, 0	0, 0, 0	0, 0, 0
0, 0, 0	0, 0, 0	0, 0, 0	0, 0, 0	0, 0, 0
1	4	2	2	3

Fig. 4. Channel to be routed

Iteration	Process	Channel illustration
Iteration 1	Pin to route: 1 PrefVert: 0; PrefHor: 1 Route: 4 Randomly chosen layer: 1 **Pin is routed.**	1 3 4 1 2 1, 0, 0 1, 0, 0 1, 0, 0 1, 0, 0 0, 0, 0 0, 0, 0 0, 0, 0 0, 0, 0 0, 0, 0 0, 0, 0 1 4 2 2 3
Iteration 2a	Pin to route: 4 PrefVert: 1; PrefHor: -1 Route: 5 Randomly chosen layer: 1 Layer 1 is occupied by net '1', so route on layer 2. Random direction: Vertical Pin not routed.	1 3 4 1 2 1, 0, 0 1, 0, 0 1, 4, 0 1, 0, 0 0, 0, 0 0, 0, 0 0, 0, 0 0, 4, 0 0, 0, 0 0, 0, 0 1 4 2 2 3
Iteration 2b	Pin: 4 PrefVert: 1; PrefHor: -1 Route: 5 Random direction: Horizontal **Pin is routed.**	1 3 4 1 2 1, 0, 0 1, 0, 0 1, 4, 0 1, 0, 0 0, 0, 0 0, 0, 0 0, 4, 0 0, 4, 0 0, 0, 0 0, 0, 0 1 4 2 2 2
Iteration 3	Pin: 1 PrefVert: -1; PrefHor: 0 Route: 2 Randomly chosen layer: 1 **Pin is routed.**	1 3 4 1 2 1, 0, 0 1, 3, 0 1, 4, 0 1, 0, 0 0, 0, 0 1, 0, 0 0, 4, 0 0, 4, 0 0, 0, 0 0, 0, 0 1 4 2 2 3

Fig. 5. Illustration of the random routing algorithm

Fitness Function

Contrary to existing approaches, the different objectives can be treated separated by MOEA. Since this eliminates the need to determine appropriate weights for aggregating the different objectives, it is now possible to specify the individual specifications instead of the aggregated functions used as described in the review. Two variations of the algorithm, in terms of the number of objectives to be minimized, are implemented. MOEA2 consists of two objective functions, while MOEA3 has three objective functions. The objective functions used in MOEA2 are:

$$f_{wirelength} = Sum\ \ of\ \ wire\ \ length\ \ in\ \ channel \tag{5}$$

$$f_{vias} = Sum\ \ of\ \ vias\ \ in\ \ channel \tag{6}$$

In addition to the above objective functions, MOEA3 has a third fitness function,

$$f_{track} = Number\ \ of\ \ tracks\ \ in\ \ channel \tag{7}$$

Diversity Preservation

The use of niche sharing [10, 12, 22] prevents genetic drift and helps promote the discovery of the whole Pareto front by the population. The basic idea is that individuals in a particular niche should share the available resources and the fitness value of an individual is degraded as the number of individuals in its neighbourhood increases. Niching can be performed in either the decision variable domain or the objective domain. The choice between niching in decision variable domain or objective domain depends on what is desired for the underlying problem. In general, niche sharing is achieved using a sharing function. Let d be the Euclidean distance between x and y. The neighbourhood size is defined in terms of d and specified by the so-called niche radius σ_{share}. The sharing function is defined as follows

$$sh(d) = \begin{cases} 1-(d/\sigma_{share})^{\alpha} & \text{if } d<\sigma_{share} \\ 0 & \text{otherwise} \end{cases} \tag{8}$$

And the niche count function is defined with the help of sharing function,

$$nc(x) = \sum\nolimits_{y} sh(dist(x,y)) \tag{9}$$

The niche radius σ_{share} is a key parameter that affects MOEA's effectiveness. In this paper, the niche radius is obtained from extensive simulation studies.

Genetic Operators

Crossover: A one-point cut is used in this algorithm. Two parents (P_1, P_2) are first chosen. If the two parents have different channel width, the one with a smaller channel width is adjusted by adding tracks randomly until the channel widths for both parents are equal. Next, a cut, X_i is randomly chosen with $2 \le X_i \le X_{term} - 2$, where X_{term} is the number of columns in the channel, as shown in Fig. 6 (a), (b). The left-hand-side of P_1 is combined with right-hand-side of P_2 to create offspring1. Likewise, the right-hand-side of P_1 is combined with the left-hand-side of P_2 to create offspring2. As some pins will be disconnected or connections crossed during this process, all such pins and connections are deleted from the offsprings, as shown in Fig. 6 (c). Thereafter, the re-route algorithm is applied on the offspring channels to route them completely as seen in Fig. 6 (d).

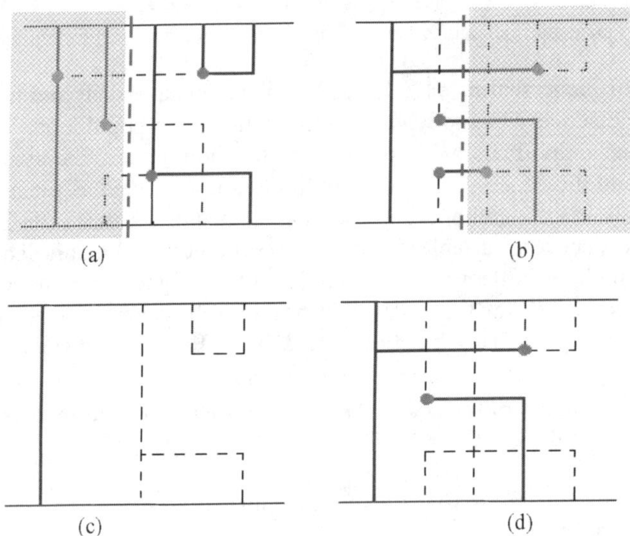

Fig. 6. Crossover operation

Multimode Mutation: Three different mutation operators are used in this algorithm. This because the CRP consists of many constraints and using merely one mutation operator will not provide a good 'exploration' of the search space. Here, the main objectives of the mutation operators are to explore the possibilities of other solutions, as well as to guide the search towards a more optimal solution space. The mutation operators used are shown in Table 1 below. Both Mutation_1 and Mutation_2 operations provide an exploration of the search space through new routing solutions provided by the operators. The Mutation_3 is used to reduce the channel width and in the process helped to guide the search towards a more optimal solution space, especially in terms of reducing the number of tracks used.

Table 1. Multimode mutation operators and their procedures

Operation	Procedures
Mutation_1	In this mutation, k nets are chosen, where $k = \frac{1}{3}N$ and N being the total number of nets in the channel. The nets chose are deleted from the channel and then the re-route algorithm is applied on the channel to produce a new complete solution.
Mutation_2	In this mutation, a point X_i is chosen along the length of the channel, such that $3 \leq X_i \leq X_{term} - 3$, where X_{term} is the total number of columns in the channel. Next, a random choice is made either to delete all connections to the left or right X_{term}. After which, the re-route algorithm is applied to the channel to create a new complete solution.
Mutation_3	In this mutation, a random track T_j is chosen, such that $1 \leq T_j \leq T_{total} - 1$, where T_{total} is the total number of tracks in the channel. The chosen track is then removed from the channel. All broken connections between any two pins of each net are deleted, while those that remain connected are left intact. The re-route algorithm is then applied to the channel to produce a new solution.

Channel Update: Redundancies in the channel will inevitable arise during the optimization process. While these redundancies have no positive contribution to the overall channel, their presence will incur unnecessary channel width. Therefore, the channel update mechanism, a local search strategy is developed to remove redundancies found in the channel. Specifically, channel update is applied to every new offspring and checked for redundant rows. Subsequently, all uncovered redundant rows will be removed from the offspring.

Program Flowchart

The flowchart of the algorithm can be seen in Fig. 7. An initial population of feasible solutions is created when the algorithm was first executed. This was done using a random router as detailed in Section 3.2.2, which created the individuals through randomized routing, so as to maintain diversity in the initial population. Following which, each of the individuals fitness is assessed. Each individual in the population is assigned a fitness value, which forms the basis of evaluation on the individual. The Pareto based ranking approach [5] that assigns the smallest rank values to all non-dominated individuals in the population and the rest of the individuals are ranked according to the number of individuals dominating them is adopted here. Therefore, the rank of an individual p in a population is given by

$$rank(p) = 1 + q \tag{10}$$

where q is the number of individuals that dominate the individual p based on the above criteria.

Individuals are selected into the mating pool by means of tournament selection where a random number (N) of individuals are chosen at each time and the fittest among these individuals is chosen to become a parent is adopted here [3, 11]. The selection criteria is based on Pareto ranking and niche count will be used in the event of a tie. Archive updating criteria was based on non-dominance criteria. In every generation, all the non-dominated solutions were updated into the archive population, whose size was dynamic. Likewise, archived solutions that are dominated are removed. In the event where the archive size reaches a predetermined limit, batch pruning which is based on niche count is employed. The process of fitness evaluation, ranking, crossover, mutation and archive update was repeated until the maximum generation is reached.

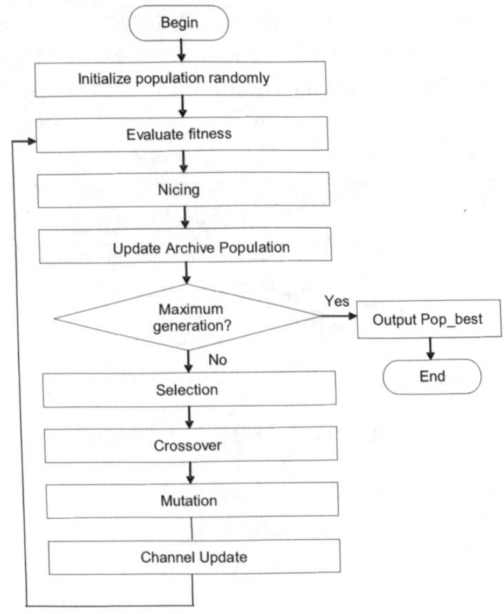

Fig. 7. The program flowchart of MOEA

4 Performance Comparison

The section begins with the listing of standard benchmark problems used and parameters settings. Following which, comparison of results obtained by using the proposed MOEA for two objectives channel routing problem can be found in section 4.2. Section 4.3 analyses the results for three objectives channel routing problem. Section 4.4 shows the comparison of different variations of the algorithm made against the best found in literature.

4.1 Simulation Setup

Five standard benchmark problems, Joobbani's test channels (Joo6_12, Joo6_13 and Joo6_16), Yoshimura-Kuh Channel (Yos), and Burstein's Difficult Channel (Burstein) are used to test the algorithm. Ten runs are performed for each algorithm setting on every test problem in order to study the statistical performance, in terms of consistency and robustness of the proposed algorithm. The algorithm is implemented in C++ programming language and the simulations were performed using an Intel Pentium4 2.4 GHz computer. Ten independent runs using ten different fixed seeds are used to simulate each benchmark test data. A summary of the simulation parameters are shown in Table 2.

Table 2. The simulation parameters

PARAMETER	VALUES
Population Size	150
Archive Size	20
Generation Size	120
Seeds used (10 runs)	99, 1, 1003, 153, 645, 475, 135, 6, 8, 127
Initial number of rows	7
Crossover probability	0.9
Mutation_1 rate	0.2
Mutation_2 rate	0.2
Mutation_3 rate	0.2
Niche radius	0.1 (normalized search space)

4.2 Two-objective Optimization (MOEA2)

The optimization results from the proposed MOEA are shown in Fig. 8 (a)-(e) while a tabulation of their variances is found in Table 3 for each benchmark test case respectively. The figures show the plots of netlength against vias for the final and archive populations belonging to different test cases. The attainment surfaces obtained by MOEA2 are represented by the black line joining non-dominated solutions.

Due to the inherently lower pressure exerted by the routing algorithm to increase the number of vias, it can be seen from the figures that the number of vias lying along the attainment surface are minimized close to optimum. As a result of which and due to the discrete nature of the problem, there are relatively few points lying along the non-dominated front for smaller test sets. Nonetheless, the trade-off curves do show that there indeed exist some form of confliction between the objectives of minimizing netlength and vias, as a decrease in netlength inevitably resulted in an increase in number of vias used. Hence this supports the use of MOEA to solve this class of problems.

Interestingly, the optimization managed to optimize benchmark, Yos, to a single point as shown in Fig. 8(e). The optimized channel solution for this benchmark has 4 tracks, 67 units of netlength and 1 via. This could imply that in the case of benchmark Yos specifically, the problem could be treated as a single objective optimization. While the objectives of the Channel routing problems may be conflicting in nature, the degree of confliction differs depending on the difficulty of the individual problem and it is possible to find a single optimal solution in some cases. Yet, there is no way to find out whether a problem is easy to solve or not before optimization is performed.

In the case of benchmark, Yos, suppose four tracks and an infinite number of layers could be used in the channel. The lower bound for netlength can be found to be 63 units, while the number of via used is 0. However, only 2 layers are allowed in this implementation, thus a netlength of 67 units and one via is considered to be very close to the lower bound of 63 units with zero via. As such, it could be almost impossible to find a solution that required zero via when netlength is increased; or fewer units of netlength when vias are allowed to increase. This could explain why only one solution point is obtained by the optimization.

Table 3. Variance values for the simulation results of MOEA2

TEST CASES	VIA	NETLENGTH
Joo6_12	0	0
Joo6_13	1.7333	7.1111
Joo6_16	0.6222	1.6
Burstein	0.1	0
Yos	0	0

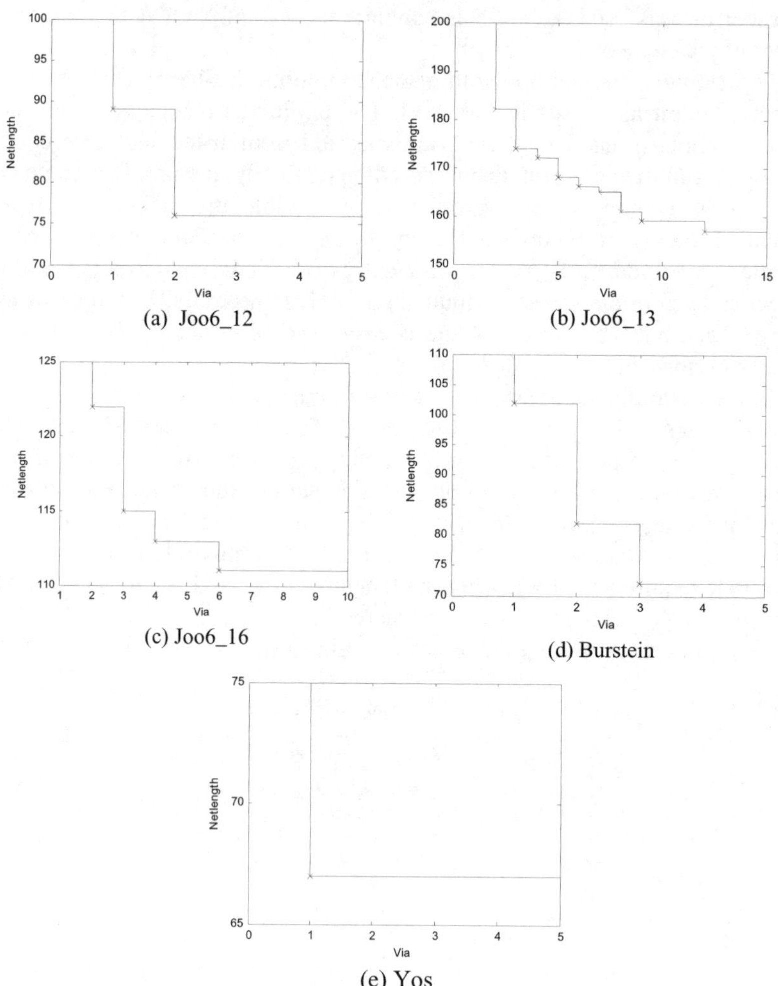

Fig. 8. Attainment surface of MOEA2 for the various benchmark problems

Channel Update

In order to evaluate the effects of channel update, two different MOEA settings are used in this section. The first setting do not allow channel update for all subsequent generations after the initial population, though update is performed before updating the archive of the population. The second setting updates all individuals in the population. Fig. 9 (a)-(e) show the box-plots of the best 10 results obtained for each of the benchmarks from 10 separate runs each.

The effects of channel updating are observed to be significant through experimental runs. It can be observed that in general, results with channel update outperformed those obtained without channel update, in terms of minimization of the objectives. Besides that, the comparison of variance of each objective shows that in general, there is a higher variance for results obtained without channel update. Hence, this shows that there is less consistency and reliability of performance if channel update is not performed.

It may appear that Channel update will exert exceptionally high pressure to reduce channel width, resulting in a loss of diversity in the population and possibly pre-mature convergence of population. However, it can be observed from the Fig. 8 that this concern is unfounded because less superior solutions did appear in the final solutions, thus showing that diversity had not been compromised. The diversity of population might be maintained due to the process of reproduction, where mating between a superior individual and one that is less superior in terms of channel width, may result in a degradation of the offspring produced and thus preventing the population from converging to a single solution.

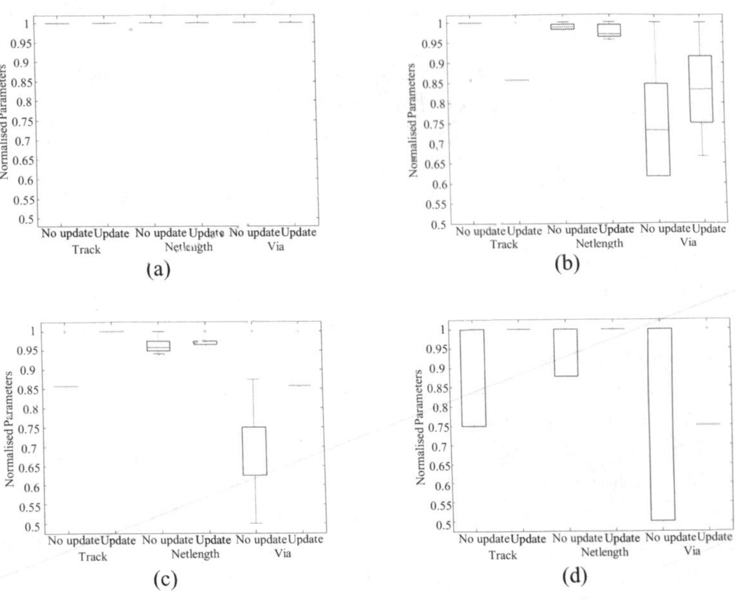

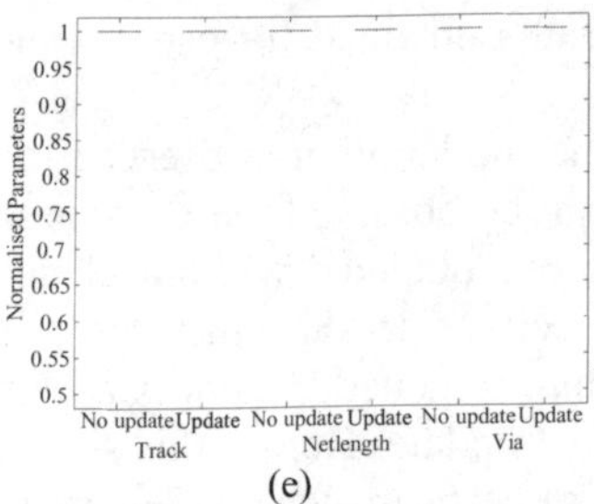

(e)

Fig. 9. Comparison of performance with and without channel update

4.3 Three-objective Optimization (MOEA3)

The algorithm was tested for its performance in the optimization of 3 objectives in MOEA3. The plots for the normalized non-dominated solutions are shown in Fig. 10. In the figures, all three objectives of minimizing netlength, via and tracks are considered. The normalized value for each of these objectives belonging to each non-dominated solution is represented by a node and the nodes representing each solution are linked together by a black line.

From the graphs, it can be observed that the lines linking via and netlength crossed one another. This shows that the improvement in the objective of reducing the number of vias inevitably resulted in an increase in the netlength. In same way, it is clear that the objective of minimization the number of vias conflicts with the objective to reduce netlength. However, the lines linking netlength and track do not cross one another in the figures. This implies that the objectives to minimize netlength and number of tracks are non-conflicting in nature. This is logical as fewer tracks used should reduce the length of wires used as well, unless there are redundant cyclical connections in the channel. From here, we can deduce that the objectives of minimizing tracks and netlength are non-conflicting generally.

As the number of tracks and netlength have been shown to be non-conflicting, it is sufficient to plot the graphs of netlength against via to show the attainment surface of the benchmarks problems obtained through the use of MOEA3. The graphs are shown in Fig. 11 (a)-(e). Similarly to the case of MOEA2, only a single point can be found for benchmark problem, Yos.

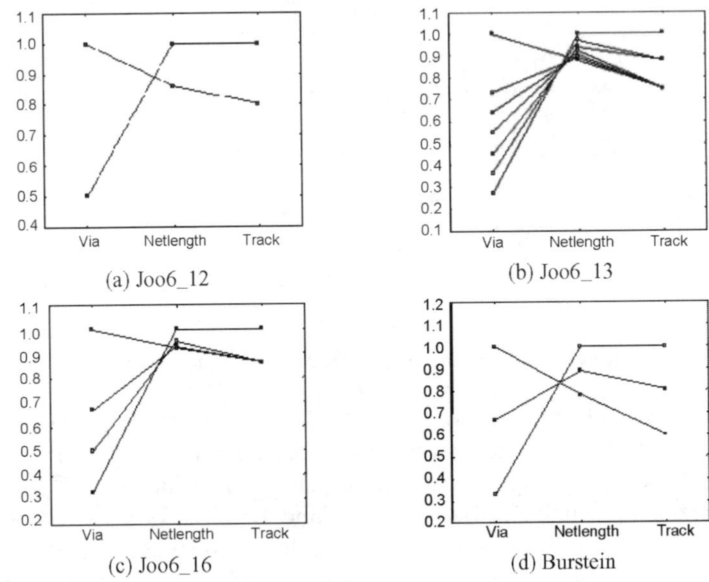

Fig. 10. Normalized non-dominated solutions for MOEA3

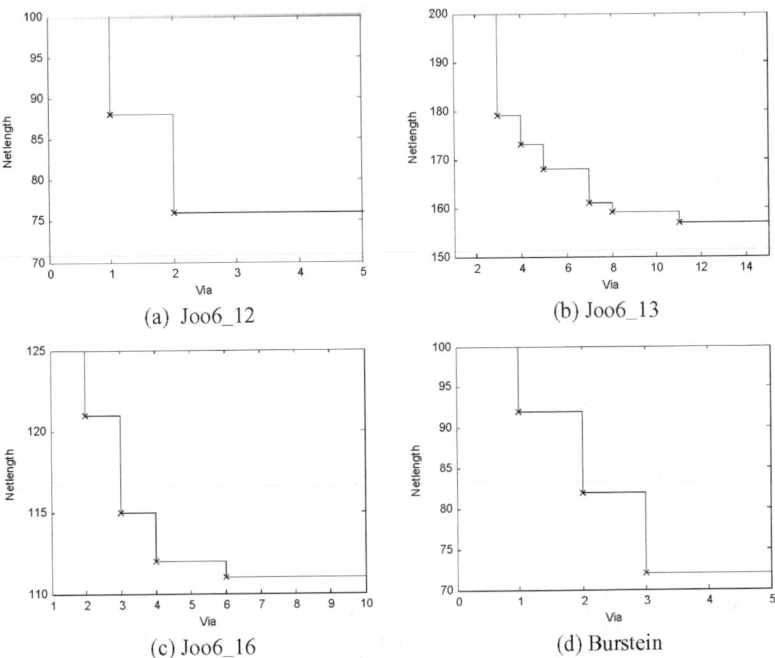

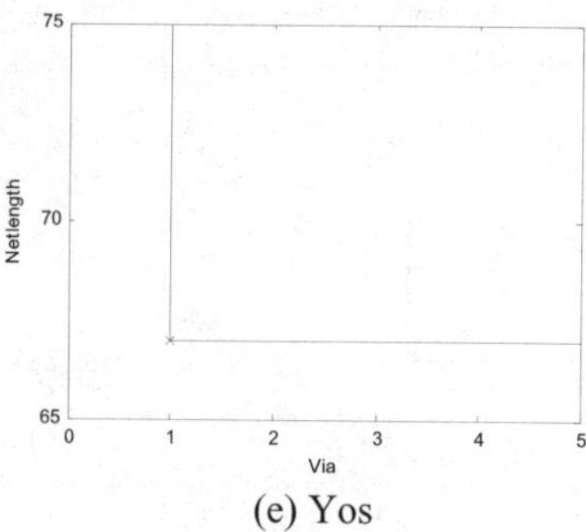

(e) Yos

Fig. 11. Attainment surface of MOEA3 for the various benchmark problems

The non-dominated solutions of both MOEA2 and MOEA3 are listed in Table 4. The first and second number represents the number of via and netlength respectively. Comparison on performance for the objectives of minimizing netlength and via is made to find out if performance of the two algorithms differs. It is observed that the solutions obtained in both optimizations are comparable. The solutions in MOEA3 do not show much deterioration in the two objectives (via and netlength) comparing to results in MOEA2. With this, it can be shown that it is possible to extend the proposed algorithm to solve any N-objectives channel routing problems without making any compromise on the solutions obtained.

Table 4. Comparison of non-dominated solutions from MOEA2 and MOEA3

Benchmark	MOEA2	MOEA3
Joo6_12	<1, 89>, <2, 76>	<1, 88>, <2, 76>
Joo6_13	<2, 182>, <3, 174>, <4, 172>, <5, 168>, <6, 166>, <7, 165>, <8, 161>, <9, 159>,<12, 157>	<3, 179>, <4, 173>, <5, 168>, <7, 161>, <8, 159>, <11, 157>
Joo6_16	<3, 115>, <4, 113>, <6, 111>	<3, 115>, <4, 112>, <6, 111>
Burstein	<1, 102>, <2, 82>, <3, 72>	<1, 92>, <2, 82>, <3, 72>
Yoshimura-Kuh	<1, 67>	<1, 67>

4.4 Comparison with Best Known Results in the Literature

Comparison of solutions obtained from MOEA2 and MOEA3 have been made against solutions reported in other researches in Table 5 according to the authors' best knowledge. From the results, it can be seen that the present work is able to yield results that are better than the best that has been known in literature. In fact, for Burstein's Difficult Channel, the channel width is reduced to 3, when the best solution found previously needed 4 tracks. As a result of which, a significantly better solution is found for Burstein's Difficult Channel.

In addition, it can be observed that the optimal number of vias used by the present work is significantly reduced for most of the solutions. This supports the assertion made about the routing algorithm proposed; that it has the advantage of finding solutions in the search space that would yield better performance in terms of the number of vias used. This also shows that in the case of the highly constrained CRP, the importance of the routing algorithm must not be overlooked as it influences the quality of solutions found.

5 Conclusions

This research has presented a novel approach to solving the channel routing problem through the use of a Multi-Objective Evolutionary Algorithm (MOEA). A novel method of random routing has also been proposed to avoid solutions that are 'expensive' in terms of the number of vias used. The algorithm is tested on five well-known benchmark problems. Analyses based on channel routing benchmark problems showed that the results obtained from the proposed algorithm are consistent, as variance in the best results obtained from ten independent runs for each test case is found to be insignificant. This confirmed the reliability of the proposed algorithm, which has also been shown to be robust as it is able to solve problems of varying sizes and difficulties. The results also show that it is possible to extend the algorithm to solve N-objective channel routing problems.

This work has confirmed the multi-objectivity nature of the channel routing problem. Comparisons have been made between the best results obtained for the different benchmark problems with those found in literature. It has been observed that the solutions obtained from the proposed approach are very competitive and in some cases significantly better than others. This suggests that the treatment of CRP as a multiobjective optimization problem to be solved by MOEA has yielded better results than previous methods that reduced the problem to single objective optimization.

Since the proposed algorithm is shown to be effective in solving two layer channel routing problems (CRP), it can be extended to an N-lay[illegible] CRP. An additional objective could be included, such as the minimizat[illegible] of the number of layers used. Besides that, the algorithm can be modified to allow for pre-routed wires and obstacles, so more practical CRP can be modeled and solved using the new algorithm. It is believed that the proposed algorithm can also be applied to switchbox routing problems, since both problems are based on similar geometry, although the switchbox routing problem consists of more constraints than the CRP.

Table 5. Comparison to the best results found in literature. (*) denotes interactive while (~) denotes use of parallel GA

BENCHMARK	ALGORITHM	COL	ROWS	NETLENGTH	VIAS
Joo6_12	Monreale [6]	12	4	84	13
	Packer [30]	12	4	82	18
	Weaver [14]	12	4	79	14
	GAP [20]	12	4	79	14
	MOEA2	12	4	76	2
	MOEA3	12	4	76	2
Joo6_13	Greedy [28]	18	8	194	38
	Het-GA [23]	18	6	172	24
	Silk [21]	18	6	171	28
	Weaver [14]	18	7	169	29
	Packer [30]	18	6	167	25
	GAP(~) [33]	18	6	164	22
	MOEA2	18	6	157	12
	MOEA3	18	6	157	11
Joo6_16	Weaver [14]	11	8	131	23
	Weaver(*)	11	7	121	21

	Monreale [6]	11	7	120	19
	HGA [9]	11	6	115	16
	GAP(~) [18]	11	6	115	15
	Het-GA [23]	11	6	115	14
	MOEA2	11	6	111	6
	MOEA3	11	6	111	6
Burstein's Difficult Channel (Burstein)	Mighty	13	4	83	8
	Packer [30]	12	4	82	10
	Monreale [6]	12	4	82	10
	GAP [20]	12	4	82	8
	Het-GA [23]	12	4	82	8
	MOEA2	12	3	72	3
	MOEA3	12	3	72	3
Yoshimura-Kuh Channel (Yos)	Yos-Kuh [32]	12	5	75	21
	Monreale [6]	12	4	72	11
	HGA [9]	12	4	71	12
	GAP [20]	12	4	70	11
	Weaver [14]	12	4	67	12
	MOEA2	12	4	67	1
	MOEA3	12	4	67	1

References

[1] Burstein, M. (1986). "Channel routing," in *Layout Design and Verification*, Ohtsuki, T. (ed), New York: Elsevier Science.

[2] Burstein, M. and Pelavin, R. (1983). "Hierarchical wire routing," *IEEE Transactions on Computer-Aided Design of Integrated Circuits and Systems*, vol. 2, no. 4, pp. 223-234.

[3] Blickle, T. and Thiele, L. (1995). "A mathematical analysis of tournament selection," in *Proceedings of the 6th International Conference on Genetic Algorithms*, pp. 9-16.

[4] Davidenko, V. N., Kureichik, V. M. and Miagkikh, V. V. (1997). "Genetic algorithm for restrictive channel routing problem," in *Proceedings of the 7th International Conference on Genetic Algorithms*, pp. 636-642.

[5] Fonseca, C. M. and Fleming, P. J. (1993). "Genetic algorithm for multiobjective optimization: formulation, discussion and generalization," in *Proceedings of the 5th International Conference on Genetic Algorithms*, pp. 416-423.

[6] Geraci, M., Orlando, P., Sorbello, F. and Vasallo, G. (1991). "A genetic algorithm for the routing of VLSI circuits," in *Proceedings of Euro ASIC91*, pp. 218-223.

[7] Gerez, S. H. and Herrmann, O. E. (1989). "Switchbox routing by stepwise reshaping," *IEEE Transactions on Computer-Aided Design of Integrated Circuits and Systems*, vol. 8, no. 12, pp. 1350-1361.

[8] Groeneveld, P. (1933). "Necessary and sufficient conditions for the routability of classical channels," *The Integration, the VLSI Journal*, vol. 16, no. 1, pp. 59-74.

[9] Göckel, N., Pudelko, G., Drechsler, R. and Becker, B. (1996). "A hybrid genetic algorithm for the channel routing problem," in *Proceedings of the 1996 IEEE International Symposium on Circuits and Systems*, pp. 675-678.

[10] Goldberg, D. E. (1989). *Genetic Algorithms in Search, Optimization and Machine Learning*, Boston: Addison Wesley.

[11] Goldberg, D. E. and Deb, K. (1991). "A comparative analysis of selection schemes used in genetic algorithms," in *Foundations of Genetic Algorithms*, Gregory, J. (ed), pp. 69-93.

[12] Goldberg, D. E. and Richardson, J. (1987). "Genetic algorithms with sharing for multimodal function optimization," in *Proceedings of the 2nd International Conference on Genetic Algorithms on Genetic algorithms and their Application*, pp. 41-49.

[13] Hashimoto, A. and Stevens, S. (1971). "Wire routing by optimizing channel assignment within large apertures," in *Proceedings of the Eight Design Automation Conference*, pages 155-169, ACM/IEEE.

[14] Joobbani, R. (1986). *An Artificial Intelligence Approach to VLSI Routing*, Boston: Kluwer Academic Publishers.

[15] Lengauer, T. (1990). *Combinatorial Algorithms for Integrated Circuit Layout*, New York: John Wiley & Sons.

[16] Leong, H. W., Wong D. F. and Liu, C. L. (1985). "A simulated annealing channel router," in *Proceeding of IEEE International Conference on CAD*, pp. 226-228.

[17] Lienig, J. (1997). "A parallel genetic algorithm for performance-driven VLSI routing," *IEEE Transactions on Evolutionary Computation*, vol. 1, No. 1, pp. 29-39.

[18] Lienig, J. (1997). "Channel and switchbox routing with minimized crosstalk - a parallel genetic approach," in *Proceedings of the 10th International Conference on VLSI Design*, pp. 27-31.

[19] Lienig, J. and Thulasiraman, K. (1994). "A new genetic algorithm for the channel routing problem," in *Proceedings of the 7th International Conference on VLSI Design*, pp. 133-136.

[20] Lienig, J. and Thulasiraman, K. (1994). "A genetic algorithm for channel routing in VLSI circuits," *Evolutionary Computation*, vol. 1, no. 4, pp. 293-311.

[21] Lin, Y. L., Hsu Y. C. and Tsai F. S. (1989). "SILK: A simulated evolution router," *IEEE Transactions Computer Aided Design of Integrated Circuits and Systems*, vol. 8, no. 10, pp. 1108-1114.

[22] Mahfoud, S. W. (1995). *Niching Methods for Genetic Algorithms*, University of Illinois, Urbana-Champaign, PhD thesis.

[23] Masuda, T., Hayashi, Y., Shigchiro, Y. and Inoue, J. (2000). "A VLSI channel routing method using genetic algorithm based on the coexistence of heterogeneous populations," *IEE Japan Extended Summary*, vol. 120-C, no. 11.

[24] Prahlada Rao, B. B. and Hansdah, R. C. (1993). "Extended distributed genetic algorithm for channel routing," in *Proceedings of the IEEE Symposium on Parallel and Distributed Processing*, pp. 726-733.

[25] Prahlada Rao, B. B., Patnik, L. M. and Hansdah, R. C. (1995) "An extended evolutionary programming algorithm for VLSI channel routing", in *Proceedings of the Fourth Annual Conference on Evolutionary Programming*, pp. 521-544.

[26] Prahlada Rao, B. B., Patnaik, L. M. and Hansdah, R. C. (1994). "A genetic algorithm for channel routing using inter-cluster mutation," in *Proceedings of the first IEEE International Conference on Evolutionary Computation*, pp. 97-103.

[27] Prahlada Rao, B.B., Patnaik, L. M. and Hansdah, R.C.(1993). "A parallel genetic algorithm for channel routing problem," in *Proceedings of the IEEE 3rd Great Lake Symposium on Design Automation of High Performance VLSI Systems*, pp. 69-70.

[28] Rivest, R. L. and Fidducia, C. M. (1982). "A greedy channel router," in *Proceedings 13th Design Automation Conference*, pp. 418-424.

[29] Saymanski, T. G (1985). "Dogleg channel routing is NP-complete, " *IEEE Transactions on Computer-Aided Design*, vol. 4, no. 1, pp. 31-41.

[30] Shin, H. and Sangiovanni-Vincentelli, A. (1987). "A detailed router based on incremental routing modifications mighty," *IEEE Transactions on Computer-Aided-Design*, vol. 6, no. 6, pp. 942-955.

[31] Tan, K. C., Khor, E. F., Lee, T. H. and Sathikannan, R. (2003). "An evolutionary algorithm with advanced goal and priority specification for multi-

objective optimization," *Journal of Artificial Intelligence Research,* vol. 18, pp. 183-215.

[32] Yoshimura T. and Kuh, E. S. (1982). "Efficient algorithms for channel routing," *IEEE Transactions on Computer Aided Design of Integrated Circuits and Systems*, vol. 1, no.1, pp. 25-35.

Simultaneous Planning and Scheduling for Multi-Autonomous Vehicles

D.K. Liu and A.K. Kulatunga

ARC Centre of Excellence for Autonomous Systems
Faculty of Engineering, University of Technology, Sydney
POBox 123, Broadway, NSW 2007, Australia, {dkliu, akula}@eng.uts.edu.au

Summary. With the increasing applications of autonomous vehicles in dynamic and strictly constrained environments such as automated container terminals, efficient task/resource allocation and motion coordination (i.e., path and speed planning) of multi-autonomous vehicles has become the critical problem and have therefore been recently recognized as the key research issues by both academics and industry. This chapter addresses a generic approach for integration of task allocation, path planning and collision avoidance, which has so far not attracted much attention in the academic literature. A Simultaneous Task Allocation and Motion Coordination (STAMC) approach is presented. Two meta-heuristic algorithms, i.e. simulated annealing and ant colony, and an auction algorithm are investigated and applied. The proposed approach is able to solve the scheduling, planning and collision avoidance problems simultaneously; improve the usage of bottleneck areas; handle dynamic traffic conditions and avoid deadlock. Simulation results demonstrated the effectiveness and efficiency of this approach.

1 Introduction

An autonomous vehicle is a driverless system which can derive information about the environment from its on-board sensors, make decisions based on the information, and control it to meet the mission requirements. Examples of autonomous vehicles include automated guided vehicle (AGV) and unmanned aerial vehicles. AGVs are equipped to navigate a flexible guided path network, either unidirectional or bidirectional. AGVs have been widely used in

D.K. Liu and A.K. Kulatunga: *Simultaneous Planning and Scheduling for Multi-Autonomous Vehicles,* Studies in Computational Intelligence (SCI) **49**, 437–464 (2007)

material handling since their introduction in 1950s [33], including indoor and outdoor environments such as manufacturing, distribution, transshipment and transportation areas.

The significant increases of AGV applications in various areas, the number of autonomous vehicles employed, the dynamics of traffic conditions and the constraints in environments have resulted in a considerable increase in the complexity of autonomous vehicle task allocation and motion coordination (i.e., path and speed planning and collision avoidance) with the goal of operating autonomous vehicles efficiently and safely. Planning and scheduling of autonomous vehicles, including task allocation (also called dispatching), path and motion planning and collision avoidance, play an important role in improving the productivity, and have therefore been recently recognized as key research issues in autonomous vehicle systems by both academics and industry [46,39,6].

The problem of AGVs planning and scheduling is to have the correct AGVs to perform the tasks in the right places and times. Functionally, there are three activities of AGVs planning and scheduling: task allocation, path planning and scheduling.

1.1 Task allocation

Task allocation or dispatching refers to a rule used to select a vehicle to perform a task, which is usually a continuous and dynamic process as the tasks and idle vehicles could change instantly. When a request is received to assign a task to a vehicle, either work-centre-initiated or vehicle-initiated task allocation [13] can be conducted. In work-centre-initiated task allocation, an AGV is selected from a set of idle AGVs. Various rules, for example, random vehicle rule, nearest vehicle rule, and least utilized vehicle rule, can be employed for assigning priorities to AGVs. Some of the task allocation policies in vehicle-initiated approach include shortest travel time/distance rule, maximum outgoing queue size rule, and modified first-comes-first-served rule [41, 48]. Evaluations on the performance of those rules and policies have been conducted. For example, De Koster, et al. [11] evaluated the performance of several real-time vehicle dispatching rules in three different environments, namely a European distribution centre, a container terminal and a production site. The vehicle dispatching problem is also studied in combination with other related optimization problems such as container location assignment in the storage area [4].

Various algorithms and models have been proposed and studied for task allocation. Examples include a dynamic deployment algorithm for dispatching AGVs to containers in order to minimize the (un)loading time for a vessel [27]; an auction algorithm as a dispatching method for AGVs [28]; a dynamic model for real-time optimization of the flow of flatcars [36] which considers constraints for assignment of trailers and containers; a mixed integer linear programming model to dispatch multi-load AGVs [18], etc. Other heuristic and computational algorithms studied for application in vehicle task allocation include Markov decision processes, fuzzy logic and neural network approaches. For real life applications with a large number of vehicles, more research for advanced heuristics and optimization approaches is required [46].

1.2 Path planning

Path planning is the selection of AGV paths with the assigned tasks before movement commences. For applications in known environments such as road networks, path planning is usually solved in two steps: (1) build a graph to represent the geometric structure of an environment and (2) perform a graph search to find a connected path between the start and destination points based on certain pre-specified criteria, current traffic situation and availability of the vehicle. The efficiency of path planning is normally analyzed in terms of distance traveled or time required to complete the given task(s).

Path planning algorithms have been studied for autonomous vehicles in partially known environments, examples include D* [42], Delayed D* [15] and E* algorithm [35]. Meta-based heuristic optimization algorithms have been studied for applications in path planning, e.g. particle swarm optimization (PSO) [37] and genetic algorithm [9].

Finding shortest paths in known and dynamic environments appears to have diverging approaches. For example, Fu and Rilett investigated the dynamic and stochastic shortest path problem by modeling link travel time as a continuous-time stochastic process [16]. By allowing for variation in speed, Horn presented an algorithm to calculate an approximation of shortest travel time after studying a number of Dijkstra variant algorithms [19].

The majority of published research on shortest path planning algorithms, which are often used to find the "best" paths for road vehicles in known environments, has dealt with static road networks that have fixed topology and less constraints. Roadmap approach, which is based on the concepts of configuration space and continuous path, is one of the most common approaches to vehicle path planning in road networks. Zhan and Noon presented a comprehensive study of shortest path algorithms on 21 road networks from 10 different states in the U.S. [49]. In this study, Dijkstra-based algorithms outperformed other algorithms in one-to-one and one-to-some fastest path problems. Husdal reported that the A* algorithm dominates the research literature for static networks and A* and Dijkstra algorithms are applicable in both static and dynamic networks [21]. Evolutionary-based algorithms, e.g. genetic algorithms, have also been applied in path planning problems [45, 22, 2].

1.3 Scheduling

Scheduling determines arrival and departure times of AGVs at each path segment, pick-up and delivery point and intersection to ensure collision free movement. An AGV's schedule is generated by combining the paths of all AGVs into an overall sequence based on certain scheduling rules such as task priorities. During scheduling, some of the planned path(s) may need to be updated to achieve a conflict-free schedule with a high degree of concurrency in AGVs movement. Those updates may include changes of vehicle speeds, paths, start and finish time, etc. Scheduling normally repeats at regular intervals to cope with foreseen and/or unforeseen problems such as the addition of new tasks, traffic congestion, changes of task priority, etc.

Collision avoidance and deadlock prevention are important issues in scheduling. Most algorithms for deadlock prevention and collision avoidance use cyclic [31] or loop based approaches. There are some studies on path topologies in AGV systems [1, 14]. In most cases uni-directional movement of a vehicle along a path is assumed [31, 32]. The approach proposed in [34] deals with optimal routing for an AGV to do a job, e.g. moving from one place to another. However the path segments allocated for an AGV can not be used by other vehicles until the AGV completes the current job. The effectiveness and robustness of an incremental path planning approach proposed by [43] were proved by comparing with the complete path planning method. A Petri net based deadlock prevention method [20] was proposed for an AGV system. This approach can not be used for a complex or difficult planning problem due to its high computational cost.

1.4 Simultaneous approaches

A few research studies have been conducted for integrated scheduling problems. The problem of integrated scheduling of various types of handling equipment at an automated container terminal in a dynamic environment was investigated [30]. In this research, an optimization based Beam Search heuristic and several dispatching rules were presented. An extensive computational study is carried out [illegible] investigate the performance of these solution methods under different scen[illegible]s, with the conclusion that the Beam Search heuristic performs the best on [illegible]rage, but that some of the relatively simple dispatching rules perform alm[illegible] well. Bish et al. [5] modeled a transportation system for container termi[illegible] which uses the assumptions of constant vehicle velocity and uni-directio[illegible] vehicle movement. An algorithm presented in [24] does fleet sizing and [illegible] routing for container transportation based on Tabu search method. This al[illegible] initially starts up with the lower bound of fleet size and increases it un[illegible] criteria are satisfied. The study in [38] gives collision free routing for AGVs in a bi-directional flow path layout. Simultaneous machine and AGV scheduling in flexible manufacturing system is introduced by Ulusoy et al. [44] in order to minimize the makespan. A genetic algorithm is studied for this simultaneous scheduling problem [45]. Hussain et al. [12] presented a generic approach to modeling integrated scheduling problems with a finite horizon such that capacity, parking space and release time constraints are met. An evolution based algorithm with domain-specific operators is proposed to solve the integrated mission assignment and path planning problem for multiple unmanned ground vehicles in a dynamic environment [22].

1.5 Issues and problems

Research on autonomous vehicle planning and scheduling in known environments such as container terminals has been attempting to provide more realistic and versatile solutions over the last decade. However, the following issues still remain unresolved:

(1) Planning and scheduling issues are often studied separately. The integration of task allocation, path planning and motion coordination has not been studied [46]. This integration forms a very challenging problem. It is expected that this integration is able to increase the efficiency of planning and scheduling, and productivity.

(2) Efficient management of congested and bottleneck areas which cause the most difficulties in planning and scheduling. In a time-dependent dynamic environment, considering the limitations on space and vehicle paths, the planned shortest paths of autonomous vehicles have significant effect on the path planning of their successors. However, most of the existing shortest path algorithms plan the shortest paths without considering the issues of traffic congestion and opposite direction collision in a long and bi-directional flow path where only one vehicle is allowed to pass each time. Avoiding congestion, deadlock and delay becomes highly significant [46].

(3) Most of the previous studies focus on selection of shortest path for a single autonomous vehicle with the consideration of computational cost [8]. Some reported approaches claimed that the variations in speed and time delays at nodes are allowed, however, those changes in speed and waiting time are normally allowed in the scheduling stage after path planning. That is to say, the planned paths can not be changed. As a result, the changes in speed and waiting time can only contribute to the solution in scheduling. The integration of task allocation, path and motion planning for a team of autonomous vehicles is expected to provide the optimal or

near optimal solution for the whole system. A generic approach is required to solve the integrated problem.

(4) How path and motion planning algorithms accommodate dynamic traffic conditions. Practically, the traffic condition related to the movement of autonomous vehicles changes from time to time. Some studies have mentioned the necessity of link weight change [40]. However, those studies focus on how the changes of connection weights among nodes affect the existing planed shortest paths [47], instead of how the planed paths affect the changes of connection weights and how the changed weights affect the successive vehicles' paths and speeds.

To the best of our knowledge, the integration of task allocation, path planning and motion coordination for multiple autonomous vehicles in dynamic and strictly constrained environments has not been well studied, and no generic approach can be used to solve the integrated problem. In this research, a Simultaneous Path and Motion Planning (SiPaMoP) [29] approach is presented (Section 2). This generic approach coordinates the motion of multiple autonomous vehicles in dynamic and strictly constrained environments by conducting vehicles' path and speed planning and collision avoidance simultaneously, which can efficiently manage congestion and bottleneck areas, avoid collisions among vehicles and between vehicles and obstacles, handle dynamic changes in traffic conditions and environments.

Based on the SiPaMoP approach, a Simultaneous Task Allocation and Motion Coordination (STAMC) approach is further presented (Section 3). Integration of task allocation with the SiPaMoP approach provides a way to dispatch vehicles by taking path and motion planning into account, and to maximize the productivity. Two meta-heuristic algorithms, i.e. simulated annealing and ant colony, and an auction algorithm are investigated and applied in this research. Simulation studies and comparison of the performance of those three algorithms are conducted. The results demonstrated the effectiveness and efficiency of those algorithms in simultaneous autonomous vehicle planning and scheduling.

2 Simultaneous Task Allocation and Motion Coordination

2.1 The architecture of the STAMC

As mentioned above, traditional centralized approaches deal with task allocation, path planning, motion planning and collision avoidance separately as shown in Fig.1. The proposed simultaneous approach, i.e. STAMC shown in Fig.2, allocates tasks to vehicles by integrating vehicle path and motion planning and collision avoidance into one stage. Dynamic traffic conditions and environment changes are taken into account with the mission and commands in the stage.

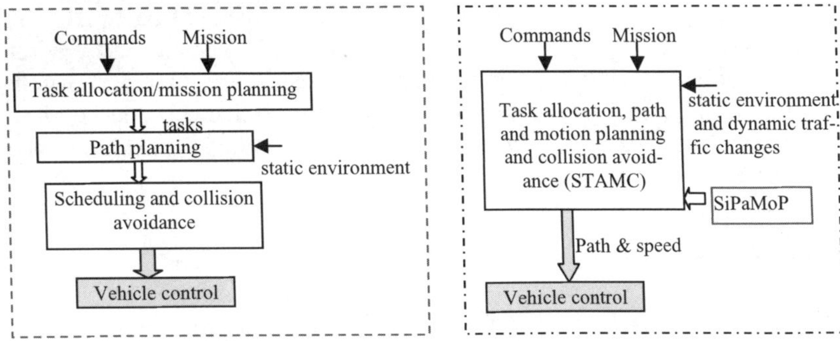

Fig. 1. Traditional sequential approach

Fig. 2. STAMC approach

The task allocation problem for multiple autonomous vehicles consists of a certain multiple vehicles (N_v) and a number of tasks (N_J) to be carried out by the vehicles. Solving such a problem amounts to making discrete choices such that an optimal solution is found among a finite or a countable number of alternatives. This problem is proved to be a NP hard combinatorial optimization problem in that it is difficult to find an optimal solution without use of an essentially enumerative algorithm, and that computational time increases exponentially with the problem size. This problem becomes more difficult to solve when a large number of vehicles are used in dynamic and strictly constrained environments. Many algorithms have been investigated to solve the problem. A simulated annealing (SA) algorithm has been proved as one of the algorithms [25, 26].

The objective of vehicle task allocation and motion coordination is to minimize the maximum travel time of vehicles (makespan), or the total travel time of all vehicles (vehicle utilization):

$$\begin{aligned} &\min.\quad MS = \max_{s=1,2,\dots,N_v} \sum_{i=1}^{B_s} t_{s,i} \quad or \\ &\min.\quad TT = \sum_{v=1}^{N_v} \sum_{i=1}^{B_s} t_{s,i} \end{aligned} \tag{1}$$

where, MS represents makespan, TT is the total travel time of all vehicles and N_v is the total number of vehicles. $t_{s,i}$ is the travel time of vehicle s performing task i, which is formulated in Section 2.2. B_s is the number of tasks allocated to vehicle s, which should satisfy:

$$\sum_{s=1}^{N_v} B_s = N_J \tag{2}$$

Eq. (2) implies that the allocated tasks to all vehicles should be equal to the total number of tasks, N_J. Other constraints are task, vehicle and environmental condition dependant, mainly including a task's start time and completion time, requirement of safety distance between vehicles, node condition (staticly locked nodes or dynamicly locked nodes), vehicle maximum speed and capacity, task priority, etc.

2.2 Motion coordination

Motion coordination of multiple autonomous vehicles is conducted by the Simultaneous Path and Motion Planning (SiPaMoP) method [29]. The SiPaMoP is used as a module in the proposed STAMC approach (Fig.2), which coordinates vehicles' motion (path and speed) and avoids collision simultaneously with the objective of minimizing the travel time of vehicles and avoiding collisions with static and moving obstacles in a dynamic and strictly constrained environment. A vehicle's path and motion are designed as:

$P_{s,i} : [p_{s,1}, p_{s,2}, \dots, p_{s,k_{s,i}}]$: path segments of vehicle s for performing task i. The length of a path segment, $p_{s,j}$ is $S_{s,j}$. The time for vehicle s to perform task i is defined as:

$$t_{s,i} = \sum_{j=1}^{k_{s,i}} \frac{S_{s,j}}{v_{s,j}}\bigg|_{v_{s,j}\neq 0} + \sum_{j=1}^{k_{s,i}} t_w\bigg|_{v_{s,j}=0} + t_{load} + t_{unload} \tag{3}$$

where, t_w represents waiting time, t_{load} is the loading time and t_{unload} the unloading time. $V_{s,i} : [v_{s,1}, v_{s,2}, \dots, v_{s,k_{s,i}}]$ is the average speed of vehicle s from the first path segment to the last link $k_{s,i}$ in the path for performing task i.

Based on the tasks allocated, changes in the environment and the location of static and moving obstacles, the SiPaMoP approach generates the path $P_{s,i}$ and speed $V_{s,i}$ simultaneously for each vehicle s (s=1, 2,…, N_v) by minimizing the operational time for a set of tasks. The SiPaMoP method is described below.

Assume nodes i, j and k are nodes connected serially from node i, j and then to k, and the weight for each connection, for example, connection between nodes i and j, is the travel time for a vehicle at the given speed or allowed maximum speed. While planning a path and speed for a vehicle to undertake a task, the SiPaMoP method will change the connection weight (travel time) between nodes i and j in the way of:

$$w_{i,j}^{d} = w_{i,j} + \Delta w \tag{4}$$

where $w_{i,j}$ is the connection weight which is equivalent to the travel time between these two nodes; $w_{i,j}^{d}$ is the new connection weight between nodes i and j. Δw is a parameter determined by

$$\Delta w = \begin{cases} 0 & \text{if node } j \text{ is free} \\ T + \Delta T & \text{otherwise} \end{cases} \tag{5}$$

where, T is the time difference between two relevant vehicles when they approach the same node j, ΔT is the minimal time difference between two vehicles according to safety requirements.

For example, as shown in Fig.3, vehicle V1 is traveling from node d to a via intermediate nodes c, j and b. The SiPaMoP algorithm is planning vehicle V2's path from node h to node l, and finds that V1 and V2 will pass the same node j within $\pm\Delta T$. So the algorithm will automatically change the connection weight between node i and node j by letting $w_{i,j}^{d} = w_{i,j} + T + \Delta T$. As a result of this weight increase between nodes i and j, V2 will either change its path, for example, travel from h to l via i, b (or c) and k, or reduce its travel speed from i to j to allow V1 to pass node j safely. This change in connection weight will automatically be removed after V2 leaves node j towards node k, which allows other vehicles passing the link without a change in speed if there is no collision in this link.

This dynamic change of connection weight between nodes i and j can also solve the collision problem in the case that V1's path is from node d to l via c, j and k, and V2 travels from node h to l. In this case, V2 will pass node j after V1 with a minimum time difference of ΔT.

The search mechanism of the SiPaMoP is based on the Dijkstra algorithm. Other search mechanism such as A* can be easily adapted. By changing the connection weight, the SiPaMoP method is able to vary the coming vehicle's path and/or speed in one of the following three ways depending which one gives the least travel time:

(1) Changing a vehicle's travel speed: this way keeps the same path as that planned without weight change but changes the vehicle's travel speed from node i to node j. As we assume the paths planned previously have higher priority than the current path being planned, and vehicles are running with their maximum speed when there is no collision on their path, the change in weight will only slow down the coming vehicle.

(2) Waiting until other vehicles passing node j. This happens normally in the bottleneck areas.

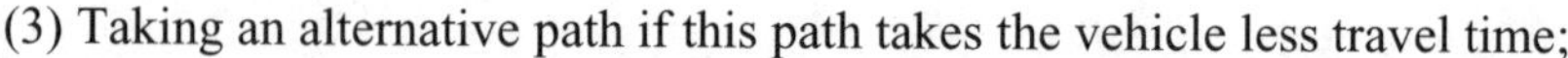

(3) Taking an alternative path if this path takes the vehicle less travel time;

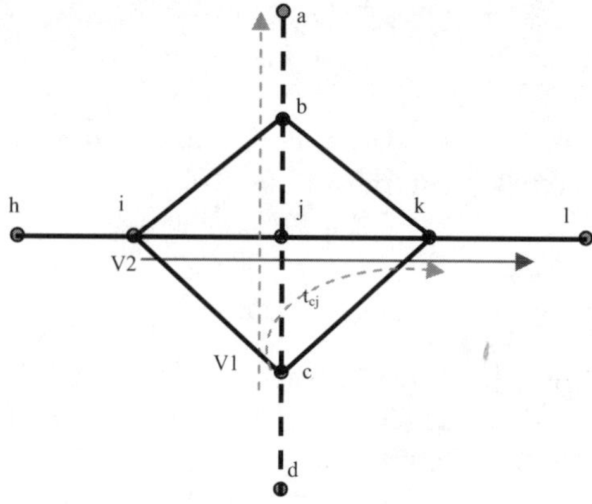

Fig. 3. A bidirectional path network

2.3 Simulation studies on the SiPaMoP

The indoor environment shown in Fig.4 is mapped as a bi-directional flow path network with 186 nodes and a number of links. The crosses (x) in the map represent the nodes where vehicles can access. The lines are the connections among nodes and represent possible paths vehicles have to follow. An autonomous vehicle is supposed to move from node to node following the link between nodes. It can be seen from the map that there are several bottleneck areas which affect the system productivity significantly. Traffic congestion is unavoidable in these areas. This situation becomes worse when more autonomous vehicles are employed. The management of these areas is a crucial issue in path and motion planning for a team of vehicles.

Four case studies are conducted below to demonstrate the performance of the SiPaMoP method under the assumptions of that all transportation demands with origin and destination are known; vehicle V1 is planned first, then vehicle V2; no target time applies to the transportation order; and all changes of traffic condition and the environment are known.

Case 1. Two vehicles collide and one changes its travel speed. In this case, vehicle V1's task is to travel from node 88 to node 55, and vehicle V2 from node 68 to node 107. The paths for the two vehicles are first planned by applying Dijkstra algorithm independently without considering other vehicles, which means that the two paths obtained may not be collision free. The planned paths are: V1 starts from node 88 to node 55 via node 69, and V2 starts from node 68 to 107 via nodes 69, 88 and 109 (Fig.5a and Table 1). Those two vehicles collide at node 69.

By applying the proposed SiPaMoP method to plan the two vehicles' paths, the two vehicles have the same paths (Fig.5b) as those obtained by Dijkstra algorithm, but the collision is avoided in the planning stage by slowing down V2 when it travels from nodes 68 to 69. After V1 leaves node 69, V2 resumes its speed to the normal value. The total travel time of V1, 7.4 (unit time), does not change, but V2 increases its travel time to 25.2 (unit time) from 20.2 (unit time) which is increased by 5 (unit time) (Table 1).

Fig. 4. An indoor environment (left) and its path network (right) (Unit: cm)

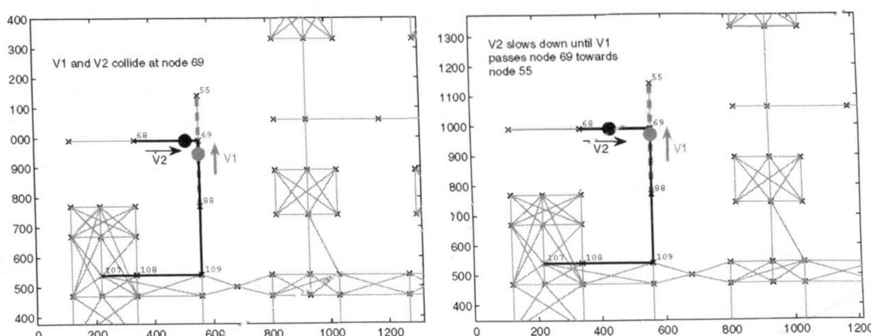

Fig. 5. Case 1; vehicles V1 and V2 collide at node 69. (a) Paths obtained by Dijkstra approach (left), (b) Paths obtained by the SiPaMoP approach (right)

Table 1. Case 1: changing vehicle travel speed: paths and travel time (unit time)

Vehicle 1								Vehicle 2							
Dijkstra only				SiPaMoP				Dijkstra only				SiPaMoP			
Start node	End node	Node time	Total time	Start node	End Node	Node time	Total time	Start node	End node	Node time	Total time	Start node	End node	Node time	Total time
88	69	4.4	4.4	88	69	4.4	4.4	68	69	4.4	4.4	**68**	**69**	**9.4**	**9.4**
69	55	3.0	7.4	69	55	3.0	7.4	69	88	4.4	8.8	69	88	4.4	13.8
								88	109	4.6	13.4	88	109	4.6	18.4
								109	108	4.4	17.8	109	108	4.4	22.8
								108	107	2.4	**20.2**	108	107	2.4	**25.2**

Case 2. Two vehicles collide and one changes its path. In this case, vehicle V1's task is to travel from node 3 to node 89, and vehicle V2 from node 112 to node 38. Similar to Case 1, the paths for the two vehicles are first planned by applying Dijkstra algorithm. As shown in Fig.6a and Table 2, the path of V1 starts from node 3 to node 89 via nodes 4, 21, 37, 57 and 71, and the path of V2 starts from node 112 to 38 via nodes 90, 71, 57 and 37. The two vehicles collide between nodes 37 and 57.

With the application of the proposed SiPaMoP method, vehicle V1 does not change its path and travel speed (Fig.6b), but the collision is avoided in the planning stage by changing V2's path between nodes 112 and 57 (Fig.6b). V2 does not change its speed. With this change of path, V2 travels a longer path than the one obtained by Dijkstra algorithm in order to avoid the collision with V1. As a result, V2's travel time increases from 18.3 (unit time) to 35 (unit time) (Table 2). The change in path, in this case, is better than changing speed because the two vehicles collide in a bottleneck area and V2 needs to wait at least 18 (unit time) (the travel time of V1 from node 21 to node 71 plus the safety distance between the two vehicles).

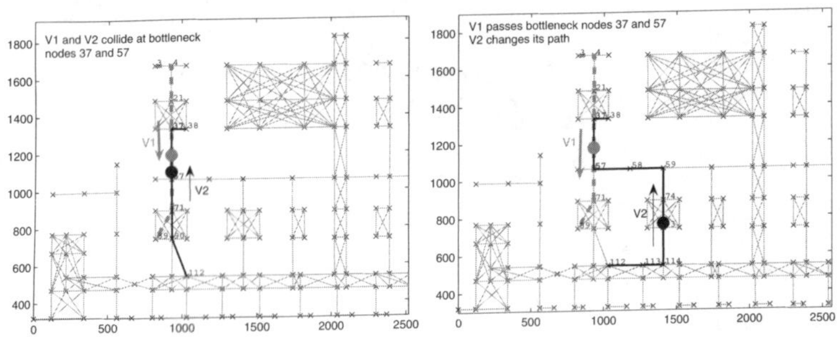

Fig. 6. Case 2: changing path: (a) Paths obtained by Dijkstra approach (left); (b) Paths obtained by the SiPaMoP approach (right)

Table 2. Case 2: changing path: paths and travel time (unit time)

Vehicle 1								Vehicle 2							
Dijkstra only				SiPaMoP				Dijkstra only				SiPaMoP			
Start node	End node	Node time	Total time	Start node	End node	Node time	Total time	Start node	End node	Node time	Total time	Start node	End node	Node time	Total time
3	4	2.2	2.2	3	4	2.2	2.2	112	90	4.5	4.5	112	**113**	4.8	4.8
4	21	3.8	6.0	4	21	3.8	6.0	90	71	3.0	7.5	113	**114**	2.8	7.6
21	37	3.0	9.0	21	37	3.0	9.0	71	57	3.4	11.0	114	**93**	4.0	11.6
37	57	5.4	14.4	37	57	5.4	14.4	57	37	5.4	16.3	93	**74**	3.0	14.6
57	71	3.4	17.8	57	71	3.4	17.8	37	38	2.0	**18.3**	74	**59**	3.4	18.0
71	89	3.7	**21.6**	71	89	3.7	**21.6**					59	**58**	4.6	22.6
												58	57	5.0	27.6
												57	37	5.4	33.0
												37	38	2.0	**35.0**

Case 3. Two vehicles collide and one vehicle waits until the other passes safely. In Case 3, vehicle V1 travels from node 50 to node 18, and vehicle V2 from node 35 to node 48. The paths for the two vehicles obtained by applying Dijkstra algorithm are: V1 starts from node 50 to node 18 via node 34 and V2 starts from node 35 to 48 via nodes 34, 33, 54 and 49. Those two vehicles collide at node 34 (Fig.7a and Table 3). By applying the proposed SiPaMoP method, vehicle V1 and V2 do not change their paths and travel speeds (Fig.7b), but V2 delays its travel by 4.2 (unit time) in order to avoid a collision with V1. With this extra waiting time, V2's travel time increases from 20.7 (unit time) to 24.9 (unit time) (Table 3). In this case, waiting at V2's start node is better than decreasing V2's speed because of the requirement of safety distance between the two vehicles.

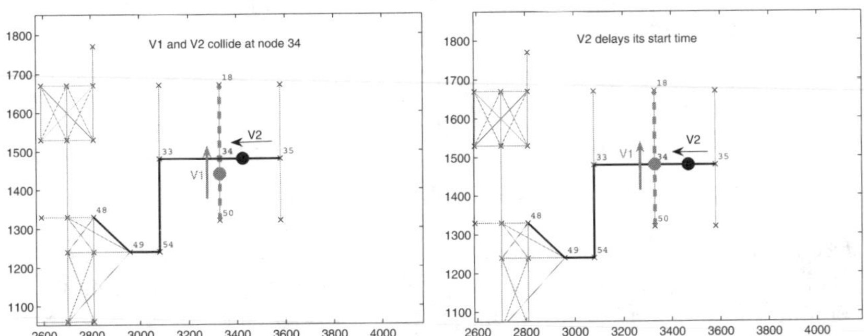

Fig. 7. Case 3: waiting: (a) Paths obtained by Dijkstra approach (left); (b) Paths obtained by the SiPaMoP approach (right)

Table 3. Case 3: waiting: paths and travel time (unit time)

Vehicle 1								Vehicle 2							
Dijkstra only				SiPaMoP				Dijkstra only				SiPaMoP			
Start node	End node	Node time	Total time	Start node	End node	Node time	Total time	Start node	End node	Node time	Total time	Start node	End node	Node time	Total time
50	34	3.2	3.2	50	34	3.2	3.2	35	34	5.0	5.0	**35**	**34**	**8.8**	**9.2**
34	18	3.8	**7.0**	34	18	3.8	**7.0**	34	33	5.0	10.0	34	33	5.0	14.2
								33	54	4.8	14.8	33	54	4.8	19.0
								54	49	2.4	17.2	54	49	2.4	21.4
								49	48	3.5	**20.7**	49	48	3.5	**24.9**

Case 4. 13 tasks and four vehicles. This case study is to show the scalability and efficiency of the SiPaMoP approach in complex applications. There are 13 tasks allocated to four vehicles with vehicle V1 allocated four tasks and each of the other three vehicles allocated three tasks (Table 4). The vehicles' current positions and tasks' start positions and destinations are also listed in Table 4. Except for that the first four tasks start from time zero, all other tasks start immediately after their predecessor tasks [39].

Table 4. Case 4: 13 tasks and four vehicles

Vehicle	Vehicle current position	Task No	Start node	Destination node	Start time
V1	176	1	176	56	0
V2	3	2	3	172	0
V3	135	3	135	25	0
V4	143	4	143	74	0
V1	56	5	56	174	19.18
V2	172	6	172	20	60.20
V3	25	7	25	53	21.61
V4	74	8	74	18	36.59
V1	174	9	174	142	45.91
V2	20	10	20	68	94.72
V3	53	11	53	170	48.46
V4	18	12	18	20	84.71
V1	142	13	142	2	92.39

For the purpose of comparison, Dijkstra algorithm is applied to plan the paths for the four vehicles undertaking their allocated tasks, but without consideration of collisions. Those paths obtained by Dijkstra algorithm are not collision free, but the makespan obtained for completing all the tasks is 151.8 (unit time) which should be the ideal target for the SiPaMoP method. The closer the makespan obtained by SiPaMoP to the target time, the better the SiPaMoP's performance. When the SiPaMoP method is applied, all the collisions are avoided in three ways (i.e. changes in path, travel speed or waiting) automatically in the planning stage depending on which one can reduce the completion time of all tasks. The makespan obtained by SiPaMoP method is 158.6 (unit time) which is very close to the target value of 151.8 (unit time).

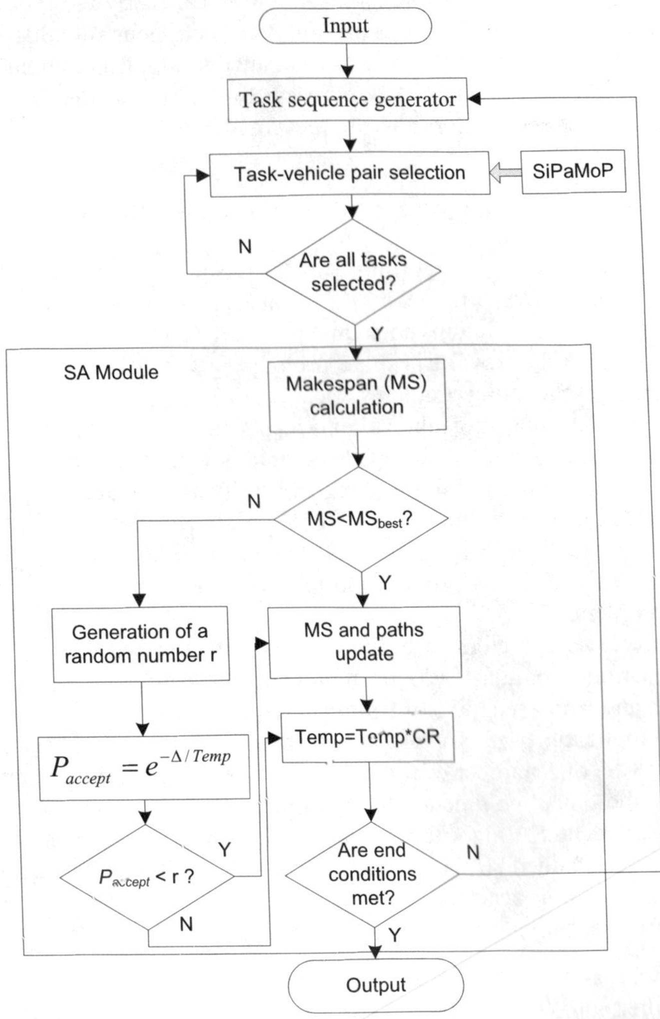

Fig. 8. Flow chart of the SA based STAMC approach

3 Simulated Annealing, Ant Colony and Auction Algorithms based STAMC

The SiPaMoP discussed above is a component in the simultaneous task allocation and motion coordination approach and is called by the task allocation module as shown in Fig.2. In this section, two meta-heuristic algorithms, i.e. simulated annealing and ant colony, and an auction algorithm are investigated for task allocation in the STAMC approach. Simulation studies and comparisons among those three algorithms are presented below.

3.1 Simulated annealing algorithm based STAMC

Simulated Annealing (SA) algorithm was first used for combinatorial optimization problem by Kirkpatrick et al. [23]. This algorithm has proven its ability to find near optimal solutions to many NP-hard combinatorial optimization problems such as the travel salesman problem, graph partitioning, quadratic assignment and scheduling problems, etc.

The simulated annealing algorithm accepts not only solutions with improved cost, but to a limited extent the solutions with deteriorated cost. This feature gives the algorithm hill climbing capability and the ability to avoid being trapped in the local optimum. Initially the probability of accepting inferior solutions is large. This probability is then reduced with the search process proceeding. SA is effective, robust and relatively easy to implement for optimization problems.

There are several factors to be considered with the application of SA: a concise description of a task allocation problem, a quantitative objective function and an annealing schedule of temperatures.

In the application of SA to autonomous vehicle task allocation, the makespan (MS) of a number of tasks, i.e. the maximum travel time among all vehicles, is the objective function to be minimized in the SA based task allocation module. The SA algorithm compares the current makespan (MS) with the best value obtained so far (MS_{best}). If the current MS is less than MS_{best} then the current MS is accepted, otherwise a random number r ($0<r<1$) is generated and the accepting probability (P_{accept}) is calculated:

$$P_{accept} = e^{-\Delta/Temp} \tag{6}$$

where, Δ is the difference between MS and MS_{best} and Temp is the current temperature of the annealing process. Next step is to reduce the annealing temperature by multiplying the Temp with the cooling rate (CR) set initially. In most cases CR is close to 1. In this research, we use the combination of minimum annealing temperature and the number of annealing cycles (j) as the convergence criteria. Fig.8 shows the flow chart of the SA based STAMC.

In the STAMC approach, the SA algorithm and SiPaMoP method conduct task allocation and collision free path and motion planning simultaneously. Task allocation is to find the tasks-vehicle pairs in an order such that the makespan of all available tasks is minimized. The makespan changes with different task orders. In the task-vehicle pair selection, the SiPaMoP module is called for each pair in order to calculate collision free shortest path from the vehicle present node to the task start node and to task destination node. After all the tasks are allocated, the SA decides whether to accept the allocation or not. If yes, this allocation is the best so far. This allocation process continues until the stop criteria are satisfied.

In the case study below, assuming there are 8 tasks (8T) and 4 vehicles (4V) running in the environment shown in Fig.4. The 8 tasks, including their start and destination nodes, are listed in Table 5, and the 4 vehicles' start positions are listed in Table 6. All the tasks are assumed to have equal priority. Task allocation process starts with randomly selected task allocation sequence. Values of the SA parameters in this case study are: start temperature =100, cooling rate CR = 0.9 and stop temperature $T_0 = 1$.

Table 5. Eight tasks' start and destination nodes

Task No	Start node	Destination node
1	177	153
2	43	83
3	113	115
4	91	148
5	166	172
6	142	138
7	85	33
8	4	76

Table 6. Four vehicles' initial positions

Vehicle No	Start node
1	174
2	171
3	77
4	167

The simulation results of task allocation obtained by the proposed SA based STAMC method are listed in Table 7 and Fig.9. For the 8T-4V problem, vehicle V1 is assigned to tasks 4 and 5, V2 task 7, V3 tasks 3, 6 and 2, and V4 tasks 1 and 8. The Gantt chart of the allocated task-vehicle pairs are shown in Fig.9. The planning order or the priority order of the 8 tasks is finally determined to be 1, 7, 3, 6, 4, 5, 2 and 8, which results in the makespan of 45.96 (unit time). The processing time of each task is also listed in the last column in Table 7.

Table 7. Simulation results obtained by SA based STAMC method

Task No	Start node	Destination node	Allocated vehicle	Planning order	Processing time (unit time)
1	177	153	V4	1	15.02
7	85	33	V2	2	45.34
3	113	115	V3	3	10.90
6	142	138	V3	4	15.30
4	91	148	V1	5	19.62
5	166	172	V1	6	12.26
2	43	83	V3	7	19.49
8	4	76	V4	8	30.94

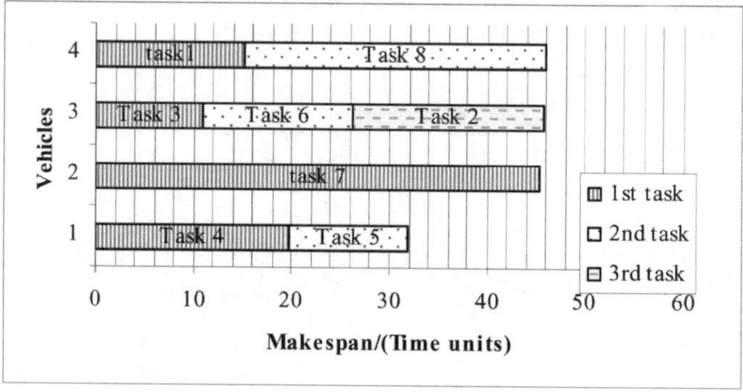

Fig. 9. Task allocation results obtained from SA based STAMC approach

3.2 Ant colony algorithm based STAMC

Ant colony algorithms were inspired by the observation of real ant colonies. Ants are social insects that live in colonies and whose behavior is directed more to the survival of the colony as a whole. An important and interesting behavior of ant colonies is their foraging behavior, and, in particular, how ants can find the shortest paths between food sources and their nest. These behaviors are modeled mathematically as a meta-heuristic algorithm where artificial ants update pheromones while traveling.

Ant colony algorithms were proposed by Colorni et al [10] as a multi-agent approach to solve difficult combinatorial optimization problems such as the traveling salesman problem and the quadratic assignment problem. There is currently ongoing activity in the scientific community to extend and apply ant-based algorithms to many different discrete optimization problems.

When an ant algorithm is applied to task allocation for multi-autonomous vehicles, an ant represents a vehicle and starts from its start node (depot). The first task of each ant is allocated randomly, then each ant selects the next task from the available task list until all tasks are selected. Each ant's total travel time is calculated based on its selected tasks, planned routes and travel speeds. The maximum traveling time out of all the ants is the makespan (MS).

For the selection of tasks that are not yet allocated, two aspects should be taken into account: how good the previous task-vehicle pairs are and how promising the next task selection is in general. The first information is stored in the pheromone trails τ_{ij} associated with each task-vehicle pare, whereas the second is the heuristic function. The measure of desirability, called visibility, is denoted by η_{ij}.

In an ant colony optimization algorithm (ACO), each ant selects its next task based on the probability of selection. The probability distribution p_{ij} is calculated by:

$$p_{ij} = \frac{(\tau_{ij})^{\alpha}(\eta_{ij})^{\beta}}{\sum_{u \in T_l}(\tau_{iu})^{\alpha}(\eta_{iu})^{\beta}} \tag{7}$$

where τ_{ij} is equal to the amount of pheromone of selecting task j after task i. The value of η_{ij} is defined as the inverse of the complete traveling time which is the sum of the transient time for the vehicle from its current position to the start node of the task, and the task processing time (vehicle travel time from the task start node to end node). The probability distribution is biased by parameters α and β that determine the relative influence of the trails and the visibility, respectively. Allocated tasks are removed from the task list T_l.

In order to improve the solutions, the pheromone trails of ants must be updated to reflect the ants' performance and the quality of the solutions found. This update is a key element to the adaptive learning technique of ACO and helps in improving the subsequent solutions. The update is conducted by reducing the amount of pheromone elements in the pheromone table of each task-ant combination in order to simulate the natural evaporation of pheromone and to ensure that no one task-vehicle combination becomes dominant. This is done by the following equation:

$$\tau_{ij} = (1-\lambda)\tau_{ij} + \lambda\tau_0 \tag{8}$$

where λ is a parameter that controls the speed of evaporation and τ_0 is the initial pheromone value. In our algorithm τ_0 is the inverse of the completed tour cost (travel time) of each ant.

The probability calculations are based on travel time instead of travel distance used in other applications such as TSP problems. The travel time comes from the SiPaMoP module. Each ant calculates the probability to select the next task based on Eq. (7). The task, which gives highest probability, will be selected. Each ant selects one task at a time until all the tasks in the task list are selected. After all tasks are allocated, the makespan can be calculated and the best makespan so far can be updated accordingly. This process repeats until the fixed number of cycles is reached.

The simulation example in Section 3.1 is also used in the ACO based STAMC. Values of the ACO parameters in this case study are: $\alpha = 3$, $\beta = 5$, $\lambda = 0.7$. The simulation results of task allocation by the ACO based method are listed in Table 8 and Fig.10. For the 8T-4V problem, vehicle V1 is assigned to tasks 2 and 6, V2 to tasks 1 and 8, V3 to tasks 3 and 4, and V4 to tasks 5 and 7. The Gantt chart of the allocated task-vehicle pairs are shown in Fig.10. The planning order or the priority order of the 8 tasks is finally determined as 2, 1, 3, 5, 6, 8, 4 and 7. The processing time of each task is also listed in the last column in Table 8. The makespan obtained from this ACO based approach is 59.94 (unit time) which is the completion time of V4.

Table 8. Simulation results obtained by ACO based STAMC method

Task No	Start node	Destination node	Allocated vehicle	Planning order	Processing time (unit time)
2	43	83	V1	1	37.48
1	177	153	V2	2	19.66
3	113	115	V3	3	10.90
5	166	172	V4	4	14.32
6	142	138	V1	5	10.25
8	4	76	V2	6	30.94
4	91	148	V3	7	16.44
7	85	33	V4	8	45.62

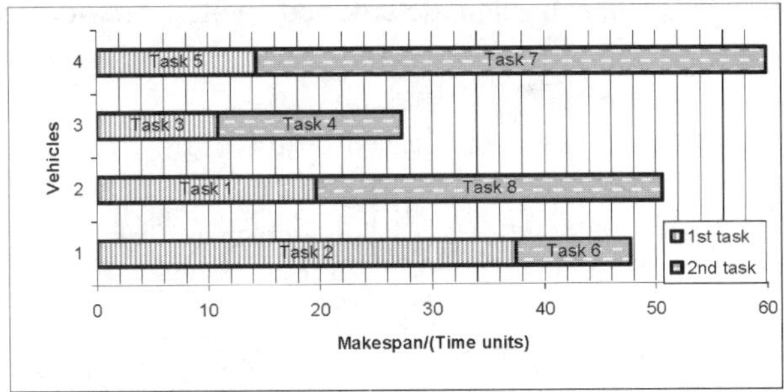

Fig. 10. Task allocation results obtained from ACO based approach

3.3 Auction algorithm based STAMC

An Auction algorithm (AA), which is derived from a typical auction process has been used in many applications such as assignment, transportation, shortest path search problems and recently for mobile robots applications [3, 7, 17]. For task allocation process in the STAMC approach, an adapted version of the auction algorithm, called first cycle auction process is used due to its ability to produce results quickly.

The main steps of the auction based simultaneous task allocation and path and motion planning algorithm are: (1) initially, the task sequence is generated randomly by the task generator; (2) The first task is broadcast to all autonomous vehicles to allow everyone to place a bid for the task; (3) After every vehicle has offered a bid, a winner is determined and allocated the task; (4) The second task is then broadcast, followed by bidding from vehicles and selecting winner. This auction process continues until all tasks are allocated.

For every vehicle, the calculation of a bid is based on the travel time to complete the current broadcast task and any previously allocated (as the winner) tasks. Once the travel time has been calculated it is used to post a bid (e.g. $B_{1,j}$ to $B_{i,j}$ in Fig. 11). The traveling time and collision free paths are obtained using the SiPaMoP algorithm. A winner is then determined based on the lowest travel time to perform the task. In the completion time calculation for an autonomous vehicle, previous task commitments of this vehicle are also considered. This will balance the load of vehicles. For example, if a previous task is allocated to a particular vehicle, then there will be fewer tendencies for the same vehicle to win the next task. In addition, load balancing of the vehicles can be achieved partially. After all tasks of the current task sequence are allocated, the makespan is calculated. This process continues for a fixed number of cycles with a different task sequence generated randomly in each cycle.

The best task sequence is then determined, which provides the minimum makespan.

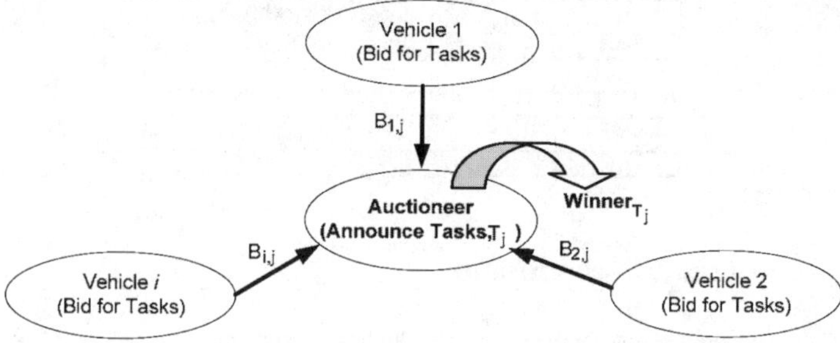

Fig. 11. Auction Process

Table 9. Simulation results obtained by AA based STAMC method

Task No	Start node	Destination node	Allocated vehicle	Planning order	Processing time (unit time)
8	4	76	V1	1	35.37
7	85	33	V2	2	45.34
3	113	115	V3	3	10.90
1	177	153	V4	4	15.02
2	43	83	V3	5	23.42
4	91	148	V4	6	15.04
5	166	172	V5	7	12.26
6	142	138	V3	8	10.25

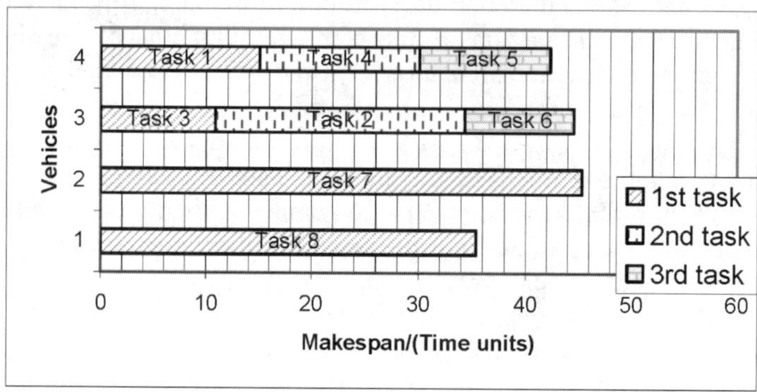

Fig. 12. Task allocation results obtained from AA based approach

The simulation example in Section 3.1 is also used in the simulation study of the AA based STAMC. The simulation results of task allocation by the AA based method are listed in Table 9 and Fig.12. For the 8T-4V problem, vehicle V1 is assigned to task 8, V2 to task 7, V3 to tasks 3, 2 and 6, and V4 tasks 1, 4 and 5. The Gantt chart of the allocated task-vehicle pairs are shown in Fig.12. The planning order of the 8 tasks is optimized as 8, 7, 3, 1, 2, 4, 5 and 6. The processing time of each task is also listed in the last column in Table 9. The makespan obtained from this AA based approach is 45.34 (unit time) which is the completion time of vehicle V2.

3.4 Comparison and discussion

In the STAMC approach studied above, task allocation is based on three algorithms namely ant colony optimization (ACO), simulated Annealing (SA) and auction Algorithm (AA). The three algorithms were applied for the same simulation case, i.e. 8 tasks and 4 vehicles. The same parameter values are used in this simulation. It can be seen from the discussions above that the task allocation results are different, but their makespans are similar (Table 10). The makespans obtained from SA and AA are 45.96 (unit time) and 45.34 (unit time), respectively, but the one from ACO is 59.94 (unit time) which is about 4 (unit time) more than the other two. The vehicle utilizations (i.e. the total travel time of all 4 vehicles for performing the 8 tasks) are also given in Table 10. Vehicle utilization in SA based STAMC (168.87 (unit time)) is very close to that of the AA base approach (167.6 (unit time)), the ACO based approach results in more vehicle utilization (185.6 (unit time)).

Table 10. Comparison of Makespans and vehicle utilization of the three methods (unit time)

Parameter	Approach		
	SA	ACO	AA
Makespan (unit time)	45.96	59.94	45.34
Vehicle utilization (unit time)	168.87	185.60	167.60

Since a task processing time includes two components, namely the travel time from a vehicle's initial node to a task's start node and the time taken from the task start node to its destination node, task processing time varies even if the same task is allocated to different vehicles. This is clearly visible from task 8, where SA and ACO methods take approximately 30 (unit time) to process it and AA method takes 35 (unit time) to process the same task. In addition to this, the processing time of the same task can also change when the vehicle performs it at different time due to traffic congestion.

In order to extensively test the performance of the proposed STAMC method for multi-autonomous vehicle task allocation and motion coordination, more simulation studies are conducted and the results are shown in Table 11. Two task-vehicle pairs, i.e. 20 tasks and 5 vehicles (20T-5V) and 40 tasks and 10 vehicles (40T-10V), are used. In each task-vehicle pair, there are six cases, each case has different tasks and vehicle positions. The number of runs of each algorithm for each task-vehicle pair is given in the second column of Table 11. Out of the makespans calculated in all the runs in each case of each pair, the minimum value is selected as the makespan for the given task-vehicle-case. It can be seen from Table 11 that the proposed STAMC is scalable and the makespans obtained from the three algorithms are similar.

Table 11. Makespan comparisons of ACO, SA and AA methods

Tasks-Vehicles	No. of runs	ACO	SA	AA
20T-5V-case 1	400	130.1	124.0	125.7
20T-5V-case 2	400	107.3	112.6	108.7
20T-5V-case 3	400	137.8	141.1	141.1
20T-5V-case 4	400	118.8	112.6	106.1
20T-5V-case 5	400	94.4	95.2	89.5
20T-5V-case 6	400	110.4	105.2	105.6
40T-10V-case 1	200	108.3	105.4	100.0
40T-10V-case 2	200	119.0	109.4	107.0
40T-10V-case 3	200	121.1	127.4	112.0
40T-10V-case 4	200	133.3	115.5	116.0
40T-10V-case 5	200	124.1	119.7	115.0
40T-10V-case 6	200	125.3	107.6	110.0

4 Conclusions and Remarks

In this research, a simultaneous path and motion planning (SiPaMoP) method is first presented for multi-autonomous vehicle collision-free path and motion planning. This method is demonstrated to be able to efficiently coordinate the motion of multiple autonomous vehicles in strictly constrained environments by conducting vehicle path and motion planning and collision avoidance simultaneously. It can also efficiently manage congestion and bottleneck areas, avoid collisions and handle dynamic changes in traffic conditions and environments.

A simultaneous task allocation and motion coordination (STAMC) approach is then studied. Integration of task allocation with the SiPaMoP method provides a way to dispatch vehicles by taking collision-free path and motion planning into account. Two metaheuristic algorithms, i.e. simulated annealing and ant colony, and an auction algorithm are investigated for task allocation in the STAMC approach. Simulation studies and performance comparison of the three algorithms demonstrated the effectiveness and efficiency of those algorithms in autonomous vehicle planning and scheduling.

Comparison of the STAMC to the sequential approach (Fig.1) and dispatching rule based approaches, and application of the proposed method in a fully automated container terminal with over 15 autonomous vehicles are currently being conducted. Combination of this approach with the scheduling and optimization of other machines such as cranes is another important issue and will be studied.

5 Acknowledgements

This work is supported by the ARC Centre of Excellence programme, funded by the Australian Research Council (ARC) and the New South Wales State Government, Australia. The authors like to acknowledge Professor G. Dissanayake and Dr. X. Wu for their contributions to this research.

References

[1] Asef-Vaziri A, Laporte G (2005) Loop based facility planning and material handling. European Journal of Operational Research, 164: 1-11

[2] Baker B, Ayechew M (2003) A genetic algorithm for the vehicle routing problem. Computers and Operations Research, 30: 787-800

[3] Bertsekas DP, Castanon DA (1989) The Auction Algorithm for Transportation Problems. Annals of Operations Research, 20: 67-96

[4] Bish EK, Leong TY, Li CL, Cheong Ng JW, Simchi-Levi D (2001) Analysis of a new vehicle scheduling and location problem. Naval Research Logistics 48: 363-385

[5] Bish EK, Chen FY, Leong YT, Nelson BL, Ng WC, Simchi-Levi D (2005) Dispatching vehicles in a mega container terminal. OR Spectrum 27: 491-506

[6] Boddy MS, Bennett BH, Isle BA, Isle RA (2004) NASA Planning and Scheduling Applications: Emerging Technologies and Mission Trends. Final Report, 28/03/2004, Adventium Labs, USA

[7] Botelho SC, Alami R (1999) M+: A scheme for multi-robot cooperation through negotiated task allocation and achievement. Proceedings of the IEEE International Conference on Robotics and Automation (ICRA'99) Detroit, Michigan, pp1234-1239

[8] Buriol LS, Ressende MGC, Thorup, M (2003) Speeding up Dynamic Shortest Path Algorithms. At&T Labs Research Technical Report TD-5RJ8B

[9] Chiba R, Ota J, Arai T (2004) Integrated design for AGV systems using cooperative co-evolution. Proceedings of IEEE/RSJ International Conference on Intelligent Robots and Systems, September 28-October 2, 2004, Sendai, Japan, pp3791-3796

[10] Colorni A, Dorigo M, Maniezzo V (1991) Distributed Optimization by Ant Colonies. Proceedings of the First European Conference on Artificial Life (ECAL 91), pp134-142

[11] De Koster RBM, Le-Anh T, Van der Meer JR (2004) Testing and classifying vehicle dispatching rules in three real-world settings. Journal of Operations Management 22:369-386

[12] Ebben M, Van der Heijden M, Hurink J, Schutten M (2004) Modeling of capacitated transportation systems for integral scheduling. OR Spectrum 26: 263-282.

[13] Egbelu PJ, Tanchoco JMA (1984) Characterization of automatic guided vehicle dispatching rules. International Journal of Production Research 22: 359-374.

[14] Farling BE, Mosier CT, Mahmoodi F (2001) Analysis of automated guided vehicle configurations in flexible manufacturing systems. International Journal of Production Research, 39: 4222-4239

[15] Ferguson D, Stentz A (2005) The Delayed D* algorithm for efficient path replanning. Proceedings of the 2005 IEEE International Conference on Robotics and Automation Barcelona, Spain, pp2057-2062

[16] Fu L, Rilett LR (1996) Expected shortest paths in dynamic and stochastic traffic networks. Transportation Research, Part B: Methodological 32: 499-516

[17] Gerkey BP, Mataric MJ (2001) Sold! Auction methods for multi-robot coordination. IEEE Transactions on Robotics and Automation, 18: 758-768

[18] Grunow M, Günther HO, Lehmann M (2004) Dispatching multi-load AGVs in highly automated seaport container terminals. OR Spectrum 26: 211-235

[19] Horn MET, Efficient modelling of travel in networks with time-varying link speeds. CSIRO Mathematical and Information Sciences Technical Report CMIS 99/97 http://www.cmis.csiro.au/Mark.Horn/

[20] Hsieh S, Kang TP (1998) Developing AGVs Petri Net Control Models from Flowpath Nets. Journal of Manufacturing Systems, 17: 237-250

[21] Husdal J (2000) Fastest path problems in dynamic transportation networks. University of Leicester, UK http://www.husdal.com/mscgis/research.htm

[22] Hussain T, Montana D, Vidaver G (2004) Evolution-based deliberative planning for cooperating unmanned ground vehicles in a dynamic environment. Proceedings of the Genetic and Evolutionary Computation Conference (GECCO)

[23] Kirkpatrick S, Vecchi MP (1983) Optimization by Simulated Annealing. Science, 220- 4598: 671-680

[24] Koo PH, Lee WS, Jang DW (2004) Fleet sizing and vehicle routing for container transportation in a static environment. OR Spectrum 26: 193-209

[25] Kulatunga AK, Liu DK, Dissanayake G (2004) Simulated Annealing Algorithm based Multi-Robot Coordination. Proceedings of the 3rd IFAC Symposium on Mechatronic Systems, Sydney, Australia, September, 2004, pp411-416

[26] Kulatunga AK, Liu DK, Dissanayake G, Siyambalapitiya SB (2006) Ant colony optimization technique for simultaneous task allocation and path planning of autonomous vehicles. Proceedings of the IEEE International Conference on Cybernetics and Intelligent Systems (CIS), 7-9 June, 2006, Bankok, Thailand, pp823-828

[27] Leong C Y (2001) Simulation study of dynamic AGV-container job deployment scheme. Master of science, National University of Singapore

[28] Lim JK, Kim KH, Yoshimoto K, Lee JH (2003) A dispatching method for automated guided vehicles by using a bidding concept. OR Spectrum 25: 25-44
[29] Liu DK, Wu X, Kulatunga AK, Dissanayake G (2006) Motion Coordination of Multi-Autonomous Vehicles in Dynamic and Strictly Constrained Environments. Proceedings of the IEEE International Conference on Cybernetics and Intelligent Systems, 7-9 June 2006, Thailand, pp204-209
[30] Meersmans PJM, Wagelmans APM (2001) Dynamic Scheduling of Handling Equipment at Automated Container Terminals https://ep.eur.nl/handle/1765/137
[31] Moorthy RL, Hock-Guan W (2000) Deadlock prediction and avoidance in an AGV system. Master of science, Sri Ramakrishna Engineering College, National University of Singapore
[32] Moorthy RL, Hock-Guan W, Ng WC, Chung-Piaw T (2003) Cyclic deadlock prediction and avoidance for zone-controlled AGV system. International Journal of Production Economics, 83: 309-324
[33] Müller T (1983) Automated Guided Vehicles. IFS (Publications) Ltd. Springer-Verlag, K/Berlin
[34] Oboth CB, Karwan MR (1999) Dynamic conflict-free routing of automated guided vehicles. International Journal of Production Research, 37: 2003- 2028
[35] Philippsen R, Siegwart R (2005) An interpolated dynamic navigation function. Proceedings of the 2005 IEEE International Conference on Robotics and Automation Barcelona, Spain, April 2005, pp3793-3800
[36] Powell WB, Carvalho TA (1998) Real-time optimization of containers and flatcars for intermodal operations. Transportation Science 32:110-126
[37] Qin Y, Sun D, Li N, Cen Y (2004) Path planning for mobile robot using the particle swarm optimization with mutation Operator. Proceedings of the 2004 International Conference on Machine Learning and Cybernetics, Aug. 2004, pp2473-2478
[38] Qiu L, HsuW-J (2001) Scheduling of AGVs in a mesh-like path topology: a case study in a container terminal, Technical Report CAIS-TR-01-35, School of Computer Engineering, Nanyang Technological University
[39] Qiu L, Hsu WJ, Huang SY, Wang H (2002) Scheduling and Routing Algorithms for AGVs: a survey. International Journal of Production research 40: 745-760
[40] Seifert RW, Kay MG, Wilson JR (1998) Evaluation of AGV routeing strategies using hierarchical simulation. International Journal of Production Research 36: 1961-1976
[41] Srinivasan MM, Bozer YA, Cho M (1994) Trip-based material handling systems: Throughput capacity analysis. IIE Transactions 26: 70-89.
[42] Stentz A (1994) Optimal and efficient path planning for partially-known environments. Proceedings of the IEEE International Conference on Robotics and Automation, San Diego, CA, USA, May 1994
[43] Taghaboni-Dutta F, Tanchoco JM (1995) Comparison of dynamic routing techniques for automated guided vehicle system. International Journal of Production Research, 33: 2653-2617
[44] Ulusoy GU, Bilge Ü (1993) Simultaneous scheduling of machines and automated guided vehicles. International Journal of Production Research 31: 2857-2873

[45] Ulusoy GU, Sivrikaya SF, Bilge U (1997) A genetic algorithm approach to the simultaneous scheduling of machines and automated guided vehicles. Computers & Operations Research, 24: 335-351

[46] Vis IFA (2006) Survey of research in the design and control of automated guided vehicle systems. European Journal of Operational Research 170: 677-709

[47] Wagner D, Willhalm T, Zaroliagis C (2003) Dynamic Shortest Paths Containers. Theoretical Computer Science, 92 http://www.elsevier.nl/locate/entcs/volume92.html

[48] Yamashita H (2001) Analysis of dispatching rules of AGV systems with multiple vehicles. IIE Transactions 33: 889-895

[49] Zhan FB, Noon CE (1996) Shortest Path Algorithms: An Evaluation using Real Road. Transportation Science, 32: 65-73

Scheduling Production and Distribution of Rapidly Perishable Materials with Hybrid GA's

David Naso, Michele Surico, and Biagio Turchiano

Dipartimento di Elettrotecnica ed Elettronica, Politecnico di Bari (Italy), Email naso@poliba.it

Summary. This paper considers the problem of scheduling the production and distribution activities of a network of plants supplying rapidly perishable materials. The main challenges are (1) addressing simultaneously several interrelated scheduling and routing problems, and (2) finding solutions that take into account the amount of cooperation/interaction necessary between the various plants. We propose a strategy that combines genetic algorithms and fast schedule construction heuristics for job scheduling and truck routing. The effectiveness of the approach is evaluated against other methods used in industrial practice on a challenging large-scale case study.

1 Introduction

Supply networks (SN's) are cooperative organizations of partially autonomous production and distribution centers. The activities of SN's involve a number of large-scale, interrelated assignment, scheduling and routing problems. Due to the well-known combinatorial nature of these problems, operating SN's is very challenging, especially when the supplied goods have an extremely short lifespan, as in the case of ready-mixed concrete (RMC) considered in this chapter. Since the product is rapidly perishable, it must be produced on demand and delivered within the time window specified by customers according to their construction plans. Clearly, it is not possible to produce in advance a certain amount of the good to cope with possible demand peaks, and therefore production and distribution activities must be necessarily synchronized and optimized on a daily basis. The perishable nature of the RMC is not the only additional complication of the supply problem. Another significant difficulty is related to the fact that orders are often much larger than the capacity of a single truck, and thus they must be split in long sequences of concatenated deliveries, which must also be coordinated to avoid interruptions between truck unloads. Both a delay between two consecutive unload operations, and a delay that exceeds the lifespan of the

D. Naso et al.: *Scheduling Production and Distribution of Rapidly Perishable Materials with Hybrid GA's*, Studies in Computational Intelligence (SCI) **49**, 465–483 (2007)

transported RMC are extremely undesirable events that involve significant economical and environmental costs. Thus, most of the optimization efforts are dedicated to minimize the likelihood of these two types of events, while maintaining the adequate levels of resource utilization and profit.

In spite of the fact that mobile communication technologies make it possible to share detailed information about distributed operations (e.g. customer orders, production schedules, location and status of delivering vehicles) in real time, the time needed by general-purpose mathematical programming solvers to find satisfactory solutions for such large-scale problems is generally too long to make these tools really useful for SN managers. Therefore, industrial companies tend to rely on skilled operators that work out scheduling decisions basing on their experience [1, 10], and plan the activities on short time horizons considering each production plant separately from the other ones. In this way, they renounce to the potential benefits of longer-term and larger-scale optimization to achieve a reduced risk of delayed delivery [20].

This chapter presents a meta-heuristic algorithm to address the RMC production and distribution problem in an efficient way. Our dynamic scheduling approach combines the following complementary tools:

1) A detailed mathematical programming model of the large-scale logistic problem that unambiguously specifies the decision variables, the technical requirements, and the constraints that must be fulfilled at each single stage of the supply.

2) A set of fast constructive heuristics that use the mathematical model to find a fully feasible solution for the entire SN starting from a given assignment of a predefined subset of problem variables.

3) A stochastic search engine based on Genetic Algorithms (GA's), and designed to optimize the subset of problem variables used by the constructive heuristics to build feasible solutions. In other words, the GA performs the optimization of the predefined objective function by continuously interacting with the constructive heuristics until the stopping criterion is met.

The integration of these tools forms a global meta-heuristic approach that achieves an effective tradeoff between exploration (the GA exploring various regions in the solution space) and exploitation mechanisms (local heuristics concentrating the search into a specific region of the search space).

With respect to our earlier work described in [13], the approach presented here overcomes some significant limitations, allowing us to extend it to the case of production centers (PC's) with more than one dock, trucks with different speed, and operation slack times of different length. The meta-heuristic tool is very effective, and can be used to address scenarios in which several hundred delivery jobs (up to 800 in our numerical experiments) must be scheduled and coordinated so as to meet product demands on a relatively wide geographical area. The chapter is organized as follows. Section 2 overviews the related literature, discussing the analogies between the RMC problem and other scheduling and routing problems. Section 3 summarizes the essential aspects of the mathematical model, introducing the essential objectives and constraints. Section 4 describes the two components of the search algorithms, i.e. the master GA engine, and the subordinate constructive heuristics. Section 5 summarizes the numerical investigation based on a large scale SN with over 10 nodes and hundreds of delivery jobs, and Section 6 is dedicated to the conclusive remarks.

2 Literature overview

A comprehensive overview of the RMC supply problem can be found in [20]. The supply process can be essentially viewed as a joint production-distribution problem, and thus related to various research areas of industrial automation and operation research. The RMC job loading stage can be modeled as a scheduling problem on parallel machines with earliness/lateness objectives, which is proven to be NP-hard (see e.g., [15]). Reference [2] considers the problem of scheduling a single RMC batch plant so to satisfy delivery time constraints. The work assumes that an unlimited fleet is available, and that each load has to be delivered at a certain desired time, while delays are addressed with a linear earliness/tardiness penalty. Reference [8] devises a scheduling algorithm composed of a timing algorithm that computes the optimal start time of each job, and a sequencing strategy to determine the processing order based on a GA. A similar approach is described in [16] and [12].

Similarly, the delivery stage of RMC can be viewed as a particular vehicle routing problem known as Multi-Depot Multi-Vehicle Routing Problem with Time Windows (MDVRPTW for brevity). The MDVRPTW is NP-hard, and even finding a feasible solution in the case of a fleet of fixed size can be an NP-complete problem [17]. Several heuristic approaches providing satisfactory solutions for various MDVRPTW problems in acceptable times have been recently surveyed [6, 21, 9], and significant results have been achieved also using GA's [18, 19], and other stochastic approaches [3]. Most literature considers an unlimited number of vehicles, although recent progresses on the case of a limited fleet have also been reported (see [7] and the included references). Routing problems in the specific context of RMC supply are analyzed in [10]. In particular, that paper considers both the carriers for the delivery of the concrete to customers' sites, and the pumps necessary to unload the cement from the trucks, and focuses on a Tabu Search algorithm for their solution. Studies on dynamic vehicle routing with variable travel times are also emerging (an updated survey is in [4]). Reference [1] integrates the scheduling and the routing problems. The paper considers the RMC supply from a single PC equipped with a fleet of vehicles with identical capacity. Truck loading and unloading times are assumed to be equal for all the jobs, so that the production scheduling is reduced to a permutation problem, and is solved with a GA. An extensive discussion on the technical differences between these recent approaches and our meta-heuristic tools is available in [13] and omitted here for brevity.

3 A Model for the RMC production and distribution

A SN for RMC consists of a consortium of P independent and distributed PC's or depots ($p \in \{1,...,P\}$ is the depot index) located in a given geographical area. Each depot has a number of loading docks (D_p) where the RMC is loaded on the trucks for its delivery. Each dock d can service only one truck at a time, but the docks of a PC work concurrently. Thus, we define the total number of loading docks:

$$D = \sum_{p=1}^{P} D_p \, . \tag{1}$$

Every day the SN has to process a set of R requests or orders from different customers ($r \in \{1,...,R\}$ is the request index), whose construction sites are spread over a certain geographical area. An order r consists of a customer-specified delivery time window $[\tau_r^{ED}, \tau_r^{LD}]$ (earliest and latest delivery time), a required amount Q_r, and a delivery location.

The SN is also equipped with a fleet of K trucks ($k \in \{1,...,K\}$ is the truck index) to deliver the product to customers' construction site. Some PC's in the SN do not own trucks, and thus rely on the fleet of the other SN nodes for the delivery of their products. Large demands exceed the capacity of one truck, and must be split in a number of sub-demands simply called *jobs* ($i_r \in \{1,...,N_r\}$ is the job index and N_r is the total number of jobs composing demand r). As mentioned, these jobs have to be synchronized, because the overall unloading process at the customer site must be uninterrupted, so as to guarantee the homogeneity of the placed fluid and the final mechanical properties of the solidified concrete. Each job is produced at a dock by mixing water with dry components directly when the product is being loaded on the truck (production is simultaneous with loading), and each dock can load one truck at a time. Loading is a non-preemptive operation encompassing a fixed setup time and a loading time interval that depends on the loading rate of the dock and on the size of the load. The latter is, in turn, limited by truck capacity.

In case none of the nodes of the SN is able to produce a certain (fraction of) demand with the requested characteristics, the production of the residual jobs is outsourced to external suppliers. Similarly, the SN can hire additional trucks to deliver jobs that cannot be handled by the internal fleet. Clearly, outsourcing production and hiring further vehicles involve additional costs, and are performed only when necessary. Moreover, the need of additional resources is not only related to the amount of requests, but also to the actual utilization of internal resources, which is also influenced by the effectiveness of cooperation between the various partner nodes.

Let us define a task of a truck as the delivery of a job to its destination, and introduce the task index $m \in \{1,..., M_k\}$, where M_k is the last task of truck k. As Fig.1 shows, the generic task m of truck k is made up by several phases, characterized by the following times:

1) The loading waiting time (LWT_{km}) before the job load starts.

2) The loading time for job i of demand r (LT_{ir}), influenced by the loading rate of the dock and by the load size.

3) The travel time from the source depot of job i_r to the destination (SDT_{ir}).

4) The unloading waiting time (UWT_{km}): Each truck is normally scheduled to arrive slightly in advance, and thus it must wait either for the initial placement of customer unloading pump or for the end of the previous unload.

5) The unloading time (UT_{ir}), depending on the unloading rate of the customer and on the load size.
6) The travel time from the customer to the PC of its next task (DST_{ir}).

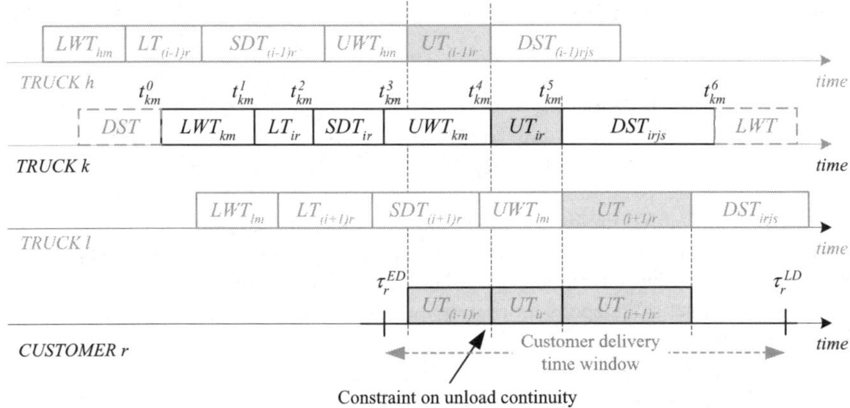

Fig. 1. Activities of truck k and their synchronization with those of other trucks at the delivery site.

Each phase has a starting and ending time described by the following notation:

t^0_{km} start of the *m*-th task for truck *k*;

t^1_{km} start of the loading phase;

t^2_{km} end of the loading phase and start of the trip toward the customer;

t^3_{km} end of the outward trip and start of the waiting time for the unloading phase;

t^4_{km} end of the waiting time and start of the unloading phase;

t^5_{km} end of the unloading phase and start of the trip toward the next source depot;

t^6_{km} end of the trip toward the source depot of the next task.

Each truck *k* executes one task after another, and the end of one task corresponds exactly to the start of the next one, i.e. it holds:

$$\forall k \in \{1,...,K\}\ , \forall m \in \{1,...,M_k - 1\}$$

$$t^0_{k(m+1)} = t^6_{km}\ . \tag{2}$$

A full solution for our problem involves the definition of all the production activities at the PC's and distribution activities of the trucks. A solution can be univocally described by a set of integer and real-valued decision variables. The first integer decision variable K_o defines the number of additional trucks that must be hired to distribute all the produced jobs. Then, a set of binary variables is used to indicate assignment decisions. Such variables are defined as follows:

$X_{ikm} \in \{0,1\}$ if the job i_r is assigned to truck k as m-th task, X_{ikm}=1, otherwise X_{ikm}=0.
$Y_{id} \in \{0,1\}$ if job i_r is produced at the dock d, Y_{id} = 1 otherwise Y_{id} = 0.
$Y_{io} \in \{0,1\}$ if the production of job i_r is outsourced, Y_{io} =1, otherwise Y_{io} =0.

Finally, real-valued timing variables indicate the start time of truck tasks. As the various truck activities are executed in sequence, their start and end times can be easily related to each other (see Fig.1). Thus, the activity scheduling can be specified with a minimal set of independent timing variables. In particular, for each truck k, our model uses the previously defined truck waiting times LWT_{km} and UWT_{km}, and the start of the first task t_{k0}^{0} as further non-negative decision variables.

To obtain a fully feasible solution, the decision variables must fulfill a large number of constraints, describing conditions as correctness or feasibility of the assignments, precedence, non-overlap, synchronization and timeliness of activities. The extensive list and formal specification of all problem constraints can be found in [13], and thus omitted for brevity.

The decision variables LWT_{km} and UWT_{km} are also key-factors for the robustness of every feasible solution. Essentially, LWT_{km} and UWT_{km} play the role of slack times between the activities executed by trucks. From the viewpoint of resource utilization, LWT_{km} and UWT_{km} should be as short as possible, so as to minimize truck idle times. On the other hand, longer waiting times make the overall schedule more tolerant to the frequent, inherently unpredictable delays occurring during transportation. For example, if a truck is scheduled to arrive at the PC 10 minutes before the actual start of next-job loading, any arrival delay shorter than the 10 minutes slack will not affect the subsequent parts of the schedule. In this paper, the desired slack times are computed as a function of the length of truck route, i.e. we assume that the waiting time must be larger than a predefined number of seconds per km of travel.

To sum up, it can be noted that there are three main factors making the considered problem particularly challengig: (1) the typical combinatorial complexity of routing and scheduling problems, (2) the high number of constraints deriving from the perishable nature of RMC, and (3) the conflicting nature of the cost minimization and delay-tolerance maximization objectives.

The objective function of our problem is computed as the sum of three terms:

$$C = C_{transport} + C_{waiting} + C_{extra}. \tag{3}$$

The transportation costs account for the distances traveled by all the trucks of the fleet, including the initial and final trips from and to the base locations. It is obtained by multiplying the total distance traveled by all trucks with an average cost per kilometer. The waiting costs account for all the truck waiting times. They are obtained by multiplying the total waiting time of all scheduled trucks (waiting before loading and before unloading) with a penalty for each minute of idle waiting. The extra costs include all the additional costs related to outsourced production, hired trucks, and payments for overtime of truck drivers. Among the three addends in (3), $C_{transport}$ and C_{extra} are computed using estimates of the actual production, outsourcing or hiring costs available for the geographical area considered in the case studies, while the penalty costs related to wait times C_{wait} are chosen heuristically basing on ranges suggested by expert managers and a number of preliminary simulations.

4 The Genetic Algorithm and the Constructive Heuristics

The mathematical model of the SN could be directly used in a general purpose mathematical programming solvers. However, as mentioned, the long computational times and the lack of flexibility strongly limit the utility of these tools in the considered context. A viable and effective alternative to cope with problem size can be achieved by distributing the optimization tasks between different heuristic tools. In particular, our approach is based on the combination of a GA and a set of Constructive Heuristic Procedures (CHP). The GA works as the master algorithm that defines a set of assignment and priority decision variables, while the CHP is a subordinate constructive tool that builds a fully legal schedule starting from the subset of decision variables specified by the GA. More precisely, this paper describes two variants of our meta-heuristic tool devised to address slightly different versions of our production and distribution problem. The first variant consider the case in which it is strongly preferred (unless for exceptional reasons) that all the jobs of a single demand are produced at the same PC. This situation is particularly common in industrial practice, because it is a means to guarantee the homogeneity of the raw materials used for a given demand, and also because when fractions of demands are supplied by different centers, higher levels of synchronization between the operations of the autonomous PC's are necessary. This first type of SN will be hereafter referred to as *Low-Cooperation SN* (LCSN). The second variant of our meta-heuristic tool applies to the SNs in which jobs of the same demand can be assigned to different PC's without particular restrictions. Sharing the production of large orders makes it possible to obtain an improved overall resource utilization, and accept requests that could be impossible for a single PC. This second type of SN will be hereafter called High-Cooperation SNs (HCSNs). The following subsections describe the main components of our search engine separately.

4.1 The Genetic Algorithm

Every GA requires the preliminary definition of a coding strategy that transforms a generic solution of the problem into a string of symbols chosen from a pre-specified alphabet. The strings should also be devised so as to let the recombination operators (crossover and mutation) obtain new solutions with relatively simple manipulations. The large number of variables in instances of realistic size (e.g., 20-60 requests, 400-800 jobs, 5-10 docks and 50-120 trucks) and the peculiarities of the considered problem make the use of a standard GA inappropriate, due to the excessively long search times. Moreover, conventional strategies to address constraint satisfaction in evolutionary computation cannot be easily applied to the RMC SN's. In fact, due to the large number of constraints, approaches based on penalty functions require extremely long search times before the GA identifies the regions of the solution space containing fully feasible solutions. On the other hand, coding strategies that guarantee that each chromosome (generated by crossover or mutation) is feasible, or repair in some way the unfeasible chromosomes after they have been generated, would lead to prohibitive computational costs.

To overcome this problem, the task of our GA is to process and optimize only a subset of the decision variables, while the remaining unassigned variables are subsequently handled by the CHP. It is important to mention here that the proposed combination of tools is slightly different from the way meta-heuristic or memetic algorithms are usually defined [22, 23, 24]. In general, memetic algorithms use a first (global) search algorithm as a tool to pass good initial solutions to a faster or more accurate second (local) search algorithm, which returns the improved solutions to the first one. In such a type of hybridization, the use of one tool (the GA or the local search tool) is not strictly necessary for the operation of the other one (in other words, the GA and the local search tool could be used separately). On the contrary, in our strategy, the GA explicitly relies on (and cannot work without) the CHP to obtain a complete solution. The CHP, in turn, requires the set of variables determined by the GA to perform its task. This coupling strategy allows us to obtain a GA that is significantly faster than a penalty-based conventional GA in obtaining feasible solutions, and that does not require complex chromosome repair strategies, as the GA always processes, compares and combines fully-legal solutions obtained by the CHP.

Customer's Request-to-Depot Assignment						Priority of request in schedule construction					
r_1	r_2	r_3	r_4	r_5	r_6	p_1	p_2	p_3	p_4	p_5	p_6
1	3	2	1	2	2	4	5	6	1	3	2

Fig. 2. A single chromosome.

To illustrate the basic operations of the GA, let us first focus on the chromosome encoding (see Fig. 2). The chromosome consists of two separate parts, both containing R (the number of demands) elements. Fig. 2 illustrates a simple example with $R = 6$ (in our experiments R ranges between 50 and 100). For the LCSN variant, the first part defines the assignment of demands (requests) to the PC's. More specifically, the chromosome defined by the GA specifies the PC where each request has to be produced. Then, a dedicated algorithm (described in the next subsection) splits the demands in jobs and assigns them to the docks of the PC. In this variant, each gene is an integer between 1 and P (the number of PC's): the l–th gene of this first part of the chromosome indicates the PC to which request r_l is assigned. For instance, in the chromosome represented in Fig. 2, requests r_1 and r_4 are assigned to PC 1, r_2 to PC 3 and all the remaining ones to PC 2. In the case of the HCSN variant, the first half of the chromosome directly assigns requests to the single docks. Then, a specific algorithm (described in the next subsection) reexamines the jobs assignment and, if necessary, redirects some of them to other docks in order to obtain a fully feasible production plan. In this case, each gene is an integer between 1 and D (the number of docks): the l–th gene of this first part of the chromosome indicates the dock to which request r_l is assigned. The second part of the chromosome (which has identical meaning for both the LCSN and the HCSN) establishes the order in which the R requests will be considered in the construction of the complete schedule of the production chain. The l-th gene in this second part indicates the demand that will be considered at the l-th step of the scheduling construction procedure. Thus, demand corresponding to p_1 will be considered first, followed by demand corresponding to p_2, and so on. For instance, in the chromosome reported in Fig. 2, request r_4 (represented by the integer 4) corresponds to p_1 and will be the first one to be allocated in the scheduling plan. The next demand in the order of priority is r_5 (represented by integer 5 and corresponding to p_2), followed by request r_6 (corresponding to p_3), and then r_1, r_3, and finally r_2. Clearly, the second part of the chromosome can be any permutation of the sequence of integers 1,2,…, R. It should be underlined that the scheduling priority is not a variable of the mathematical model, but rather a seed that specifies to other constructive heuristics how to handle the remaining decision variables. It is evident that the chromosome structure is particularly transparent, as it essentially specifies which PC must process the request (or at least the largest part of the request), and the priority of the request in the activity of the PC. In this way, if necessary, a plant manager can easily assign requests to specific PC's (fixing a priori one or more genes in the left-hand part of the chromosome) or to impose higher priority to specific requests (fixing their position in the right-hand part) without needing to handle/readjust the whole schedule of activities by hand.

Once a new chromosome is available (e.g., as the output of a crossover or a mutation), it is passed to the CHP, which is in charge of determining (1) a complete, fully feasible schedule of all the SN's activities, (2) the corresponding overall cost of the solution (see eq. (3)) returned to the GA fitness as fitness of the chromosome.

Fig. 3 provides a generic pseudo-code for the GA. The included functions can be summarized as follows:

1) `random_pop`: randomly generates a population of n_{pop}=100 chromosomes.

2) `fitness_eval(Pop())`: for each chromosome in the population, this function runs the selected CHP to obtain a feasible schedule and evaluate the associated fitness.

3) `select(Pop())`: selects the mating pool of the population solutions using tournament selection [11].

4 – 5) `crossover(Pop())` and `mutation(Pop())`: generate a new population of chromosomes by applying crossover to 50% and mutation to 10% of randomly selected chromosomes in the mating pool. Both operators randomly select a point in the chromosome, and apply a different operator if the selected point is in the first or in the second part of the string, always producing legal offspring solutions. Technical

details on these operators and on their implementation can be found in [13] and in [11, 14, 5], respectively, and are omitted here, for brevity.

6) `Pop(i) ∪ p_best` the GA completes the new population by adding the elitist individual p_best .

7) `WHILE ... END WHILE`: Iterate steps 2-6 for the available computation time (usually 200 iterations).

```
/* Genetic Algorithm Startup */
i = 1;
Pop(1) = random_pop
fitness_eval(Pop(1))
i = 2;

/* main loop of the GA */

WHILE terminating_condition == false

    Pop(i) = select(Pop(i-1));
    Pop(i) = crossover&mutation(Pop(i));

    Fitness_eval(Pop(i))

    Pop(i)=Pop(i) ∪ p_best
    i=i+1;

END WHILE
```

Fig. 3. Pseudocode for the main optimization loop.

4.2 Constructing a complete schedule from a chromosome

Two different CHP's are used to solve the LCSN's and the HCSN's problems. Both versions need a chromosome as input, and return a fully legal schedule and the associated value of the cost function (3). As an extensive description of the HCNS is available in [13], here we concentrate on the essential details, and on the differences between the two variants. The CHP is composed of two modules. The first one is devoted to production scheduling, while the second addresses truck routing. Dock scheduling involves the assignment of each job to a dock, and the definition of feasible start and end times for each job loading. To perform this task, the CHP assumes that a truck is always available at the dock to load the job (truck routing is considered in a subsequent stage), and executes a sequence of operations that can be basically distinguished in three sequential phases.

Phase 1. The CHP considers each request following the priority order assigned in the chromosome. For each request, it checks if the distance between delivery location and assigned PC permits the end of the unloading before the RMC set (otherwise the request is redirected to other PC's, considered in order of shortest distance from the customer location, and the chromosome is changed accordingly). The jobs of the request with highest priority (e.g., r_4 in Fig. 2) are scheduled on the first dock of the assigned PC at their ideal start times, i.e., those that make the unloading of first job start exactly at customer-specified earliest delivery time. The CHP tries to find loading windows for the jobs of the second and subsequent requests in the same way, avoiding overlaps previously assigned jobs. If overlaps are detected, several attempts are performed to overcome the conflict. First, the CHP tries to assign the job to another dock of the same PC, and if this attempt is unsuccessful it runs a forward-backward search algorithm that tries to insert the conflicting jobs in the idle time windows of all the docks of PC, without altering the schedules of the jobs with higher priorities. The jobs that are still unscheduled at the end of this procedure will be reconsidered in a subsequent phase. In the case none of the jobs composing a demand can be scheduled on the assigned PC, the entire demand is reassigned to another PC of the SN, considered in order of shortest distance from the customer, and the chromosome is changed accordingly.

Phase 2. the CHP reconsiders the unscheduled jobs, and attempts to place them on other PC's redirecting all the unscheduled jobs of a request r to the same PC to minimize the amount of coupling between PC's. The depots are considered in order of shortest distance from the customer's site to minimize the distance between PC and customer.

Phase 3. The main peculiarity of this last attempt is that the CHP now is allowed to force job insertion by moderately shifting backward in time some of the already scheduled jobs. This operation may significantly alter the already constructed part of the schedule, and may also involve undesirable effects (e.g., determining excessively long waiting times before the unloading of some job). For this reasons, the result of this operation is accepted only if the consequent additional cost due to the increased waiting times is lower than the cost of outsourcing the unscheduled job under consideration. Finally, all the jobs unscheduled at the end of the third phase must be outsourced.

Once the assignment solution encoded in the chromosome is converted in a feasible loading sequence for each PC, the fleet of trucks must be assigned to jobs (setting the values of decision variables X_{ikm}) and routed from PC's to customer sites and vice-versa to pickup and deliver loads. Basically, the truck scheduling must guarantee that a truck assigned to a job is available at the loading dock of the supplying PC at the scheduled loading start time. The procedure consists of the following two phases.

Phase 4-Phase 5. Both phases perform the same sequence of operations, with the only difference that Phase 4 uses the trucks owned by a PC to deliver jobs of the same PC, and then Phase 5 deals with all the jobs that remained unassigned at the end of Phase 4. The jobs are examined in order of start time. To illustrate the truck allocation procedure, let us divide the trucks available to deliver a job i into two sets: the first one is made up by the vehicles that have not left their base location since the beginning of the working day (hereinafter defined as type 1); on the contrary, the trucks that have already completed some transport operations, and are able to reach the assigned PC before the loading start time of job i, compose the second set (type 2).

The objective of the LCSN CHP is to minimize the number of used trucks. For this reason, it first tries to assign trucks of type 2. When multiple trucks of type 2 are available, the LCSN CHP sorts them giving higher priority to those with the latest expected arrival time at the depot (first ranking criterion) and with shortest distance from PC (second ranking criterion). Finally, the job is assigned to the first vehicle in rank. This hierarchical criterion was preferred to other combinations of routing rules because it provided the most satisfactory results in a preliminary comparison of alternative routing strategies. Finally, if no truck is available to deliver a given job, the procedure requests the hiring of a truck. The complete scheme of the proposed algorithm is illustrated in Fig. 4.

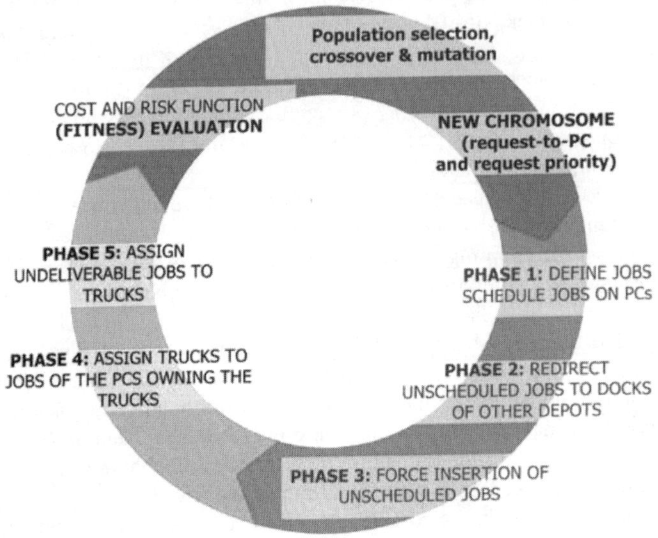

Fig. 4. The main structure of the proposed search algorithm.

The main difference between the CHP's used for the LCSN and HCSN cases is related to job redirection. In particular, the redirection of jobs to docks of other PC's is allowed without limitations for the HCSN case, while it determines additional penalty costs in the CHP for LCSN.

Finally, it is worth noting that the proposed encoding scheme well lends itself to interpretation and manual interventions by plant managers. In particular, one or more genes in the left-hand part of the chromosome can be directly assigned by plant managers in order to force requests to be produced at specific PC's. Similarly, plant managers may decide to assign higher priority to specific requests by fixing their position in the right-hand part of the chromosome.

5 Case study

The SN considered in this paper describes an hypothetical consortium of 10 independent PC's (with a total of 15 loading docks) spread over a large area in the southern Italy. Customers have construction sites located in the area surrounding the SN, as illustrated in Fig. 5. Only the PC's with more than one dock have their own fleet of trucks (a total of 120 units). We consider a set of instances describing typical workdays of such a SN, with daily production satisfying 60 requests (over 800 jobs, see Fig 6. for an example of a docks Gantt diagram). The values of the main parameters of the SN and of the GA are summarized in the Tables 1 and 2.

Table 1. Cost parameters in normalized units (NU's)

cost for each Km of travel of the trucks	10
penalty for idle time	15
loss of income for m^3 of concrete to outsource	2000
cost of an hired truck	10000
extra pay for truck drivers' overtime minute	5

Table 2. Configuration of the genetic algorithm

population size (randomly generated)	100
termination condition (generations)	200
crossover rate	50%
mutation rate	10%

The GA's parameters have been chosen after an extensive preliminary tuning phase. The average behavior of the GA with the configuration in Table 2 is illustrated in Fig. 7.

We compare the results obtained by the two variants of our meta-heuristic tool with a reference constructive heuristics implementing the typical constructive criteria used by plant expert managers. This comparison complements the analysis provided in [13], where a preliminary version of the HCSN CHP variant is compared with several other scheduling strategies on 250 instances differing in size and complexity. With respect to the data used in [13], the SN and the instances considered here are significantly larger and more challenging. We focus on 50 reference instances having two different levels of complexity, referred to as *normal* and *hard*, according to the amount of conflicts between the customer requests.

Tables 3 and 4 highlight that the variant for HCSN can significantly outperform the overall results obtainable with the strategy for LCSN in terms of both average and standard deviation (STD) of the final costs. On the other hand, it could be underlined again that the HCSN requires more complex information and communication infrastructures, necessary to guarantee activity synchronization in an inherently stochastic (truck delays) environment.

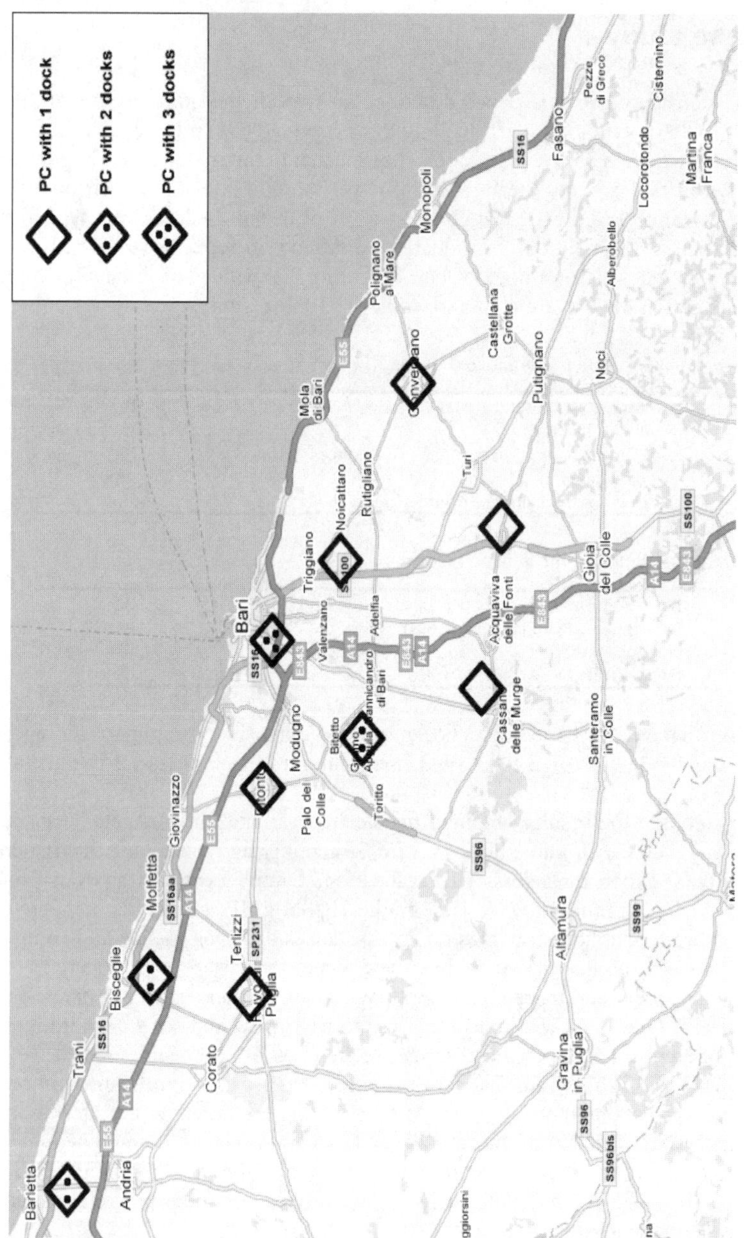

Fig. 5. Case Study: map of the area with PC locations.

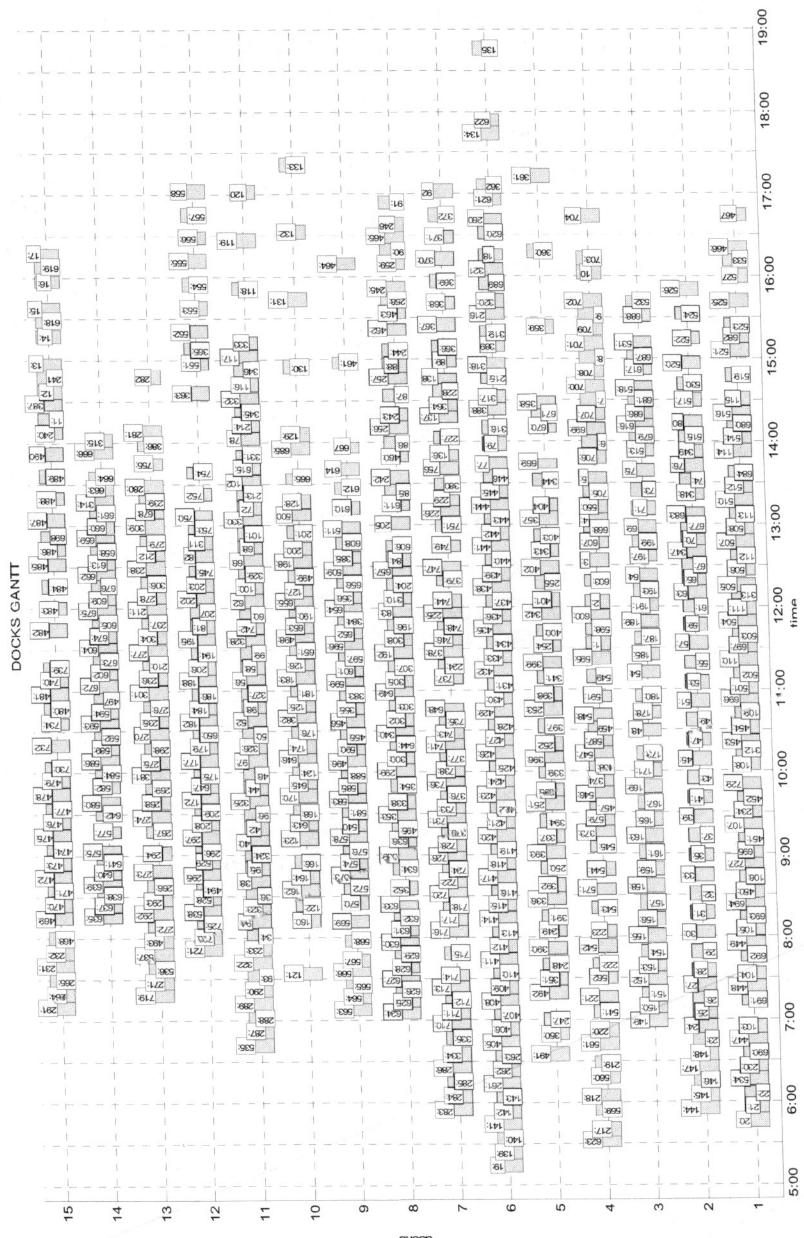

Fig. 6. Example of Gantt diagram for the production scheduling at the docks.

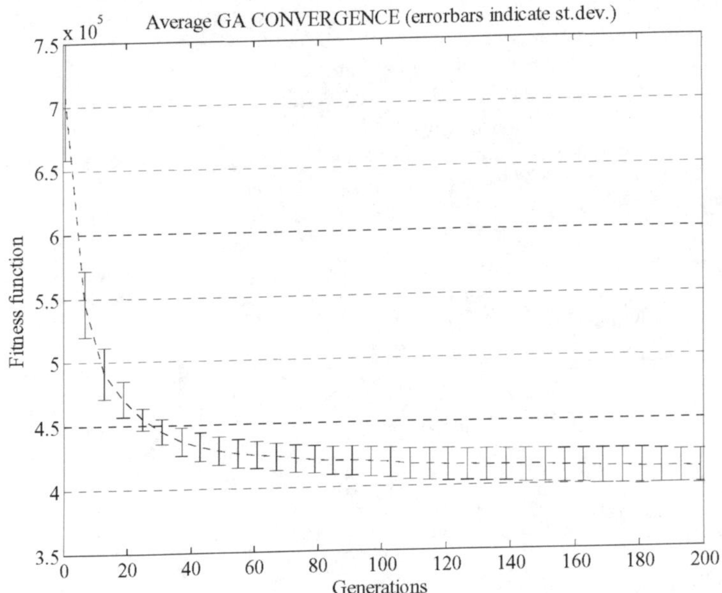

Fig. 7. Convergence rate of the GA.

Table 3. Average characteristics of the solutions found by the two variants on a set of 25 instances of *normal* difficulty

CHP→	LCSN		HCSN	
↓Cost Factors	avg	std	avg	std
Empty trips (Km)	18087.86	453.92	18016.71	331.15
Loaded trips (Km)	19805.57	1490.57	18276.86	1047.12
Waiting time (min)	23055.00	1234.22	23264.86	783.28
Number of hired trucks	28.29	2.67	27.14	1.29
Number of outsourced jobs (m^3)	1.14 (9.86)	0.35 (2.76)	0.00 (0)	0.00 (0)
Total cost (NU's)	1027379	34834	983309	30294
Algorithm Exec. Time (min)	19.84	0.74	21.03	0.91

Table 4. Average characteristics of the solutions found by the two variants on a set of 25 instances of *hard* difficulty

CHP→	LCSN		HCSN	
↓Cost Factors	avg	std	avg	std
Empty trips (Km)	18087.86	453.92	18016.71	331.15
Loaded trips (Km)	19805.57	1490.57	18276.86	1047.12
Waiting time (min)	23055.00	1234.22	23264.86	783.28
Number of hired trucks	28.29	2.67	27.14	1.29
Number of outsourced jobs (m^3)	1.14 (9.86)	0.35 (2.76)	0.00 (0)	0.00 (0)
Total cost (NU's)	1027379	34834	983309	30294
Algorithm Exec. Time (min)	19.84	0.74	21.03	0.91

In a second campaign of numerical experiments, we compare the two variants of our meta-heuristics with a reference constructive algorithm proposed in [13]. Essentially, this algorithm constructs the schedule by assigning jobs to the PC closest to customer's site (Shorter Distance, SD), and assigning tasks to the candidate truck that has the Smallest Idle Time (SIT). Conflicts and overlaps are handled with a series of constructive rules similar to those described for the CHP. The effectiveness of the solutions obtained by this algorithm, hereafter referred to as SD-SIT, depends on the particular characteristics of the considered instance. In any case, it represents a significant reference for comparison, as it has been shown that it can achieve very satisfactory solution in many reference scenarios derived from industrial data.

Table 5. Average components of the cost function for the LCSN and the SD-SIT on a set of 100 instances (*normal* and *hard* difficulties)

CHP→	SD-SIT		LCSN	
↓Components	avg	std	avg	std
Empty trips (Km)	15582.44	422.26	14984.84	490.094
Loaded trips (Km)	10983.15	309.64	16199.27	1351.114
Waiting time (min)	22141.55	617.87	16943.73	996.402
Hired trucks	21.98	0.46	19.08	1.75
Number of outsourced jobs	11.80	0.31	5.45	0.17
(m^3)	(115.34)	(1.72)	(52.955)	(1,38)
Total cost (NU's)	1048309	36885	862757	27689
Search Execution Time (min)	1.04		17.46	

Table 6. Average components of the cost function for the HCSN and the SD-SIT on a set of 100 instances (*normal* and *hard* difficulties)

CHP→	SD-SIT		HCSN	
↓Components	avg	std	avg	std
Empty trips (Km)	16474.13	444.76	15622.42	354.893
Loaded trips (Km)	11716.04	325.34	15819.79	1028.566
Waiting time (min)	23530.44	641.03	18055.4	745.2005
Number of hired trucks	16.285	0.44	10.41	1.00
Number of outsourced jobs (m^3)	11.80	0.31	4.90	0.16
	(115.34)	(1.72)	(46.34)	(1.22)
Total cost (NU's)	1028388	37619	782033	23284
Search Execution Time (min)	1.08		18.63	

Tables 5 and 6 compare the LCSN and the HCSN CHP with the SD-SIT strategy on two distinct sets of 100 instances (50 *normal* and 50 *hard*). The tables clearly show that our meta-heuristics always outperform the SD-SIT. In particular, while the SD-SIT is effective in optimizing the amount of truck travel, the two CHPs obtain a better distribution of production activities between PC's (thus allowing a substantial reduction of the outsourced production) and use a lower amount of hired trucks. On the other hand, being a relatively simple constructive heuristics, the SD-SIT is remarkably faster (15-18 times) in terms of execution time. Nevertheless, the search times of the evolutionary approaches LCSN and HCSN is still acceptable even for problems with several hundreds of jobs, and the amount of cost saving obtained fully counterbalances the relatively increased computational demand.

6 Conclusions

In this chapter, we have considered the challenging problem of coordinating the production and distribution activities of a network of PC's for the supply of rapidly perishable goods. The proposed meta-heuristic approaches are able to provide highly effective solutions with the desired tradeoff between production costs and ability to tolerate small delivery delays. We considered two different types of SN's (independent centers with restricted cooperation, and highly cooperative network). The numerical results show that in both cases the proposed approach can find effective solutions. It is important to remark that the proposed approach could be easily extended to other types of SN's by changing the problem specific heuristics used in the CHP.

Long term research involves the investigation of innovative paradigms based on distributed optimization, in which enhanced reactivity and fault tolerances are achieved by distributing the optimization task between various decision nodes located at each PC of the SN.

References

1. Feng, C. W., Cheng, T. M., Wu, H. T.: Optimizing the schedule of dispatching RMC trucks through genetic algorithms. Automation in Construction, Vol. 13, issue 3 (2004) 327 – 340
2. Garcia, J. M., Lozano, S., Smith, K., Kwok, T., Villa, G.: Coordinated scheduling of production and delivery from multiple plants and with time windows using genetic algorithms. Proceedings of the 9th International Conference on Neural Information Processing, ICONIP '02, Vol. 3 (2002) 1153 – 1158
3. Gendreau, M., Laporte, G., Séguin, R.: A tabu search heuristic for the vehicle routing problem with stochastic demands and customers. Operations Research, Vol. 44, issue 3 (1996) 469 – 477
4. Ghiani, G., Guerriero, F., Laporte, G., Musmanno, R.: Real-time vehicle routing: Solution concepts, algorithms and parallel computing strategies. European Journal of Operational Research, Vol. 151 (2003) 1 – 11
5. Ishibuchi, H., Yoshida, T., Murata, T.: Balance between genetic search and local search in memetic algorithms for multiobjective permutation flowshop scheduling. IEEE Transactions on Evolutionary Computation, Vol. 7 issue 2 (2003) 204 – 223
6. Laporte, G., Gendreau, M., Potvin, J. Y., Semet, F.: Classical and modern heuristics for the vehicle routing problem. International Transactions in Operational Research, Vol. 7, issue 4 – 5 (2000) 285 – 300
7. Lau, H. C., Sim, M., Teo, K. M.: Vehicle routing problem with time windows and a limited number of vehicles. European Journal of Operational Research, Vol. 148 (2003) 559 – 569
8. Lee, C. Y., Choi, J. Y.: A genetic algorithm for job sequencing problems with distinct due dates and general early-tardy penalty weights. Computers & Operations Research, Vol. 22, issue 8 (1995) 857 – 869
9. Marinakis, Y., Migdalas, A.: Annotated Bibliography in Vehicle Routing. Operational Research — An International Journal, Vol. 2, (2003) 32 – 46
10. Matsatsinis, Nikolaos F.: Towards a decision support system for the ready concrete distribution system: A case of a Greek company. European Journal of Operational Research, Vol. 152, issue 2 (2004) 487 – 499

11. Michalewicz, Z.: Genetic Algorithms + Data Structures = Evolution Programs. 3rd edn. Springer-Verlag, Berlin Heidelberg New York (1996)
12. Min, L., Cheng, W.: A genetic algorithm for minimizing the makespan in the case of scheduling identical parallel machines. Artificial Intelligence in Engineering, Vol. 13, issue 4 (1999) 399 – 403
13. D. Naso, M. Surico, B. Turchiano, U. Kaymak, "Genetic algorithms for supply chain scheduling: a case study on ready mixed concrete", Erasmus Research Institute of Management – Report ERS-2004-096-LIS, to appear on *European Journal of Operation Research,* 2006 or 2007
14. Nearchou, A. C.: The effect of various operators on the genetic search for large scheduling problems. International Journal of Production Economics, Vol. 88, issue 2 (2004) 191 – 203
15. Pinedo M.: Scheduling: theory, algorithms, and systems. Prentice-Hall, Englewood Cliffs, New Jersey (1995) Chap. 4, 86
16. Serifoglu, F. S., Ulusoy, G,: Parallel machine scheduling with earliness and tardiness penalties. Computers & Operations Research, Vol. 26, issue 8 (1999) 773 – 787
17. Solomon, M. M.: Algorithms for the Vehicle Routing and Scheduling Problems with Time Window Constraints. Operation Research, Vol. 35, issue 2, (1987) 254 – 265
18. Tan, K. C., Lee, L. H., Ou, K.: Hybrid genetic algorithms in solving vehicle routing problems with time window constraints. Asia-Pacific Journal of Operational Research, Vol. 18, issue 1 (2001a) 121 – 130
19. Tan, K. C., Lee, L. H., Zhu, Q. L., Ou, K.: Heuristic methods for vehicle routing problem with time windows. Artificial Intelligence in Engineering, Vol. 15, issue 3 (2001b) 281 – 295
20. Tommelein, I. D., Li, A.: Just-In-Time Concrete Delivery: Mapping Alternatives for Vertical Supply Chain Integration. Proceedings of the Seventh Annual Conference of the International Group for Lean Construction IGLC-7, University of California, Berkeley, California, (1999) 97 – 108
21. Toth, P., Vigo, D.: Models, relaxations and exact approaches for the capacitated vehicle routing problem. Discrete Applied Mathematics, Vol. 123, issue 1 – 3 (2002) 487 – 512
22. H. Ishibuchi, T. Yoshida, and T. Murata: Balance between genetic search and local search in memetic algorithms for multiobjective permutation flowshop scheduling. IEEE Trans. Evol. Comput., vol. 7 (2003), pp. 204 – 223
23. N. Krasnogor, J. Smith: A tutorial for competent memetic algorithms: model, taxonomy, and design issues. IEEE Transactions on Evolutionary Computation, Vol. 9, Issue 5 (2005), 474 – 488
24. Y. S. Ong, M. H. Lim, N. Zhu, K. W. Wong: Classification of adaptive memetic algorithms: a comparative study. IEEE Transactions on Systems, Man and Cybernetics, Part B. Vol. 36, issue 1(2006) 141 – 152

A Scenario-based Evolutionary Scheduling Approach for Assessing Future Supply Chain Fleet Capabilities

Stephen Baker[1], Axel Bender[2], Hussein Abbass[3], and Ruhul Sarker[4]

[1] Defence Science and Technology Organization Land Operations Division P.O. Box 1500, Edinburgh SA 5111, Australia. steve.baker@dsto.defence.gov.au
[2] Defence Science and Technology Organization Land Operations Division P.O. Box 1500, Edinburgh SA 5111, Australia. axel.bender@dsto.defence.gov.au
[3] Defence and Security Applications Research Centre, Australian Defence Force Academy, University of New South Wales, Canberra, ACT 2600, Australia. abbass@itee.adfa.edu.au
[4] Defence and Security Applications Research Centre, Australian Defence Force Academy, University of New South Wales, Canberra, ACT 2600, Australia. ruhul@cs.adfa.edu.au

When assessing land vehicle fleet capabilities in off-shore military operations, it is necessary to develop and apply a reliable mechanism for creating and testing different fleet structures. In this paper, the problem of optimizing vehicle fleet mixes in the settings of military deployments is introduced, the complexity of this problem is discussed and a fleet-optimization system that optimizes multiple objectives while satisfying the system constraints is introduced. The system is based on a set of heuristics that can answer a number of key questions required for long-term capability planning such as utilization of current fleet, mix of different vehicle and modular units for a given scenario, and overall fleet structure. It is evolutionary and uses a relaxed version of the concept of Pareto-dominance to identify the set of different options available for possible use by military analysts. The effectiveness of our system through a case study based on a simple example dataset is demonstrated. The novelty in this work is that we optimize a problem where scheduling is tightly coupled with routing and bin packing. In addition, land-based operations are also constrained with different standard operating procedures that complicate the scheduling problem. This paper is a first attempt towards solving this complex problem.

S. Baker et al.: *A Scenario-based Evolutionary Scheduling Approach for Assessing Future Supply Chain Fleet Capabilities,* Studies in Computational Intelligence (SCI) **49**, 485–511 (2007)

1 Introduction

Vehicle fleets of large organizations, such as land-based military forces, significantly contribute to operational effectiveness in all battlefield operations systems. For example, let us consider a military context. Here, one objective from the organization's point of view is its ability to negotiate diverse terrain quickly and under adverse environmental conditions. Another objective is to build the capacity to sustain prolonged and sizeable operations. Both these objectives depend heavily on the force's core transport capability. Designing and optimizing such a capability in a manner that meets operational requirements and, at the same time, is cost efficient therefore is of utmost importance to any land-based military force.

The strategic significance of optimizing military vehicle fleets is matched by the complexity of the problem. Owing to its large-scale and combinatorial complexity, determining and optimizing the mix of a large heterogeneous transport fleet carries a heavy computational burden because of two reasons. The first reason is the large-scale hard-to-decompose nature of the problem. The second is the combinatorial complexity of the problem. As shown in this chapter, the problem has a coupled interaction between three NP-complete sub-problems; these are: scheduling, routing, and bin-packing. The realization of any of these problems in an acceptable time frame is very unlikely. Thus heuristic or approximation methods are often employed [2, 19].

While the subject of fleet mix optimization has been studied in the context of commercial fleets [9] and some military airlift fleets [1], little research effort has been focussed on the problem domain of purely land-based military vehicle fleets. The military land environment includes a number of operational constraints many of which have unique military dimensions. Amongst them are threats, risks and concomitant protection requirements, terrain limitations, load compatibility, maintenance requirements, task connectivity, convoy requirements, crew restrictions, tasking time windows, and occupational health and safety requirements. Additionally, military vehicle fleets confront a multitude of tasks, subsets of which may be isolated to a particular industry sector in the commercial domain such as in the case of routine transport of personnel. In this regard, examinations of fleet mix problems in the literature often are made with reference to one particular industry sector.

The nature of the future military vehicle fleet examined in this paper can be described as 'modularized'. Incorporating both truck and trailer assets, the concept of a modularized vehicle fleet sees a basic truck and/or trailer combination configurable to a task specific variant, suitable for a particular payload or function, by the addition of an appropriate 'module'. Vehicles of a general cargo functional type, for example, should be configurable by adding to a base vehicle chassis a flat rack (a type of demountable truck tray to carry cargo loads), ISO container (shipping container compliant with standards promulgated by the International Organization for Standardization, ISO), bulk liquid tank, dump or tip-truck module. Other vehicle functional types also employ this modularized capability. The concept of utilizing modules is designed to divorce the payload task functionality from the mobility and manoeuvrability functions of the vehicle chassis. Additionally, the notion is that with common interfaces, system components can be designed so that vehicles can easily and quickly swap modules to meet contemporary mission requirements.

As indicated, the modularized concept is applicable to both trucks and trailers as shown by current day examples in Figure 1. As seen in this figure, allied with the modularized vehicle fleet concept is the adoption in cargo carrying vehicles of load handling systems that are organic to trucks (known as integral load handling systems, ILHS) and allow loading/unloading of modules from trucks and trailers without the need for other material handling equipment, such as forklifts and cranes.

Fig. 1. On left: Modular Vehicle with ISO Module and Integral Load Handling System[7]. On right: Modular Vehicle with Integral Load Handling System and Modular Trailer both with Flat Rack Modules[8].

Quite apart from the strategic significance and complexity of the fleet mix issue, it is this modularized concept that adds a novel dimension to the problem addressed in this paper. While previous analysis have addressed container movement for example, we are not aware of such a modularized concept being examined, and also note that the wide use of trailer assets in military fleets introduces a further range of issues that have received relatively narrow investigation in the literature.

The Modularized Fleet Mix Problem (MFMP) being examined in this paper therefore can be stated as:

A deployed military force has a range of mobility tasks to be undertaken utilizing a heterogeneous modularized vehicle fleet incorporating truck and/or trailer operations. The deployed military force along with its vehicle fleet is distributed among many locations in an area of operation. Each mobility task is characterized as requiring a number and range of modules to be moved between locations, to meet a priority for movement, within time window constraints. Each truck and trailer type has characteristics in terms of its ability to carry a particular range and number of modules and its ability to move across particular terrain classifications at particular speeds. The problem is to select trucks, trailers and modules assets to provide the best fleet outcomes. The fleet mix options are to be assessed against several efficiency and effectiveness criteria.

In this Chapter some of the approaches and methods that have been applied to vehicle fleet optimization problems are reviewed. The heuristics-based solver system is then presented followed by applying it to a simple case study. In the final section future work is outlined.

2 Fleet mix problems and approaches

Most of the fleet mix problems described in the literature also address, simultaneously, fleet management problems other than fleet optimization. Amongst these are vehicle assignment, vehicle routing, and/or vehicle scheduling within a network of customers or demand locations, and they are often referred to as, or based upon, the classical Vehicle Routing Problem (VRP), Vehicle Scheduling Problem (VSP) and combinations or variants thereof. The survey by Bodin *et al* [9], despite its age, continues to be cited as one of the most comprehensive undertaken in this area of interest. Along with this survey, Desrochers et al [11] highlight the great variety of problem types examined and the principal characteristics that differentiate them.

As Fisher and Jaikumar [12] point out 'literally person-centuries have been devoted to developing a sophisticated theory' for problems in this domain. As such, resource routing, scheduling and assignment represent huge fields of research endeavor where the application of heuristics and meta-heuristics is prominent. Ruiz *et al* [16] also note that while solution approaches based on exact methods have been applied for reasonably sized problems, generally only basic versions of such problems, such as the VRP or VSP, are considered.

The fleet mix problem presented in this Chapter is most closely aligned with the VSP, which can be stated in the following manner: *Complete a range of tasks, originating from multiple depots, by employing appropriate movement assets within established time windows so as to optimize such aspects as the number of vehicles of various types, fixed and variable costs, and vehicle capacity utilization.* Some recent examination of VSP can be found in Park [18], Dell'Amico *et al* [10], and Ferland and Michelon [6]. In this area the literature demonstrates the application of a wide range of heuristic and meta-heuristic solution approaches.

However, despite this interest in vehicle operations, few applications of combined truck and trailer fleet mix problems can be found in the literature. Those few areas of investigation are generally aligned with either the Vehicle Routing Problem with Trailers (VRPT), the Truck and Trailer Routing Problem (TTRP), or the Truck and Trailer Vehicle Routing Problem (TTVRP).

Semet and Taillard [4] consider a VRP that includes the use of trailers under accessibility constraints. Semet [3] similarly models a related problem called the 'partial accessibility constrained VRP'. As an extension of the VRP, Semet categorizes customers as either 'trailer-customers' and therefore accessible by either a truck or a truck-trailer combination, or as 'truck-customers' and therefore accessible by a truck only.

Gerdessen [7] examines a similar extension of the classical VRP entitled the VRP with trailers (VRPT) to determine optimal truck and trailer combinations. This problem is based on the consideration that manoeuvring problems may be encountered at certain customer sites. As a result it considers unhitching trailers at various parking sites in order to visit some 'difficult' customers with an easily manoeuvrable truck only.

Chao [5] considers a related problem identified as the Truck and Trailer Routing Problem (TTRP). The core of the problem is as presented by Gerdessen, however, a number of important assumptions are relaxed. The VRPT studied by Gerdessen differs from Chao's TTRP in that all customers have unit demand, customers are assigned manoeuvring times instead of customer types, each customer location can be used as a trailer parking place and each trailer is parked exactly once. In their heuristic solution approaches both Gerdessen and Chao develop construction and improvement algorithms. Chao, however, applies a solution construction method and a tabu search improvement heuristic together with the deviation concept found in deterministic annealing to solve the TTRP. In a later article Scheuerer [17] also addresses the TTRP and proposes two construction heuristics for this problem and a tabu search heuristic.

The aforementioned problem constructs involving trailer operations belong to the domain of vehicle routing. With the addition of time windows, Tan *et al* [8] introduce an element of scheduling when examining the Truck and Trailer Vehicle Routing Problem (TTVRP). In addition to time windows, unlike the TTRP, the TTVRP:

- requires vehicles to visit designated trailer exchange points for picking up the correct trailer types depending on the tasks to be undertaken,
- models trucks as essentially prime movers with no organic carrying capacity (i.e. trucks do not operate independently of trailers), and
- allows for the outsourcing of tasks that are not routed by sub-fleets in the TTVRP.

Tan *et al* [8] propose a hybrid multi-objective evolutionary algorithm incorporating specialized genetic operators, variable length chromosome representation and a local search heuristic to find the Pareto optimal routing solution.

In [15], a heuristic was developed for the multi-period multi-commodity transportation problem. In this problem, commodities are to be transported from a number of sources to a number of destinations. This is only part of the problem we are dealing with in this chapter because it does not consider the path taking from one place to another, and it does not consider the many constraints we will explain later on.

In the military domain, recent attempts to build a computerized scheduling system include the work in [13], where the concept of multi-agent systems is proposed to model military scheduling problems. In a more recent publication by these authors [14], the authors focus on the dynamic scheduling problem and the dynamic load balance between agents in terms of processing time.

3 Scenario-Based Military Vehicle Fleet Scheduling Approach

The problem being introduced in the introduction is more complex than what can be found in the literature, and consists of a number of coupled sub-problems. Before the proposed approach is discussed, the problem is stated formally. In the discussion presented from now on, the trailers are ignored in order to simplify the problem formulation. Including trailers in the approach does not pose a fundamentally new problem, however it increases combinatorial complexity.

Assume a command structure (see Figure 2) defined as a tree-graph , where U are the units in the command structure and O is the set of directed-edges. It provides information on which unit is in command of which other unit(s). Each unit is divided into a set of sub-units (capability groups). Overall, the command structure defines the ownership of vehicles; vehicles owned by a particular unit cannot be used by any other units. However, by having defined a command structure changes to vehicle ownership can be made when needed. For instance, a vehicle owned by a unit may be allowed to be used by any other unit who is in command of the former unit. This enables the examination of the effect of changes to vehicle ownership on fleet mix optimization, by "switching" between organic (situation-based, location-based and dynamic ownership) and hierarchical fixed ownership. The command structure described here also allows for the study of other ownership philosophies, such as centralized fleet management or network-enabled ownership in which a unit closest to a vehicle can use it independent of ownership issues.

Assume a nodal structure (*i.e.* a model of a geographical region) represented as a graph, $G(N, E)$, N is the set of nodes, while E is the set of directed edges. The nodal structure represents physical locations, where nodes represent places and edges represent roads. There are three different types of nodes: a physical location, a dummy node, and an exchange point as shown in Figure 3. Dummy nodes are used when there are multiple roads between two nodes. An exchange point is a physical location agreed-on by the units as a mid-point for vehicles to meet to exchange materials. For example, assume a need arises to transport materials from two nodes, N1 and N2. An exchange point E1 between these two nodes is used such that vehicles move from both N1 and N2 to meet at E1 at a given point of time to exchange materials. In

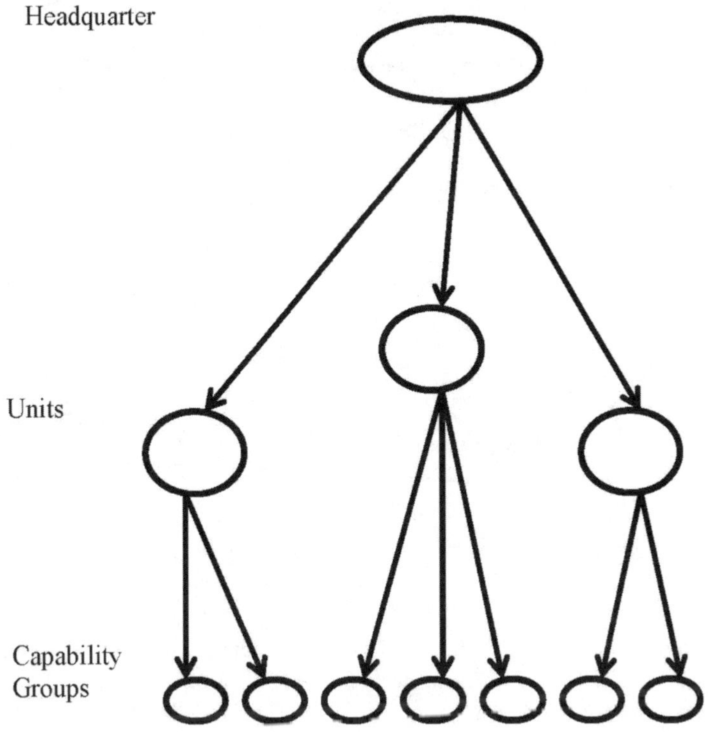

Fig. 2. A simple visualization of a simple tree-based command structure.

this case, vehicles from N1 (N2) will move from N1 (N2) to E2 and back to N1 (N2) respectively.

A capability group can exist in one physical node only at any point of time, while a single physical node can have many capability groups. Each node has a number of properties including the size of the node, the average length between any two points within the node, the maximum day and night speed of internal roads and the unique set of sub-units located at this node. Since some tasks can be internal within a node, this node-specific information is necessary for scheduling internal tasks. Each edge has a number of parameters associated with it, including the mobility criterion of the road, the length, width, and maximum load (for example, when there is a bridge, the load is limited), and maximum day and night speed of vehicles travelling on the road.

Assume the existence of V^k vehicles, where k represents the vehicle type; thus it also represents the combination of modules that the vehicle chassis can carry. Also assume the existence of M^r modules, where r is the module type which also represents the set of materials that can go with each module. Further assume the existence of T^w_{di} tasks, where w is the time window in

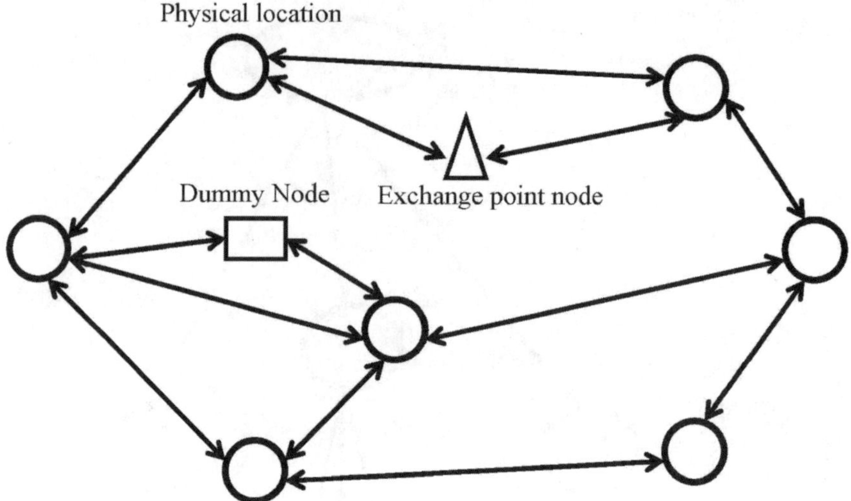

Fig. 3. A simple nodal structure.

which a task needs to be fulfilled, d is the duration of the task, and i is a unique task ID. Let C^k be the cost of a vehicle of type k, L^k be the length of a vehicle of type k, and C^r be the cost of a module of type r. The cost in this simple formulation can be seen as the acquisition cost thought in our tool, it covers other cost factors such as expected preventive-maintenance cost during the expected lifetime of an asset, and operation cost. The problem is to identify a mixture of vehicles and modules to fulfil the tasks such that

1. the cost is minimum;

$$\min \sum_k C^k V^k + \sum_r C^r M^r$$

2. the mixture is balanced (a balance between different vehicle types); $|k|$ in the equation is the cardinality of the set k while $\overline{V}$ is the average number of vehicles of all types.

$$\min \frac{\sum_k (V^k - \overline{V})^2}{|k|}$$

3. the lane meter (a measure that describes the space a vehicle occupies in a strategic sealift vessel) is minimum.

$$\min \sum_k L^k V^k$$

Typically, the three objectives may exist in conflict, i.e. fleet options that are optimal when assessed against one of the objectives may not optimize the other two objectives. The mixture balance is calculated through the variance of the fleet mix vector.

The vehicle is assumed to return to the origin after fulfilling the task. In the current version of the model, a working day is eight hours for a driver. The daylight period is from 6am to 6pm and the night time is the rest of the day. A fixed loading and unloading time for the vehicle is assumed.

When solving this problem, time is of the essence. Given that the size of operation imposes a huge number of tasks, it is not convenient to allow the solver to generate many (sometimes any) infeasible solutions. Thus, the solver is designed around the concept of generate-mix-improve.

The current solver is designed as presented in Figure 4. The system is characterized by a unique and modular structure which provides the system with the required flexibility to evolve over time with minimal changes. An 'oracle' agent is introduced that interacts with three main agents:

1. the route agent;
2. the task agent; and,
3. the combination agent.

The oracle coordinates the synchronization among these three agents. It provides the main internal interface for exchanging data and information between the three agents as well as the recombination operation of Stage 3 of the heuristic presented later in this Chapter.

The distributed multi-agent system is written in Java. Figure 4 depicts the overall design of this multi-agent system. Some of these agents undertake pre-processing operations when they first get activated. Some of these pre-processing operations are inefficient when the problem size is small as they have a large fixed-cost. However, when the problem size becomes large, the pre-processing operations save a huge amount of time. The pre-processing operations associated with each agent are described when we explain each agent below. The overall system flow chart is depicted in Figure 5.

The first operation takes place in the route agent. This agent is responsible for providing a route for a given task, vehicle, trailer if needed, and modules if needed. It gets initialized by generating all possible routes between any two nodes in the network. This is an exponential list as the average degree per node in the network increases. However, given that a typical military network tends to have a small average degree per node, the complexity of this process is not that expensive. The fixed inputs to the route agent is the start node,

Constraint Category	Constraints
Tasks	material types material quantity material volume early start time and date latest finish time and date duration priority source/origin of the task destination intermediate deliverables and local demands frequency for doing the task preferred vehicle preferred module
Modules	payload in volume payload in kilograms the combination of materials
Trailers	valid combinations of modules mobility criterion (restriction on suitable terrains) payload in volume for fixed trailers payload in kilograms for fixed trailers maintenance parameters
Vehicles	valid combinations of modules driver skill level mobility criterion (restriction on suitable terrains) vehicle ownership (whether only owners can assign tasks or not) payload in volume for fixed vehicles payload in kilograms for fixed vehicles fuel capacity dimensions crew size required maintenance parameters speed limits load/uploading time
Routing	Road capacity Road mobility criterion Road risk level Road length Road width day and night speed
Operating procedures	Convoy requirements for certain tasks Minimum and maximum number of vehicles on a given road
Drivers	maximum working hours in a day maximum working hours without a stop rest time

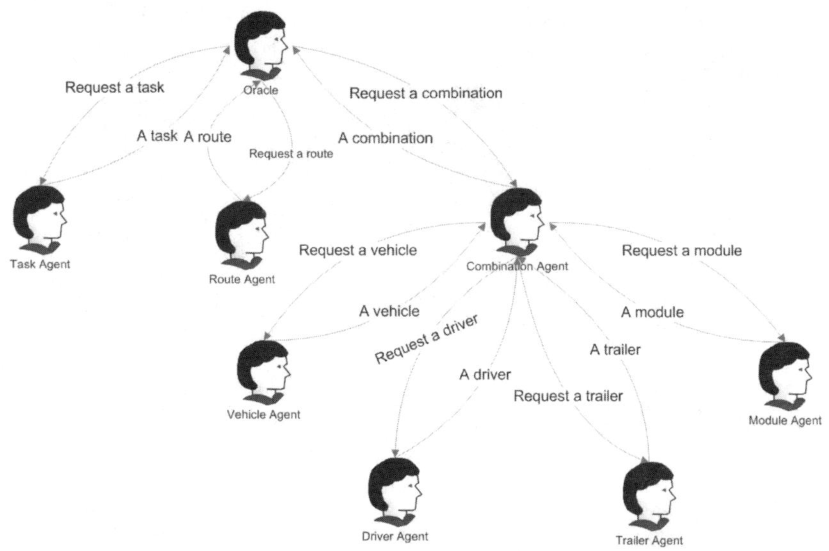

Fig. 4. The multi-agent system in the solver of the scenario-based military vehicle fleet scheduling problem.

end node, task, and proposed vehicle. Optional inputs include the proposed trailer and/or modules associated with the vehicle. The route agent is then responsible for deciding which of these different routes is suitable for a task (*i.e.* satisfies the constraints). The routes agent uses the following criteria to decide on which route is most suitable for the given task–vehicle combination:

- mobility criterion (the worst mobility criterion of the route which, in the military domain, is based on the quantifiable assessment of road surface, terrain conditions, slopes, obstacles, *etc*),
- the total distance of the route, and
- the maximum load that can be transported through this route.

The combination agent oversees the decision on forming a vehicle-trailer-module combination given a task. The pre-processing associated with this agent is that it enumerates the possible list of modules that can be carried with each vehicle or trailer type. For each combination, the total payload in kilograms and volume, and the mobility criterion of the vehicle (or vehicle–trailer combination) are calculated.

The multi-agent system works by message-passing between agents. Each message is tagged with a unique ID to overcome deadlocks. It works in a similar way as a ticket reservation system, where the resource is either free, reserved, or scheduled. A reserved resource for one operation is not available for any other operation until it gets unlocked.

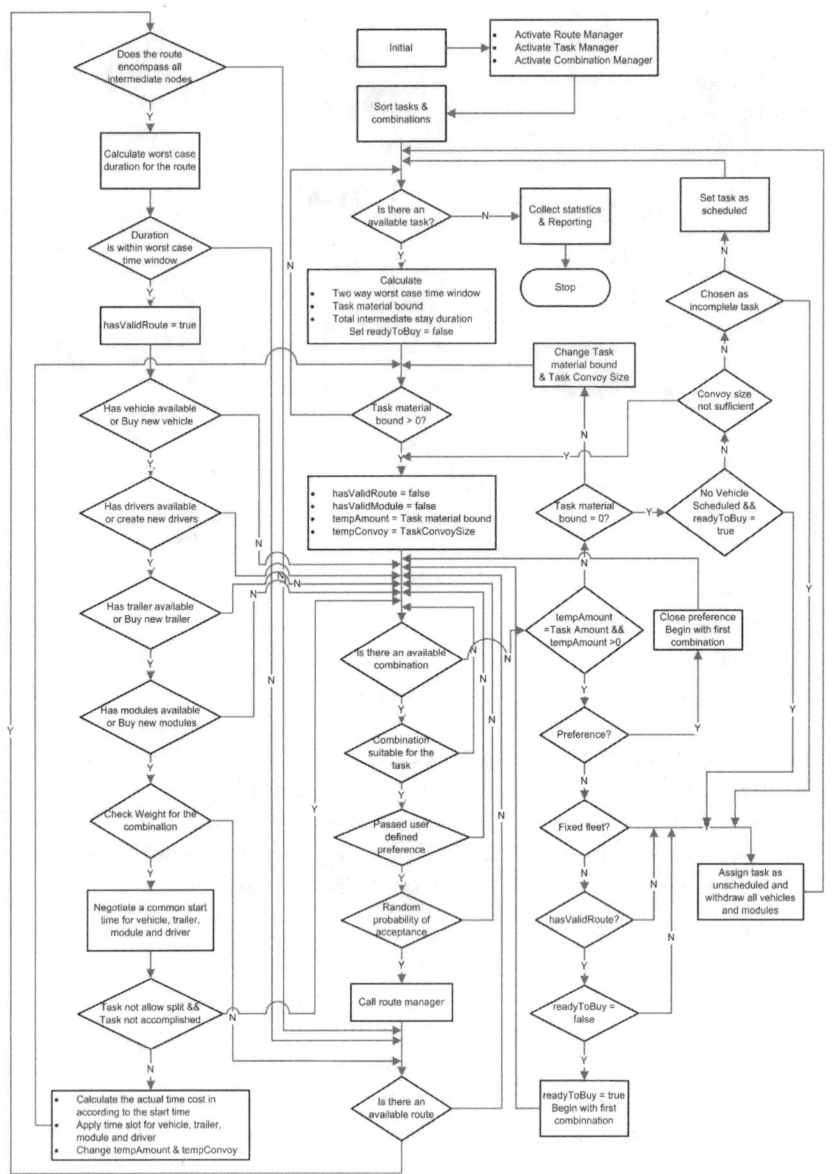

Fig. 5. A holistic view of the oracle.

The task agent is responsible for choosing a task for the schedule agent to schedule. When a task is selected for scheduling, the task agent orders the tasks using a random sequence (to overcome the problem of fixed bias discussed later in this section) of the following set of criteria:

- Descending order on earliest start time,
- Descending order on latest finish time,
- Sorting the origins according to ID,
- Sorting the destination according to ID,
- Sorting on priority from important to less important with 1 being the highest priority,
- Descending or ascending order on weight,
- Descending or ascending order on volume

In the nodal structure, each node is given a unique ID. The ID is generated by the system from left to right, top–down. This represents some sort of a linear encoding for the nodal structure and is used as a criterion above to sort tasks based on flow of origin/destination.

Once the task agent has sorted the tasks, a task is selected and its characteristics are passed on (by the oracle agent) to the vehicle agent so that the vehicle agent can select a suitable vehicle for the task. If there is no suitable vehicle available to perform a task, then the vehicle agent has access to an (unlimited) pool of vehicles in order to 'create' the required vehicle. The vehicle selection process uses the following constraints:

1. The default start time for a task is 6 am.
2. A driver (also a vehicle) cannot operate more than ten hours a day.
3. Every five hours, the driver/vehicle must task a break for 30 minutes.
4. If the sum of the duration of the trip and the time the vehicle has been operating so far without a break is greater than the maximum allowed time to operate without a break, the driver will take a break before commencing the task
5. The mobility criterion of a vehicle needs to be suited to the road. Here we assume that a vehicle with a mobility criterion VM can be used for a road with mobility criterion RM if VM $\leq$ RM.
6. The class of supply for the vehicle is consistent with the materials to be transported. The heuristic is configured to prefer a light vehicle over a water tanker if the amount of water to be transported is less than 1000lt since it is not cost effective to use a water tanker for a small amount of water.
7. If the vehicle is an existing one (it has been created before), the owner of the vehicle must reside at the location from where the task originates.
8. Volume, pax, and weight of tasks must be less than those of the vehicles; otherwise the task will be split into smaller tasks.

The overall scheduling heuristic comprises three stages. In the first stage, a single feasible solution is generated by calling the multi-agent system. Each agent in the system makes a decision using the set of criteria explained above. The heuristics associated with each agent are explained above. The generated solution is guaranteed to satisfy all constraints. If during the generation procedure one or more of the constraints cannot be satisfied - for example, because of a lack of a suitable vehicle - the system takes specific actions to satisfy all constraints - for example, buying a suitable vehicle. In the second stage, the first stage is called a number of times to create a number of feasible solutions by shuffling some criteria in the heuristics. In the third stage, the solutions go through selection and recombination, and new solutions evolve. The new solutions are used as input to the first stage in order to repair them should they not satisfy all of the constraints. The main algorithm is given below.

Algorithm 1 *A pseudo code of the system*

```
    Initialize the sorted task list, SList, to empty
  Stage 1:
    For each task T in the sorted task list
      Until T is completed (T.quantity>0)
        While there is combination available based on T
          Select one combination from C according to the heuristic
          While there is a route available based on T and C
            Select one route R according to the heuristic
            While there is a set of modules based on T,C,R
              Select a set of modules M according to the heuristic
              Break;
              While there is a vehicle based on T,C,R
                Obtain vehicle V according to the heuristic
                Reset the available time of V and M
                Break;
              End While
            End While
          End While
        End While
      End Until
    End For
    Let the created schedule be S0
    If the size of SList is < maximum size
      Add S0 to SList
    Else if S0 dominates any other schedule S1 ∈ SList
      Replace S1 with S0
    Else if S0 is non-dominated when compared to all schedules ∈ SList
      Add S0 to SList
    End if
  Stage 2:
    While (time-elapsed < time-available)
      Shuffle the orders for sorting
      Call Stage 1
    End while
  Stage 3:
    While (time-elapsed < time-available)
```

```
        Select 2 schedules S1 and S2 from SList
        Let S3 = Operator(S1,S2)
        If the size of SList is < maximum size
            Add S3 to SList
        Else if S3 dominates any other schedule S0 ∈ SList
            Replace S0 with S3
        Else if S3 is non-dominated when compared to all schedules ∈ SList
            Add S3 to SList
        End if
    End while
```

In the described approach, we minimize three different objectives. The first objective is budget or the total purchasing cost of the fleet. The second objective is the cost of strategic lift expressed in lane meters: imagine that all vehicles are lined-up in a single line, the strategic lift 'bill' would be the length of this line. The third objective is the variance of the fleet composition; this reflects how uniform the fleet is. It is not desirable to have a fleet just with heavy vehicles since it does not provide enough flexibility and operational maneuverability. We define non-uniformity as the variance of vehicle numbers of a given type compared to the average number of vehicles irrespective of type. For instance if we have a fleet of 30 light, 30 medium, and 30 heavy vehicles, then the average number of vehicles per type is 30 and the variance is 0, which describes a (very) uniform fleet mix.

One issue that deserves a discussion here is why the fleet variance was chosen over other objectives such as completing tasks on time. Fleet variance as an objective actually subsumes many other objectives in the sense that if the fastest way to complete the list of tasks is by having a fleet of small vehicles only, this is a potential extreme point on the Pareto front. Thus, using the fleet variance as an objective allows for solutions to complete tasks quickly as well as solutions where a balance in capability is achieved. In general, when we optimize future fleets, we should keep in mind that the future is usually hard to predict; thus a balance in fleet mix is needed. In scenarios where small vehicles cannot operate, heavy vehicles would be needed while in other scenarios small vehicles can be preferred.

Once the set of criteria that a heuristic follows gets initialized, a problem of fixed bias may arise. Take, for example, a heuristic which always prefers a large vehicle over a small one. This heuristic will only generate a fleet with all vehicles being large. We may not worry so much about this fixed bias since in stage 3, solutions get selected and combined. However, we use a parameter called "acceptance rate" to test whether or not this type of bias creates a diversity problem for stage 3. The "acceptance rate" parameter represents the probability a decision made by an agent will be accepted. If the agent's decision is rejected, the agent will need to re-generate another decision. If no such alternative decision is available, the agent will need to back-track and the original decision is chosen.

4 The Evolutionary computation Setup

We use a parameter called "population size" to signify the maximum number of solutions generated by the scheduling algorithm before selection pressure is applied. If this parameter is set, say, to 20, all first 20 solutions will be accepted regardless of their objective values. Once 20 solutions have been generated, any additional new solution will be added to the population only if it is non-dominated or it dominates any of the solutions in the population. In the latter case, the new solution replaces the dominated solution. This is the selection strategy we used because it is important for our end user to see the full scale of non-dominated options available. This is because many qualitative objectives are hard, if not impossible, to include in the model.

We use this strategy because of the nature of our problem. Let us assume that the objectives are not in conflict; thus there might be a single unique optimal solution. This solution, from a practical sense, may not necessarily be the best solution. There can be many other qualitative ways to judge on the quality of solutions from a military perspective. The process to come up with proper measures for the quality of the fleet can be too complex that makes the cost for evaluating a single fleet too expensive.

The fundamental question in this work is what is the smallest fleet size to perform a certain scenario. Therefore, a chromosome is defined as a string of integers representing the fleet composition for each capability group. Assuming that we have k different type of vehicles, r different type of modules, l different type of trailers, and m capability group, the chromosome length is $k \times r \times l \times m$.

We use an arithmetic crossover operator for this exercise. We intend in future work to experiment with different crossover operators. The arithmetic crossover operator works as follows: assume two chromosomes $p1$ and $p2$, then the child c is generated through $c = \alpha \times p1 + (1 - \alpha) \times p2$, where α is randomly chosen from a uniform distribution $U(0, 1)$. The new generated fleet is used as an input to the multi-agent system and get repaired if it is not sufficient to undertake the tasks (assuming that we are allowed to 'create' or buy more vehicles). This repair operator is used as some sort of corrective mutation. As a result, we did not use a mutation operator, although it would be interesting to try to do so in the future.

5 Validation

In this section, we validate the model by comparing it against two integer linear programming models solved using the branch and bound technique. The answers were obtained on the same computer with the following specifications: CPU: Intel Centrino 1.8G Hz; Memory: 521MB; OS: Windows XP professional SP2.

We simplify the problem to a two-node model, where all tasks start at one node and ends at the other. There is only one mobility criterion dictating a constant speed of 80Km/h. The length of the road is 40Km, which forces the time needed to deliver a task and return to base to be one hour.

In the first model, we assume that all tasks start together; thus the problem is reduced to the following very simple integer linear programming model:

Let x_{jk} be the number of vehicles of type j undertaking task k. Let C_j be the cost of buying one vehicle of type j, $Capacity_j$ be the capacity of a vehicle of type j, and Q_k the capacity required by task k. The model is

$$\min \sum_j C_j \times P_j$$

S.T.

$$\sum_k x_{jk} \leq P_j$$

$$\sum_j Capacity_j \times x_{jk} \geq Q_k$$

$$x_{jk} \geq 0; \quad integer$$

In this model, the problem is reduced to a very simple case where only the number of vehicles need to be determined in this simple environment and all tasks will be undertaken simultaneously. We assume that there are 3 different vehicles: Heavy with capacity 1000 Kg, Medium with capacity 500 Kg, and small with capacity 100 Kg. We assume 10 tasks, 3 with capacity 1000 Kg each, 3 with capacity 500 Kg each and 4 with capacity 100 Kg each.

The second model tests the scalability of traditional optimization in a simple situation. We allow tasks to be undertaken in different time slots. In fact, the problem is fully decomposable but we do not use this decomposition, neither in our system nor in the traditional optimization model, as it is the result of our simplification and cannot be achieved in more complicated situations. Overlapping the data will complicate the mathematical model unnecessarily, when the objective of this section is merely to test some aspects of our system. We assume that we have ten different time slots and each task can be done in anyone of them or being split as well. The second model is as follows:

$$\min \sum_j C_j \times P_j$$

S.T.

$$\sum_t \sum_k x_{jkt} \leq P_j$$

$$\sum_t \sum_j Capacity_j \times x_{jkt} \geq Q_k$$

$$x_{jk} \geq 0; \quad integer$$

In both models, we use three different cost functions. The costs used are listed below:

Cost function	Heavy Vehicle	Medium Vehicle	Small Vehicle
1	400000	225000	50000
2	350000	212500	50000
3	250000	187500	50000

The first mathematical model has 30 variables while the second has 300 variables. We used the LINGO software to solve the integer linear programming model. The LINGO code for the second example is shown in the appendix. The same code can be used for the first example by setting $t = 1$. One can also notice that there is a third constraint being added to limit the number of vehicles of any type to 10 for numerical reasons in LINGO to establish an upper bound on the search space.

Both the optimization model and the proposed system reached the optimal solution. For the first model, both LINGO and the proposed system took less than one second (running time rather than CPU time) to solve the problem for all cost functions. The results of the second model were surprising. The proposed system took between one and three seconds to solve each problem, while LINGO took 13 seconds on the first cost function, 36 seconds on the second cost function and LINGO complained when solving the third cost function. When we investigated the problem further, we found that when the equality component of the inequalities is fully satisfied, it resulted in an ill-matrix in LINGO. To overcome this, if the reader changes the *Capacity* in the third last line of the code to 1001 instead of 1000 and 501 instead of 500, LINGO will run fast and will give solutions in a similar speed as our algorithm. We attach LINGO code so that the reader can try for him/herself these difficulties which was surprising for us.

6 Case study

In this section, we present an example with three different vehicle types: heavy (mobility criterion 3, day speed 60 km/hour, night speed 40 km/hour), medium (mobility criterion 2, day speed 80 km/hour, night speed 50 km/hour) and light (mobility criterion 1, day speed 80 km/hour, night speed 50 km/hour). There are also three different module types which can carry up to 1000, 500 and 100 Kg of materials. We use three different cost functions to demonstrate the functionality of the system. The cost functions are plotted in

Figure 6 with the nodal structure in Figure 7. The three different purchasing cost for the vehicles are (in dimensionless units): L: 50,000, M: 225,000, H: 400,000; L: 50,000, M: 212,500, H: 350,000; and L: 50,000, M: 187,500, H: 250,000.

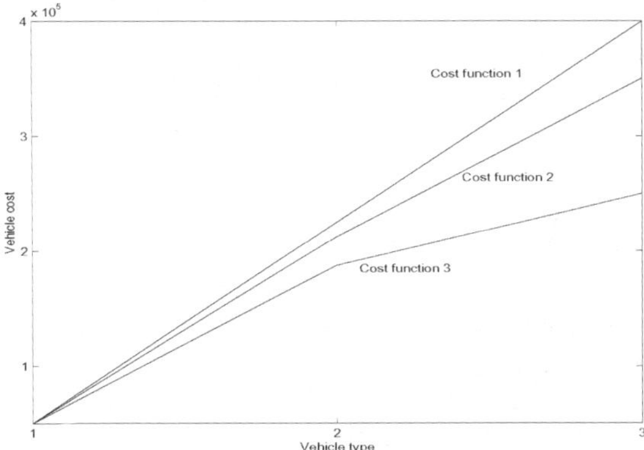

Fig. 6. The three different cost functions for the truck types.

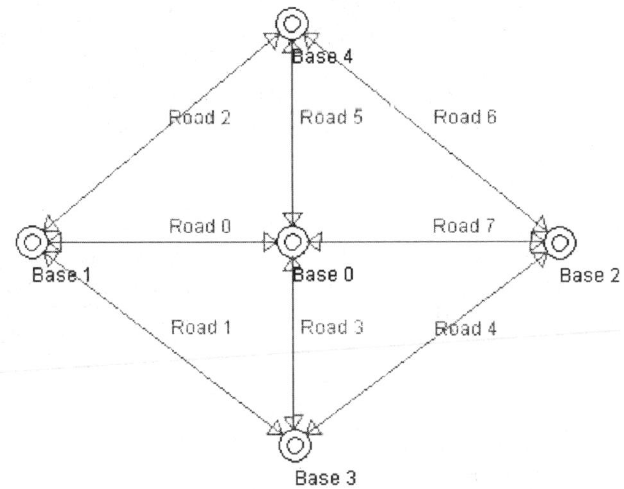

Fig. 7. The nodal structure.

The nodal structure has the following characteristics

Road Name	From	To	Mobility Criteria	Length	Width
Road 0	Base 1	Base 0	3	25	10
Road 1	Base 1	Base 3	4	50	10
Road 2	Base 1	Base 4	4	55	10
Road 3	Base 3	Base 0	3	30	10
Road 4	Base 3	Base 2	4	60	10
Road 5	Base 4	Base 0	3	30	10
Road 6	Base 4	Base 2	4	55	10
Road 7	Base 0	Base 2	3	35	10

There are two materials, food and water. The early start time for all tasks is the start of the day and the latest finish time is eight hours afterwards.

There are 114 daily tasks included in the test, divided into 6 different groups for validation purposes. The six groups are: [(Water 1000 + Food 1000) * 1]; [(Food 1000) * 1, (Water 1000) * 1]; [(Water 500 + Food 500) * 2]; [(Food 500) * 2, (Water 500) * 2]; [(Water 100 + Food 100) * 4]; [(Food 100) * 3, (Water 100) * 3]. Each group is assigned to a different base-base combination. These combinations, ordered according to groups, are: Base1-Base2; Base1-Base3; Base1-Base4; Base3-Base2; Base3-Base4; and Base4-Base2 respectively.

We experimented with four different "acceptance rate" levels: 1, 0.99, 0.95, and 0.80. We also experimented with three different population sizes: 10, 25, and 50.

The convention for presenting the results is as follows. Each figure consists of 24 sub-figures divided into a consecutive group of 6 figures. The top three figures in each group depict all solutions that have been generated in stage 2 (marked with circles) and stage 3 (marked with black dots) in the algorithm. All solutions shown in the figures are valid solutions. For solutions generated in stage 3, those solutions are repaired for feasibility before get evaluated and presented in the figure. The bottom three figures in each group depict the final population at the end of the run.

Representative figures with different cost functions for population size 10 are shown in Figures 8, 9, 10. By scrutinizing these figures, one can notice the following:

1. With a smaller acceptance rate, more distinct solutions get generated in stage 2 as can be seen by the increase in the number of circles in all graphs. With acceptance rate 1, very few distinct solutions get generate in stage 2 and mostly these solutions would have high variance, indicating the effect of the bias in the heuristic. However, the wide spectrum of solutions that get generated by using low acceptance rate, can also be generated through the recombination operator. This can be noticed by comparing the population corresponding to acceptance rate of 1 with the population corresponding to acceptance rate of 0.8. The two figures are almost iden-

tical except that most of the solutions with the former acceptance rate get generated in stage 3.

2. Surprisingly, the argument of the usefulness of using a lower acceptance rate as a replacement for the recombination operator does not hold true as we can see when we compare the final population corresponding to acceptance rate of 0.95 and acceptance rate of 0.8. It seems that the final population corresponding to the latter acceptance rate has less spread than the former. This may entail that a very low acceptance rate may deteriorate the solution quality. However, the recombination operator in conjunction with a high acceptance rate would perform consistently well.
3. The population size seems to have no effect on all acceptance rates with the first cost function. One needs to interpret this remark carefully since our definition of population size is not the traditional evolutionary computation definition in the sense that we use a varying population size. The population size makes a difference only in the first x number of solutions that get generated, whether they will enter into a competition or not. As such, this initial pressure has less effect in this problem when the rate of change in the cost function is almost constant. However, when we look at the second and third cost functions, one can notice that the higher population size, the more variations we get in the final population. This variation could have been redundancy (duplicate solutions) in the population, however after a closer examination to the populations we found that these are legitimate variations (unique and mostly non-dominated solutions).
4. The cost functions, on the other hand, had a notable impact on the spread and variety of solutions that we obtained. It is clear when comparing across the different cost functions that the smaller the rate of change in the cost function (the slop of each function in Figure 6), the less variability of solutions one obtains. One may be biased when comparing the sub-plots showing the total cost since it is natural that the change in cost would change these sub-plots. However, the phenomena that the cost function has an impact on the variability of solutions being generated can particularly be noticed when comparing the sub-plots of lane meter against variance.
5. The computational cost for generating these solutions was very low. A valid solution that satisfies all constraints takes less than a second to be generated on a Pentium 4 PC. We should emphasis also that we paid a lot of attention in the design and coding of the multi-agent system to optimize the code and speed up the processes.

7 Conclusion

Applying the presented solver system to the simple case study illustrates that (1) schedule recombination (Stage 3) results in improvement in the fleet mix

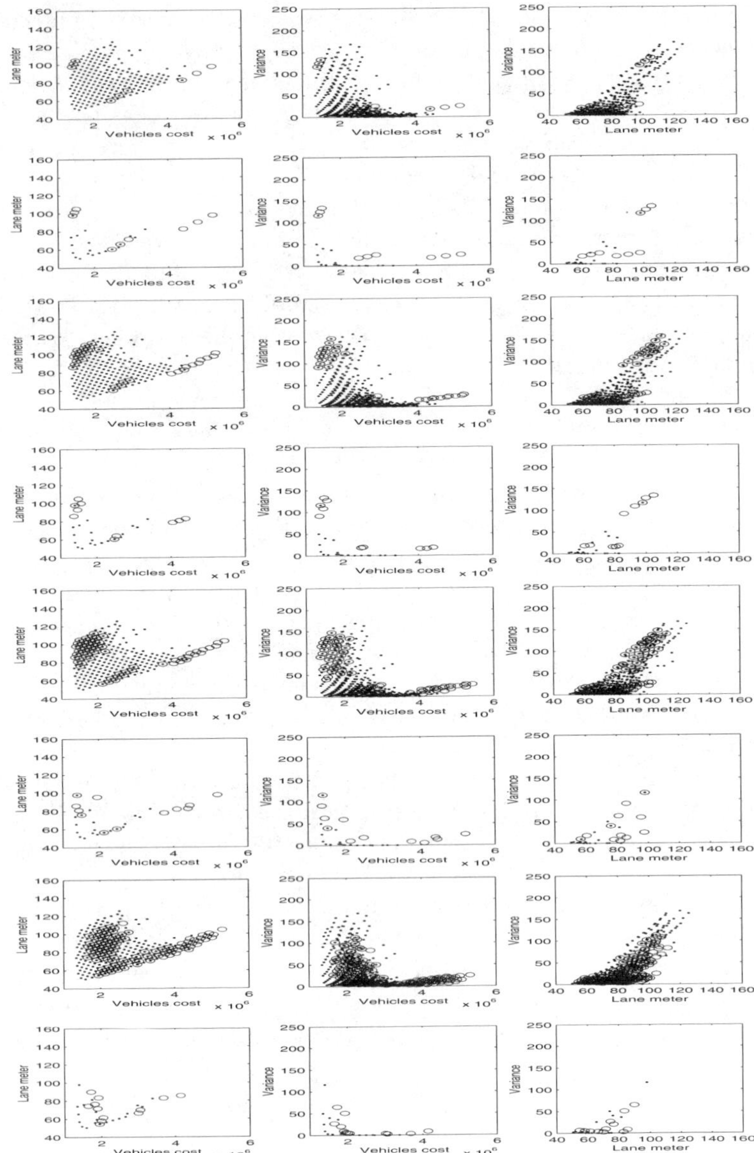

Fig. 8. Cost function 1 with different acceptance rates and population size 10. Figures are grouped with a group size of 6. Groups from top to down corresponds to acceptance rates of 1, 0.99, 0.95, and 0.80 respectively. Circles and dots represent solutions generated in stage 2 and 3 respectively.

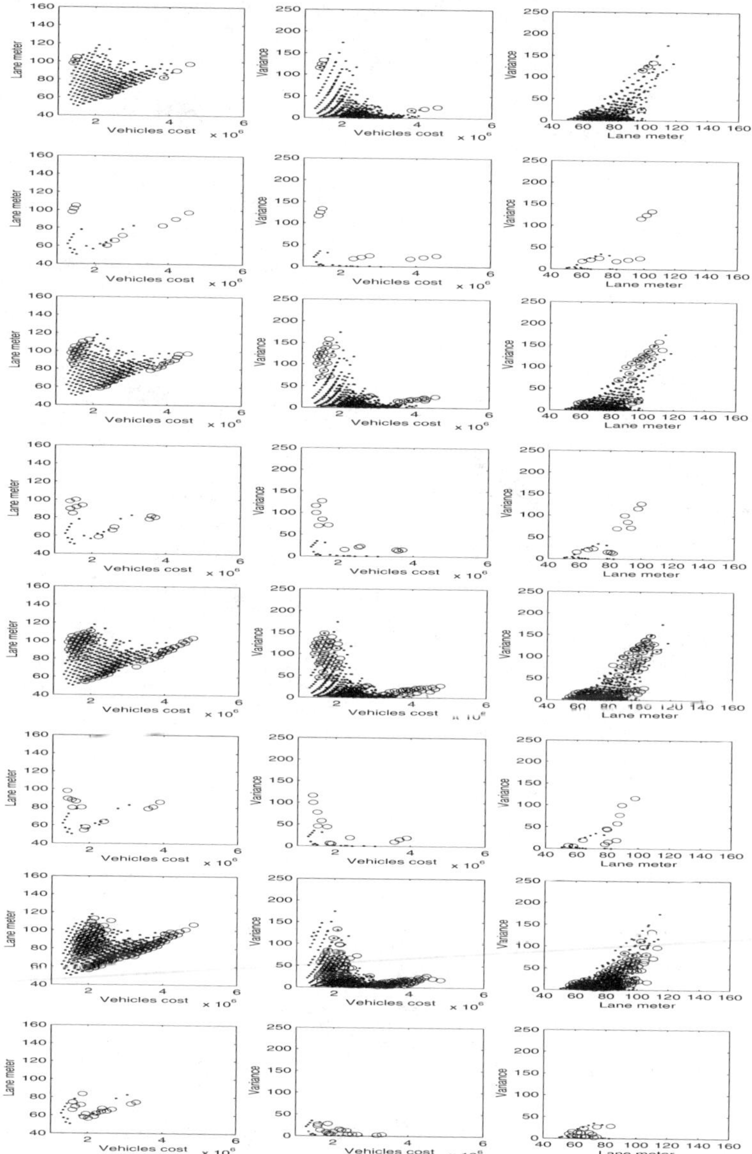

Fig. 9. Cost function 2 with different acceptance rates and population size 10. Figures are grouped with a group size of 6. Groups from top to down corresponds to acceptance rates of 1, 0.99, 0.95, and 0.80 respectively. Circles and dots represent solutions generated in stage 2 and 3 respectively.

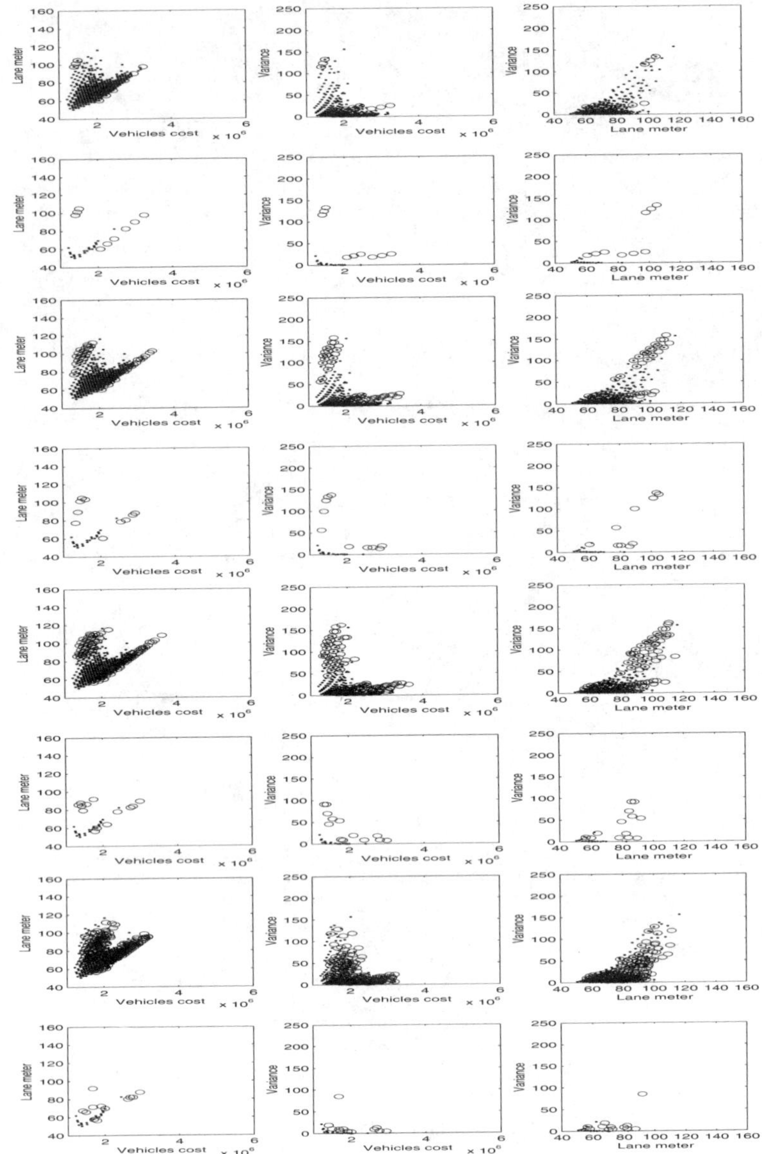

Fig. 10. Cost function 3 with different acceptance rates and population size 10. Figures are grouped with a group size of 6. Groups from top to down corresponds to acceptance rates of 1, 0.99, 0.95, and 0.80 respectively. Circles and dots represent solutions generated in stage 2 and 3 respectively.

solutions found in Stage 2, and (2) the solver system produces fleet options with great efficiency at low computational cost. This instils confidence that the developed method can be applied to more complex examples. However, examining examples is not equivalent to proofing the effectiveness of the approach.

We are currently working on refinements to the heuristics employed in the three stages of the solver system, and the definition of optimization objectives that provide a "good" spread for the Stage 2 seed solutions upon which improved solutions are generated during Stage 3. We have also started to investigate the structure of the fleet option landscape in order to obtain a better understanding of the accuracy of the optimization approximation used, and to evaluate computational effort required to find accurate approximation.

In future work, we will continue with the development of the solver system to enable the optimization of vehicle fleets that operate in dynamically changing military environments characterized, in parts, by enemy disruption and high uncertainty of demands in supply classes, such as ammunition. We expect that the evolutionary computation method will change to deal with the dynamics. Our previous work on dynamic environments in general reveal that identification of linkage and building blocks is a stable approach to deal with changing environments. In this problem, it would be interesting to define what a building block is; especially with the tight coupling of scheduling, routing and bin-packing.

8 Acknowledgements

The paper is written as unclassified materials of a DSTO-ADFA project on Defence Fleet-mix Problem. The project is funded by the Australian Department of Defence, Land Operations Division (LOD) of the Australian Defence Science and Technology Organization (DSTO) awarded to Hussein Abbass and Ruhul Sarker of the Australian Defence Force Academy, University of New South Wales, Canberra. The authors like to thank Dr Yin Shan and Mr Qi Fan for their effort in modifying the codes, system implementation and experimentation. The authors also like to thank the management of LOD, DSTO for allowing them to submit the paper.

9 Appendix

Algorithm 2 *The LINGO Code for Example 2*

```
MODEL: SETS:
    VEHICLE: CAPACITY, COST, NUM;
    TASK: REQD;
    PERIOD;
    LINKS (VEHICLE, TASK, PERIOD): XVAR;
```

```
ENDSETS
MIN = SUM (VEHICLE(J): COST(J) * NUM(J));
@FOR (TASK(K):
  @SUM ( PERIOD (T):
    @SUM(VEHICLE(J):CAPACITY(J)*XVAR(J,K,T)))>=REQD(K));
@FOR (VEHICLE(J):
  @FOR ( PERIOD (T):
    @SUM ( TASK (K): XVAR(J,K,T))<= NUM(J)));
@FOR (VEHICLE(J):
  @FOR ( PERIOD (T):
    @FOR (TASK (K): XVAR(J,K,T) <= 10)));
@FOR (VEHICLE(J):
  @FOR ( PERIOD (T):
    @FOR (TASK (K): @GIN(XVAR(J,K,T)))));
DATA:
   VEHICLE = H M S;
   COST = 250000 187500 50000;
   TASK = T1 T2 T3 T4 T5 T6 T7 T8 T9 T10;
   PERIOD = P1 P2 P3 P4 P5 P6 P7 P8 P9 P10;
   CAPACITY = 1000 500 100;
   REQD = 1000 1000 1000 500 500 500 100 100 100 100;
ENDDATA END
```

References

1. Morton D.P., Rosenthal R.E., and Weng L.T. Optimization modelling for airlift mobility. Technical report, United States Navy, Naval Postgraduate School, Monterey CA, 1995.
2. Baita F., Pesenti R., Ukovich W., and Favaretto D. A comparision of different solution approaches to the vehicle scheduling problem in a practical case. *Computers & Operations Research*, 27(13):1249–1269, 2000.
3. Semet F. A two-phase algorithm for the partial accessibility constrained vehicle routing problem. *Annals of Operations Research*, 61(1):45–65, 1995.
4. Semet F. and Taillard E. Solving real-life vehicle routing problems efficiently using tabu search. *Annals of Operations Research*, 41(4):469–488, 1993.
5. Chao I.M. A tabu search method for the truck and trailer routing problem. *Computers & Operations Research*, 29(1):33–51, 2002.
6. Ferland J.A. and Michelon P. The vehicle scheduling problem with multiple vehicle types. *Journal of the Operational Research Society*, 39(6):577–583, 1988.
7. Gerdessen J.C. Vehicle routing problem with trailers. *European Journal of Operational Research*, 93(1):135–147, 1996.
8. Tan K.C., Chew Y.H., and Lee L.H. A hybrid multi-objective algorithm for solving truck and trailver vehicle routing problems. *European Journal of Operational Research*, 2005. Corrected Proof, In Press.

9. Bodin L.D., Golden B.L., Assad A.A., and Ball M.O. Routing and scheduling of vehicles and crews: The state of the art. *Computers & Operations Research*, 10(2):63–211, 1983.
10. Dell'Amico M., Fischetti M., and Toth P. Heuristic algorithms for the multiple depot vehicle scheduling problem. *Management Science*, 39(1):115–125, 1993.
11. Desrochers M., Lenstra J.K., and Savelsbergh M.W.P. A classification scheme for vehicle routing and scheduling problems. *European Journal of Operational Research*, 46(3), 1990.
12. Fisher M.L. and Jaikumar R. A generalised assignment heuristic for vehicle routing. *Networks*, 11:109–124, 1981.
13. D. Montana, J. Herrero, G. Vidaver, and G. Bidwell. A multiagent society for military transportation scheduling. *Journal of Scheduling*, 3(4):225–246, 2000.
14. D. Montana, G. Vidaver, and T. Hussain. A reconfigurable multiagent society for transportation scheduling and dynamic rescheduling. In *International Conference on Integration of Knowledge Intensive Multi-Agent Systems (KIMAS)*, 2005.
15. KL Poh, KW Choo, and CG Wong. A heuristic approach to the multi-period multi-commodity transportation problem. *Journal of the Operational Research Society*, 56:708–718, 2005.
16. Ruiz R., Maroto C., and Alcaraz J. A decision support system for a real vehicle routing problem. *European Journal of Operational Research*, 153(3):593–606, 2004.
17. Scheuerer S. A tabu search heuristic for the truck and trailer routing problem. *Computers & Operations Research*, 2005. Corrected Proof, In Press.
18. Park Y.B. A solution of the bicriteria vehicle scheduling problems with time and area-dependent travel speeds. *Computers & Industrial Engineering*, 38(1):173–187, 2000.
19. Park Y.B. A hybrid genetic algorithm for the vehicle scheduling problem with due times and time deadlines. *International Journal of Production Economics*, 73:175–188, 2001.

Evolutionary Optimization of Business Process Designs

Ashutosh Tiwari, Kostas Vergidis and Rajkumar Roy

School of Applied Sciences, Cranfield University,
Cranfield, MK43 0AL, U.K.

Summary. Business process redesign and improvement have become increasingly attractive in the wider area of business process intelligence. Although there are many attempts to establish a qualitative business process redesign framework, there is little work on quantitative business process analysis and optimization. Furthermore, most of the attempts to analyze and optimize a business process are manual without involving a formal automated methodology. Business process optimization can be classified as a scheduling problem, expressed as the selection of alternative activities in the appropriate sequence for the available resources to be transformed and thus satisfy the business process objectives. This chapter provides an overview of the current research about business process analysis and optimization and introduces an evolutionary approach. It demonstrates how a business process design problem can be modeled as a multi-objective optimization problem and solved using existing techniques. An illustrative case study is presented to demonstrate the results obtained through three multi-objective optimization algorithms. It is shown that multi-objective optimization of business processes is a highly constrained problem with fragmented search space. However, the results demonstrate a successful attempt and highlight the directions for future research in the area.

1 Introduction

Business process redesign is inherently linked to the scheduling problems. One of the main topics of research in scheduling has been the optimal allocation of resources to tasks. Business processes can be analyzed using

A. Tiwari et al.: *Evolutionary Optimization of Business Process Designs*, Studies in Computational Intelligence (SCI) **49**, 513–541 (2007)

similar perspective. The design and management of business processes is a key factor for companies to effectively compete in today's volatile business environment. By focusing on the optimization and continuous improvement of business processes, organisations can establish a solid competitive advantage by reducing cost, improving quality and efficiency, and enabling adaptation to changing requirements. Multi-objective optimization of business processes can result in novel approaches and more efficient ways of business process improvement as more than one optimization criteria can be selected and satisfied concurrently.

The rest of the chapter is organized as follows: Section 2 introduces the basic concepts of business processes and examines the basic formal definitions. Section 3 presents existing performance analysis techniques for business processes and section 4 investigates current optimization approaches for formal business process models. Section 5 presents the authors' approach towards evolutionary multi-objective optimization of business processes and section 6 introduces the results of this approach. Finally, section 7 discusses the potential directions for future research in the area.

2 Formal Approaches to Business Processes

For a business process design to be optimized, the model construction methodology plays a decisive role. This section provides an introduction to the 'business process' concept and presents the most common modeling notions to provide a familiarization with the basic ideas. The focus is to the business process modeling techniques that allow formal optimization. This section begins with a discussion on how different authors perceive business processes and how these are related to scheduling.

2.1 Business Processes and Scheduling

An overview of most common business process definitions is provided in [16] suggesting that most of the attempts to define a business process are inadequate and confined to a mechanistic viewpoint of the process. Most of these definitions ignore the human side resulting in static simplified representations of business processes [18]. According to [5], business processes are complicated and thus more difficult to be fully specified. However, the 'mechanistic' definitions of business processes bring them closer to scheduling problems thus making a range of successful approaches already applied to scheduling, available.

There are many authors influenced by this perspective and provide business process descriptions that are more indicative of scheduling problem descriptions. For example, Hammer and Champy [9] suggest that a process is a 'set of partially ordered activities intended to reach a goal'. A process is thus a specific ordering of work activities across time and place with a beginning, an end and clearly identified inputs and outputs: a structure for action. A similar perception [3] also suggests that a business process can be perceived as a network of tasks. This approach is in line with scheduling, as it is concerned with resource allocation to tasks and justifies the attempt in [28] to map scheduling problems using a technique (i.e: Petri-nets) that has been already used for business process modeling.

Another attempt to define business processes and relate them to scheduling is found in [2] describing a business process as 'the combination of a set of activities within an enterprise with a structure describing their logical order and dependence, whose objective is to produce a desired result'. Also, in [11] the business process model described is quite similar to the Resource-Constraint Project Scheduling Problem (RCPSP). An inadequate attempt to define and model a business process could have a significantly negative effect on people's productivity by restraining their ability to gain expertise and apply innovative approaches while carrying out a business process.

2.2 Business Process Modeling

Business process modeling gives a snapshot of what is perceived at a point in time, in terms of a process that takes place in a business environment. Process models are currently best used to represent the internal elements of business processes; for example the activities needed and their dependencies, the dataflow, the roles and actors involved, and the goals. Along with the large number of attempts to define business processes, there is an abundance of techniques for capturing and modeling business processes. Aguilar-Saven [2] provides an overview of the most commonly used business process modeling techniques.

According to [17], business process improvement is dependent on an *insight* in the structure of business processes and their relations. This insight can be obtained by creating business process models that clearly and precisely represent the *essence* of the business organization. Business process modeling is itself a complex interdisciplinary and time-consuming process that involves judgments based on domain knowledge and experience, due to the multifaceted and dynamic nature of organizations [23]. The main objective of process modeling is the high-level specification of processes thus a process model should be properly defined, analyzed, verified, and refined to all of its aspects including structure, data flow, roles and constraints [24]. It is also important to identify the uses or purposes of the model when undertaking modeling of any kind in order to select the most appropriate technique [2]. Phalp [22] also highlights *notation* and *method* as two important parameters that need to be taken into account when modeling a business process as the model's general purpose and specific characteristics are significantly influenced by these two.

Taking the above into consideration, we can apply a number of criteria to classify the various modeling approaches. Frameworks and classifications of business process models can be found in relevant literature based on the purpose of the model, the notation or the structure ([2], [14]). We introduce a classification of business process models according to their formality, i.e. their ability to represent a process in a mathematically correct and rigorous way. The result of this classification is two groups of process models. On one hand are the so-called traditional or diagrammatic methods of process modeling, e.g. flowcharts or IDEF models and on the other hand, the business process modeling approaches that can be formally (i.e. mathematically or algorithmically) analyzed and verified. The next section introduces three representative formal business process modeling approaches. We choose to focus on these models because performance analysis and optimization – that is our main areas of focus- can be applied almost exclusively to this second group.

2.3 Formal Modeling Approaches

The most frequently recognized shortcoming of process modeling is the lack of analysis tools. Owing to the qualitative and static nature of most process models, mathematical techniques are difficult to apply. In order to make the process modeling methodologies more attractive, formal techniques for analysis of process models are required [30]. Formal process models are the ones in which process concepts are defined rigorously and precisely, so that mathematics can be used to analyze, extract knowledge from and reason about them. An advantage of formal models is that they can be verified mathematically, can be proved that they are consistent, and have or lack certain properties [15]. Although there are a number of formal modeling approaches, the majority of the business reengineering community uses simple diagrammatic modeling techniques [19]. This affects the analysis of business processes restricting it to simple inspection of the business process diagrams thus the conclusions are mostly heavily dependent upon the skill of the modeler.

Luttighuis *et al.* [17] recommends formal methods for detailed analysis of process over the diagrammatic modeling. The same perspective is also highlighted in [2] when suggesting that the analysis of business processes needs models that present both the dynamic and functional aspects of the process and also sophisticated mechanisms that qualitative analysis of static diagrammatic models cannot offer. Although formal methods can provide significant benefits to business process modeling by introducing new perspectives, there is a lack of formal methods to support the actual design of business processes [11]. This is mainly because design elements and constraints on process designs are hard to characterize in a formal way amenable to analytical methods. The qualitative nature of process designs explains the difficulty of 'parametric' models of business processes [26]. Three representative formal business process models found in the literature are discussed below.

2.3.1 Petri-nets

Petri-nets is a formal graphical process modeling language. According to [10], Petri-nets help describe the semantics of process control flow, including basic branch and join rules, as well as more complicated synchronization scenarios. Petri-nets are an established tool for modeling and analyzing processes that has been widely recognized. They can be used as a design language for the specification of complex workflows and also Petri-net theory provides powerful analysis techniques that can be used to verify the correctness of workflow procedures – they can be used for both qualitative and quantitative analysis of workflows and workflow systems [29]. A Petri-net is a directed graph that uses as main constructs places, transitions, tokens and arcs.

- *Places:* drawn as a circle, a place is a stopping point in a process, the attainment of a milestone.
- *Transitions:* a transition is a rectangle that represents an event or action.
- *Tokens:* A token is a black dot residing in a place representing the current state of the process. During the execution of the process, tokens move from place to place.
- *Arcs:* An arc is a link from a transition to a place or a place to a transition.

Van der Aalst [29] supports that Petri-nets have a series of advantages that played a key role in their establishment. Although they support diagrammatic process modeling, they also provide formal semantics. Moreover, apart from process events, they record the various states of the process providing a more holistic view on the process flow. Also, they offer an abundance of analysis techniques that can be used to evaluate the *performance* of the modeled business process by calculating e.g. the estimated throughput of a process, the average throughput time of a job, the estimated occupation rate, etc. [27]

However some attempts to use Petri nets in practice reveal two serious drawbacks. First of all, there is no data concept and hence the models often become excessively large, because all data manipulation has to be represented directly in the net structure. Secondly, there are no hierarchy concepts, and thus it is not possible to build a large model via a set of separate sub-models with well-defined interfaces [2]. For these reasons Petri-nets have been extended supporting constructs like time and color.

2.3.2 Business Process AI-based Language

The second approach to formal business process modeling comes from [15]. The proposed business process modeling methodology is constructed with an Artificial Intelligence (AI) programming language thus ensuring the formality of the proposed process model.

The methodology begins with the definition of business process objectives. The output is a detailed formal specification of a business process that achieves those objectives. This perspective is established and confirmed by the logical assumption that a process model cannot be represented by a single model but as a set of various sub-models that capture the business process from different viewpoints. There are five interconnected sub-models specified to formally describe different aspects of the business process are:

- *organizational sub-model,* describing the actors that participate in the process, their roles, their responsibilities and their capabilities,
- *objectives and goals sub-model,* describing what the process and its actors try to achieve,
- *process sub-model,* describing how the process will achieve those goals,
- *concepts sub-model,* describing non-intentional entities, and
- *constraints sub-model,* describing factors limiting what the enterprise and its components can do.

Each of these models consists of various concepts that are formally described by the declarative logical language *L* [7]. *L* is used to introduce appropriate constructs, write axioms and capture process semantics. The emphasis is on *actor* and *role* concepts as each *role* involves a set of responsibilities and actions that need to be carried out by an *actor*. *L* is used here to precisely define the relationships between various concepts.

Using the same methodology, other sub-model concepts like *goals* and *actions* are specified. This approach ensures the formality of the concepts although it becomes increasingly complicated as the concepts to be modeled incorporate other concepts. This attempt can be discouraging because 'it is a lot of work to create and maintain a formal business process and also retain its consistency' [15]. Other disadvantages of this proposal lie in the use of complex mathematical notation that might put off the business analyst and the skills of AI programming that are essential but rarely found in an average manager.

2.3.3 Scheduling-based Mathematical Formulation of Business Processes

The third approach to formal business process modeling comes from [11] and it is related to mathematical definition of business processes. This modeling approach is linked with three different optimization approaches that will be thoroughly discussed later on this chapter. A business process is described using a mathematical model with an objective function which can portray any business process objective e.g. cost. The objective function is minimized or maximized by the optimization algorithm.

The main concepts used in the process design are activities and resources. A business process is perceived as a sequence of activities. These activities use some resources and produce others to be used by the following activities until the goal resources are produced. Resources are the physical or information objects which flow through the system. Activities are transformation steps which use resources as inputs and produce new ones as outputs. Both activities and resources are represented as sets. Each process begins with some input resources and produces a desired set of output resources. Each activity has two parameters: one for its starting time and another for its execution duration. The input resources of this activity must be available before the activity starts and the output resources must be produced after the activity has been executed. The time that a resource becomes available is another parameter critical to process feasibility.

The particular modeling approach gives flexibility to the authors, to apply different optimization methods. It is also close to scheduling problem definition as it explicitly involves activities and resources to a business process design. The process model is adjusted and further defined in each of the different optimization attempts. This approach will be more extensively discussed later as it forms the basis for our evolutionary optimization of business process designs.

3 Business Processes Performance Analysis

Qualitative performance analysis of business processes can only occur to formal business process models. It is the only kind of process analysis that can contribute to process optimization by identifying the process bottlenecks that can be optimized. A business process might be correct (verified) and also produce the expected outcome in a given context (validated) but it could still have redundant steps or not satisfactory performance in terms of cost effectiveness, duration, resource allocation etc. Before introducing the existing business process optimization approaches in the next section, this section provides a brief introduction to different process analysis attempts in literature that approach the subject from different perspectives and can be considered as a significant first step towards business process improvement.

3.1 The Fuzzy Logic Approach

Zakarian [30] attempts to model and quantify a business process using a combination of fuzzy logic and rule-based reasoning. The main motivation for this approach is to model efficiently the uncertain and incomplete information of process variables that exist in most of the traditional modelling techniques. The starting point of this approach is a business process illustrated with an IDEF3 model. This model is used as the basis for quantification and performance analysis of the business process. IDEF3 is quite popular and widely used modelling method in the business process context. One of the major advantages of IDEF3 representation is its simplicity and its descriptive power. The essence of IDEF3 methodology is its ability to describe activities and their relationship at various levels of detail, because an initial model includes parent activities that can be decomposed into lower level activities. These models are also easy to extend. IDEF3 offers several important characteristics for successful process representation:

1. process description in the form of activities,
2. structure of the underlying process, and
3. flow of objects and their relationship.

In the first step of the process presented in [30], IF-THEN fuzzy rules are extracted from the IDEF3 model and thus the linguistic variables are defined. The linguistic variables, which owing to their nature contain incomplete or uncertain information, fall into two categories: input and output variables. In the formal model, which consists of IF-THEN fuzzy rules, the input variables appear in the IF part, while the output variables are found in the consequent THEN part. In order to understand the author's approach an example with two fuzzy rules is demonstrated with the assumption that the rules were extracted from a simple IDEF3 model:

- *Rule 1:* IF product stock is low THEN Marketing department informs customers about urgent orders.
- *Rule 2*: IF shipping products' weight is heavy THEN postage costs are high.

The linguistic variables for *rule 1* are 'low' and 'urgent' and for *rule 2* 'heavy' and 'high' and they need to be quantified. The quantification will take place by assigning possibility distributions to the linguistic variables. Possibility distribution is the definition of the lower and upper limits of a fuzzy set using numbers. For example, 'low' product cost can be defined by the fuzzy set as being lower than 1,000 product items. The possibility distribution is precisely defined as the result of a process called 'de-fuzzification'. De-fuzzification is about obtaining a crisp value for each of the output fuzzy set variables (e.g. 'high post costs' can be de-fuzzified to 100 dollars).

The main outcome of this methodology is that having started from an IDEF3 model containing incomplete information about its linguistic variables, a set of specific IF-THEN rules is extracted based on the model's flow. The linguistic variables of these rules are then categorised into fuzzy sets which are de-fuzzified by assigning precise boundaries. The significance of this procedure is that the process is accurately executed and its output can be quantified and predicted. Performance analysis can then occur by combining different values for each variable to estimate the various process outputs.

This methodology was applied to a real industry process problem (film decomposition process) and succeeded in calculating process outputs for different values of the linguistic variables of the process. This provides the opportunity to perform performance analysis of the process output for different scenarios. This modelling and analysis approach, being co-operative (or supplementary) to an IDEF3 process model, manages to represent and quantify information that is usually not applicable to diagrammatic process models. The effectiveness of quantification is apparent as it can be used to combine a number of values for each participating variable and to produce potential process outputs helping the business analyst to track which process scenario is more beneficial.

3.2 Quantifying Role Activity Diagrams

Role activity diagrams (RADs) are based on a graphical view of the process from the perspective of individual roles, concentrating on the responsibility of roles and the interactions between them [12]. Roles are abstract notations of behaviour describing a desired behaviour within the organisation. RADs are object-state transition diagrams used in object-oriented models. They describe how a role object changes state as a result of the occurring actions and interactions.

Phalp [21] attempts to quantify RADs thus create a measure for performance analysis. The main objective of this approach is to minimise the coupling between the various roles in a Role Activity Diagram. Coupling is defined as the interaction between two or more roles in order to complete a task [20]. Coupling can be captured and visualised within RADs. The starting point for this coupling measure is the sum of actions and interactions of a particular role. This measure is called 'coupling ratio' (C_pF_X) measuring the correlation between actions (independent activities of a role) and interactions (involvement of another role) in terms of percentage. If the ratio is high, it means that the coupling is large and the business process needs further improvement.

A major benefit of the coupling ratio is that it enables comparison between roles of different sizes [21]. Reducing coupling allows roles to become more autonomous because they do not need to synchronise. This has a major effect on the business process as it improves the performance in a way that each role performs its tasks more quickly and with less opportunity to delay since there is independence from redundant interactions. This approach enables business process performance analysis by quantifying one feature, i.e. role coupling. Reducing coupling in a process, directly results in the process becoming more straight-forward, faster and – in general terms- improved through the analysis and inspection of the coupling ratio.

3.3 Performance Analysis through a Query Language

The last business process performance analysis approach is workflow related. Workflow is a similar concept to business process and many authors are using these two keywords interchangeably. In [1] an approach is proposed for the performance evaluation of automated business processes based on the measurement language WPQL (Workflow Performance Query Language). The adoption of a Workflow Management System (WfMS) to automate a business process gives the opportunity to collect real execution data continuously, from which exact information about the process performance can be obtained. On the one hand, such data can be used for monitoring, work balancing and decision support. On the other, execution data can feed simulation tools that exploit mathematical models. Through the simulation, it is possible to obtain an assessment of the current process performance and to formulate hypotheses about possible reengineering alternatives.

Although formal languages have been exploited in order to define the concept of process, the use of formal languages to handle the problem of performance evaluation of workflow has received not as much coverage [1]. A language is an application tool that could enable the writing of queries against a WfMS in order to compute measures about given workflow entities. The main benefit of such a tool is the expressive power of a programming language to define and evaluate new performance measures. The basic idea of the WPQL language is based on the following steps:

1. Define a new measure
2. Select workflow entities to measure
3. Apply the measure to the selected entities

The WPQL also offers a standard set of performance measures as well as the facilities to define new ones. WPQL is a specialized language, its focus is on the measurement of workflow related quantities and its constructs have been introduced to ease the handling of concepts such as process or workflow participant.

4 Scheduling-based Business Process Optimization Approaches

After reviewing the business process modeling techniques and performance analysis approaches, it is time to examine the not-as-many business process optimization attempts. Process improvement – often used as an umbrella term for performance analysis and business process optimization- is one of the most significant motivations for process modeling. According to [25], large organizations are attempting to map their processes for two main reasons: One is to acquire a realistic knowledge of the current situation and flow of activities within the organization and second to efficiently improve those processes thus meeting the organization goals. The prerequisites for business process optimization are: (i) the business processes should be correctly designed, (ii) their execution should be supported by a system that can meet the workload requirements, and (iii) the (human or automated) process resources should be able to perform their work items in a timely fashion [8]. The same would stand for any other scheduling-type of problem.

The quality of a business process is defined by many, often conflicting criteria such as costs, duration, or quality of output. A business process optimization approach should clearly define and specify how optimization is perceived and which aspects of the process are going to be optimized. Not many optimization techniques found in literature are suitable for business processes. Due to their qualitative nature, process designs are hard to characterize in a formal way amenable to analytical methods and thus improve them in a measurable way. There is also a lack of tools for identifying the bottleneck areas in these models. Their qualitative nature also explains the difficulty of developing 'parametric' models of business processes. For these reasons, there is a lack of algorithmic approaches for the optimization of business processes [26]. In this section some optimization techniques found in literature are discussed.

4.1 Mathematical Programming Formulation

As mentioned previously, there is an attempt in [11] to optimise a business process using three different approaches that are examined thoroughly in this section. Having already mentioned the modelling approach towards business processes it becomes easier to understand the perspective towards the model's optimization.

The formulation of the process design as a mathematical problem is close to scheduling problems. A linear objective function and a number of constraints are used to describe the problem and cover all of its aspects. The objective function –which can illustrate a process objective e.g. cost- is minimised or maximised according to the goal of optimization. The constraints describe and ensure the feasibility of the process in a mathematically formal way. A constraint, for example, can restrict the use of a particular resource before it is produced within the process. The main concepts used in the process design are activities and resources. The mathematical constraints regarding the activities and resources can be grouped into two major categories:

1. constraints related to input and output resources of each activity and
2. constraints regarding the sequence and timing of resources and activities.

Each process begins with some input resources (I_{glob}) and produces a desired set of output resources (O_{glob}). The participating activities should be sequenced in such a way that they use some resources as inputs and then produce resources that can be used as inputs by other activities until the desired output is produced. Constraints that belong to the first group make sure that input resources are available by activities to use and that the final set of output resources is eventually produced. The second group of constraints checks the time sequence of activities and resources. Each activity has a starting time p and an execution duration d. The input resources of this activity must be available before p and the output resources must be produced in $p+d$ time. The time that a resource becomes available is q and is critical to the feasibility of the process.

In order to formally set the constraints, a number of variables and arrays that bind together the resources and activities are being introduced. This increases the complexity of the process model but also ensures its mathematical formality. It also makes the model more flexible as a constraint can be eliminated to simplify a particular aspect of the model or extra constraints can be inserted to further shape the model. The total number of constraints in the final mathematical is thirteen. According to experiments [11] the mathematical approach produced satisfying results but high execution times.

4.2 Direct Branch and Bound Method

Apart from mathematical programming, branch and bound method is another way of optimising a business process [11]. Optimization problems that have a significant number of binary variables are often solved using branch and bound algorithms as they are easy to implement. Moreover, these algorithms provide a bound on the optimal objective during execution, meaning that they can be stopped when the potential improvement is small, without solving the problem to full optimality. During the branch and bound procedure, a search tree is generated and in every branching step another activity is added to the tree constructing the process design i.e. at the first level, the first activity to be included in the design is selected, at the second level the second activity and so on. When a feasible solution is found then the particular node is no longer expanded if it contains a feasible solution. In order to implement a branch and bound strategy, two particular problems are being examined [11]:

1. The selection of a node to be expanded.
2. The computation of bounds on the objective value at each node.

In terms of the first problem, two strategies attempt to restrict the set of activities considered for branching: (i) The forward strategy that starts from global inputs using resources available to link activities and produce the global output and (ii) the backward strategy that starts from global outputs and tries to generate initial resources. Computational experiments confirm that the forward strategy is considerably more efficient. These experiments show that the branch and bound algorithm turned out to be considerably fast, solving most of the problems quickly and yielding better results than other approaches.

4.3 Genetic Algorithms

Genetic Algorithms (GAs) has been a popular method that has been developed and successfully applied to complex problems in a variety of areas. A significant advantage of GAs is that they maintain a population of possible solutions to reach feasibility and that makes them more powerful. Another significant advantage is their extendibility to optimise a problem with more than one objective. Multi-objectivity makes genetic algorithms a flexible and truly beneficial methodology that can be applied to any optimization problem.

In order to find an optimal solution, a genetic algorithm imitates the process of natural evolution. It works on a large number of solutions in parallel, where each solution corresponds to an individual in a population. Each solution is represented by an appropriately coded string. Initially, a set of randomly generated solutions is produced. Then, superior solutions are selected to form a new population. The selection probability depends on the objective function value. The resulting selected individuals are then selected for mating. A crossover operation exchanges information between two individuals. Finally, a mutation operation changes the values of randomly chosen bits. This process continues until some pre-defined termination criteria are fulfilled.

The business process design that is proposed in [11] has a significant number of constraints. There are two different approaches to deal with constraints in the GAs optimization approach. In the first approach, a penalty term for constraint violation is added to the original objective function. The second approach modifies the genetic operators to limit the search space to feasible solutions. This approach is appropriate if feasible changes can easily be determined. Nevertheless, during the performance tests, genetic algorithms show weak performance. Their main problem is the feasibility maintenance in this business process design problem. Our approach –presented on the next section- attempts to resolve these problems by modifying the business process model appropriately and then applying selected evolutionary algorithms to achieve multi-objective optimization.

5 Scheduling-based Evolutionary Multi-objective Optimization

This section introduces our approach towards business process optimization. A formally defined business process model is optimized using evolutionary algorithms. The utilization of evolutionary techniques provides the additional ability of optimizing the problem under multiple criteria. The framework to be presented next, aims to introduce a methodology for applying multi-objective optimization algorithms to business process models. It consists of two main stages: The first stage of the framework is the business process model definition and the second stage involves the application of the evolutionary algorithms to a test business process model for optimization results generation.

5.1 Business Process Model

The first stage of the optimization framework is the business process model specification. The model has a mathematical basis to ensure formality, consistency and rigor. The business process model is limited to a series of mathematical constraints that define the feasibility boundaries of the business process and a set of objective functions that consist of the various business process objectives. Representing a business process using a formal mathematical model guarantees the construction of consistent and rigorous business processes following a formally correct, repeatable and most importantly, verifiable approach [15].

The business process model consists of two key concepts: activities and resources. Apart from the resources that are generated within the process, the business process design has two sets of resources, the initial (I_{glob}) and the final (O_{glob}) resources. The initial resources are available at the beginning of the business process, while the final resources form the business process final output. The resources flow through the process and belong to two categories: physical and information resources. The activities, on the other hand, are perceived as the transformation steps within the process that use some resources as inputs and produce others as outputs. In a feasible process design, all the activities are in a defined sequence, the availability of resources is adequate and most importantly the final (output) resources are being produced by the participating activities.

For the business process model to be optimised, it is necessary to define the process objectives (e.g. process duration) as well as the input variables (e.g. activity duration). The business process model receives as input variables the participating activities and their starting times, whereas the aim is to produce an optimised process in terms of minimising two objectives, the process duration and cost. For each process design, there is a library of candidate activities with attributes such as activity duration and activity cost. For an optimised business process design to be produced, a set of activities that generate minimum business process cost and duration needs to be selected. It is important to note that the framework works independently of the number of objectives and their type. Process duration and are cost chosen as the two objectives for business process improvement. The complete mathematical model is the following:

$$f_1(P) = \max(q_j) \to \min, \forall j : b_j \in go_j$$

$$f_2(P) = \sum u_{i1} x_i \to \min$$

s.t.

1. $x_i \le r_{ij}, \forall i, j : b_j \in I_i, b_j \in B_P,$
2. $x_i \le y_j, \forall i, j : b_j \in I_i, b_j \in B_I,$
3. $go_i + \sum_i r_{ij} \le M \cdot gi_j + \sum_i t_{ij} x_i, \forall j : b_j \in B_P,$
4. $y_j \le gi_j + \sum_i t_{ij} x_i, \forall j : b_j \in B_I,$
5. $y_j \ge go_j,$
6. $p_i \ge q_j - M(1 - x_i), \forall i, j : b_j \in I_i,$
7. $q_j \le p_i + \delta_i + M(1 - x_i), \forall i : b_j \in O_i$
8. $q_j \ge p_i + \delta_i - M(1 - x_i) - M(1 - \lambda_{ij}), \forall i : b_j \in O_i,$
9. $\lambda_{ij} \le x_i, \forall i, j : b_j \in O_i,$
10. $\sum_{i : b_j \in O_i} \lambda_{ij} \ge \sum_i r_{ij} + og_i - M(1 - y_j), \forall j : B_P, gi_j = 0,$
11. $\sum_{i : b_j \in O_i} \lambda_{ij} \ge 1 - M(1 - y_j), \forall j : b_j \in B_P, gi_j = 0,$
12. $x_i \in \{0,1\}, \forall i,$
13. $\lambda_{ij} \in \{0,1\}, \forall i, j : b_j \in O_i.$

The mathematical model of business process defines the optimization objectives with two objective functions and ensures the business process consistency and feasibility with thirteen constraints. Further objectives can be added with extra functions. The process model appears to be complicated in contrast to the simplistic approach of the business process consisting only of activities and resources. A brief description of the mathematical model's main features can provide a good understanding of its functionality. The mathematical model consists of a number of binary variables and binary matrices that have an impact on the production of feasible process designs since they result to a highly fragmented search space. The first objective function (f_1) of the model calculates the duration of the business process. The total duration for a feasible process equals the time the last resource that belongs to global outputs is produced. The second objective function (f_2) calculates the business process cost as the sum of costs of all participating activities.

Table 1. Main Parameters in Mathematical Model

Parameter	Explanation
u_{i1}	Cost of execution for activity a_i.
x_i	Binary variable that indicates whether a candidate activity a_i participates in the business process design.
y_j	Binary variable that indicates whether resource b_j is or becomes available in the business process.
$t_{i,j}$	Matrix of binary variables that link the activities with their output resources.
r_{ij}	Matrix of binary variables that indicate if a unit of physical resource b_j is available for use by activity a_i.
gi_j & go_j	One-dimensional binary constants that indicate which resources belong to global inputs and/or global outputs.
M	Large constant indicating that physical resources contained in the set of global inputs are available in unlimited amounts.
p_i	Starting time of activity a_i.
q_j	Time when the resource b_j becomes available.
δ_i	Duration of activity a_i.
λ_{ij}	Binary variable indicating that activity a_i is used to create resource b_j.
I_i / O_i	Sets of input/output resources of activity a_i.
B_P / B_I	Set of physical / information resources b_j.

The mathematical model constraints ensure that the model produces feasible business processes by examining different aspects of the business process model. Table 1 provides an explanatory legend for all the mathematical model variables and parameters and table 2 provides a short description of each constraint of the mathematical model. It is also important to highlight two features of the business process model. The mathematical model consists of many discrete binary variables that significantly increase the complexity of even a simple process design as the search space for feasible solutions is highly fragmented. Another feature of the business process model is that although it is simple to conceive and understand, it is highly constrained when it comes to formal mathematical definition. This can create serious difficulties in locating the optimum solutions since even feasible solutions are hard to be produced. The concepts that describe the business process and its mathematical model are inspired by [11]. Our aim is to extend the model to multi-objectivity and optimise it using multi-objective evolutionary algorithms.

5.2 Test Problem Construction

This section describes the construction of a business process design test problem. Five different test process designs were constructed for assessing their capability to be optimised by the optimization algorithms. We will describe analytically the construction of one of these problems.

The test problems utilised, have an increasing number of activities participating in each process design. The results demonstrate that this is a significant cause in increasing the problem complexity and has a serious impact on the quality of results. Each of the problems has a fixed predefined number of participating activities in the process. The initial and final resources of the business process are given. Remember that for each process design there is a library of candidate activities that can potentially participate in the process.

A business process design is optimised when the activities selected along with their starting times produce a business process with minimum duration and cost in contrast with any other combination of candidate activities. The case study discussed here is based on the generic business process model, it is called ActivitiesST4 and it involves four participating activities. The library of candidate activities contains 10 activities that can be alternatively used in various combinations of four. For ActivitiesST4 design to be optimised, the four activities selected to participate must minimise the total process cost and duration.

Table 2. Explanation of Constraints

1. $x_i \le r_{ij}, \forall i, j : b_j \in I_i, b_j \in B_P$

All input physical resources of an activity must be available ($r_{ij}=1$) at some stage of the process if the activity is participating ($x_i=1$).

2. $x_i \le y_j, \forall i, j : b_j \in I_i, b_j \in B_I$

All input information resources (y_j) of an activity must be available at some stage of the process if the activity is participating ($x_i=1$).

3. $go_i + \sum_i r_{ij} \le M \cdot gi_j + \sum_i t_{ij} x_i, \forall j : b_j \in B_P$

The output physical resources -final or not- must not exceed the sum of initial and produced -during the process.

4. $y_j \le gi_j + \sum_i t_{ij} x_i, \forall j : b_j \in B_I$

An information resource (y_j) can be available either at the beginning of the process -as initial resource (gi_j) - or as an output resource of a participating activity.

5. $y_j \ge go_j$

A resource (y_j) cannot be part of the output without first being available at some stage of the process (go_j).

6. $p_i \ge q_j - M(1-x_i), \forall i, j : b_j \in I_i$

In terms of time, a participating activity must start (p_i) only after the time that all its input resources have become available.

7. $q_j \le p_i + \delta_i + M(1-x_i), \forall i : b_j \in O_i$

8. $q_j \ge p_i + \delta_i - M(1-x_i) - M(1-\lambda_{ij}), \forall i : b_j \in O_i$

In terms of time, an output resource must become available exactly when the generating activity has been completed ($q_j=p_i$).

9. $\lambda_{ij} \le x_i, \forall i, j : b_j \in O_i$

A non-participating activity ($x_i=0$) cannot have output resources ($\lambda_{ij}=1$).

10. $\sum_{i:b_j \in O_i} \lambda_{ij} \ge \sum_i r_{ij} + og_i - M(1-y_j), \forall j : B_P, gi_j = 0,$

When a physical resource does not belong to initial resources, it must be produced in greater or equal amounts to the required resource inputs.

11. $\sum_{i:b_j \in O_i} \lambda_{ij} \ge 1 - M(1-y_j), \forall j : b_j \in B_P, gi_j = 0$

Each physical resource that does not belong to initial resources but appears in the output of a participating activity must be produced at least once.

12. $x_i \in \{0,1\}, \forall i$

The variable x (indicating participating activities) must be binary.

13. $\lambda_{ij} \in \{0,1\}, \forall i, j : b_j \in O_i$

The variable λ (indicating output resource j of activity i) must be binary.

The process design sketch of ActivitiesST4 problem is demonstrated in figure 1. Process optimization depends on the input parameters:

1. The appropriate activities need to be selected from the library and combined according to their duration and cost attributes and
2. The activities' starting times need to be properly calculated in order for the process outputs to be produced as early as possible and thus minimise the total process duration.

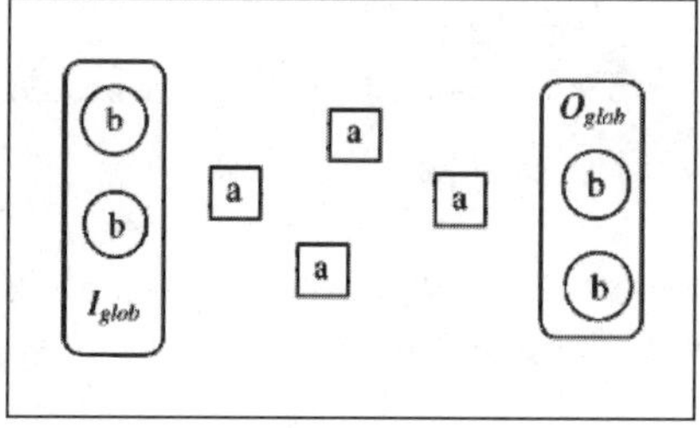

Fig. 1. ActivitiesST4 initial process design

The process design of figure 1 can be described as follows: There are two global input resources to start the process. These two resources together with the two global outputs are considered as constants. The system variables of the problem are the four participating activities and their starting time attribute. This means that the optimization algorithms attempt to meet the optimization objectives by defining a set of four activities (from a library of 10 alternatives) and the starting time for each of them. All the potential activities are stored in a built-in library and the algorithms can select any four activities. The four potential activities of the process design must be combined in a way that the given process output resources are produced. The optimization criteria are the minimisation of process duration and cost.

The other four test problems have identical structure as the one described. Their major differences are in the number of activities that participate in the business process design and the size of the library of candidate activities. The test process designs range from simple ones (e.g. ActivitiesST2 with two participating activities and a library of 10) to more complex ones (e.g. ActivitiesST5 with 5 activities in the process design and a library of 20 alternatives). As is mentioned earlier in this section, although the different test problems give the impression of having minor differences in the process size and library, these have a significant impact on the optimization performance as is demonstrated on the next section.

5.3 Experimental Results

This section describes the experimental results for the test problem of the previous section. Three popular evolutionary algorithms that allow multi-objective optimization have been selected to optimise the business process model. The optimization algorithms that were selected are NSGA2, SPEA2 and MOPSO. These algorithms attempt to optimize the process designs by selecting different sets of activities and defining their starting times. Non-dominated Sorting Genetic Algorithm II (NSGA2) is non-dominated, sorting-based, multi-objective evolutionary algorithm [6]. NSGA2 has been quite popular and has been applied to many problems on a number of research areas. Strength Pareto Evolutionary Algorithm II (SPEA2) is another elitist evolutionary algorithm with a fine-grained fitness assignment strategy, a density estimation technique, and an enhanced archive truncation method [31]. SPEA2 has also been quite popular and used in a variety of optimization problems. Multi-Objective Particle Swarm Optimization (MOPSO) is different from most evolutionary computation techniques as it is an extension of the Particle Swarm Optimization (PSO) method. MOPSO is demonstrating better performance in problems that have continuous search space [13]. Since more than one optimization methods are applied to the business process model, the opportunity of comparing the performance of the different algorithms in the particular problem context is appealing. NSGA2 and SPEA2 are being demonstrated in a number of papers and although similar they battle each other in quality of results in different subject domains. MOPSO on the other hand has never been used in such a constrained problem.

All the five test problems are optimised with each of the evolutionary algorithms. To evaluate the results a metric is also introduced which demonstrates the algorithms' performance. The 'success ratio' is the opposite of error ratio [6] and is calculated as the percentage of generated solutions that belong to the Pareto optimal front against the total number of solutions. The equation of the success ratio is:

$$s_R = \frac{no_of_solutions \in P^*}{total_solutions} \%$$

The numerator of the success ratio holds the number of generated solutions that belong to P* (Pareto optimal front) while the denominator holds the total number of generated solutions.

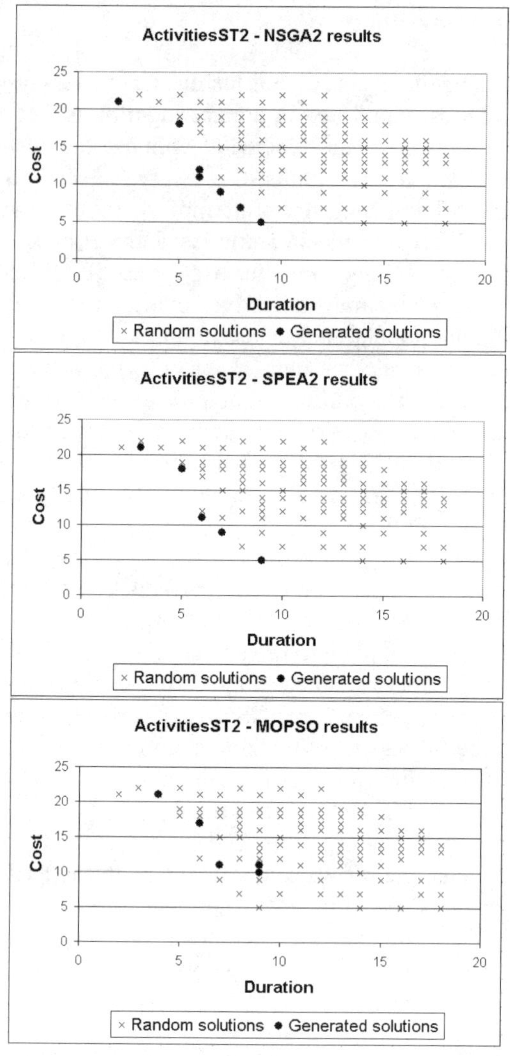

Fig. 2. Generated solutions for ActivitiesST4 by the optimization algorithms

In this particular context, the success ratio (s_R) calculates the percentage of the *near-Pareto* optimal solutions that the optimization algorithm has generated because being a real-life situation the actual Pareto optimal front is not known for the test problem. To acquire a picture of the search space 10,000 random solutions were created and the feasible solutions amongst these were identified and plotted. Therefore, near-Pareto optimality of a solution in this case is defined with respect to the large set of randomly generated solutions such as the ones demonstrated in figure 2. A solution generated by an algorithm is considered here as near-Pareto optimal if it is non-dominated with respect to the set of these random solutions.

The test problems are incorporated in KEA toolbox [4] an optimization platform that uses (among others) NSGA2, SPEA2 and MOPSO algorithms to optimise user-defined problems. In order to produce the results, each of the optimization algorithms was executed 30 times with different random seed values. Most of these 30 runs produced similar results. The results presented here belong to one of those runs. The graphs in figure 2 demonstrate the solutions that each of the optimization techniques generated for the test process design. These solutions consist of feasible business processes with minimised process duration and cost. The graphs depict the process duration and cost values for both the random population and the optimised. The dotted points represent the solutions of each technique while the ‘x points’ the random solutions. Each graph demonstrates the results for ActivitiesST4 process design by NSGA2, SPEA2 and MOPSO algorithms.

The success ratio was used to evaluate the results that the optimization algorithms produced. Figure 3 demonstrates the success ratio percentages for all the five test problems. For ActivitiesST2 process design, both NSGA2 and SPEA2 performed very well, unlike MOPSO that identified only 40% of the near-Pareto optimal solutions. SPEA2 also produced very good results for ActivitiesST3 problem, while NSGA2 gave a satisfactory number of optimum solutions. Nevertheless the algorithms’ performance drops significantly with the addition of an extra activity in ActivitiesST4. MOPSO performs poorly as apart from the first test problem it does not seem to be able to locate optimum solutions. Moving to test problems with bigger activity libraries, NSGA2 produced satisfactory results for ActivitiesST4(20) only, while for ActivitiesST5 problem none of the algorithms was able to locate solutions near the Pareto front. The average success ratio for both NSGA2 and SPEA2 is approximately 40%, while for MOPSO is only 8%.

Before the results are further discussed, the features of the search space need to be highlighted once more as they seriously influence the quality of the results. The mathematical model of the business process designs consists of discrete binary variables that increase the optimization complexity even for a simple process design as the search space for feasible solutions is highly fragmented. Also the business process models are highly constrained having 13 constraints to check for every possible set of solutions, decreasing the performance of the algorithms.

The optimization algorithms have a difficult task even to produce sets of feasible solutions. Given the complex nature of the business process design problem, the overall performance of NSGA2 can be characterised as good and can be attributed to its elitism. As NSGA2 archives the optimum solutions of each generation and compares them with the ones it produces, it manages to preserve the identified feasible solutions. SPEA2 is also an elitist algorithm that provides bigger spread of the solutions. It also preserves feasible individuals through generations and that justifies its satisfactory results. On the other hand, MOPSO seems to have a serious problem in successfully defining the guide that combines the two objectives. The algorithm demonstrates poor performance as the solutions that are generated are not near to the Pareto optimal front for most of the test problems. This supports the claim that MOPSO has best performance in problems with continuous search space [13] which is not the case here.

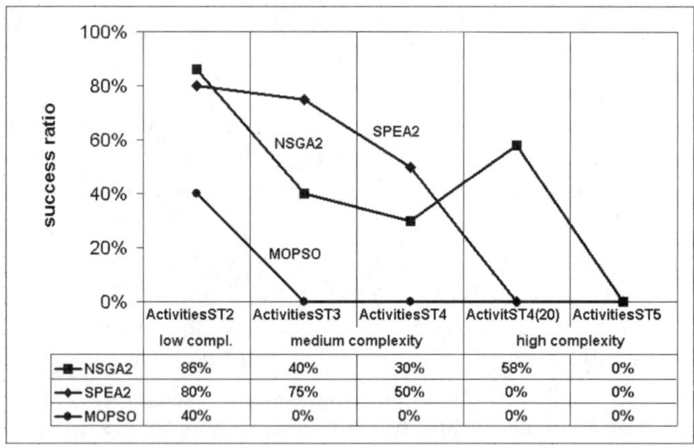

	ActivitiesST2	ActivitiesST3	ActivitiesST4	ActivitST4(20)	ActivitiesST5
	low compl.	medium complexity		high complexity	
NSGA2	86%	40%	30%	58%	0%
SPEA2	80%	75%	50%	0%	0%
MOPSO	40%	0%	0%	0%	0%

Fig. 3. Evaluation graph based on success ratio

Figure 3 also shows that as the complexity of the problems increases, the performance of the optimization algorithms declines significantly. The simplest of test problems (ActivitiesST2) is handled well by all three algorithms. Moving to medium complexity problems, SPEA2 provides better results while NSGA2 hits back on high complexity problems with 58% success on one of the problems. On average performance NSGA2 holds the best position with slightly better results that SPEA2 which has also performed above 40% on average. This enhances the view of SPEA2 and NSGA2 behaving very similar on various problems [31]. Many applications of the NSGA2, SPEA2 and MOPSO are not as successful in dealing with large dimensional problems and extremely disconnected Pareto fronts.

6 Discussion of Results

This section discusses the practical implications of the business process optimization framework, along with its limitations. As mentioned in a previous section, there are not many optimization techniques for business processes. Many of the techniques only provide quantifiable measures from diagrammatic process models.

However, the test problems demonstrated that the proposed framework is capable of successfully applying multi-objective optimization to various business process designs. The ability to produce an overall 40% of optimum solutions provides a good set of optimised alternative business processes with trade-offs in process duration and cost. This gives the capability to the business analyst to select a business process from a range of near-Pareto optimal solutions according to decision making priorities. This extends the approach beyond a single objective. The results are indicative but also promising, and future research can lead to better quality results.

During the development of the multi-objective optimization methodology a number of limitations were unveiled. The first limitation originates from the mathematical model of the business process. The mathematical model focuses on activities and resources as its two main concepts and it ignores the participating (physical or mechanical) actors. This consequently results in what is criticised as 'a mechanistic viewpoint of business processes' [16]. Also it is more difficult for a formal business process modelling technique to capture the roles of the participants than a diagrammatic approach which visualises the flow of the process.

Another limitation lies in the selection of the five process designs as test problems. In order to better assess the optimization techniques used, an approach with a scalable range of problems was selected. Evaluating the algorithms' performance using a larger series of problems can better demonstrate the algorithms' behaviour providing a better and more apparent performance overview. The last limitation is linked with the evaluation metric that did not take into account the diversity of the generated solutions.

Formal business process optimization techniques can significantly contribute to the wider area of business process improvement in a number of ways. Firstly, an analytical method which takes into account the entire range of possible designs might produce process designs that are overlooked or cannot be conceived by a human designer. Secondly, by optimizing a process for different design criteria, the inherent trade-offs and the sensitivity of results to variations in design parameters will become more transparent with an analytical method. This can help a designer in identifying those parameters that are most important in achieving the desired goals. Therefore, business process optimization based on mathematical or algorithmical techniques can contribute significantly to introducing new perspectives and approaches.

7 Directions for Future Research

Future research in this area should focus on building more complete process models, testing more complicated process designs and exploring more efficient metrics. The construction of a business process model that can cover more aspects of a 'closer to real world' business process can be very useful for effective business process optimization. Business processes in real world have patterns such as feedback loops or decision points. Modelling and optimizing these aspects can prove a complicated process with increased complexity. Future research should also focus on selecting the most appropriate techniques for business process multi-objective optimization from a wider set of techniques and algorithms, and thus locating more accurately the most suitable optimization method. To improve the optimisation results of similar problems, there is a need to develop novel initialization and recombination schemes instead of relying to existing evolutionary approaches. Again, scheduling domain can provide techniques that have been successfully applied to solve complex scheduling problems of similar nature.

8 Conclusions

This chapter presented business process re-design as a problem of similar nature to scheduling. Business process optimization was perceived as the combination and sequencing of resources and alternative activities. After examining the business process concept in a generic sense and examining a selection of business process modeling, performance analysis and optimization techniques, we presented a framework for applying multi-objective optimization to business processes. By developing a formal business process model and orienting it to multi-objectivity, the generation of optimized business processes was facilitated.

The business process optimization problem is unique because of its highly constrained nature and the fragmented search space that have a significant impact on locating the optimum solutions. It is shown that state-of-the-art multi-objective optimization algorithms, such as NSGA2 and SPEA2, produce satisfactory results by managing to generate and preserve optimal solutions. This provides a number of alternative optimised process designs for the business analyst to decide the trade-offs between the different objectives. The results presented here demonstrate that principles of scheduling could be effectively applied for optimization of business processes. This work is encouraging for further research in the area of business process multi-objective optimization.

References

[1] Abate AF, Esposito A, Grieco N, Nota G (2002) Workflow Performance Evaluation Through WPQL. In: Proceedings of the 14th International Conference on Software Engineering and Knowledge Engineering, Vol. 27. ACM Press, New York, pp 489-495

[2] Aguilar-Saven RS (2004) Business process modeling: Review and framework. Int J of Production Economics 90: 129-149

[3] Aiello R, Esposito A, Nota G (2002) A Hierarchical Measurement Framework for the Evaluation of Automated Business Processes. Int. J. of Software Eng. and Knowledge Eng. 12: 331-361

[4] Bartz-Beielstein T, Mehnen J, Naujoks B, Schmitt K, Zibold D (2004) KEA - A software package for development, analysis and application of multiple objective evolutionary algorithms, <http://www.isf.maschinenbau.uni-dortmund.de/veroeff/documents/2004_mehn_kea_a.pdf> viewed February 2, 2006

[5] Curtis B, Kellner M, Over J (1992) Process modeling. Communications of the ACM 35: 75-90

[6] Deb, K.: Multi-objective optimization using evolutionary algorithms. John Wiley & Sons, New York (2001)

[7] Enderton, H.B.: A Mathematical Introduction to Logic. Academic Press, New York (1972)

[8] Grigori D, Casati F, Castellanos M, Dayal U, Sayal M, Shan M.C. (2004) Business Process Intelligence. Computers in Industry 53: 321-343

[9] Hammer M, Champy J (1993): Re-engineering the Corporation: A manifesto for business revolution. Harper Business, New York

[10] Havey M (2005) Essential Business Process Modeling. O'Reilly, U.S.A

[11] Hofacker I, Vetschera R (2001) Algorithmical approaches to business process design. Computers & Operations Research 28: 1253-1275

[12] Holt AW, Ramsey HR, Grimes JD (1983) Coordination systems technology as a programming environment. Electrical Communication 57: 307-314

[13] Kennedy J, Eberhart R (1999) The Particle Swarm: social adaptation in information-processing systems. In: Corne D, Dorigo M, Glover F, (Eds.): New Ideas in Optimization. McGraw-Hill, Cambridge, pp 379-388
[14] Kettinger WJ, Teng JTC, Guha S (1997) Business Process Change: A Study of Methodologies, Techniques and Tools. MIS Quarterly 21: 55-80
[15] Koubarakis M, Plexousakis D (2001) A formal framework for business process modeling and design. Information Systems 27: 299-319
[16] Lindsay A, Downs D, Lunn K (2003) Business processes - attempts to find a definition. Information and Software Technology 45: 1015-1019
[17] Luttighuis PO, Lankhorst M, van de Wetering R, Bal R, van den Berg H (2001) Visualizing Business Processes. Computer Languages 27: 39-59
[18] Melao N, Pidd M (2000) A conceptual framework for understanding business processes and business process modeling. Information Systems Journal 10: 105-129
[19] Miers D (1994) Use of tool and technology within a BPR initiative. In: Coulson-Thomas C (ed.): Business Process Re-engineering: Myth and Reality. Elsevier Science, North-Holland, Amsterdam, pp 142-165
[20] Ould MA (1995) Business Processes: Modeling and Analysis for Reengineering & Improvement. John Wiley, New York
[21] Phalp K, Shepperd M (2000) Quantitative analysis of static models of process. The Journal of Systems and Software 52: 105-112
[22] Phalp K (1998) CAP framework for business process modeling. Information and Software Technology 40: 731-744
[23] Reyneri C (1999) Operational building blocks for business process modeling. Computers in Industry 40: 115-123
[24] Sadiq W, Orlowska EM (2000) Analyzing process models using graph reduction techniques. Information Systems 25: 117-134
[25] Smith H (2003) Business process management - the third wave: business process modeling language (bpml) and its pi-calculus format. Information and Software Technology 45: 1065-1069
[26] Tiwari A (2001) Evolutionary computing techniques for handling variables interaction in engineering design optimization. PhD Thesis, SIMS, Cranfield University, Cranfield, UK
[27] van der Aalst W.M.P, van Hee K.M (1995) Framework for Business Process Redesign. In: JR Callahan (ed.) Proceedings of the Fourth Workshop on Enabling Technologies: Infrastructure for Collaborative Enterprises. IEEE Computer Society Press, Berkeley Springs, pp 36-45
[28] van der Aalst WMP (1996) Petri Net Based Scheduling. OR Spectrum 18: 219-229
[29] van der Aalst WMP (1998) The Application of Petri-Nets to Workflow Management. Journal of Circuits, Systems and Computers 8: 21-66
[30] Zakarian A (2001) Analysis of process models: A fuzzy logic approach. The Int J of Advanced Manuf. Technology 17: 444-452
[31] Zitzler E, Laumanns M, Thiele L (2001) SPEA2: Improving the Strength Pareto Evolutionary Algorithm. TIK Report Nr. 103, Computer Engineer-

ing and Networks Lab (TIK), Swiss Federal Institute of Technology (ETH) Zurich

Using a Large Set of Low Level Heuristics in a Hyperheuristic Approach to Personnel Scheduling

Peter Cowling and Konstantin Chakhlevitch

Modeling Optimisation Scheduling And Intelligent Computing (MOSAIC) Research Centre, Department of Computing, University of Bradford, Bradford BD7 1DP, UK

Summary. A *hyperheuristic* is a high-level search method which manages the choice of low level heuristics, making it a robust and easy to implement approach for complex real-world problems. We only need to develop new low level heuristics and define the objective functions in order to apply a hyperheuristic to an entirely new problem. Although hyperheuristic methods require limited problem-specific information, their performance for a particular problem is determined to a great extent by the quality of low level heuristics used. This chapter addresses the question of designing the set of low level heuristics for the problem under consideration. We construct a large set of low level heuristics by using a technique which allows us to "multiply" partial low level heuristics. We apply hyperheuristic methods to a trainer scheduling problem using commercial data from a large financial institution. The results of the experiments show that simple hyperheuristic approaches can successfully tackle a complex real-world problem provided that low level heuristics are carefully selected to treat various constraints. We examine experimentally how the choice of different sets of low level heuristics affects the solution quality.

1 Introduction

Given its economic importance, there is continuing research interest in solving real-world personnel scheduling problems. The purpose of personnel scheduling is to allocate the available workforce to timeslots and locations and to assign particular tasks to each member of staff optimising

P. Cowling and K. Chakhlevitch: *Using a Large Set of Low Level Heuristics in a Hyperheuristic Approach to Personnel Scheduling*, Studies in Computational Intelligence (SCI) **49**, 543–576 (2007)

various measures such as worker quality of life, workforce utilisation and service quality. However, real-world scheduling problems require increasingly complex models and finding optimal solutions may require prohibitive amounts of computer time. Heuristic methods are often used in practice, which produce solutions of acceptable quality in reasonable time. Various metaheuristic approaches have been developed and successfully applied for different personnel scheduling problems. Recent examples include fast local search and guided local search algorithms applied to British Telecom's workforce scheduling problem [24]; a simulated annealing approach for shift scheduling problems [23]; tabu search applied to audit staff scheduling [10]; different approaches to tackle a nurse rostering problem, specifically tabu search with strategic oscillation [11], genetic algorithms [1], and memetic algorithms and their hybrids with tabu search [2].

Although metaheuristics and especially their hybrids have proved to be quite efficient for solving some real-world scheduling problems, their application is usually dependent on the problem domain. Specific metaheuristic approaches designed to effectively solve a particular problem may not be applicable or may produce very poor solutions for other problems or even for the other instances of the same problem. Metaheuristics incorporate information specific for the problem and require expertise both in the problem domain and in heuristic methods. Therefore, metaheuristics are often quite expensive to implement [7]. For that reason, development of general domain-independent heuristic search techniques has received increased attention from researchers. These new approaches have recently become known as *hyperheuristics* and their development is described by Burke et al. [3] as "an emerging direction in modern search technology".

The term "hyperheuristic" was introduced in [7], as an approach that manages the choice of which low level heuristic method should be applied at any given time, depending upon the characteristics of the region of the solution space currently under exploration, and the history of each low level heuristic. This means that a hyperheuristic does not search directly for a better solution of the problem but instead it looks for a method by which a promising solution can be obtained. A hyperheuristic requires limited domain-specific information which is concentrated in the set of low level heuristics and the objective function(s). Low level heuristics usually represent simple local search neighbourhoods or the rules used by a human expert for constructing solutions.

Several hyperheuristic approaches have been presented over recent years. Gratch and Chien in [14] develop a general adaptive problem-solving approach which automatically acquires domain-specific information and selects well-suited heuristic method from a given set. In [20], Randall and Abramson develop a general metaheuristic based solver for combinatorial optimisation problems. Their solver uses a linked list representation of the problem which enables rapid prototyping of solution heuristics for different problems. Nareyek in [19] presents a method which learns how to select promising heuristics during the search process using different weight adaptation mechanisms. The method is tested on two optimisation problems, yielding promising results.

Genetic algorithm (GA) based hyperheuristics use indirect chromosome coding, i.e. such that the chromosome represents a sequence of heuristics for solving the problem rather than a solution. The indirect GA evolves the heuristic choice in order to find combinations of heuristics that lead to improved solutions of the problem. Fang, Ross and Corne [12] use an indirect GA for solving open shop scheduling problems. They use a set of simple dispatching rules as the basic blocks for chromosome, and the sequence of rules in a chromosome specifies the method for schedule construction. The GA developed in [22] evolves constraint satisfaction strategies for examination timetabling problems. Hart and Ross [17] develop a GA-based heuristic combination method for solving dynamic job shop scheduling problems. The chromosome in the GA encodes the combination of two methods: an algorithm for constructing a set of operations to be scheduled and a heuristic which selects an operation from the set. Heuristics represent simple priority rules. Such an approach outperforms other heuristic combination methods for benchmark problems. The method of evolving heuristic choice is successfully implemented in [18] for a com plex real-world problem of scheduling chicken catching and transportation. Cowling, Kendall and Han [6] introduce a robust hyper-GA approach which can easily be reimplemented for different problems. Each individual in the hyper-GA population encodes a sequence of low level heuristics and indicates which heuristics to apply and in what order. The hyper-GA is applied to a trainer scheduling problem and produces strong results. The approach is further advanced in [15] to allow the length of the chromosome to change adaptively during the search.

Ross et al. [21] use a learning classifier system based on evolutionary algorithm (EA) to learn a solution process for various instances of the bin-packing problem. Their method determines an order in which simple heuristics should be applied to solve particular problem instance. The approach produces optimal solutions for most instances whereas the heuristics applied separately are not able to achieve an optimum very often.

Burke, Kendall and Soubeiga [4] investigate the robustness of tabu search based hyperheuristic applied to multiple instances of timetabling and rostering problems. The choice of low level heuristics at any time is based on their ranks. The ranks of low level heuristics are dynamically adjusted during the run of hyperheuristic using reinforcement learning techniques similar to those employed by Nareyek [19]. A tabu list of low level heuristics with poor recent performance is maintained in order to avoid their selection too soon. The approach produces results comparable to those of problem-specific metaheuristics for both problems.

A hyperheuristic approach based on statistical ranking of low level heuristics is proposed by Cowling, Kendall and Soubeiga in [7]-[9]. They introduce a choice function for low level heuristics which accumulates information about their recent performance. The choice function represents the weighted sum of three components which contain the information regarding recent performance of each low level heuristic, information about recent effectiveness of consecutive pairs of low level heuristics, and the amount of time since each heuristic was last called. The weights of the components are selected empirically [7] or automatically adjusted as the search progresses [8], [9]. The authors present hyperheuristics which employ different techniques for selection of low level heuristics based on the values of the choice function. The approaches are successfully tested on different real-world personnel scheduling problems.

In this chapter we study the performance of a range of hyperheuristics applied to a real-world personnel scheduling problem, focusing particularly on the question of designing the set of low level heuristics for the problem under consideration. Although hyperheuristic methods use limited problem-specific information, it appears intuitively likely that their performance for a particular problem is determined to a significant extent by the low level heuristics used. In this chapter we investigate experimentally whether this intuition is correct. We split simple local search neighbourhoods into "event selection" and "resource selection" rules, then use software engineering techniques to "multiply" these approaches. Hence by writing only 27 pieces of code (where only 5 are substantially different) we are able to quickly produce 95 low level heuristics. Such an approach allows us to effectively treat various constraints of the problem by having several low level heuristics which can deal with each constraint. We examine how the choice of different subsets of low level heuristics affects the outcome of applying a hyperheuristic.

The contribution of this chapter is twofold. Firstly, it further investigates the application of hyperheuristic methodology for solving real-world optimisation problems. We present a group of "peckish" [5] hyperheuristics where simple greedy and random approaches are effectively combined in order to achieve a good balance between intensification and diversification of the search in the space of low level heuristics. We also study the use of metaheuristic approaches at a higher level for managing the choice of low level heuristics. The previous work in this area has employed GAs (e.g. [6], [15]). Here we develop hyperheuristics based on the concepts of the variable neighbourhood search [16] and tabu search [13] metaheuristics. Our tabu search based hyperheuristics differ from that presented in [4] in tabu list contents and the way in which the tabu list is maintained. Secondly, unlike classical problems such as open shop, job shop and bin-packing problems where various priority rules and simple heuristics are known from the literature and can be effectively combined (as in [12], [17] and [21]), for hard real-world problems usually there is no obvious choice of low level heuristics. The set of problem-specific heuristics can be developed intuitively or based on the rules used by human experts to construct solutions. In this chapter we introduce a scheme for designing a set of low level heuristics. The approach is based on the following simple idea. In order to improve (or to modify) the current solution at any time, we have to answer two questions: which part or component of the solution should be a subject of change (*what* question) and what changes should be applied to it (*how* question). The idea is implemented by creating two separate sets of simple rules for event selection (*what*) and resource selection (*how*) respectively. The choice of rules allows us to treat various problem constraints. The rules from two sets are combined ("multiplied") resulting in a large collection of low level heuristics. This approach is easy to implement and could be used for a wide range of other real-world scheduling problems.

The rest of the chapter is organised as follows. In Section 2 we formulate the trainer scheduling problem which belongs to a wide class of personnel scheduling problems. Section 3 describes the set of low level heuristics, in Section 4 we introduce hyperheuristic approaches, and Section 5 presents and analyses the results of detailed experiments. Section 6 concludes the chapter.

2 The trainer scheduling problem

The trainer scheduling problem arises in a large financial institution which regularly organises training for its personnel. The problem involves assigning a number of training courses (events) to a limited number of training staff, locations, and timeslots as shown in Figure 1. Each event has a numerical priority and the travel of each trainer is penalised depending on the distance from the home location of the trainer to the location where the event is conducted. The objective is to maximise the total priority for scheduled events while minimising the total travel penalty for the training staff. A previous project by one of the authors developed the scheduling decision support system currently in use by the financial institution; therefore we have excellent access to user expertise and problem data.

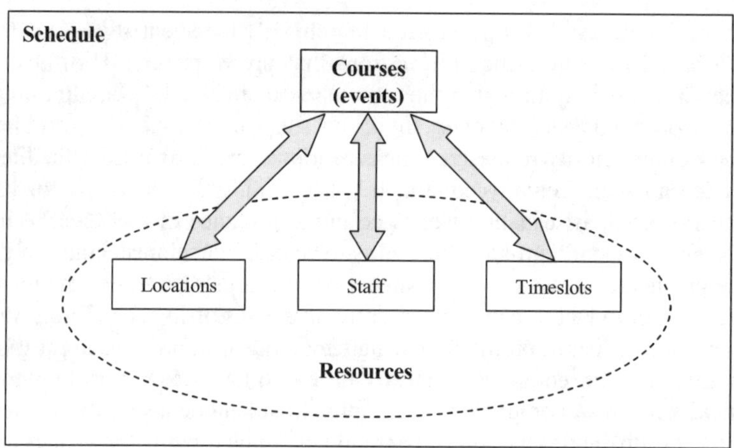

Fig. 1. The trainer scheduling problem.

The problem has a number of constraints:

- Each event can be delivered only by trainers from a limited pool who are competent in the particular topic of the event;
- Each event can be delivered at only one location from the limited list of possible locations for the event;
- Each location has a limited number of rooms and the rooms differ in types and capacities. Each event requires a specified number of rooms each of which satisfies one of the capacity and type requirements for the event;
- Each event can start only within a given time window;
- Trainers are not available on pre-booked holidays. Part-time trainers work only on certain days of the week;
- Each trainer can deliver courses for at most a specified proportion of his/her available timeslots (with the remaining time used for preparation);
- The events are compound, i.e. each event consists of one or more parts (elements). There are complex space/time/resource relationships between the elements of the event. The first element of each event should start within the time window for the event; start times of subsequent elements may be shifted relative to the start of the first element and may have their own time windows. Each element has its

own requirements for rooms and trainers. Different elements of an event may require the same trainer, chosen from the pool of possible trainers. The duration of each element is given and preemptions are not allowed.

The mathematical model for a much simplified version of the trainer scheduling problem is presented in [6]. The authors consider simple events which do not consist of any elements and many constraints from the real-world version of the problem (considered in this work) are relaxed in their model. Note that development of integer programming formulations for real-world problems with a complex structure and a large number of non-trivial constraints is often quite tricky and burdensome task and a resulting model can become very complex and difficult to follow. For our problem, a mathematical model would be too complex to either solve or offer an insight into the problem and we do not present it here. Therefore, in what follows we only define the objective function.

Let $S = \{s_i\}$ be the set of possible schedules for the problem, $E = \{e_j\}$, $j = 1, \dots, n$ be the set of events, where n is the number of events to be scheduled, and $G_i \subseteq E$ be the set of scheduled events in schedule s_i. Let w_j be the priority of event e_j and p_{ij} be the travel penalty associated with event e_j in schedule s_i ($p_{ij} = 0$ if event e_j is not scheduled). Our objective is to maximise the total priority of scheduled events minus total penalty imposed on travel of training staff to the events' locations:

$$f = \max_i \sum_{j:e_j \in G_i} (w_j - p_{ij}).$$

We use real datasets provided by the financial institution, each representing the training delivered by about 50 training staff over a period of 3 months in 16 different locations, with around 200 events to be scheduled. Prior to the use of a commercial decision support system (developed by one of the authors), it required about 9-person days of regional manager time to produce an acceptable quarterly schedule.

3 Low level heuristics

The design of the set of problem-specific low level heuristics for the problem is important for the quality of the outcome of applying a hyperheuristic. We use the following approaches to choosing our low level heuristics.

1. Since our primary objective is to maximise the total priority of scheduled events, we need low level heuristics which insert events which are not yet scheduled.

2. We have to include heuristics concerned with decreasing travel penalties.
3. We need a range of low level heuristics to resolve conflicts in trainer, room or timeslot requirements.
4. Our low level heuristics should recognise and deal with difficult-to-schedule, tightly constrained events, ideally scheduling them early in the process.

Taking into account the above considerations we propose below a scheme for low level heuristics. Each low level heuristic represents the combination of an event selection rule and a resource selection rule. Both categories of selection rules are divided into two subsets: rules for events which are not yet scheduled and rules for events which are already in the current schedule.

3.1 Event selection rules

We use 5 rules for selection from the list of not yet scheduled events:

ESn1. Select event at random;

ESn2. Select event with the highest priority;

ESn3. Select event with the smallest number of possible trainers;

ESn4. Select event with the smallest number of possible locations;

ESn5. Select event with the smallest number of feasible trainer-location-timeslot combinations.

For already scheduled events we consider the following 7 rules:

ESs1. Select event at random;

ESs2. Select event with the highest priority;

ESs3. Select event with the widest time window;

ESs4. Select event with the highest travel penalty,

ESs5. Select event with the largest number of possible trainers;

ESs6. Select event with the largest number of possible locations;

ESs7. Select event with the largest number of feasible trainer-location-timeslot combinations.

Note that in rules *ESn3 – ESn5* and *ESs3 – ESs7* we randomly select one event from the top 30% of events in the whole list of not scheduled or already scheduled events sorted according to the relevant criterion. This allows diversity in the solution process preventing selection of the same events too often which can lead to wasted CPU time. On the other hand, it ensures reasonable focus on difficult parts of the schedule. In rules *ESn2* and *ESs2* we simply randomly select one of the events with the highest priority, since in this case the number of events with the same priority is sufficiently large, as for the datasets used events are put into only five categories according to their priority.

The selection criteria *ESn3 – ESn5* for not scheduled events are chosen in such a way as to always attempt to insert the most constrained events in the schedule, whereas rules *ESs3 – ESs7* select scheduled events with the largest number of possible resources. We are not interested in rescheduling events with very limited resources because it is very unlikely that alternative available resources exist for such events. At the same time, the first two rules (similar for both categories) ensure that there are no events which are completely ignored.

3.2 Resource selection rules

For not scheduled events we suggest 5 resource (timeslot/trainer/location/rooms) selection rules:

RSn1. Select the first found combination of available resources. If no such combination exists, select resources available after unscheduling the first found conflicting event;

RSn2. Select the best combination of available resources in terms of travel penalty. If no such combination exists, select resources available after unscheduling a conflicting event which leads to the lowest penalty for the inserted event;

RSn3. As *RSn2*, but the conflicting event with the smallest value of (*priority – penalty*) is considered;

RSn4. As *RSn2*, but the conflicting event with the lowest priority is sought;

RSn5. As *RSn2*, but the conflicting event with the highest travel penalty is chosen.

When the conflicting event here is unscheduled it frees up resources necessary for the selected event to be scheduled. Note that after scheduling the selected event instead of some conflicting event, we immediately attempt to reschedule the conflicting event. If there are no possible resources, we include the event into the list of not scheduled events.

In order to move scheduled events (e.g. to reduce travel penalties), we should find available resources which are different to the currently scheduled ones. We introduce 10 rules of resource selection:

RSs1. Select the first found combination of available resources;

RSs2. Select the best combination of available resources in terms of travel penalty;

RSs3. Select a possible location with the lowest travel penalty for unchanged trainers/timeslots;

RSs4. Select a possible location randomly for unchanged trainers/timeslots;

RSs5. Select the next available timeslot for unchanged location/trainers;

RSs6. Select possible trainers randomly for unchanged location/timeslots;

RSs7. Select possible trainers with the lowest travel penalty for unchanged location/timeslots;

RSs8. Select possible trainers with the maximum number of available timeslots for unchanged location/timeslots;

RSs9. Select possible trainers with the minimum number of scheduled events for unchanged location/timeslots;

RSs10. Select possible trainers with the maximum number of timeslots remaining under their workload limits for unchanged location/timeslots.

Selection of new resources for the events which have already been scheduled aims to achieve two targets. First, moving these events can improve the schedule reducing travel penalties. Second, freed up resources can be used for other events which are not yet scheduled.

3.3 Low level heuristics

Combining event selection rules with resource selection rules for each category of events, we construct a set of 95 low level heuristics (5*5=25 heuristics for not scheduled events and 7*10=70 heuristics for scheduled events). Although the actual number of low level heuristics is large, such an approach is quite easy to implement. Since different heuristics can use the same event or resource selection rules as their components, we create only 27 pieces of code representing different event/resource selection mechanisms, and only 5 of these pieces of code are substantially different to each other.

For implementation purposes and easier reference low level heuristics are numbered from 1 to 95. The numbers from 1 to 25 are assigned to low level heuristics for not scheduled events and the numbers from 26 to 95 correspond to low level heuristics for already scheduled events. Given an event selection rule *i* and a resource selection rule *j* it is easy to calculate a low level heuristic's number (as shown in Figure 2) and vice versa. We shall refer to any particular low level heuristic by its number in further discussion.

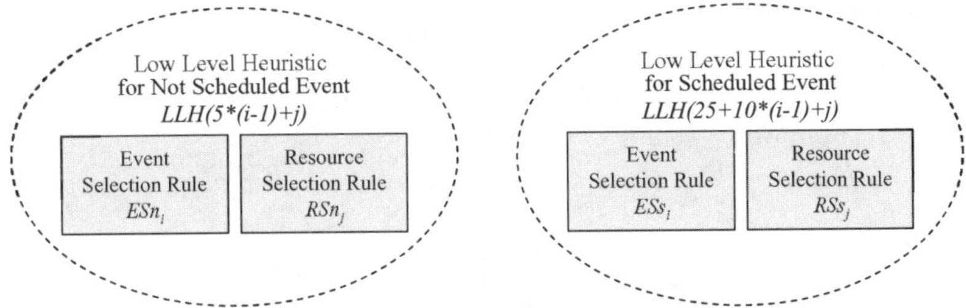

Fig. 2. Low level heuristics.

4 Hyperheuristics

In this section we present 3 groups of hyperheuristic approaches which can be easily applied with only minor modifications to a very wide range of optimisation problems given a set of problem-specific low level heuristics and an objective function.

The first group includes simple random and greedy methods.

- Hyperheuristic ***Random*** selects a low level heuristic randomly from the whole set at each iteration and applies it, repeating this process until some stopping condition is true.
- Hyperheuristic ***Greedy*** selects and applies at each iteration the best low level heuristic from the set. The "best" heuristic means either the heuristic providing the greatest improvement to the objective function or the heuristic leading to the smallest deterioration (or yielding zero improvement) if there are no improving heuristics. Here and later

through the text we consider improvements upon the current objective value, not upon the best value found so far. Ties are broken randomly.

- Hyperheuristic ***Greedy-Improvement*** uses only improving low level heuristics.

To combine random and greedy methods, a group of so-called peckish hyperheuristics is introduced. The term "peckish" is used in [5] where population initialisation algorithms for evolutionary timetabling containing features of both greedy and random methods are considered. In our peckish hyperheuristics we experiment with various degrees of greediness and randomness aiming to achieve better balance between intensification and diversification components of the search process than in pure greedy or random approaches. The general scheme for peckish hyperheuristic can be represented as follows:

Repeat
CandidateList ← *{"good" low level heuristics};*
If (CandidateList ≠ *Ø)*
Select a low level heuristic at random from CandidateList and apply it;
Else
Select a low level heuristic using another random method and apply it;
Until (Stopping condition is true)

In this pseudocode, "good" low level heuristics may mean only improving heuristics or the top n low level heuristics from the set of 95 (not necessarily improving). In the former case, "another random method" may involve random selection of low level heuristics either from the whole set or from the subset of the best n low level heuristics; in the latter case, it is not applicable since the candidate list is always non-empty. The range of peckish hyperheuristics is given below.

- Hyperheuristic ***Peckish1*** randomly selects a low level heuristic at each iteration from the candidate list of low level heuristics which improve the current solution. If the candidate list is empty, a low level heuristic is selected randomly from the whole set of low level heuristics.

- Hyperheuristic ***Peckish2*** randomly selects a low level heuristic from the candidate list which contains the n best (not necessarily improving) heuristics from the whole set of 95 low level heuristics. The candidate list size n plays an important role here: it determines how "greedy" and how "random" the hyperheuristic is – an increase in the candidate list size adds randomness and a decrease in it makes the hyperheuristic greedier.

- Hyperheuristic ***Peckish3*** combines the features of the previous two methods. At each iteration, it attempts to form the candidate list of only improving heuristics and if such a list is not empty, randomly selects a low level heuristic from it. Otherwise, random selection from the subset of the best n non-improving heuristics is applied.

- Hyperheuristic ***Peckish4*** employs ideas related to Variable Neighbourhood Search [16]. The candidate list contains only improving low level heuristics. Variable n indicates how many non-improving low level heuristics will be considered for random selection if the candidate list is empty. The value of n is initially set to 1. If at some iteration there is an improving low level heuristic(s), the hyperheuristic selects either one of the improving heuristics randomly (***Peckish4-v1***) or the best one (***Peckish4-v2***) and the value of n is reset to 1. Otherwise, a low level heuristic is randomly selected from n best non-improving heuristics and the value of n is then incremented for the next iteration. The dynamic changes of variable n provide intensification/diversification of the search.

Hyperheuristics from the third group are based on the tabu search meta-heuristic [13].

- Hyperheuristic ***TabuHeuristic*** employs a tabu list of recently called heuristics. The size of the tabu list is fixed and set to some prespecified value l. The algorithm greedily selects the best low level heuristic at each iteration. If such a heuristic leads to an improved objective function value, it is always selected and released from the tabu list if there; a non-improving heuristic is chosen only if it is not in the tabu list and immediately becomes tabu after its application.

- Hyperheuristic ***TabuEvent*** is similar to TabuHeuristic except that it is more conventional in that the tabu list holds recently selected events.

- We also consider hyperheuristics ***TabuHeuristicAdaptive*** and ***TabuEventAdaptive*** which are analogous to the previous two but allow the tabu list size to change adaptively as the search progresses. We decrease the tabu list size when the current solution keeps improving and increase it when we are having trouble in finding a better solution. After empirical evaluations of different adaptation schemes we have adopted the following one: increase the tabu list size by one in the case of non-improvement and decrease it by $\lfloor current\ size/10 \rfloor$ *4 when an improvement occurs discarding the oldest members of the list as necessary, where $\lfloor x \rfloor$ denotes the largest integer which is less than or equal to x. The maximum tabu list size is also specified to prevent unlimited growth of the tabu list.

For all our hyperheuristics except ***Random*** we do not consider low level heuristics producing no changes to the current schedule (although we do consider low level heuristics which leave the objective value unchanged). This prevents "dummy" iterations and ensures that the solution is always perturbed at each iteration.

5 Experiments and results

The problem model, all low level heuristics and hyperheuristics have been implemented in Microsoft Visual C++ and the experiments have been conducted on a Pentium 4 1600MHz PC with 640MB RAM running under Windows 2000. We use real data from two datasets which represent two problem instances of different dimensions and with different degrees of difficulty. ***Dataset1*** contains 224 events, 53 training staff, 16 locations and 37 rooms. In ***Dataset2*** there are 147 events, 54 trainers, 16 locations and 39 rooms. The scheduling period is 3 months in both cases. In spite of a significantly smaller number of events, ***Dataset2*** is more difficult than ***Dataset1***. The majority of the events from ***Dataset2*** have tight time windows and very restricted lists of possible trainers and locations whereas the constraints in ***Dataset1*** are not so strict.

5.1 Initial schedules

Initial schedules are constructed using two greedy heuristics. In the first approach we take events one by one in a random order and select the first available combination of possible resources for each event (if any exists) until all the events have been tried. The possible trainers and locations are considered in the order in which they appear in the dataset. The second approach first sorts the events in descending order of their priority. Then each group of events with the same priority is randomly permuted and the final ordered list of events is created. Events are taken from the list one by one; all available combinations of possible trainers and locations are considered for each event and the combination yielding the lowest travel penalty is selected for scheduling.

We use two initial solutions of different quality for our experiments. The "poor" initial solution is the worst one among 10 solutions obtained by applying the first greedy method and the "good" initial solution is the best one out of 10 runs of the second greedy heuristic. The good initial solution is far superior to the poor one in terms of the total priority of scheduled events, the total penalty and the number of events not scheduled.

5.2 Upper bounds

In order to assess the quality of the solutions obtained by our approach, we need to compare them with either optimal solutions or solutions constructed manually by a human expert. However, due to the inherent complexity of the problem under study, the optimal solutions for non-trivial instances cannot be obtained using manual or enumerative techniques. Manually constructed schedules are not available. Hence in our experiments we will compare the results with the upper bounds of the optimal values obtained via relaxing some of the problem constraints.

Before calculating upper bounds, we analysed the results of multiple runs of hyperheuristic ***Random*** applied to both problem instances in order to check whether all the events could be scheduled. Since the random approach is very fast, we could easily allow it to perform a large number of iterations. Comparing the results of multiple runs, we found that there were always a few events which were not scheduled. Moreover, similar events (however, not always the same) repeatedly appeared in the lists of events not scheduled after each run of the hyperheuristic. Further analysing the input data, we have identified the groups of events which share conflicting trainers, locations, and timeslots. We proved that such conflicts could not be resolved and at least one event from each group of conflicting events is impossible to schedule. Therefore, the maximum possible number of scheduled events has been found.

To calculate the upper bound of all possible schedules, assume that all the events are scheduled except those which are impossible to schedule due to irresolvable conflicts. Select from each group of conflicting events the event with the lowest priority and form the list of events which are not considered. The total priority for the rest of the events is calculated giving the optimal value for the total priority. Then we relax the constraints on availability of trainers and rooms, rooms' types and capacities, and on starting times for the events. For each event we identify the location/trainer(s) combination yielding the minimum travel penalty. If the lowest possible travel penalty for the event exceeds its priority (which can happen for some low priority events), we exclude such an event from further consideration and reduce the total priority value accordingly. Subtracting the sum of the lowest possible travel penalties for all events which can be scheduled with positive (*priority – penalty*) from the total priority, we obtain an upper bound on the objective value.

Note that the upper bounds calculated in such a way, while useful in our experiments, are not particularly tight since a large number of constraints are ignored.

5.3 Experiments

We performed 10 runs for each hyperheuristic described in Section 4, starting from the same initial solution. Both poor and good initial solutions are tried in separate experiments for both problem instances. We test different values of parameters *n* and *l* for hyperheuristics ***Peckish2*** and ***Peckish3*** and for tabu list based hyperheuristics respectively. The stopping condition for each run is the number of iterations which is set to 500.

In addition to experiments with the whole set of 95 low level heuristics, we conduct separate experiments with subsets of 3, 5, and 10 low level heuristics. The heuristics for such small sets are selected randomly from the whole set. We ensure that in each subset approximately a quarter of the heuristics deal with not scheduled events and the remaining heuristics contain the rules for already scheduled events.

5.4 Results

5.4.1 The whole set of low level heuristics.

The average performances of hyperheuristics for both datasets and the whole set of 95 low level heuristics are presented in Tables 1-4. We use the following notation in the tables:

PR_{best} – the highest total priority among 10 schedules obtained;

PR_{av} – average total priority over 10 runs of a hyperheuristic;

PN_{best} – the lowest total penalty among 10 schedules obtained;

PN_{av} – average total penalty over 10 runs of a hyperheuristic;

F_{best} – the highest objective value among 10 schedules obtained;

F_{av} – average objective value over 10 runs of a hyperheuristic;

NS_{best} – the minimum number of not scheduled events among 10 schedules obtained;

NS_{av} – average number of not scheduled events over 10 runs of a hyperheuristic;

CPU_{av} – average running time of the hyperheuristic in seconds (over 10 runs);

I_{best} – average iteration number (over 10 runs of a hyperheuristic) when the solution with the best objective value is found for the first time (in fact, a number of solutions with the same objective can be obtained during one run).

Table 1. Performance of hyperheuristics on ***Dataset1*** starting from good initial solution.

Hyperheuristic	PR_{best}	PR_{av}	PN_{best}	PN_{av}	F_{best}	F_{av}	NS_{best}	NS_{av}	CPU_{av}	I_{best}
Initial solution	181235		824		180411		16			
Random	**185410**	184670	960	1198	184118	183472	**5**	7	39.37	146
Greedy-I	184335	184323	868	882	183447	183441	9	10	25.10	7
Greedy	185385	**185080**	816	877	184493	184203	6	7	1452.59	408
Peckish1	**185410**	**185205**	908	959	184473	**184246**	**5**	**5**	1567.91	357
Peckish2(10)	184385	184378	852	862	183533	183516	7	7	1628.28	440
Peckish2(25)	185385	184578	832	876	184465	183702	6	7	1580.33	430
Peckish2(40)	185385	184763	864	958	184357	183805	6	8	1539.50	258
Peckish3(10)	185385	184780	832	868	184497	183912	6	7	1605.66	357
Peckish3(25)	185385	**185080**	832	884	184493	184196	6	7	1507.53	397
Peckish3(40)	185385	184973	836	878	184485	184094	6	7	1412.58	388
Peckish4-v1	185385	184683	824	861	184497	183822	6	7	1551.34	428
Peckish4-v2	185385	184880	828	869	184493	184011	6	7	1608.55	396
TabuH(10)	185385	184783	828	865	184489	183917	6	7	1602.29	412
TabuH(30)	185385	184878	800	860	184497	184017	6	7	1543.00	394
TabuH(60)	185385	184983	836	870	**184508**	184113	6	7	1526.00	382
TabuE(10)	185385	184980	**796**	867	184501	184113	6	7	1510.22	411
TabuE(30)	185385	184878	828	863	184496	184014	6	7	1491.60	383
TabuE(60)	185385	184975	800	861	184497	184114	6	7	1437.43	402
TabuE(100)	185385	185075	824	865	**184508**	**184210**	6	7	1418.07	405
TabuHA(30)	185385	184780	816	**859**	184497	183921	6	7	1648.09	412
TabuHA(45)	185385	184875	**792**	861	184489	184014	6	7	1597.33	366
TabuHA(60)	185385	184778	808	860	184493	183918	6	7	1519.37	408
TabuEA(30)	185385	184680	804	**845**	**184517**	183835	6	7	1569.42	407
TabuEA(45)	185385	**185080**	820	872	184501	**184208**	6	7	1587.26	421
TabuEA(60)	185385	**185083**	812	868	184505	**184215**	6	**6**	1491.22	388
TabuEA(100)	185385	184783	**792**	**852**	**184517**	183931	6	7	1480.62	417
Upper bound	185385		622		184763		6			

We also use short names for some hyperheuristics: ***Greedy-I*** for ***Greedy-Improvement***, ***TabuH*** and ***TabuE*** for ***TabuHeuristic*** and ***TabuEvent*** respectively, ***TabuHA*** and ***TabuEA*** for ***TabuHeuristicAdaptive*** and ***TabuEventAdaptive***.

Notably good values in each column are highlighted in bold font. Note that very small differences between values in the tables represent practically very significant differences in trainer inconvenience due to additional travel, or additional low priority scheduled events.

From Tables 1-4 we can observe that our hyperheuristics perform well for both starting solutions. As expected, random and greedy hyperheuristics (***Random*** and ***Greedy-Improvement***) produce the worst results. Random frequently selects low level heuristics which lead to deterioration of the objective function and all the solutions are quite poor in terms of travel penalty. Even long runs of ***Random*** (300000 iterations and 10 hours of CPU time) generate solutions with very high total travel penalties and therefore with lower objective values. ***Greedy-Improvement*** stops too early when there is no any improving heuristic at some iteration.

Table 2. Performance of hyperheuristics on ***Dataset1*** starting from poor initial solution.

Hyperheuristic	PR_{best}	PR_{av}	PN_{best}	PN_{av}	F_{best}	F_{av}	NS_{best}	NS_{av}	CPU_{av}	I_{best}
Initial solution	168360		3040		165320		24			
Random	185410	184405	2576	2667	182834	181738	5	6	34.33	444
Greedy-I	184410	184005	1272	1438	183058	182567	6	7	378.65	122
Greedy	185410	185110	1048	1102	184362	184008	5	5	1380.96	435
Peckish1	185410	184710	1036	1124	184266	183586	5	6	1659.12	467
Peckish2(10)	185410	185110	1136	1248	184178	183862	5	5	1659.29	492
Peckish2(25)	185410	185110	1352	1442	183994	183668	5	5	1726.89	495
Peckish2(40)	185410	184605	1576	1682	183814	182923	5	6	1716.05	494
Peckish3(10)	185410	185108	**1028**	1137	184286	183971	5	5	1771.24	439
Peckish3(25)	185410	184810	**988**	**1094**	184346	183716	5	6	1749.45	467
Peckish3(40)	185410	184510	**948**	**1095**	184346	183415	5	6	1648.88	473
Peckish4-v1	185410	185010	1068	1136	184342	183874	5	5	1719.59	470
Peckish4-v2	185410	185210	1032	1115	184358	184095	5	5	1525.66	460
TabuH(10)	185410	185210	1040	1110	**184370**	184100	5	5	1478.13	472
TabuH(30)	185410	185210	1056	1112	184322	184098	5	5	1492.37	453
TabuH(60)	185410	185110	**1020**	**1100**	184346	184010	5	5	1443.72	473
TabuE(10)	185410	**185310**	1068	1124	184342	184186	5	5	1469.56	437
TabuE(30)	185410	**185408**	1056	1106	184354	**184302**	5	5	1438.45	418
TabuE(60)	185410	185210	1064	1103	184342	184107	5	5	1397.09	449
TabuE(100)	185410	185210	1056	**1096**	184354	184114	5	5	1360.32	458
TabuHA(30)	185410	185210	1036	1129	**184374**	184081	5	5	1406.52	450
TabuHA(45)	185410	**185310**	1064	1121	184346	184189	5	5	1465.93	453
TabuHA(60)	185410	185110	1032	1112	**184378**	183998	5	5	1424.55	469
TabuEA(30)	185410	185210	1080	1122	184330	184088	5	5	1404.00	464
TabuEA(45)	185410	**185310**	1036	1105	**184374**	**184205**	5	5	1419.11	472
TabuEA(60)	185410	**185310**	1068	1108	184342	**184202**	5	5	1396.10	447
TabuEA(100)	185410	185208	1060	1115	184350	184093	5	5	1450.29	448
Upper bound	185385		622		184763		6			

The rest of the hyperheuristics produce similar high-quality results. This provides evidence that a very rich set of low level heuristics can be expected to perform well. There is no clear champion among the hyperheuristics, although the tables suggest that ***TabuEvent*** hyperheuristics and their adaptive versions perform better than others. The reason is probably that the tabu lists in these hyperheuristics help to avoid cycles, forbidding selection of events for which resources have been recently changed. Since we have a rich set of low level heuristics, each individual objective and each constraint can be treated by a group of several low level heuristics capable of resolving various conflicts. The large number of low level heuristics also allows us to consider many different events at each iteration which increases the chances of selecting the "right" event whose scheduling or rescheduling leads to greater improvements of the solution and may guide the search in a promising direction.

For peckish hyperheuristics, those with greater degrees of randomness (***Peckish1*** and parameterised peckish hyperheuristics with greater values of parameters) usually produce better results for ***Dataset1*** and a good initial solution than their more greedy competitors (see Table 1). This follows from the fact that the travel penalties for scheduled events in initial solution are already quite low and reducing them as well as inserting new events into the schedule becomes more difficult.

Scheduling a new event often requires a few rescheduling steps making the penalties higher in order to release the necessary resources. It is unlikely that hyperheuristics with a dominating greedy component would apply low level heuristics which significantly worsen the current solution by increasing the penalties (preferring zero-improving low level heuristics instead) while hyperheuristics employing more randomness could accept these moves. Therefore the latter hyperheuristics are more effective in terms of inserting "difficult" events into the schedule and the solutions constructed by them generally have a higher total priority. Since the total priority component dominates in the objective function, it is fair to assume that the average performance of the hyperheuristic is determined primarily by the total priority of the solutions. The figures from columns PR_{av} and F_{av} in Tables 1-4 support this statement: the best average objective values are achieved when the average priority values are the highest.

Table 3. Performance of hyperheuristics on ***Dataset2*** starting from good initial solution.

Hyperheuristic	PR_{best}	PR_{av}	PN_{best}	PN_{av}	F_{best}	F_{av}	NS_{best}	NS_{av}	CPU_{av}	I_{best}
Initial solution	72654		648		72006		21			
Random	77199	75759	697	765	76424	74994	8	9	8.41	365
Greedy-I	74289	74289	661	661	73628	73628	12	12	9.68	11
Greedy	**78289**	77009	633	658	**77578**	76351	7	9	742.71	373
Peckish1	**79199**	77018	646	683	**78418**	76335	**6**	**8**	367.77	342
Peckish2(10)	78189	76589	623	649	77500	75940	8	10	579.83	374
Peckish2(25)	77389	76549	**620**	658	76643	75891	7	9	653.98	358
Peckish2(40)	**78289**	77014	660	711	77544	76303	7	9	666.91	401
Peckish3(10)	78189	76799	623	**648**	77495	76152	8	9	574.38	423
Peckish3(25)	78189	76589	**618**	649	77486	75941	8	10	494.29	293
Peckish3(40)	77199	76604	633	652	76554	75952	8	9	406.44	340
Peckish4-v1	77199	76790	623	**641**	76554	76149	8	9	612.82	319
Peckish4-v2	78189	**77420**	633	658	77512	**76762**	8	8	775.75	353
TabuH(10)	77189	76699	627	**636**	76556	76063	9	9	644.78	265
TabuH(30)	78189	**77419**	633	676	77512	**76743**	8	9	578.34	338
TabuH(60)	78199	76604	633	672	77501	75932	7	9	471.72	310
TabuE(10)	78189	**77399**	633	667	77510	**76732**	8	9	838.40	377
TabuE(30)	78189	77109	623	656	77494	76453	8	9	677.18	316
TabuE(60)	**79189**	77219	633	674	**78450**	76545	7	9	655.69	366
TabuE(80)	**78289**	**77421**	633	661	**77592**	**76760**	**7**	**8**	565.43	393
TabuHA(30)	78189	77329	633	665	77496	76664	8	9	593.63	303
TabuHA(45)	78189	77099	**620**	655	77496	76444	8	9	468.67	303
TabuHA(60)	77289	76740	633	**647**	76650	76093	8	9	464.06	349
TabuEA(30)	78189	77119	635	653	77512	76466	8	9	476.36	432
TabuEA(45)	78189	77209	635	661	77510	76548	8	9	508.69	289
TabuEA(60)	78189	77019	625	**648**	77485	76371	8	9	538.41	322
TabuEA(80)	78189	77209	633	665	77496	76544	8	9	436.69	313
Upper bound	79389		656		78733		5			

Table 4. Performance of hyperheuristics on ***Dataset2*** starting from poor initial solution.

Hyperheuristic	PR_{best}	PR_{av}	PN_{best}	PN_{av}	F_{best}	F_{av}	NS_{best}	NS_{av}	CPU_{av}	I_{best}
Initial solution	65799		834		64965		23			
Random	76974	74847	706	814	76065	74033	7	10	7.31	404
Greedy-I	76199	75240	731	755	75401	74484	9	11	23.05	31
Greedy	78199	**77609**	671	720	77474	**76889**	7	8	335.14	390
Peckish1	78199	77089	652	**698**	77457	76391	**6**	**7**	311.37	376
Peckish2(10)	78199	76749	**626**	703	77469	76046	7	8	348.48	464
Peckish2(25)	78199	77097	661	729	77453	76367	7	8	322.71	430
Peckish2(40)	77299	76436	697	739	76554	75697	7	9	366.44	390
Peckish3(10)	78199	77308	655	706	77500	76603	7	8	327.59	366
Peckish3(25)	78199	77409	655	**698**	77469	76711	7	8	330.06	373
Peckish3(40)	78199	77339	647	705	77486	76634	7	8	313.05	398
Peckish4-v1	78199	77209	651	709	77484	76500	7	8	324.37	401
Peckish4-v2	78199	77198	711	725	77478	76473	7	8	320.62	389
TabuHc(10)	78199	77399	665	710	77468	76689	7	8	319.39	333
TabuH(30)	78199	77489	**646**	704	77474	76785	7	8	344.08	369
TabuH(60)	78199	77299	705	726	77474	76573	7	8	358.25	422
TabuE(10)	78199	77409	653	**698**	77480	76711	7	8	334.62	381
TabuE(30)	**79199**	77399	657	703	**78430**	76696	**6**	8	341.06	379
TabuE(60)	78199	77499	665	**693**	77486	**76806**	7	8	343.90	378
TabuE(80)	**79199**	**77699**	655	709	**78418**	**76990**	**6**	8	349.89	407
TabuHA(30)	78199	77399	666	717	77484	76682	7	8	337.81	414
TabuHA(45)	78199	**77788**	655	714	77474	**77074**	7	8	357.96	373
TabuHA(60)	78199	77499	671	709	77490	76790	7	8	328.59	315
TabuEA(30)	78199	77399	671	711	77466	76688	7	8	326.10	375
TabuEA(45)	78199	77399	**630**	**695**	77472	76704	7	8	353.26	415
TabuEA(60)	78199	77499	661	718	77468	76782	7	8	346.14	362
TabuEA(80)	**79199**	77199	**646**	704	**78422**	76496	**6**	8	354.14	391
Upper bound	79389		656		78733		5			

From Table 1 we can also conclude that tabu list based hyperheuristics with large tabu list sizes are preferable for this instance. The larger tabu list size serves a similar role to the randomness in the peckish approaches and means that the probability of selecting low level heuristics with negative improvement is higher. This can be a factor when starting from a good initial solution as it was mentioned above.

In contrast, in case of a poor initial solution (Table 2) hyperheuristics with a larger greedy component are quite consistent and produce slightly better outcomes. Since there are more opportunities to improve the travel penalties and scheduling new events is easier, improving low level heuristics appear in many iterations. The greediest hyperheuristics would likely accept the most improving low level heuristics at each iteration providing a faster growth of the objective value. Hence low to moderate values for the tabu list sizes are also desirable.

For ***Dataset2***, however, the moderate values of the parameters for all groups of hyperheuristics are preferred and more balanced hyperheuristics are generally better (see Tables 3-4). The reason is probably that the events in the ***Dataset2*** are tightly constrained and less possibilities exist to manipulate with the available resources. As a result, a good balance between greediness and randomness is needed. The only exception is ***TabuEvent***, where it seems that this hyperheuristic works better when the tabu list contains approximately half of the events from the dataset.

As it is discussed in subsection 5.2, for some low priority events the travel penalty may exceed priority for any possible combination of locations and trainers. In fact, there is only one such an "unprofitable" event in each dataset and including it in the schedule worsens the objective value. Hyperheuristics usually ignore the unprofitable event and it does not appear in the best schedule generated (except ***Random*** and ***Peckish1***) unless it has already been included in initial schedule. In the latter case, hyperheuristics do not remove this "unprofitable" event from the schedule and the resulting objective value is slightly worse than it could be without this event being scheduled. In our experiments, good initial schedules for both datasets do not contain unprofitable events while in poor initial solutions these events are scheduled. Therefore, the numbers of not scheduled events in Tables 2 and 4 are often less by 1 than in Tables 1 and 3 due to extra unprofitable events in the schedules. For "easy" ***Dataset1*** these numbers are often even less than the value specified in the upper bound (see Table 2 and ***Random*** and ***Peckish1*** hyperheuristics in Table 1).

It would be desirable to control the appearance of unprofitable events in the schedule from the point of view of the current objective criterion. It could be easily implemented by including into the initial greedy heuristic the mechanism of identifying such events (simple check similar to that used in calculating the upper bound) and marking them as "redundant" to prevent low level heuristics from dealing with them. Alternatively, we may include a low level heuristic which removes unprofitable events. It is worth noting that the problem owners prefer to see these events in the schedule, so the human decision maker can decide whether they remain.

From Tables 1-4 we can notice a clear tradeoff between the total priority and total penalty in the objective function. The best solutions in terms of priority are not the leaders in terms of penalty and vice versa. Indeed, each additional event scheduled incurs extra penalty; such a penalty can be quite considerable when a hyperheuristic inserts a "difficult" event at a late iteration, although the gain yielded by the priority of the inserted event usually exceeds the loss incurred by a higher penalty. In general, the larger the number of events scheduled, the higher the total penalty of the solution. For ***Dataset2*** the total penalties of some solutions are lower than the penalty of the upper bound because some "expensive" (in terms of penalty) events have been left unscheduled by certain hyperheuristics.

For ***Dataset1*** the best solutions are usually found after 400 iterations which indicates that given more time the hyperheuristics may construct even better solutions thereby closing the gap to the upper bound. This is especially true when starting from a poor initial solution since the penalties for some events can be further reduced. For ***Dataset2*** the number of iterations before the best solution is found rarely exceeds 400 and we can not expect much improvement even with extra time available. A single run of each hyperheuristic requires about 25 minutes of CPU time for ***Dataset1*** and 5-10 minutes for ***Dataset2***.

5.4.2 Behaviour of hyperheuristics

Although the performance of hyperheuristics is quite similar for good and poor initial solutions in terms of objective value, the structure of schedules obtained is different, as is the behaviour of the hyperheuristics. The typical behaviour of a hyperheuristic is shown in Figure 3. When a hyperheuristic starts from a good initial solution, there are frequent improvements in the first 20-30 iterations when the hyperheuristic schedules the "easiest" not scheduled events and further improves penalties for some events. Then a sequence of iterations yielding zero improvements in the objective function follows until the next low level heuristic call which results in an improved objective, and so on. The current objective value deteriorates only rarely and then only during the runs of ***Peckish1*** or parameterised hyperheuristics when the candidate list size or tabu list size is large enough. The improvements become more and more rare towards the end of the run and usually represent reductions in travel penalties. It is not easy to schedule the remaining "difficult" not scheduled events because in order to put such events into a schedule we need to change the resources for some conflicting events which often requires an increase of the travel penalties. But due to the fact that travel penalties for scheduled events in a good initial solution are already set to the good values by the initial greedy heuristic and because of the certain degree of greediness in our hyperheuristics, the choice of low level heuristics at each iteration is usually limited to those yielding zero improvement. In spite of such limitations, hyperheuristics are still able to schedule the majority of the events before the stopping condition is met, although each insertion of a new event requires a long sequence of iterations with zero improvements. The final schedule is almost always attractive in terms of travel penalty and priority when assessed by users but may contain some not scheduled events when compared to the upper bound.

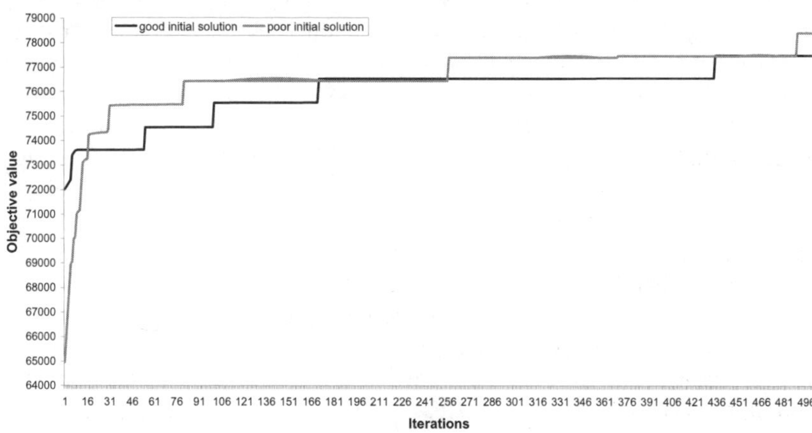

Fig. 3. Performance of hyperheuristic ***TabuEventAdaptive(80)*** on ***Dataset2*** starting from different initial solutions.

In a poor initial solution the travel penalties are often quite high. In order to achieve better values for penalties, new combinations of trainers and locations are selected which may differ significantly from the current ones. This ability to vary the trainers and the locations makes it possible to schedule new events more quickly and easily. There are a lot of improving iterations among the first 100-200 iterations, then improvements become more rare and low level heuristics with zero improvements start to play their part in the search process (see Figure 3). The schedules obtained contain more scheduled events on the average than when starting from a good initial solution and therefore the total priority is higher. The travel penalty for these schedules is generally worse than if we start from a good initial solution, but since total priority of scheduled events is weighted more highly than travel penalties, the schedules generated starting from poor initial solutions are generally better on the average.

5.4.3 Long runs of hyperheuristics

The best schedules out of each group of 10 runs are usually constructed starting from a good initial solution. We have also run the hyperheuristics for a much longer time (10000 iterations) to have a better idea about the quality of the solutions obtained so far and the magnitude of further improvements possible. For ***Dataset1*** and good initial solution different hyperheuristics consistently achieve very similar outcomes being able to schedule all events which may be profitably scheduled and make a few additional penalty improvements. The best objective value after 10000 iterations is slightly better than the best one obtained during 500 iterations: for ***Dataset1*** it is 184537 against 184517 which means lower penalties for 3-5 events. On the other hand, in case of a poor initial solution, long runs of hyperheuristics are not able even to "catch" the best values in Table 1 despite multiple improvements in travel penalties during the runs. Although the gap in total travel penalty is reduced comparing to that of between the values in Tables 1 and 2, it still remains quite noticeable. For ***Dataset2*** more significant improvements can be achieved in longer runs since there is larger number of not scheduled events after 500 iterations. Hyperheuristics are able to schedule most of them making the total priority of schedules higher. However, the outcomes are not as consistent as for ***Dataset1*** and the best result for 500 iterations (78450 in Table 3) is never beaten (possibly because it at or very close to the optimal). Note that such long runs are time consuming: for ***Dataset1*** each run requires about 10 hours of CPU time against 25 minutes for 500 iterations and for ***Dataset2*** about 2 hours instead of 10 minutes.

5.4.4 Summary of results for a large set of low level heuristics

Table 5 summarises the best results produced by hyperheuristics for the whole set of low level heuristics. The figures represent the deviation in percent from the upper bound averaged over 10 runs of hyperheuristics, so smaller numbers represent better solutions. Three different values of the parameters for hyperheuristics employing candidate lists and tabu lists have been tested and those providing the most consistent results for corresponding hyperheuristics are shown in Table 5. Hyperheuristic ***TabuEvent*** produces its best outcomes when the tabu list is allowed to contain up to approximately half of all events in the dataset denoted by N/2 in the table. Table 5 demonstrates the high quality of solutions when a large collection of low level heuristics is used. For "easy" ***Dataset1*** the difference from the upper bound is well under 1% and for "difficult" ***Dataset2*** it does not exceed 3.6%, for all hyperheuristics except simple random and greedy.

5.4.5 Small sets of low level heuristics

The results of the previous subsection indicate that a large collection of low level heuristics is very effective. We conducted additional experiments with smaller sets of low level heuristics. The main purpose is to check whether a "good" subset of low level heuristics can produce the results of similar quality as for the whole set. Although random choice of low level heuristics for subsets can not guarantee that the "best" low level heuristics will be selected, such an approach may provide an idea about the difference in results of our hyperheuristics for different subsets, thus confirming the existence of strongly and poorly performing low level heuristics in a large set.

The average performance of hyperheuristics for a reduced set of 10 randomly selected low level heuristics is presented in Table 6. The results are still good enough although not as good as for the whole set of low level heuristics. The hyperheuristics nearly always perform better starting from the good initial solution rather than a poor one; the only exception is ***TabuHeuristicAdaptive*** for ***Dataset2***.

Table 5. Average performance of hyperheuristics for the set of 95 low level heuristics (deviation in % from upper bound).

Hyperheuristic	Dataset1		Dataset2	
	good init. solution	poor init. solution	good init. solution	poor init. solution
Initial solution	2.35	10.52	8.53	17.48
Random	0.69	1.63	4.74	5.96
Greedy-I	0.71	1.18	6.47	5.38
Greedy	0.30	0.40	3.01	2.33
Peckish1	0.27	0.63	3.03	2.96
Peckish2(25)	0.57	0.59	3.60	2.99
Peckish3(25)	0.30	0.56	3.54	2.56
Peckish4-v1	0.50	0.47	3.27	2.82
TabuH(30)	0.40	0.35	2.52	2.46
TabuE(N/2)	0.29	0.34	2.49	2.20
TabuHA(45)	0.40	0.30	2.90	2.10
TabuEA(45)	0.29	0.29	2.76	2.56

Table 6. Average performance of hyperheuristics for the set of 10 low level heuristics (deviation in % from upper bound).

Hyperheuristic	Dataset1		Dataset2	
	good init. solution	poor init. solution	good init. solution	poor init. solution
Initial solution	2.35	10.52	8.53	17.48
Random	0.50	1.61	4.69	7.06
Greedy-I	1.81	4.53	7.81	13.71
Greedy	0.75	0.94	4.35	4.70
Peckish1	0.43	1.31	4.37	4.80
Peckish2(3)	0.63	1.49	4.28	5.44
Peckish3(6)	0.57	1.24	3.76	4.61
Peckish4-v1	0.43	1.11	3.75	4.79
TabuH(3)	0.66	1.12	4.11	5.08
TabuE(60)	0.64	1.19	4.18	4.87
TabuHA(4)	0.57	1.14	4.57	4.54
TabuEA(30)	0.71	1.17	4.36	4.87

The best randomly generated combinations of 10 low level heuristics construct solutions comparable to that of the large set, while the running time is less. It is important to notice that the best subsets of low level heuristics are different for ***Dataset1*** and ***Dataset2***, which indicates that the performance of low level heuristics depends on the problem instance. The worst selection of 10 low level heuristics produces a schedule of relatively poor quality. Choosing an appropriate subset of low level heuristics from a large superset is an interesting research direction which we will pursue in the future.

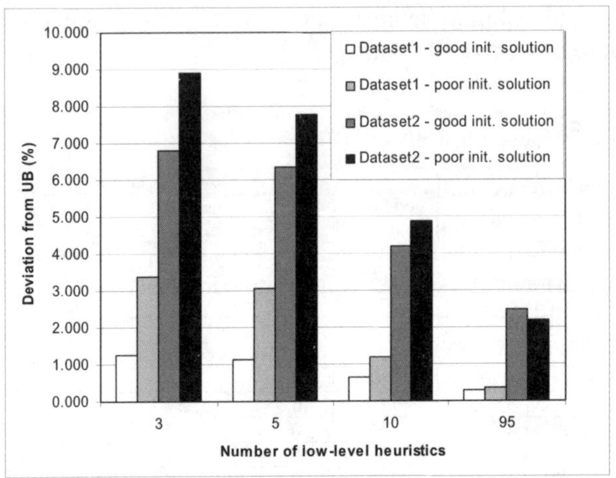

Fig. 4. Average performance of hyperheuristic ***TabuEvent*** on different sets of low level heuristics.

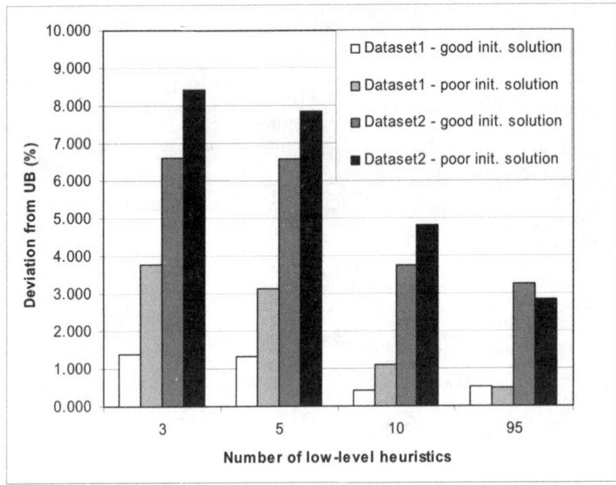

Fig. 5. Average performance of hyperheuristic ***Peckish4-v1*** on different sets of low level heuristics.

The relative performance of hyperheuristics ***TabuEvent*** and ***Peckish4-v1*** on the sets of 3, 5, 10, and 95 low level heuristics is shown in Figure 4 and Figure 5. We have selected these hyperheuristics as representatives of peckish and tabu list based groups which perform well on all sets. Figures 4 and 5 show the advantages of our approach of "multiplying" event selection rules and resource selection rules for the trainer scheduling problem especially for the problem instance with tight constraints. The figures lend support to the claim that more low level heuristics is generally better.

5.4.6 Performance of individual low level heuristics

In this subsection we analyse how individual low level heuristics contribute to the solution process. Such an analysis is important from two points of view. First, it would help us to make a conclusion about the predictability of the performance of low level heuristics for a given problem instance. Second, it might lend further support to the idea that having a large, diverse set of low level heuristics is advantageous.

Low level heuristics for not scheduled events (e.g. Figures 6a and 6b) attempt to insert new events into the schedule, which improves the current solution in terms of priority (although travel penalty can be sometimes improved in parallel if another scheduled event is moved). Low level heuristics for already scheduled events (e.g. Figures 6c and 6d) reduce travel penalties by changing event locations or trainers. We observe frequent positive improvements in the first 30-50 iterations when there are not scheduled events which are relatively easy to schedule. We can expect that low level heuristics for not scheduled events could be called by the hyperheuristic frequently at that time because of the greater magnitude of improvements than for their penalty-improving counterparts. After these "easy gains", primarily iterations with zero and negative improvements occur until the stopping condition is met, punctuated by a small number of improving iterations. It appears that each of the low level heuristics in Figures 6a-6d is able to find different, significant changes at different points in the search. Similar behaviour is seen for many other low level heuristics. Whilst it is hard to pin down when or why a particular low level heuristic will produce such an improvement, having a diverse range of low level heuristics makes these improvements much more common than for a smaller collection.

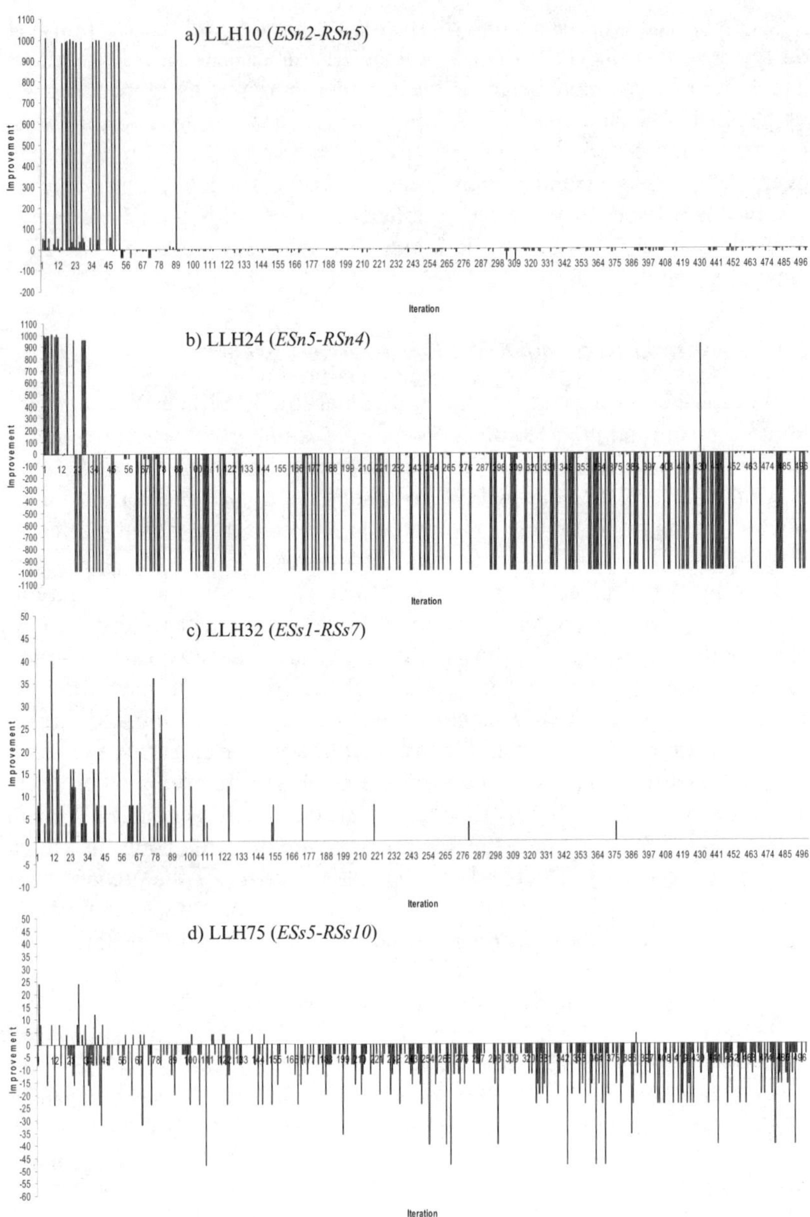

Fig. 6. Performance of low level heuristics during 500 iterations (hyperheuristic ***Peckish4***, ***Dataset1***, poor initial solution).

Table 7. The most and the least frequently applied low level heuristics over 10 runs of hyperheuristic ***Peckish4-v1*** for ***Dataset2*** using a poor initial solution.

Most frequently applied LLH			
LLH number	Event selection method	Resource selection method	Number of calls
For not scheduled events			
LLH13	ESn3	RSn3	19
LLH3	ESn1	RSn3	18
LLH6	ESn2	RSn1	15
LLH4	ESn1	RSn4	14
LLH8	ESn2	RSn3	13
LLH11	ESn3	RSn1	13
LLH24	ESn5	RSn4	13
LLH7	ESn2	RSn2	12
LLH12	ESn3	RSn2	12
LLH14	ESn3	RSn4	12
For already scheduled events			
LLH50	ESs3	RSs5	265
LLH47	ESs3	RSs2	215
LLH92	ESs7	RSs7	202
LLH94	ESs7	RSs9	198
LLH53	ESs3	RSs8	191
LLH52	ESs3	RSs7	187
LLH93	ESs7	RSs8	182
LLH90	ESs7	RSs5	180
LLH32	ESs1	RSs7	158
LLH46	ESs3	RSs1	136
LLH87	ESs7	RSs2	131
LLH27	ESs1	RSs2	118
LLH51	ESs3	RSs6	112
LLH42	ESs2	RSs7	105
LLH31	ESs1	RSs6	104
LLH91	ESs7	RSs6	103
LLH26	ESs1	RSs1	102
LLH34	ESs1	RSs9	100
LLH54	ESs3	RSs9	99
LLH33	ESs1	RSs8	95

Least frequently applied LLH			
LLH number	Event selection method	Resource selection method	Number of calls
For not scheduled events			
LLH18	ESn4	RSn3	9
LLH23	ESn5	RSn3	9
LLH16	ESn4	RSn1	8
LLH21	ESn5	RSn1	8
LLH22	ESn5	RSn2	8
LLH1	ESn1	RSn1	7
LLH10	ESn2	RSn5	7
LLH19	ESn4	RSn4	6
LLH20	ESn4	RSn5	5
LLH17	ESn4	RSn2	4
For already scheduled events			
LLH80	ESs6	RSs5	28
LLH76	ESs6	RSs1	20
LLH70	ESs5	RSs5	19
LLH40	ESs2	RSs5	16
LLH49	ESs3	RSs4	13
LLH79	ESs6	RSs4	13
LLH89	ESs7	RSs4	10
LLH69	ESs5	RSs4	7
LLH29	ESs1	RSs4	6
LLH60	ESs4	RSs5	5
LLH39	ESs2	RSs4	2
LLH59	ESs4	RSs4	2
LLH38	ESs2	RSs3	1
LLH88	ESs7	RSs3	1
LLH95	ESs7	RSs10	1
LLH28	ESs1	RSs3	0
LLH48	ESs3	RSs3	0
LLH58	ESs4	RSs3	0
LLH68	ESs5	RSs3	0
LLH78	ESs6	RSs3	0

Table 7 contains the statistics for those low level heuristics which are the most and the least frequently called over 10 runs of hyperheuristic ***Peckish4-v1*** applied to ***Dataset2*** starting from a poor initial solution. The total numbers of calls are given for the top 10 low level heuristics for not scheduled events and for the best 20 low level heuristics for already scheduled events. Not surprisingly, the latter low level heuristics are applied much more often since a lot of rescheduling steps are usually required until the resources necessary for scheduling a new event become available. Comparing two parts of Table 7, we can observe that the difference in numbers of calls for low level heuristics for not scheduled events is not very significant: all of them are more or less effective. This may follow from the fact that early in the search the list of not scheduled events in our experiments is quite short and the number of conflicting events for a particular not scheduled event may be very limited (often, there is only one conflicting event). Therefore, the same event may be selected by different event selection rules and the same conflicting event may be chosen by applying different resource selection rules. However, as we have seen in Figures 6a and 6d, later in the search these low level heuristics for not scheduled events have different, complementary behaviour. For the second category of low level heuristics, clear champions and outsiders can be identified. Why does this happen? The main factor is the structure of the constraints for a particular problem instance.

Table 8. Frequency of occurrence of selection rules in applied low level heuristics over 10 runs of hyperheuristic ***Peckish4-v1*** for ***Dataset2*** and a poor initial solution.

Event selection rules		Resource selection rules	
Rule	Number of calls	Rule	Number of calls
For not scheduled events		For not scheduled events	
ESn3	67	RSn3	68
ESn1	60	RSn4	55
ESn2	57	RSn1	51
ESn5	50	RSn2	46
ESn4	32	RSn5	46
For already scheduled events		For already scheduled events	
ESs3	1284	RSs7	823
ESs7	1090	RSs8	659
ESs1	842	RSs2	647
ESs2	563	RSs9	612
ESs6	381	RSs5	591
ESs5	319	RSs6	499
ESs4	255	RSs1	495
		RSs10	353
		RSs4	53
		RSs3	2

From Table 8, which demonstrates the frequency of appearance of individual selection rules in the applied low level heuristics, we can find out, for example, that resource selection rules which involve selection of alternative locations for the events (specifically *RSs3* and *RSs4*) are not very useful. It is because the tightest constraints in ***Dataset2*** are those that restrict the number of possible locations for the events. Table 7 confirms that all low level heuristics which employ resource selection rules dealing with locations are among the least frequently called (regardless of which event selection rule is used) and some of them are never applied being unable to change the current solution (see right part of Table 7). On the other hand, ***Dataset2*** contains groups of the events with a wide time windows and big lists of possible trainers. Hence, low level heuristics which deal with such kinds of events and look for the alternative trainers or timeslots, should perform reasonably well. For example, the leading low level heuristic *LLH50* (see left part of Table 7) selects the alternative timeslot(s) for the event with a wide time window. Almost surely such a timeslot will be found at any iteration and *LLH50* can be potentially applied by the hyper-heuristic. This type of behaviour of problem instances was not apparent when the model was developed, and indeed business practices have changed repeatedly since the initial contact with the financial company. Hence having a wide range of ways to deal with varying problem data is important and useful here, and for a range of other real-world problems.

The discussion above does not suggest that there are generally good or relatively useless low level heuristics in the set. It rather persuades that there are low level heuristics which are fitted better for the particular problem instance than others and some low level heuristics are not very useful for that instance. It is difficult to predict the behaviour of individual low level heuristics for a particular dataset. However, a large, diverse set of low level heuristics provides our approach with the flexibility and robustness leading to high quality results for a wide variety of foreseen and unforeseen characteristics of instances of the problem. Although the set of "good" low level heuristics is very different for each dataset, the hyperheuristic is able to shape a collection of low level heuristics into a method tailored for a specific problem instance.

6 Conclusions

Hyperheuristics are starting to prove themselves as fast and effective methods for solving complex real-world optimisation problems. A hyperheuristic is a robust approach which manages the choice of low level heuristics and requires limited problem-specific information. We have compared a wide range of hyperheuristic approaches for a real-world trainer scheduling problem in this chapter.

The performance of hyperheuristics is determined to a great extent by the quality of low level heuristics used. We split simple local search neighbourhoods into "event selection" and "resource selection" rules and construct a large set of low level heuristics by using all possible combinations of those rules. Such an approach allows us to effectively handle various constraints of the problem and requires only limited implementation time. From our experiments, we observe that a rich set of low level heuristics can be used to construct very good solutions to a complex real-world problem. Furthermore, small subsets of low level heuristics must be carefully chosen to generate high quality results. We have demonstrated effective peckish and tabu hyperheuristics.

In further research we intend to develop methodologies for finding promising low level heuristics from a large set of candidates by learning the behaviour of a low level heuristic in a given situation. We expect that this can significantly reduce CPU times, increasing the number of iterations possible and hopefully solution quality, while still meeting the low CPU times demanded by many practical scheduling decision support systems. We believe that our methods of quickly generating a large number of low level heuristics and putting them together with hyperheuristic AI is likely to be applicable to a wide range of complex real-world planning and scheduling optimisation problems.

References

[1] U. Aickelin and K. Dowsland (2000). "Exploiting problem structure in a genetic algorithm approach to a nurse rostering problem," *Journal of Scheduling* 3, 139-153.

[2] E. Burke, P. Cowling, P. De Causmaecker, and G. Vanden Berghe (2001). "A memetic approach to the nurse rostering problem," *Applied Intelligence* 15, 199-214.

[3] E. Burke, G. Kendall, J. Newall, E. Hart, P. Ross, and S. Schulenburg (2003). "Hyperheuristics: an emerging direction in modern search technology." In F. Glover and G. A. Kochenberger (eds.), *Handbook of Metaheuristics*. Kluwer Academic Publishers, 457-474.

[4] E. Burke, G. Kendall, and E. Soubeiga (2003). "A tabu-search hyperheuristic for timetabling and rostering," *Journal of Heuristics* 9, 451-470

[5] D. Corne and P. Ross (1995). "Peckish initialisation strategies for evolutionary timetabling." In E. Burke and P. Ross (eds.), *Practice and Theory of Automated Timetabling*. Springer Lecture Notes in Computer Science 1153, Springer, 227-240.

[6] P. Cowling, G. Kendall, and L. Han (2002). "An investigation of a hyperheuristic genetic algorithm applied to a trainer scheduling problem." In *Proceedings of 2002 Congress on Evolutionary Computation (CEC2002)*, IEEE Computer Society Press, Honolulu, USA, 1185-1190.

[7] P. Cowling, G. Kendall, and E. Soubeiga (2000). "A Hyperheuristic Approach to Scheduling a Sales Summit." In E. Burke and W. Erben (eds.), *Practice and Theory of Automated Timetabling III: PATAT 2000*. Springer Lecture Notes in Computer Science 2079, Springer, 176-190.

[8] P. Cowling, G. Kendall, and E. Soubeiga (2001). "A parameter-free hyperheuristic for scheduling a sales summit." In *Proceedings of the Third Metaheuristic International Conference (MIC'2001)*, Porto, Portugal, 127-131.

[9] P. Cowling, G. Kendall, and E. Soubeiga (2002). "Hyperheuristics: a tool for rapid prototyping in scheduling and optimisation." In S. Cagoni, J. Gottlieb, E. Hart, M. Middendorf and G. Raidl (eds.), *Applications of Evolutionary Computing: Proceedings of Evo Workshops 2002*. Springer Lecture Notes in Computer Science 2279, Springer, 1-10.

[10] B. Dodin, A. A. Elimam, and E. Rolland (1998). "Tabu search in audit scheduling," *European Journal of Operational Research* 106, 373-392.

[11] K. Dowsland (1998). "Nurse scheduling with tabu search and strategic oscillation," *European Journal of Operational Research* 106, 393-407.

[12] H.-L. Fang, P. Ross, and D. Corne (1994). "A promising hybrid GA/heuristic approach for open-shop scheduling problems." In A. Cohn (ed.), *Proceedings of ECAI 94: 11th European Conference on Artificial Intelligence*. John Wiley & Sons, 590-594.

[13] F. Glover and M. Laguna (1997). *Tabu Search*. Kluwer Academic Publishers, Norwell, MA.

[14] J. Gratch and S. Chien (1996). "Adaptive problem-solving for large-scale scheduling problems: a case study," *Journal of Artificial Intelligence Research* 4, 365-396.

[15] L. Han, G. Kendall, and P. Cowling (2002). "An adaptive length chromosome hyperheuristic genetic algorithm for a trainer scheduling problem." In *Proceedings of the 4th Asia-Pacific Conference on Simulated Evolution and Learning (SEAL'02)*, Singapore, 267-271.

[16] P. Hansen and N. Mladenović (2001). "Variable neighbourhood search: Principles and applications," *European Journal of Operational Research* 130, 449-467.

[17] E. Hart and P. Ross. (1998). "A heuristic combination method for solving job-shop scheduling problems." In A.E. Eiben, T. Back, M. Schoenauer, H-P.Schwefel (eds), *Parallel Problem Solving from Nature V*. Springer Lecture Notes in Computer Science 1498, Springer, 845-854.

[18] E. Hart, P. Ross, and J. Nelson (1998). "Solving a real-world problem using an evolving heuristically driven schedule builder," *Evolutionary Computation* 6, 61-80.

[19] A. Nareyek (2003). "Choosing search heuristics by non-stationary reinforcement learning." In M. Resende and J. de Sousa (eds.), *Metaheuristics: Computer Decision-Making*. Kluwer Academic Publishers, 523-544.

[20] M. Randall and D. Abramson (2001). "A general meta-heuristic based solver for combinatorial optimisation problems," *Computational Optimisation and Applications* 20, 185-210.

[21] P. Ross, S. Schulenburg, J. G. Marín-Blázquez, and E. Hart (2002). "Hyper-heuristics: learning to combine simple heuristics in bin-packing problems." In *Proceedings of the Genetic and Evolutionary Computation Conference (GECCO 2002)*. Morgan Kauffmann, 942-948.

[22] H. Terashima-Marín, P. Ross, and M. Valenzuela-Rendón (1999). "Evolution of constraint strategies in examination timetabling." In W. Banzhaf et al. (eds.), *Proceedings of the GECCO99 Genetic and Evolutionary Computation Conference*. Morgan Kaufmann, 635-642.

[23] G. Thompson (1996). "A simulated annealing heuristic for shift scheduling using non-continuously available employees," *Computers and Operations Research* 23, 275-288.

[24] E. Tsang and C. Voudouris (1997). "Fast local search and guided local search and their application to British Telecom's workforce scheduling problem," *Operations Research Letters* 20, 119-127.

A Genetic-Algorithm-Based Reconfigurable Scheduler

David Montana, Talib Hussain and Gordon Vidaver

BBN Technologies, Cambridge MA 02138, USA
dmontana@bbn.com, thussain@bbn.com, gvidaver@bbn.com

Summary. Scheduling problems vary widely in the nature of their constraints and optimization criteria. Most scheduling algorithms make restrictive assumptions about the constraints and criteria and hence are applicable to only a limited set of scheduling problems. A reconfigurable scheduler is one that, unlike most schedulers, is easily configured to handle a wide variety of scheduling problems with different types of constraints and criteria. We have implemented a reconfigurable scheduler, called Vishnu, that handles an especially large range of scheduling problems. Vishnu is based upon a genetic algorithm that feeds task orderings to a greedy scheduler, which in turn allocates those tasks to a schedule. The scheduling logic (i.e. constraints and optimization criteria) is reconfigurable, and Vishnu includes a general and easily expandable framework for expressing this logic using hooks and formulas. The scheduler can find an optimized schedule for any problem specified in this framework. We illustrate Vishnu's flexibility and evaluate its performance using a variety of scheduling problems, including some classic ones and others from real-world scheduling projects.

1 Introduction

Most optimizing schedulers solve a limited class of scheduling problems in a single domain. In contrast, a reconfigurable scheduler can solve a wide range of different problems across a variety of domains. Using a reconfigurable scheduler, a user should be able to switch easily between scheduling, for example, taxicab pickups, athletic fields, factory machinery, classrooms, and service visits.

Although the term "reconfigurable scheduler" is new, the concept of reconfigurable scheduling has existed for decades, and a variety of reconfigurable schedulers have been created. However, the progress in this area has been slow recently. To a large extent, this reflects the view that the existing reconfigurable scheduling frameworks, such as AMPL [1] and OPL Studio [2], are as powerful as possible, and offer little room for improvement. However, we claim that our more recently developed reconfigurable scheduler, called Vishnu, is a significantly better approach that is closer to the ideal of full coverage of real-world scheduling problems.

Like other reconfigurable schedulers, Vishnu has both an optimizer and a problem representation framework, as first described in lesser detail in [3]. Unlike other

D. Montana et al.: *A Genetic-Algorithm-Based Reconfigurable Scheduler,* Studies in Computational Intelligence (SCI) **49**, 577–611 (2007)

such schedulers, the optimizer is a hybrid of a genetic algorithm and a greedy schedule builder. The genetic algorithm generates orderings of the task to schedule, and the schedule builder adds one task at a time to the schedule in that order. In both the genetic algorithm and schedule builder are logic hooks where the user can specify key pieces of the scheduling logic, such as the optimization criterion, the task durations, and amount of capacity consumed. The user specifies these using a formula language that is similar to those used in modern spreadsheet programs. For each hook, the user either specifies a formula for computing the result or accepts the default formula. Vishnu also provides additional components, such as a graphical user interface and configurable statistical tables.

Section 2 provides a brief background of reconfigurable scheduling and evolutionary scheduling, with the focus of the latter on those schedulers closest to our particular approach. Section 3 gives an overview of the Vishnu optimizer and problem representation framework, with an emphasis on a high-level view of its capabilities and philosophy. Section 4 examines Vishnu in greater detail, describing how each of the different types of constraints is implemented. Section 5 demonstrates the performance and capabilities of Vishnu on both classical scheduling problems and complex real-world problems. While the former are useful for benchmarking, the latter more accurately showcase Vishnu's power to allow quick development of solutions to complex and unique scheduling problems. Section 6 concludes the paper.

2 Background

Our work is at the intersection of two distinct threads of scheduling research: reconfigurable scheduling and evolutionary scheduling. We now discuss each of these threads.

2.1 Reconfigurable Scheduling: The Concept

A reconfigurable scheduler has two main components: a problem representation framework and an optimizer (also called a solver). The problem representation framework provides a means for a user to specify the hard and soft constraints of a scheduling problem. Hard constraints cannot be violated and determine what constitutes a legal schedule. Soft constraints are preferences that need not be satisified, but violating them causes a decrease in schedule quality; hence, they determine what constitutes a good schedule and define the optimization criterion. Generally, the framework includes a language to represent these constraints. The optimizer searches for a schedule that satisfies all the hard constraints and that optimally (or at least nearly optimally) trades off between the different soft constraints. Preferably, the optimizer can solve any problem specified in the framework. Additional components are also highly desirable. A graphical user interface allows a user to view, modify, and otherwise interact with the schedules created. A database allows multiple users to interact with the same scheduling problem. Configurable statistics gathering allows a user to create statistical tables matched to the problem.

There are some clear advantages to a reconfigurable scheduler [4]. Primarily, it greatly reduces both the time and cost of developing a scheduler for a new scheduling problem. If the nature of the scheduling problem changes, it is quick and easy to modify to incorporate these changes. A reconfigurable scheduler is reusable, so a user does not have to develop or purchase, and then learn how to use, a different scheduler for each scheduling problem. So, the existence of an effective and easy-to-use reconfigurable scheduler would make optimized scheduling available to a much larger set of users.

A reconfigurable scheduler cannot solve every scheduling problem. There will always be some type of constraint that cannot be represented in the problem representation framework. The goal for a reconfigurable scheduler is to approach the ideal, handling as many different types of real-world scheduling problems and scheduling concepts as possible. It should be capable of being easily extended to cover new concepts to ensure that it can grow towards the ideal. Furthermore, the problem representation framework should make it easy for a user to define scheduling problems, and the optimizer should perform a reasonably efficient search for a schedule.

2.2 Reconfigurable Scheduling: Previous Work

The separation of the problem representation from the schedule generation process, which is central to reconfigurable scheduling, is not a new idea. Both the mathematical programming community and constraint programming community have a history of work on problem representation languages and associated solvers.

For mathematical programming, AMPL [1] is the most popular modeling language and serves as a good representative of its class. Competitors and predecessors, such as GAMS [5], have similar functionality. AMPL allows representation of the algebraic constraints and optimization criteria used in mathematical programming. There exist multiple solvers, including CPLEX [6], that generate solutions to the problems modeled in AMPL. From the viewpoint of reconfigurable scheduling, there are two major shortcomings of the AMPL approach. First, the solvers generally cannot solve all problems representable in AMPL. For example, CPLEX can only solve problems amenable to linear programming, mixed integer programming, or convex optimization. Second, there is a limited representation capability in AMPL for logical, as opposed to algebraic, constraints. As an example, it would be hard to represent the following constraint in AMPL: "If it is later than 10:00 and a resource has already done a full hour of work executing tasks and has not rested yet, then the resource should rest for 10 minutes."

Constraint programming also has its own languages and frameworks for representing scheduling problems. Some examples are CHIP [7], Prolog III [8], and ODO [9]. The solvers associated with constraint programming are usually tree-based search algorithms. One traditional shortcoming of the constraint programming approach is an inability to express constraints involving complex algebraic expressions. A second shortcoming is that the tree search method is ineffective at global optimization. Recent developments have addressed these problems. A new constraint programming language, OPL [2], allows representation of complex algebraic constraints in addition to logical constraints. There are associated OPL solvers, such as those available in ILOG's OPL Studio product. Improvements to the tree-based search technique, such as those described in [10], have increased the optimization performance. Still, there are a variety of types of constraints that cannot be represented in OPL, and hence many real-world scheduling problems that cannot be solved. So, there is still much room for improvement.

Reconfigurable scheduling is a much more recent development in the evolutionary computation, as well as the larger metaheuristic, community. In addition to Vishnu, there have been some other efforts at making reconfigurable genetic schedulers, including [11] and [12], but these lack the generality of Vishnu.

An area closely related to reconfigurable scheduling is scheduling ontologies [13, 14]. A scheduling ontology is a set of vocabulary, concepts and relations that can be used to describe and characterize different scheduling problems. It is essentially equivalent to the problem representation framework component of a reconfigurable scheduler. Smith and Becker [13] have created a fairly extensive ontology, and we aim to create a reconfigurable scheduler that can both represent and schedule as extensive a set of scheduling concepts as possible.

2.3 Evolutionary Scheduling: Previous Work

Genetic algorithms (and, more generally, evolutionary algorithms) have achieved success in scheduling applications [15]. There are several reasons why evolutionary algorithms are well suited for most scheduling problems, including many for which traditional mathematical programming techniques are inadequate. First, they are easy to apply to almost any optimization problem, including those with complex and/or discontinuous constraints/criteria that may derail other algorithms. Second, evolutionary algorithms are good at searching large and rugged search spaces to find nearly optimal solutions; furthermore, they can find good, though suboptimal, solutions very quickly. Third, genetic algorithms, with their population-based approach, allow for easy and effective large-scale parallelization [16], and this can provide a further performance boost.

The combination of complex constraints with the fact that orderings (rather than binary or numerical values) are often the primary outputs means that the standard chromosome representation (a binary string) is often not appropriate for scheduling problems. In fact, the first use of genetic algorithms for scheduling (by Davis [17]) also introduced one of the first non-standard chromosomes. There has been a large variety of representations and genetic operators used for evolutionary scheduling. Many of these are targeted to specific scheduling problems, such as the vehicle routing problem with time windows [18]. Such an approach is necessary in order to compete in terms of performance with other techniques tuned to specific problems, whether they be classic scheduling problems or real-world applications.

However, approaches targeted to specific scheduling problems are not useful for reconfigurable scheduling because of their lack of generality. An alternative is an order-based genetic algorithm combined with a greedy schedule builder [19, 20].

This is one of the earliest approaches to evolutionary scheduling, and its advantage is its universal applicability. The order-based genetic algorithm [21, 22] was developed based on the recognition that for problems like the traveling salesman problem, the goal is to find the best ordering of N objects. Its chromosome is a direct representation of a permutation of N objects, labeled 1 through N, and its operators are designed to manipulate chromosomes of this type. Whitley [20] and Syswerda [19] developed an approach whereby an order-based genetic algorithm can be applied to more complex types of scheduling problems by adding a greedy schedule builder. The order-based genetic algorithm generates orderings of the tasks to schedule, while the greedy schedule builder translates these orderings into schedules, handling the tasks in the order in which they are presented in the chromosome. This approach is universally applicable, since it is generally easy to create a greedy schedule builder for a scheduling problem.

3 Vishnu Overview

Vishnu is more complex than the standard scheduling application because in order to achieve its generality it needs to handle many different aspects of scheduling. In this section, we provide a general overview of the approach before exploring many of the details in the next section.

3.1 The Genetic Algorithm

In our approach, the genetic algorithm generates task orderings and relies on a schedule builder to translate these into actual schedules. The genetic representation uses an order-based chromosome. Each chromosome is some permutation of the integers 1 through N, where N is the number of tasks to schedule and each number corresponds to a task.

The only novelty of our genetic representation is that it incorporates *prerequisite constraints*. If task A has task B as a prerequisite, then task B must be handled earlier in the scheduling process than task A. (Note that this does not necessarily preclude task A from being scheduled at an earlier time than task B.) The genetic algorithm enforces prerequisite constraints by only generating chromosomes with orderings that obey all such constraints. A reordering operation is applied to every chromosome produced (either by mutation and crossover or during initialization of the population) to maintain these constraints. It works by finding any task that is earlier in the chromosome than any of its prerequisites and changing its location so that it is directly after the last of its prerequisites.

The crossover operator used by the genetic algorithm is position-based crossover [19], and its operation is illustrated in Figure 1. It works as follows. A set of positions is randomly selected (which in the example of Figure 1 are positions 4, 6 and 7). The elements at these selected positions in the first parent (which in the example are the integers 4, 6 and 7) are maintained at these positions in the child. The remaining elements (which in the example are the integers 1, 2, 3 and 5) are used to fill in the

```
      *   * *
(1 2 3 4 5 6 7) crossover                   reorder
(4 7 2 5 6 1 3) ---------> (2 5 1 4 3 6 7) -------> (2 1 4 5 3 6 7)

      *   * *
(1 2 3 4 5 6 7) mutation   (2 5 1 4 3 6 7)  reorder  (2 1 4 5 3 6 7)
```

Fig. 1. The crossover and mutation operators. The *'s indicate the randomly selected positions. It is assumed that the prerequisites are such that task 4 precedes task 5 and task 2 precedes task 6.

remaining slots in the child, but will in general be at different positions in the child than in the first parent. The order of these elements in the child will be the same as their order in the second parent (which in the example means that 2 is placed in the first empty position, followed in order by 5, 1 and 3).

Also illustrated in Figure 1 is the mutation operator. It works the same as the crossover operator except without a second parent to provide the ordering for the subset of elements that are reordered in the child. Instead, the new order of the shuffled elements is randomly selected.

Each member of the initial population is generated by selecting a random ordering and then reordering the entries to obey the prerequisites constraints. The flow of operations of the genetic algorithm is shown in Figure 2.

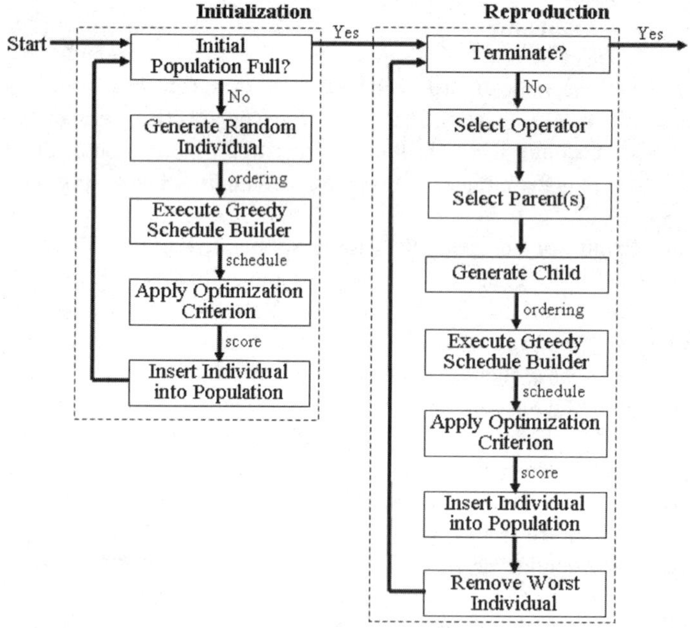

Fig. 2. The operation of the genetic algorithm.

The genetic algorithm is steady-state, which means that it generates and replaces one individual at a time rather than an entire population. The advantage of a steady-state replacement strategy is that the search generally proceeds faster, since the genetic algorithm can use good individuals as soon as they are created rather than waiting for generational boundaries. Since there are no generations, the amount of work done by the search algorithm is measured by the number of individuals evaluated. There is a uniqueness constraint to ensure that there are no two identical individuals in the population; duplicates generated by the genetic operators are discarded without being evaluated.

A fitness function is used to evaluate each individual, i.e. each task ordering. The evaluation starts by feeding the tasks to the greedy schedule builder in the specified order. The result is a schedule that obeys the required hard constraints. The optimization criterion, specified in a manner discussed in Section 4.3, then produces a numerical score representing the quality of the schedule, and hence the fitness of the individual.

There are two primary parameters for the user to specify: the population size and the number of evaluations. Increasing the population size and the number of evaluations increases the expected quality of the schedule at the expense of a longer search. As an extreme, if the user wants to execute just a greedy scheduler and bypass the genetic search, he can set the number of evaluations to be one.

Other parameters for the genetic algorithm are usually just set to their default values, as we do for all the experiments described in Section 5. The default probability of selecting mutation as the genetic operator is 0.5, with default probability also 0.5 for crossover. Parent selection is done using an exponential probability distribution, i.e. the individuals in the population are ranked and the ith best individual has selection probability that is some factor k as great as the $(i-1)^{st}$ best. This factor k is set by default to be 1-(10/popSize). There are also parameters that can specify termination criteria that are alternatives to ending after a certain number of evaluations. These include: the maximum number of duplicates (which specifies to stop the search after a certain number of duplicate individuals has been discarded), the maximum run time (which specifies to terminate after a certain amount of wall clock time has elapsed), and the maximum age of the best individual (which specifies to terminate if no progress has been made during a certain number of consecutive evaluations). By default, these parameters are set so that the number of evaluations is guaranteed to be the termination criterion.

Note that it is possible to use other optimization techniques, such as simulated annealing or tabu search, to search the space of permutations (task orderings). Which technique is best depends on the characteristics of the problem, particularly the search space size and ruggedness. Genetic algorithms are a good match for Vishnu because they display good performance over a wide range of different search space characteristics, which is important because Vishnu should be able to handle a wide variety of different scheduling problems.

3.2 Greedy Schedule Builder Overview

Our greedy schedule builder needs to be more general than those designed for specific scheduling problems, such as the active schedule generation algorithm for job-shop scheduling [23]. While the details of the greedy schedule builder are complicated, the basic idea is simple. We provide a high-level description of the algorithm with italicized concepts corresponding to some (but not all) of the logic hooks where the user specifies the logic in a manner explained below.

The schedule builder assigns tasks one at a time in the specified order. For each task, it looks for the best resource(s) and time for that task, where the goodness of a resource and time is evaluated based on a specified *greedy criterion*. The schedule builder only considers resources *capable* of performing the task. For each capable resource, it checks the legality of scheduling the task at its specified *target start time*. (The target start time can be the beginning of time, indicating to schedule the task as early as possible.) If assigning at this time would not violate any hard constraints, such as *availability* or *capacity*, then it uses this time as the sole potential assignment time for this resource. If assigning at the target start time would violate a hard constraint, then the schedule builder finds the two times closest to the target time, one earlier and one later, at which it is legal to assign the task. Because one or both of these times may not exist, this search results in 0, 1 or 2 possible assignment times for this resource. From all the possible resources and corresponding times, the schedule builder selects the best one based on the greedy criterion and schedules the task there. If no legal resource and time exist (i.e. there is no way to schedule the task without violating a hard constraint), then this task is not scheduled.

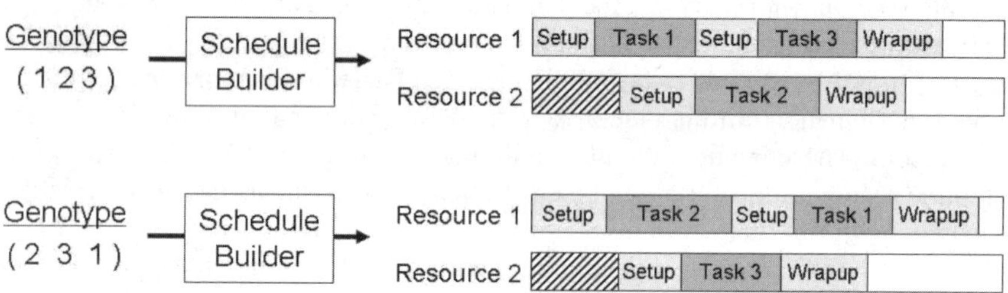

Fig. 3. The greedy schedule builder converts two different genotypes (task orderings) into two different phenotypes (schedules). The diagonally striped area indicates where the resource is unavailable.

Figure 3 illustrates the operation of the greedy schedule builder for a very simple scheduling problem. It shows how two different genotypes translate into two different phenotypes/schedules. The example makes certain assumptions about what the user has specified for the scheduling logic. These include the following default behaviors: (a) the target start time for all tasks is the beginning of scheduling window, (b) all resources are capable of performing all tasks, (c) tasks are always available, (d) there is no multitasking, i.e. a resource performs only one task at a time, and (e) each task requires only one resource. It also assumes that the greedy criterion specifies that given a choice between multiple times at which to schedule a task, the earliest is the best. For the first genotype, the greedy scheduler starts with task 1. For resource 1, the task can be scheduled at its target time. For resource 2, the task cannot be scheduled at its target start time, so the schedule builder looks forward in time to find the first available slot, which is right after the initial block when resource 2 is unavailable. (Looking backward in time yields nothing.) Since the greedy criterion says that earlier is better, the schedule builder chooses resource 1. Next, it schedules task 2. The first available slot on resource 2 is earlier than that on resource 1, so the schedule builder assigns task 2 to resource 2. Finally, using the same logic, task 3 is assigned to resource 1 in the slot after task 1. (Note that each task has three stages: setup, execution and wrapup, although one or two of these may require zero time.) For the second genotype, the schedule builder uses the same procedure, but a different order in which the tasks are scheduled leads to a different schedule. If the optimization criterion is makespan (i.e. minimize the end time of the final task completed), then the top schedule is better than the bottom one.

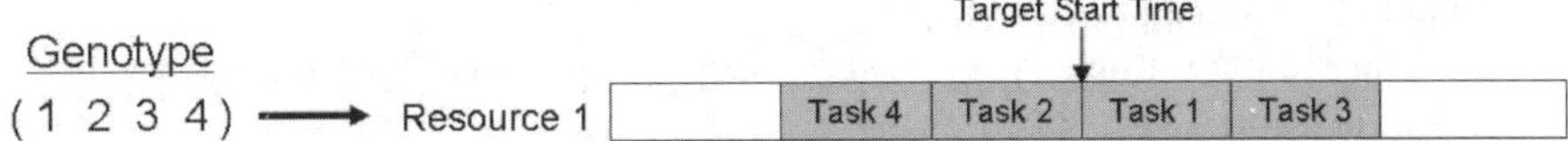

Fig. 4. Another greedy scheduler example, with different problem specifications.

Figure 4 shows what happens for a different set of problem specifications. The key differences from the prior one is: (a) the target start time is in the middle of the scheduling window rather than at the beginning, and (b) the greedy criterion is a penalty proportional to the deviation of the actual start time from the target time, with lateness penalized 1.5 times more heavily than earliness. The first task scheduled, task 1, is placed at the target time. The second task, task 2, cannot be assigned at the target time, so there are two options. Searching forward in time finds the first available slot is immediately after task 1, while searching backward finds the slot right before task 1. The greedy criterion selects the earlier slot, since the penalty is smaller. For task 3, the greedy criterion prefers the slot after task 1 rather than before task 2.

3.3 Problem Representation Framework Overview

Like the greedy schedule builder, the problem representation framework operates on a simple basic concept, with the complexity in the details. We provide an overview here and fill in the details in the next section.

The genetic algorithm and greedy schedule builder both export logic hooks where the user can specify logic specific to a particular scheduling problem. Examples of such hooks include one that allows the user to specify the task execution duration for any task, one that allows the user to specify the prerequisites for any task, and one that allows the user to specify the optimization criterion. The user defines a formula for each hook or accepts the default value for that hook.

The formula is then evaluated within context to provide the required information to the scheduling algorithm. Establishing the context primarily involves defining special variables and setting their values accordingly. The two most commonly used variables are *task* and *resource*, which refer to the particular task and/or resource about which to compute the information. All hooks provide the context variables *tasks* and *resources*, which are lists of all the tasks and resources, and *winStart*, which gives the earliest time in the scheduling window. The scheduling algorithm sets the context and evaluates the formula in that context, so all the user has to do is define the formula.

As an example, consider the Execution Duration hook, which specifies the amount of time that a resource spends executing a task and is discussed further in Section 4.3. We examine some different possibilities for formulas for this hook, starting with simple ones and then more complex ones.

- *Empty* - The default for this hook is 0, so if no formula is specified, then all tasks require 0 seconds to execute.
- **5** - This simple formula specifies that every task requires 5 seconds to execute.
- **task.duration** - This specifies that the time to execute a task is given by a field called *duration* in the task. (Note that the '.' character is the notation used to access a field inside a structure.)
- **task.distance / resource.speed** - The execution duration is the task's distance divided by the resource's speed.
- **entry (resource.durations, task.type)** - Each resource has a field *durations* that is a list of how long it takes that resource to execute different types of tasks, and each task has a numerical field that specifies its type. The task execution duration is the entry in the resource's list corresponding to the type of the task. (Note that *entry* is one of many predefined functions in Vishnu's formula language; it accesses the nth element of a list.)

4 Vishnu Details

In this section, we provide details about Vishnu. We start with a discussion of how to represent data, most notably data about the tasks and resources. We then describe the formula language and how formulas are evaluated. We next examine in-depth the different logic hooks and how the scheduling algorithm incorporates them. We conclude with examples of how to represent, and hence solve, three classic scheduling problems using Vishnu.

4.1 Scheduling Data

Vishnu provides a small number of atomic data types plus the ability to combine these atomic data types into composite data types. Some commonly used composite data types are predefined, but each scheduling problem requires the definition of new problem-specific composite data types.

The atomic data types are *string*, *number*, *boolean*, and *time*, plus a special type, *list*, which is a variable-sized set of instances of a single specified data type. The predefined composite data types include *interval*, which contains the fields start (a *time*) and end (a *time*), and *matrix*, which contains fields numrows (a *number*), numcols (a *number*), and values (a *list* of *numbers*).

The problem-specific data types are built from the atomic and predefined types. The type for a field can itself be another problem-specific type, and hence it is possible to construct arbitrarily complex data types. For each scheduling problem, one data type must be declared to represent tasks and another type specified as representing resources. Each of these two types must have one field that serves as a key, providing unique identification of instances of this type.

All data for a scheduling problem must be instances of the composite data types (including the predefined types) defined for that problem. While the primary data (i.e. what the scheduler needs to schedule) are the tasks and resources, other types of data (e.g. business rules or distance matrices) may be required to define the scheduling logic.

Examples of representation of scheduling data are given in Section 4.4 and Section 5.

4.2 Formulas: Language and Evaluation

Formulas are built from the following types of components: constants, variables, accessors, operators, and functions. Accessors provide access to the fields of a data structure using the notation '.' followed by the field name. (For example, task.id gives the id field of the data structure referenced by the variable task.) There is a fixed set of standard arithmetic (+, -, *, /) and comparison operators (=, <>, <, <=, >, >=) written using infix notation. There is also an expandable library of functions, where the syntax for invoking a function is fcnName (*arg1*, ..., *argn*).

Note that *null* means no value, and all accessors, operators and functions must handle the case when one or more of their arguments are null, usually by just returning null. A null value can be introduced into a formula by context variables such as *next* or *previous* (which are null when the task of interest is respectively the last or first task for its resource) or by functions such as *resourceFor* and *taskStartTime* (which are null if the task is not assigned to a resource).

There are too many functions to list them all here, so we provide a representative sample:

- **if** (*a*, *b*, *c*) returns the evaluation of *b* if *a* evaluates to true and returns the evaluation of *c* if *a* is false, or returns null if the third argument, *c*, is omitted.
- **list** (...) combines all of its arguments into a list.
- **interval** (*startTime*, *endTime*) returns an interval object.
- **entry** (*list*, *index*) returns the element of *list* at *index*.
- **distance** (*location1*, *location2*) returns the distance between the two locations.

- **max** (*a*, *b*) returns the maximum of the two numbers.
- **and** (...) returns true if all of its boolean arguments are true.
- **hasValue** (*a*) returns false if *a* is null and true otherwise.
- **withVar** (*varName*, *varValue*, *a*) evalutes *a* with variable *varName* bound to *varValue*.
- **mapOver** (*list*, *varName*, *a*) binds the variable *varName* to each element of *list* in succession and evaluates *a*, returning the results as a new list.
- **sumOver** (*list*, *varName*, *a*) does the same as mapOver except returning the sum of the results.
- **taskStartTime** (*task*) returns the currently assigned start time of *task* (or null if *task* is not assigned). The functions **taskSetupTime**, **taskEndTime** and **taskWrapupTime** return the other three times associated with a task assignment. The functions **formerSetupTime** and **formerWrapupTime** provide the setup and wrapup times of an already scheduled task prior to the assignment of another task that may have caused these times to change.
- **resourceFor** (*task*) returns the currently assigned resource of *task*.
- **lastTask** (*resource*) returns the last task assigned to *resource*, and **complete** (*resource*) returns this task's end time.

Note that these functions can be classified into different types including mathematical functions (max, and, distance), control functions (if, withVar), data constructors (list, interval), list operations (mapOver, sumOver, entry), and schedule accessors (taskStartTime, taskEndTime, resourceFor).

Many examples of formulas are given in Sections 4.3, 4.4 and 5.

When the formulas are originally written, a compiler transforms them into parse trees, which are what the scheduler executes to evaluate formulas. As part of the compilation process, the compiler checks that each formula is correct with respect to data types, i.e. that the formula returns the data type expected by the hook and that each argument to each function is of the expected type. Before execution of a formula, the scheduler ensures that all context-dependent variables, such as *task* and *resource*, are set appropriately.

Since the formulas are evaluated many times in different contexts, the efficiency of formula evaluation is a major aspect of scheduler performance. One important way in which we have made formula evaluation more efficient is by caching (i.e. storing) the results of formulas and recalling, rather than recomputing, the results in the future. The trick is knowing when, both in terms of which formulas and which contexts, caching results and then recalling them is valid. The validity of cached results depends on the variables and functions used in the formula. If the formula contains any schedule-dependent function (i.e. a function whose returned value depends not just on its arguments but also the current schedule, such as taskStartTime or resourceFor), then the formula needs to be re-evaluated whenever the schedule or arguments change. If the formula contains no schedule-dependent functions but refers to both task and resource variables, it needs to be evaluated once for each task/resource pair. If the formula just references the task (or resource) variable, then the formula must be evaluated just once for each task (or resource).

Hook	Default	Description
Optimization Criterion	0	Measure of quality of current schedule
Greedy Criterion	0	Measure of quality of assignment of **task** to **resource**
Target Start Time	winStart	Optimal time for **task** to begin when assigned to **resource**
Prerequisites	empty list	Tasks that must be scheduled before scheduling **task**
Execution Duration	0	Seconds required for **resource** to perform **task**
Setup Duration	0	Seconds **resource** prepares for **task** after doing **previous**
Wrapup Duration	0	Seconds **resource** cleans up after **task** before doing **next**
Breakable	false	Can a task be executed in discontinuous time intervals?
Resource Unavailable	empty list	All intervals of time when **resource** is busy
Task Unavailable	empty list	All intervals when **task** cannot be scheduled on **resource**
Capability	true	Can **resource** perform **task**?
Capacity Thresholds	empty list	Maximum capacity of each type for **resource**
Capacity Contributions	empty list	How much **task** adds to each type of capacity of **resource**
Capacity Resets	empty list	Capacity restored to **resource** by performing **task**
Multitasking Type	none	How resources perform more than one task at a time
Groupable	false	Can **task1** and **task2** can be performed in the same group?
Multiresource Reqts	empty list	Set of requirements **task** needs satisfied by its resources
Satisfied Requirements	empty list	Contribution of **resource** to requirements of **task**
Auxilliary Tasks Before	empty list	Set of auxilliary tasks to schedule before scheduling **task**
Auxilliary Tasks After	empty list	Set of auxilliary tasks to schedule after scheduling **task**

Table 1. The scheduling logic hooks, with context variables bolded

4.3 Hooks and Scheduling Logic

In this section, we discuss all of the hooks currently available in Vishnu. To define the logic for a scheduling problem, a user must associate a formula with each relevant hook or accept the hook's default. Table 1 provides a brief overview of the different hooks. Sections 4.3-4.3 discuss in greater detail the function of each hook and its role in the scheduling process. The hooks are divided into different classes to facilitate our explanation.

[Note that this set of hooks is only a current snapshot. Over time, we have gradually added new hooks, which explains why there is greater functionality available now than when we first introduced Vishnu in [3]. When we identify aspects of a scheduling problem that cannot be represented by the current hooks, we may specify a new hook to provide this functionality (along with updates to the schedule builder to utilize this hook). Careful consideration is given to ensure that a new hook is broadly applicable and not just a one-of-a-kind fix.]

Scheduler Directives

The first set of hooks we examine are those that instruct the scheduler how to execute rather than specifying problem constraints.

Optimization Criterion is the hook whose formula produces a numerical score for the current schedule. As discussed above, the genetic algorithm searches for the schedule with the smallest possible value for this score. All the different types of *soft constraints* of the problem must be aggregated into a single score, with the user (i.e. the person defining the problem) responsible for specifying how to combine them. The simplest and most common approach for combining penalties from different soft constraints is a weighted sum, but there are many other possibilities. Since this hook references the schedule as a whole rather than a particular task or resource, the only context variables are the standard ones: *tasks*, *resources* and *winStart*. A simple example is the formula that implements makespan, which scores a schedule based on the latest end time of any task:

$$\mathrm{maxOver}(tasks, \text{“}t\text{”}, \mathrm{taskEndTime}(t)) - winStart$$

Note that subtracting winStart allows the formula to return a number, which is the expected data type, rather than a time. A more complicated sample formula combines a penalty for a task being late, a penalty for the time resources spend setting up, and a penalty for each task that was not scheduled:

$$\mathrm{sumOver}(tasks, \text{“}t\text{”}, 10 * \mathrm{if}(\mathrm{taskStartTime}(t) > t.dueDate,$$
$$\mathrm{taskStartTime}(t) - t.dueDate, 0) + (\mathrm{taskStartTime}(t) - \mathrm{taskSetupTime}(t)) +$$
$$\mathrm{if}(\mathrm{hasValue}(\mathrm{resourceFor}(t)), 0, 1000000))$$

assuming that the task data type has a field *dueDate*. Other examples of schedule characteristics that might be penalized include resources working overtime or changes from a previous schedule. There should always be a formula specified for this hook except when the genetic algorithm is not used (with the user content to accept the first schedule produced by the greedy schedule builder).

Greedy Criterion is the hook whose formula produces a numerical score for any potential assignment of a task to a resource. As discussed above, it is used by the greedy scheduler to compare different possible task assignments to select the best one, and is essentially equivalent to a dispatch rule. It is often, but not always, an incremental version of the optimization criterion. Like the optimization criterion, it can combine multiple subcriteria into a single score. The context variables *task* and *resource* refer to the task and resource being assigned. A simple example for the case when earlier is better is the formula:

$$taskEndTime(task) - winStart$$

Another sample formula penalizes both any deviation from the task's target start time and an assignment to any resource other than the task's best resource:

$$\mathrm{abs}(\mathrm{taskStartTime}(task) - task.bestTime) +$$
$$\mathrm{if}(resource.name = task.bestResource, 0, 1000)$$

Other potential types of criteria include minimizing the travel (setup) time for a task or finding the least full resource.

Target Start Time tells the greedy schedule builder the optimal point on a resource's schedule to try to assign a task. As described above, for a given resource, the greedy scheduler searches forward and backward from the best time for the first legal assignment times and only considers assignments at these (at most) two times. This approach limits the number of assignment times to consider and hence the computation required. The context variables are *task* and *resource*. Often, the formula for this hook is not specified, instead accepting the default value *winStart*, which indicates to schedule the task as early as possible. An example of a case where earlier is not better is scheduling the delivery of food to a party. If the food arrives too early, it will not be fresh for the party, but it also should not arrive too late, hence making the best time n hours before the party. Note that a hard constraint on task availability (see below) may be used in conjunction to ensure that the food is never scheduled to be delivered after the party starts.

Prerequisites tells the genetic algorithm which other tasks must precede a given task in any generated task ordering, and hence will be handled earlier than this task in the greedy schedule building process. The context variable is *task*, and the formula must return a list of task names. The most common reason that task A would have task B as a prerequisite is that task A is constrained to start after task B finished, and hence needs task B scheduled to determine its own availability. However, there are other possibilities. For example, tasks A and B might need to be performed simultaneously, and task B is the more difficult of the two to schedule, so task B should be scheduled first and then task A assigned at the same time. Another example is backwards planning, where the last leg of a journey is scheduled first, with the scheduling process working backwards towards the earlier legs.

Task Durations

A second set of hooks defines the time required for a resource to perform a task. Recall that there are three stages for this process: setup, execution and wrapup. Each of these stages has a hook that tells the time in seconds required for this stage. Another hook tells whether or not task performance can be broken into multiple disconnected intervals.

Execution Duration determines the length of time spent in the execution stage when a resource performs a task. Since this potentially depends on the identity of both the task and resource, the context variables include *task* and *resource*. Some sample formulas for this hook were provided in Section 3.3. Note that during the execution stage, the task and resource both must be available, which limits the times at which the greedy scheduler can place the task.

Setup Duration computes the time a resource spends in the setup stage before performing a task. It generally depends not just on the task and resource but also on what the resource was previously doing, i.e. the previous task. Hence, the context variables include *task*, *resource* and *previous*, where *previous* references the previous task and is set to null if the task being performed is the earliest on the resource's schedule. For example, a painting machine might have a setup duration of 0 if the previous task has the same color as the current task and 2 minutes if the previous

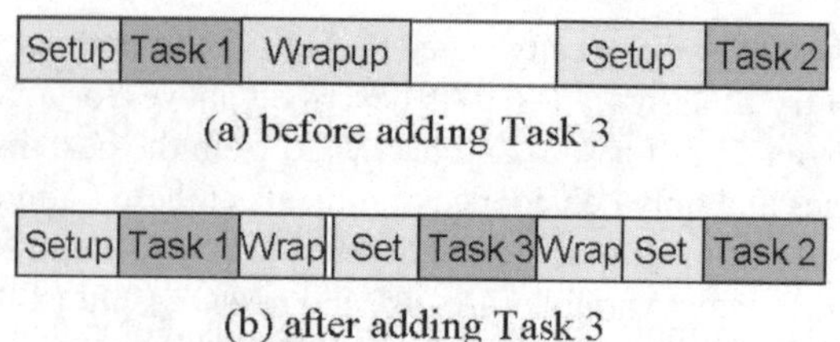

Fig. 5. The setup and wrapup durations of already scheduled tasks can change when a new task is inserted.

task has a different color, as expressed by the formula:

$$\text{if}(previous.color = task.color, 0, 120)$$

Another example is where the setup time represents travel time between the previous and current tasks and is proportional to the distance between the geographic locations of these tasks, as given by the formula:

$$\text{distance}(task.location, previous.location)/resource.speed;$$

Because of the potential dependency on the previous task, the result of this hook may need to be recomputed as the greedy scheduler searches through the resource's schedule trying to find where the task fits. In this case, it also needs to recompute the setup duration of the following task when inserting in a spot other than the end of the schedule, as illustrated in Figure 5. Note that only the resource, and not the task, needs to be available during the setup stage.

Wrapup Duration computes the time a resource spends in the wrapup stage. It can depend on the task, the resource and what the resource is doing afterwards (i.e. the next task), so the context variables include *task*, *resource* and *next*. Consider the example where if a task is the final one on a resource's schedule, then the resource needs to spend five minutes cleaning up, but otherwise no time during wrapup. The formula to capture this is

$$\text{if}(\text{hasvalue}(next), 0, 300)$$

Like Setup Duration, the value for this hook is potentially recomputed not just for each potential position of the task on the resource's schedule but also for the preceding task already on the resource's schedule, as illustrated in Figure 5.

Breakable tells whether the task performance interval can be split into discontinuous sections. For example, if a coffee break for the resources is scheduled for 10:30-10:45, a breakable task that requires an hour can start at 10:00 and be completed in two intervals, 10:00-10:30 and 10:45-11:15. This hook is just a choice of two values, yes or no, with the default being no. (There could be further development in the future of the semantics for breakable tasks, e.g. allowing specification of the conditions under which tasks can be broken and into what size chunks the task can be broken.)

Availability

Another set of hooks specifies when tasks and resources are available to be scheduled.

Resource Unavailable specifies the times at which a resource is not available to be scheduled. This is independent of the tasks being scheduled, so the only context variable is *resource*. The formula returns a list of intervals. For example, if a person works from noon to 8 PM on weekdays, then this resource is unavailable on weekends and between 8 PM and noon. Other examples are fixed break times or maintenance downtimes.

Task Unavailable specifies the times at which a task is not available to be scheduled. Unavailable intervals can be independent of how other tasks are scheduled. For example, a service call can only be scheduled when a person is around to allow entrance, or a delivery cannot be scheduled after the time when the item is needed. However, unavailable intervals can also depend on other task's assignments. For example, it is very common that one task cannot be scheduled to start earlier than the end time of another task. Another example is that a task may only be allowed to be scheduled at the same time as another. A sample formula that specifies both that the task finishes before a due date and that it must wait for the end of another task is:

$$\begin{aligned}&\mathrm{list}(\ \mathrm{interval}(task.dueDate, endTime), \mathrm{interval}(winStart,\\&\quad \mathrm{if}(\mathrm{hasValue}(task.followsTask),\\&\quad\ \ \mathrm{taskEndTime}(\mathrm{taskNamed}(task.followsTask)), winStart)))\end{aligned}$$

Capability specifies whether a particular resource is capable of performing a particular task. For example, an electrician can perform electrical wiring tasks but not plumbing tasks, and a painting robot can perform painting tasks but not welding tasks. The context variables are *task* and *resource*, and the hook returns a boolean indicating whether the resource is capable. An example formula that searches for a particular skill on a resource's list of skills possessed is

$$\mathrm{contains}(resource.skills, task.skillRequired)$$

Capacities

Capacities are hard limits on what resources can do based on accumulation of quantities over multiple tasks. There are a few hooks for specifying the functionality of capacity constraints.

Capacity Thresholds is the hook that specifies for a resource the thresholds that cannot be exceeded for each type of capacity. In general, capacity has multiple dimensions. For example, there are limits on both total volume and total weight for the cargo transported by a vehicle. As another example, an employee may have limits on both the hours worked in a day and the hours worked in a week. Therefore, the hook expects a list of numbers that are the thresholds, as in the formula

$$\mathrm{list}(resource.maxWeightCargo, resource.maxVolumeCargo)$$

The context variable is *resource*. If there are no capacity constraints, then the hook should use the default, which is the empty list.

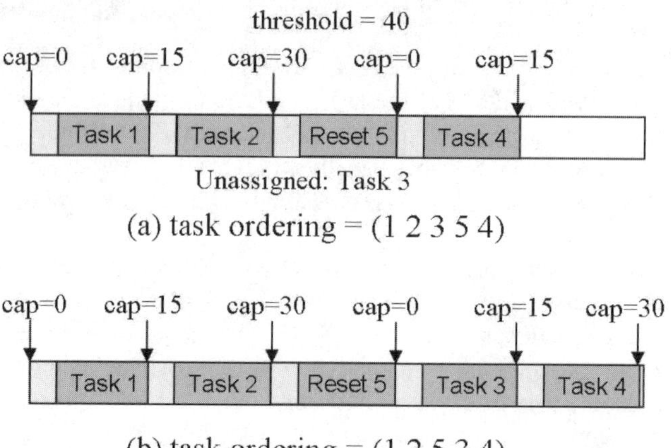

Fig. 6. This examples illustration how capacity can affect scheduling. The reset task, task 5, must be scheduled prior to other tasks to provide the additional capacity for these other tasks.

Capacity Contributions specifies the amount that a task contributes towards filling the capacity of a particular resource. A task cannot be assigned to a resource if it causes the threshold to be exceeded in any dimension. The context variables include *task* and *resource*, and formula should return a list of numbers of the same size returned by the Capacity Thresholds hook. The contributions of multiple tasks are accumulated by a simple vector sum. Note that when multitasking, i.e. a resource performing more than one task at a time, the capacities are summed and compared at a particular time, but otherwise they are accumulated over the duration of a resource's schedule. Continuing the weight and volume example, a sample formula is

$$\text{list}(task.weight, task.volume)$$

A more complex sample formula indicates that the amount of fuel required for a fuel truck to service a fuel request is the amount of fuel requested plus the amount of fuel for the truck to travel:

$$\begin{aligned}&\text{list}(task.requestedFuel + resource.consumptionRate*\\&\qquad(\text{taskStartTime}(task) - \text{taskSetupTime}(task)))\end{aligned}$$

Capacity Resets specifies when a task causes the accumulated capacity contributions of a resource to be reduced rather than increased. In many cases, the capacity contributions do not just keep on accumulating over time. At some point, an event happens that causes the accumulated counts to be reset, or partially reset. For example, with the capacity constraint that an employee can only work so many hours in a day, the total hours worked is reset to zero when a new day begins. Similarly, the total weight of a vehicle's cargo is reset to zero when all the cargo is dropped off at a central depot. The formula on this hook returns a list of quantities by which to reduce the accumulated contributions, with the stipulation that they cannot be set to less than zero. The context variables are *task* and *resource*, and the default is an empty list, indicating no reset. An example formula that resets the contributions to zero if the task is to drop off the cargo is

$$\text{if}(task.isDropoff, \text{list}(1000000, 1000000), \text{list}(0, 0))$$

Figure 6 shows the different aspects of capacity, including capacity resets, being used.

Multitasking

Multitasking is when a resource can perform multiple tasks simultaneously. There are a few hooks that determine this functionality.

Multitasking Type is a hook that has no formula but instead just a choice among three options. The first option is *none*, which means that the resource can only perform a single task at a time. This is the default and the most common option. With no multitasking, for the greedy scheduler to place a task on a resource's schedule, it needs to find a place where the time to perform the task, including the setup and wrapup, does not overlap the time to perform any other task already on the schedule. The second option is *grouped* (or batched) multitasking. In this case, a resource can perform multiple tasks simultaneously but only if all the tasks start at the same time and end at the same time. For example, consider a ship transporting cargo from one port to another. If each task is an item to transport, then the ship can perform more than one task at a time, but only if the items all depart from the same origin at the same time and travel to the same destination. The capacity constraints set the limit on how many simultaneous tasks a resource can handle. When assigning a task to a resource the greedy schedule builder can either add the task to an existing group or start a new group. The third option is *ungrouped* (or asynchronous) multitasking. In this case, the resource can perform multiple tasks simultaneously, and there is no need to synchronize the start and end times of the tasks. Hence, partial overlapping of the task performance periods is permitted as long as the capacity constraints are not violated at any time in the new task performance interval. An example of ungrouped multitasking is when a resource is actually a pool of homogeneous sub-resources, e.g. a set of N electricians or M cutting machines.

Groupable is a hook that only applies when using grouped multitasking. It specifies whether two tasks can be put in the same execution group, i.e. executed by the same resource at the same time. Context variables *task1* and *task2* provide references to the two tasks, and the formula returns a boolean. A sample formula indicates that two tasks are groupable if and only if they share the same origin and destination ports:

$$\text{and}(task1.origin = task2.origin, task1.destination = task2.destination)$$

Multiresourcing

Multiresourcing is when tasks potentially require more than one resource, and this section describes the hooks that specify this functionality. Note that when a task uses more than one resource, the greedy scheduler currently assumes that all the resources are committed to the task for the entire duration of task execution. If the scheduling logic requires finer-grain control over the times at which the different resources are busy, then auxilliary tasks (see Section 4.3) should be used instead, with the single task split into multiple tasks that are tightly coupled.

Multiresource Requirements produces a list of numbers, similar to the capacity thresholds, that enumerate what requirements the resources in aggregate must satisfy. The context variable is *task*. As an example, if a class needs one classroom, one teacher, and a certain number of teaching assistants, the formula is

$$\text{list}(1, 1, task.assistantsRequired)$$

As another example, if a resupply task requires a certain number of gallons of fuels and a certain number of rounds of ammunition, the formula is

$$\text{list}(task.fuelRequired, task.ammoRequired)$$

The default is the empty list, and hence no multiresourcing.

Multiresource Contributions specifies a list of numbers, similar to the capacity contributions, that are the contributions that a resource makes towards satisfying a task's multiresourcing requirements. As resources are added to a task, the contributions accumulate until all requirements are fully satisfied or it is proven impossible to do so. The greedy scheduler adds a new resource to a set of resources potentially satisfying a task's requirements only if it provides a non-zero contribution to at least one of the requirements not yet satisfied. The context variables are *task* and *resource*. Continuing the class scheduling example, a sample formula is

$$\text{list}(\text{if}(resource.isClassroom, 1, 0), \text{if}(resource.isTeacher, 1, 0),$$
$$\text{if}(resource.isAssistant, 1, 0))$$

Auxilliary Tasks

Auxilliary tasks provide no direct benefit from being scheduled, and hence are not included in the task ordering created by the genetic algorithm. They are helper tasks that allow the primary tasks to complete. An auxilliary task that helps a particular primary task complete is scheduled at the same point in the scheduling process as the primary task. As an example, consider trying to schedule an air marshal to monitor a certain flight from Los Angeles to New York, but there is no air marshal scheduled to be in Los Angeles at that time. Scheduling a connecting flight as an auxilliary task can put a marshal in position to perform the primary task. As discussed above, auxilliary tasks are used for the situation when a task requires a variety of resources at different times. In this case, the task is divided into multiple tasks, one of which is the primary task and the rest auxilliary. For example, a military air mission needs not just a plane and crew, but also maintenance crews and facilities, runways, airspace, etc. at different points in its execution.

Auxilliary Tasks Before provides a list of names of auxilliary tasks associated with a particular primary task that should be scheduled prior to scheduling the primary task. (Here, prior means earlier in the scheduling process, not earlier in the schedule.) The default is the empty list, and the context variable is *task*.

Auxilliary Tasks After is the same as Auxilliary Tasks Before but indicates the tasks scheduled subsequent to primary task rather than prior.

4.4 Examples - Classic Scheduling Problems

We now provide three examples of problem specifications using Vishnu. The problems are well-known and well-studied benchmarks from the operations research literature. (The OR-Library [24] is a good source of such classic problems and is available on the web at http://graph.ms.ic.ac.uk/info.html.) They are logically (though not computationally) simple, and therefore provide good examples with which to illustrate how to apply Vishnu before moving to more logically complex problems. While the Vishnu formulas can be hard to understand initially, it takes a relatively small amount of experience for a user to become acclimated to its method of problem representation.

Traveling Salesman Problem (TSP)

A salesman starts at a given city, travels to a set of other cities visiting each city once, and then returns to the starting city. The objective is to schedule the visits so as to minimize the total distance traveled.

Hook	Formula
Optimization Criterion	taskEndTime (taskNamed ("*City 1*"))
Prerequisites	if (*task.index* = 1, mapover (*tasks*, "*t*", if (*t.index* <> 1, *t.id*)))
Setup Duration	entry (*task.distances*, if (hasvalue (*previous*), *previous.index*, 1))

Table 2. Scheduling Logic for the Traveling Salesman Problem

The task data type, *city*, has fields id (a string), index (a number), and distances (a list of numbers). The resource data type, *salesman*, has the field id (a string). The data has one salesman with id = "Salesman". There are N cities. The ith city has index = i, id = "City i", and distances equal to a list containing the distance from each city to this one.

The hooks with associated formulas are shown in Table 2. The optimization criterion is the completion time, where this time is equal to the distance traveled. The prerequisities constraint indicates that the salesman cannot return to the city of origin, city 1, until he has visited every other city. The setup duration is obtained by looking up the distance from the previous city in the current city's list of distances.

Job-Shop Scheduling Problem (JSSP)

This problem was originally proposed by [25]. There are M machines and N manufacturing jobs to be completed. Each job has M steps, with each step corresponding to a different specified machine. There is a specified order in which the steps for a certain job must be performed, with each step not able to start until the previous step has ended. The objective is to minimize the end time of the last step completed.

Hook	**Formula**
Optimization Criterion	maxover (*resources*, "*r*", complete (*r*)) - *winStart*
Prerequisites	if (*task.preceedingStep* <> "", list (*task.preceedingStep*))
Execution Duration	*task.duration*
Capability	*task.machine* = *resource.id*
Task Unavailable	if (*task.preceedingStep* <> "", list (interval (*winStart*, taskEndTime (taskNamed (*task.preceedingStep*)))))

Table 3. Scheduling Logic for the Job-shop Scheduling Problem

The task data type, *step*, has fields id (string), duration (number), machine (string), and preceedingStep (string). The resource data type, *machine*, has field id (string).

The hooks with associated formulas are shown in Table 3. The optimization criterion is the makespan. If a task is not the first step in a job, its prerequisite is the step that precedes it, and it is unavailable to be scheduled earlier than the end time of this preceding task. Only the designated machine is capable of performing a task. Since there is only one choice of resource, there is no need for a greedy criterion.

Hook	**Formula**
Optimization Criterion	sumOver (*tasks*, "*t*", (taskStartTime (*t*) - taskSetupTime (*t*)) + if (hasvalue (resourceFor (*t*)), 0, 1E7))
Greedy Criterion	if (hasvalue (*next*), 0, if (hasvalue (*previous*), taskWrapupTime (*task*) - formerWrapupTime (*previous*), taskWrapupTime (*task*) - *winStart*))
Prerequisites	mapover (*tasks*, "*t*", if (*t.latest* < *task.earliest*, *t.id*))
Execution Duration	extra.serviceTime
Setup Duration	distance (*task.location*, if (hasvalue (*previous*), *previous.location*, *extra.depotLocation*))
Wrapup Duration	if (hasvalue (*next*), 0, distance (*task.location*, *extra.depotLocation*))
Task Unavailable	list (interval (*winStart*, *task.readyTime*), interval (*task.latest* + *extra.serviceTime*), *endTime*))
Capacity Contributions	*task.load*
Capacity Thresholds	*extra.capacity*

Table 4. Scheduling Logic for the Vehicle Routing Problem with Time Windows

Vehicle Routing Problem with Time Windows (VRPTW)

This is a more logically complex problem than the previous two. It is described in [26]. There are M vehicles and N customers from whom to pick up cargo. Each vehicle has a limited capacity for cargo, and each piece of cargo contributes a different amount towards this capacity. There is a certain window of time in which each pickup must be initiated, and the pickups require a certain non-zero time. Each vehicle that is utilized starts at a central depot, makes a circuit of all its customers, and then returns to the depot. The objective is to minimize the total distance traveled by the vehicles.

The task data type has fields id (string), load (number), earliest (number), latest (number), and location (xycoor). The resource data type has fields id (string) and capacity (number). A third data type has fields serviceTime (number) and depotLocation (xycoor) and has a single instance named *extra*.

The hooks with associated formulas are shown in Table 4. The optimization criterion is the sum of the distances traveled by the vehicles plus a large penalty that is proportional to the number of unassigned tasks. The execution duration is just the constant defined in *extra*. The setup duration is the distance from the location of the previous pickup, or if this is the first pickup, the distance from the depot. The wrapup duration is nonzero only if this is the last pickup, in which case it is the distance to return to the depot. Each customer is unavailable before the start of its pickup window and after the end of this window allowing the time for pickup. The capacity formulas indicate that the sum of the loads contributed by the different customers cannot exceed a vehicle's capacity.

The greedy criterion and prerequisites formulas are actually not part of the problem specification but directives that help the scheduler find a solution faster. Task B is defined to be a prerequisite for task A if the end of B's pickup window is earlier than the start of A's window. This allows the greedy scheduler to predominantly build the schedule forward in time. The greedy criterion states that if there is a way to fit a new task to schedule earlier than the last task of a resource, that is the preferred assignment. Otherwise, the preference is to find the assignment that packs the schedule as compactly as possible.

4.5 Additional Scheduling Capabilities

While problem specification and automated schedule creation are the core capabilities required by a reconfigurable scheduler, there are additional capabilities that make it more practical and generally applicable. We now provide a brief discussion of some of these additional capabilities that Vishnu possesses.

Dynamic rescheduling is the process of creating a modified schedule from an existing schedule in response to updates to the data. Many real-world scheduling problems require the ability to change the schedule "on the fly", i.e. during the process of executing the schedule. Vishnu provides a few mechanisms to support dynamic rescheduling, including sticky assignments and frozen assignments [27].

In many cases, when doing dynamic rescheduling, the scheduler should attempt to minimize the perturbation from the prior schedule, i.e. maximize *schedule stability*. A *sticky* assignment is a type of soft constraint that penalizes a new assignment for a task for differing from the previous assignment for the task. To implement sticky assignments, the scheduler provides functions priorResource, priorStartTime, etc. that provide information about the prior assignment. This allows the user to explicitly include penalties for maintaining schedule stability on the Optimization Criterion, Greedy Criterion, and Target Start Time hooks. Furthermore, tasks provide data fields that indicate the level of commitment to the current assignment, providing a priority for maintaining this assignment.

The scheduler also allows an assignment to be *frozen*, i.e. creating a hard constraint that the assignment stay the same upon rescheduling. There are two reasons why a user might want to make an assignment frozen, which is a hard constraint, rather than sticky, which is a soft constraint. First, the human scheduler may want to force a particular assignment without giving an override option to the computer. Second, it is much more efficient computationally, since tasks with frozen assignments fall automatically into place without the need to search for the best position. In contrast, tasks with sticky assignments require a scheduling decision to be made, and in general place the same computational burden on the scheduler as a new and previously unscheduled task.

Schedule display and interactive scheduling are other features that enhances Vishnu's usefulness. Vishnu automatically generates color-coded Gantt charts to display a schedule. The colors used for the different assignments and the text to display are specified by the user employing formulas of the same type used to specify the scheduling logic. The display also includes the capability to generate user-defined spreadsheet-like data tables, with the data to display also based on formulas.

Vishnu not only displays the schedule for the user but also allows the user to modify the schedule. It provides various ways for users to make their own assignments, or undo existing assignments, including drag-and-drop. After a user has made his own assignment of a task to a resource, he can make this assignment sticky or frozen so that the scheduler cannot just discard it during the next round of scheduling. In this way, the user and automated scheduler can work together to produce a final schedule.

A composable software architecture makes a reconfigurable scheduler, such as Vishnu, applicable in more situations. For certain applications, a single-user standalone scheduler is sufficient, with any sharing of schedule data done via files. However, other applications require that the scheduler be integrated as part of a larger software system. Although a discussion of the details of the software architecture of Vishnu is beyond the scope of this paper, we do note that Vishnu is composable in a variety of ways. First, it has a web-based deployment mode, in which schedules are stored in a database and are accessible via a web server. This allows shared viewing and editing of schedules across multiple locations. Second, Vishnu has been integrated as the basis for scheduling agents in a multiagent scheduling architecture [28]. This allows multiple schedulers to cooperate on producing schedules, which is important when a scheduling problem is too big and/or too heterogeneous for a single scheduler. Third, Vishnu provides standard interfaces for communication with non-scheduling software, including formats for passing data back and forth and a well-defined interface. This allows easy integration into larger software systems.

5 Evaluation

The traditional metrics for evaluating scheduling algorithms do not measure what Vishnu is trying to accomplish. Traditionally, evaluation involves selecting a set of benchmarks that are instances of the particular problem for which the scheduler was designed. The performance of the scheduling algorithm can be compared to what other algorithms achieve on the same benchmarks, both in terms of quality of solution and the time to reach the solution. This makes sense as a way to compare schedulers targeted to a specific problem.

However, this is not the right way to evaluate Vishnu. The goal of Vishnu is to make it quick and easy to develop optimized scheduling solutions to new scheduling problems. Therefore, the primary metrics should be (a) the ability to solve a wide range of problems with just reconfiguration and (b) the ease of solution development, i.e. the time and effort it takes a user to configure for a particular problem. For certain problems, particularly well-studied benchmarks, Vishnu's generality will mean that its performance based on traditional metrics will be inferior to that of problem-specific schedulers, sacrificing raw speed for flexibility and ease of solution development. For those readers familiar with software development, this tradeoff is analogous to that between high-level programming languages, such as C++ or Java, and low-level languages, such as assembly or microcode.

While there is no obvious way to measure the ease of development or range of problems solved, we can provide sample problems, anecdotes, and an analysis of the capabilities of Vishnu. Of course, traditional measures of scheduling performance still matter for Vishnu (just as for programming languages), so we additionally provide some performance numbers on a few benchmark problems to show that Vishnu passes the threshold for acceptability. Section 5.1 discusses two sample problems logically more complex than the benchmark problems as a way to show how easy it is to develop scheduler for new problems, even if they are complex. Section 5.2 provides speed and optimality performance numbers on some common benchmark problems. Section 5.3 provides a brief analysis of the capabilities of Vishnu not possessed by other reconfigurable schedulers, which allow it to solve a wider range of problems.

5.1 Sample Problems

The following two sample problems demonstrate how a Vishnu scheduler can be easily specified for problems with greater logical complexity than traditional benchmarks. These are just a representative set, and a variety of additional problems are available in the demonstration at the Vishnu web site [29].

Air Marshal Scheduling

Air marshals are people who fly on commercial airline flights and monitor them for terrorist, or other illegal, activities. There are not enough marshals to monitor all flights. Therefore, good scheduling is required to put marshals on as many flights, particularly those designated as high priority, as possible while not putting undue burdens on the marshals.

Each marshal has an airport designated as his home base. He should take a set of flights that form a circuit that eventually brings him back home. The **hard constraints** of the problem are

(a) All flights leave and arrive at their scheduled times.
(b) The marshal must be located at an airport to take a flight that flies from it.
(c) A marshal must arrive at least an hour early for a flight.
(d) A marshal can work at most 14 hours in a day, with a maximum of 8 hours spent on flights.
(e) A marshal cannot work during his time off.

The **soft constraints** are

(a) As many different flights as possible, particularly the high priority ones, should be covered.
(b) Marshals should return home at the end of the current scheduling window.

Air marshal scheduling is a variation of the air crew scheduling problem. Over time, a standard approach for air crew scheduling has emerged where the problem is split into two subproblems, crew pairing and crew assignment (or rostering) [30]. The former creates a set of circuits of flights, each of which ends at the same airport as it begins. The latter determines which crew members to assign to each route/circuit. This approach works well for large-scale problems when there are many flights and many potential crew members. However, it does not scale down well to small numbers of crew members (marshals), in which case building the routes cannot be done independent of knowledge of the crew members' work schedule.

We were able to create an optimizing solution to the air marshal scheduling problem within a day using Vishnu. This is very rapid turnaround time given the complexity of the problem. A version is shown in Table 5. A big benefit of Vishnu is the compactness of the problem representation, requiring only a small number of lines of formulas.

We now describe how the specifications in Table 5 meet each of the constraints for the problem given above, starting with the **hard constraints**.

(a) To ensure that flights only depart and arrive as scheduled requires a combination of formulas on two hooks. Execution Duration constrains each flight to extend for its scheduled length of time, while Task Unavailable constraints a flight to not start before its schedule departure or end after its scheduled arrival.
(b) To ensure that a marshal must be at an airport in order to fly from it involves two hooks. The Setup Duration hook determines at which airport the marshal is located by finding the arrival airport of the marshal's previous flight, or the marshal's home airport if the current flight is his first. If this airport is not the departure airport of the current flight, the formula on the hook evaluates to a

Hook	Formula

Optimization Criterion	sumOver (*tasks*, "*t*", if (hasValue (resourceFor (*t*)), 0,
	if (*t.priority* = 1, 1, if (*t.priority* = 2, 0.2, 0.04)))) +
	sumOver (*resources*, "*r*",
	if (lastTask (*r*).*arrivesWhere* = *resource.home*, 0, 0.5))

Greedy Criterion	*task.departsWhen* - complete (*resource*)
Prerequisites	mapover (*tasks*, "*t2*", if (*task.departsWhen* >= *t2.arrivesWhen* + entry (*t2.arrivesWhere.minConnectTime*, *task.departsWhere.index*), *t2.name*))
Execution Duration	*task.arrivesWhen* - *task.departsWhen*
Setup Duration	if (if (hasvalue (*previous*), *previous.arrivesWhere* = *task.departsWhere*, *resource.home* = *task.departsWhere*), 3600, 999999)
Resource Unavailable	resource.unavailable
Task Unavailable	list (interval (*winStart*, *task.departsWhen*), interval (*task.arrivesWhen*, *endTime*))
Capability	complete (*resource*) + entry (lastTask (*resource*).*arrivesWhere.minConnectTime*, *task.departsWhere.index*) <= *task.departsWhen*
Capacity Thresholds	list (28800, 50400)

Capacity Contributions	list (*task.arrivesWhen* - *task.departsWhen*,
	if (and (hasvalue (*previous*), not (*isReset*)), *task.arrivesWhen* -
	previous.arrivesWhen, *task.arrivesWhen* - *task.departsWhen*) +
	if (and (not (hasvalue (*next*)),
	endTime - *task.arrivesWhen* < 68400.0),
	entry (*task.arrivesWhere.travelTimes*, *resource.home.index*), 0))

Capacity Resets	if (and (hasvalue (*previous*), *task.departsWhen* - *previous.arrivesWhen* > 36000, andOver (tasksFor (*resource*), "*t*", or (capacityReset (*t*, 1.0) = 0.0, taskStartTime (*t*) >= taskStartTime (*task*)))), list (1E8, 1E8), list (0, 0))
Auxilliary Tasks Before	list (entry (entry (if (hasValue (lastTask (*resource*)), lastTask (*resource*).*arrivesWhere*, *resource.home*).*connectSchedules*, *task.departsWhere.index*).*latestConnect*, (*task.departsWhen* - *winStart*) / 1800 + 1))

Table 5. Scheduling Logic for the Air Marshal Scheduling problem

very large number, making it impossible to assign the flight to the marshal. The Auxilliary Tasks Before hook checks to see if there is a connecting flight in the case that the two airports are different. If so, it specifies to assign this flight as a way to position the marshal for the flight of interest.

(c) To ensure that marshals are at least an hour early for their flight, the Setup Duration evaluates to 3600 seconds when the marshal is at the right airport.

(d) Enforcement of the flight-time and working-hours limitations relies on the three capacity-related hooks. The limits are set to 8 hours and 14 hours respectively by the Capacity Thresholds hook. The Capacity Contributions hook specifies that the additional flight time is the length of the flight, while the additional time worked is the length of the flight if this is the first flight since a reset or, otherwise, the length of the flight plus the time spent between flights.

(e) The marshals' time off from work is protected from assignments by the Resource Unavailable hook.

The Optimization Criterion formula embodies the **soft constraints**, penalizing 1.0, 0.4 and 0.02 for each flight not covered with priority 1, 2 and 3 respectively, plus 0.5 for each air marshal not home at the end of the scheduling window. Like in the vehicle routing problem discussed above, other hooks assist in achieving the goals of the Optimization Criterion. The Greedy Criterion hook specifies that the greedy scheduler should pick the marshal for a flight that would have the shortest waiting from his last flight to this flight. The Prerequisites hook says that a flight cannot be scheduled until all flights that could possibly feed it by bringing a marshal to its originating airport are all scheduled. At that point, the greedy scheduler knows which marshals will be at the airport and therefore can make an informed decision. The Capability and Capacity Contributions hooks both help get marshals home by prohibiting them from traveling too far away without enough time remaining to get home. Specifying this as a hard constraint rather than just a greedy preference makes it much less likely that a marshal will not make it home.

Battlefield Supply Scheduling

An agile military requires that its combat units be able to fight for long periods of time without running out of supplies. To accomplish this, supply trucks drive around the battlefield, although preferably not the area of actual fighting, delivering their cargo to the vehicles of the combat units. Each unit has multiple vehicles all geographically clustered, so it is efficient for one supply truck (or a small number of supply trucks) to deliver all the supplies of a particular unit. The problem is how to schedule the deliveries of multiple supply trucks to the different combat units.

There are actually different battlefield supply scheduling problems with different constraints for each broad class of supplies. Here, we consider two different classes of supplies: fuel and ammunition (ammo). The problem specifications shown in Table 6 show the formulas separately for food and ammunition for those hooks where the formulas differ. One of the big benefits of Vishnu is that it allows easy adjustment for the idiosyncracies of variations on a single problem, as illustrated by the ease with which we can specify different problem specification, and hence schedulers, for the two types of supplies.

The **hard constraints** of the problem are

(a) Each delivery requires five minutes to execute.

(b) The supply trucks travel a time equal to the distance between the points divided by the truck's average speed.

(c) There is a fixed window of time for each delivery to occur.

(d) (ammo) There are different types of ammunition, so each request needs to be matched with the types and quantities available on a supply truck.

Hook	Formula
Greedy Criterion	*omitted for brevity*
Target Start Time	*task.desiredTime* - 300
Prerequisites	mapover (*tasks*, "*t*", if (*t.DesiredTime* ¡ *task.DesiredTime*, *t.id*))
Execution Duration	300
Setup Duration	distance (if (hasValue (*previous*), *previous.location*, *resource.initialLocation*), *task.location*) / *resource.speed* * 3600
Capability (Fuel)	or (*task.refillAmount* = 0, *task.recipient* = *resource.id*)
Capability (Ammo)	and (or (*task.refillAmount* = 0, *task.recipient* = *resource.id*), hasValue (find (*resource.loadedAmmo*, "*a*", *a.type* = *task.desiredType*)))
Task Unavailable	list (interval (*winStart*, *task.earliest*), interval (*task.latest*, *endTime*))
Cap Thresholds (Fuel)	*resource.initialGallons*
Cap Thresh (Ammo)	mapOver (*resource.loadedAmmo*, "*a*", *a.rounds*)
Capacity Contributions (Fuel)	if (*task.refillAmount* > 0, 0, withVar ("f", find (*task.fuelProfile*, "f2", and (taskStartTime (*task*) >= *f2.startTime*, taskStartTime (*task*) <= *f2.endTime*)), (*f.endValue* - *f.startValue*) * (taskStartTime (*task*) - *f.startTime*) / (*f.endTime* - *f.startTime*) + *f.startValue*))
Capacity Contributions (Ammo)	mapOver (*resource.loadedAmmo*, "*a*", if (and (*task.refillAmount* = 0, *a.type* = *task.desiredType*), max (1, min (*task.desiredQuantity*, capacityRemaining (*resource*, *task.DesiredTime*, indexOf (resourceFor (*task*).*loadedAmmo*, "*a*", *a.Type* = *task.desiredType*)))), 0))
Capacity Resets (Fuel)	*task.refillAmount*
Cap Resets (Ammo)	mapOver (*resource.loadedAmmo*, "*a*", if (*a.type* = *task.desiredType*, *task.refillAmount*, 0)))

Table 6. Scheduling Logic for the Battlefield Supply Scheduling problem

(e) (ammo) The requested amount is a maximum, and less can be delivered if the full amount is not available.

(f) (fuel) A different amount of fuel is required depending on when the fuel is delivered. The recipient continually consumes fuel while awaiting the delivery, and hence needs more fuel the later the delivery. A piecewise linear fuel profile specifies how much is required.

(g) The supply trucks themselves are restocked at fixed times and locations by fixed amounts.

The **soft constraints** are

(a) The schedule should fill as many requests as possible.

(b) For each delivery, the time should be as close as possible to the desired time with preference to earlier than later.

(c) The schedule should minimize travel time/distance of the supply trucks.

The problem specification in Table 6 satisfies the **hard constraints** as follows.

(a) The time for a delivery is specified by the Execution Duration hook.

(b) The travel time is specified by Setup Duration, with the prior location being either the location of the last delivery or, if this is the first delivery, then the truck's initial location.

(c) The delivery window is specified by Task Unavailable.

(d) (ammo) To ensure that the truck carries the right type of ammunition, the Capability formula checks that the desired type is in the truck's list of initial supplies. The quantities of the different types of ammunition are tracked using the capacities, with each dimension of the capacity corresponding to a particular type of ammunition. Capacity Thresholds indicates that the maximum of each type of supply is as given at initialization. Capacity Contributions removes the amount delivered from the truck's stock.

(e) (ammo) Capacity Contributions sets the amount delivered to be the minimum of the amount requested and the amount left on the truck but never less than one.

(f) (fuel) Each task has an associated fuel profile that is a list with each element corresponding to one piece of the piecewise linear function. Capacity Contributions finds the right piece and interpolates between the endpoints.

(g) Capacity Resets restores the inventory on the supply truck.

With respect to the **soft constraints**, for the particular application it was more important to have fast turnaround than fully optimized schedules. So, we used just a single greedy schedule generated from a random task ordering. This meant that we had no need for an optimization criterion. The Greedy Criterion is not shown in Table 6 because it is a bit too long and is not particularly instructive. However, it is simple in concept. There are four penalty terms. One penalizes additional travel time. A second term penalizes the deviation of the actual delivery time from the desired time, with a much heavier weight for being late. The third rewards putting two supply tasks from the same unit consecutively on a resource. The fourth penalizes putting two supply tasks from different units contiguously on a resource. The Prerequisites formula specifies to perform the scheduling of tasks in the order of their desired delivery times, although there is much room for randomness in the ordering with many tasks having the same desired time.

This was a successful application of Vishnu [31]. We were able to quickly build different types of supply schedulers and easily adjust them to changing specifications of the problem. We were also able to easily integrate the schedulers into a larger multiagent system for managing supplies.

5.2 Performance Results

We now provide performance numbers on some benchmark classic scheduling problems. As discussed above, Vishnu cannot always compete in terms of speed and optimality with problem-specific algorithms. However, the results on these benchmarks still provide some idea of how Vishnu will perform on other problems similar in scale that do not have existing solutions.

Problem Name	Population Size	Evaluations	Optimal Score	Median Score	Average Score	Avg Time (M:S)
JSSP-mt06	1000	5000	55	55	55	0:01
JSSP-mt10	500	10000	930	1012	1010	0:06
JSSP-mt10	5000	100000	930	982	982	1:04
JSSP-mt10	50000	1000000	930	961	962	10:21
TSP-bays29	5000	140000	2020	2028	2042	0:11
VRPTW-c101	100	200	827.3	828.9	828.9	< 0:01

Table 7. Summary of experimental results

Table 7 summarizes the results. All the experiments involved ten runs of Vishnu on the problem, reporting the mean and median scores of the resulting schedule produced and the mean time for the run to complete. In addition, the table shows the two parameters that need to be selected for the genetic algorithm, the population size and the number of evaluations performed, as well as the score for the known optimal solution to the problem. All the runs were made on a 2.8GHz Pentium 4 processor.

The data sets we used were

- Muth-Thompson 6x6 (mt06) and Muth-Thompson 10x10 (mt10) are two standard benchmarks for job-shop scheduling [25]. The former has 6 machines/resources and 6 jobs, and hence 36 tasks. The latter has 10 machines and 10 jobs, and hence 100 tasks. They are available from OR-Library (under the names ft06 and ft10).
- The bays29 traveling salesman problem is a 29-city symmetric problem available at the TSPLIB web site [32].
- The c101 vehicle routing problem with time windows is one of the Solomon benchmarks [26]. There are 100 pickups/tasks and 25 vehicles/resources. This is one of the instances where the time windows are very tight.

Job-Shop Scheduling Problem - The Muth-Thompson 6x6 problem is not a difficult problem, but it is also not trivial given the 36 tasks to schedule. Therefore, the ability of the automated scheduler to consistently find an optimal solution in under 5000 evaluations and 1 second reflects well on both the effectiveness of the genetic search algorithm (which needs to explore only a very small fraction of the search space of task orderings) and the efficiency of the greedy schedule builder (which despite its generality can still build each schedule in approximately 0.2 msec).

The Muth-Thompson 10x10 problem is much more difficult. In fact, despite much attention, there was no solution proven to be optimal until relatively recently [33]. The results show three sets of experiments with three different genetic algorithm parameters. The first set of parameters has a small population size, leading to fast convergence. The second set of parameters is a factor of ten longer before convergence than the first, and the third a factor of ten longer than the second. This illustrates two properties of our reconfigurable scheduler (and of many pure genetic algorithms). First, on difficult problems it can quickly find a reasonable solution (within

8.6% of the optimum in 6 seconds and within 5.6% of the optimum in a minute) but, even with much longer runs, may not find the optimum solution. Second, there is a tradeoff between the search time and the expected quality of the solution, which can be selected explicitly via choice of the parameters.

Traveling Salesman Problem - The 29-city problem used is a small one, but it is sufficient to show that our algorithm clearly cannot compete with custom designed algorithms. For example, the Concorde algorithm [34] completes the bays29 problem in 0.13 seconds, as compared to our algorithm taking over 11 seconds on a much faster machine. The fact that our algorithm is a genetic algorithm is not the primary issue, as a variety of competetive genetic algorithms for the traveling salesman problem attest [35]. There are two main reasons for our algorithm's shortcomings, both of which are related to the fact that the traveling salesman problem is purely a routing problem rather than a true scheduling problem (insofar as time-based constraints are not involved). First, there is a lot known about the structure of the search space (particularly when the distances are symmetric), and large performance gains can be achieved by designing an algorithm that exploits this structure. Second, there is a large software overhead, since Vishnu builds a full schedule as part of the evaluation process, while a particular route can be evaluated just by summing the distances.

Vehicle Routing Problem with Time Windows (VRPTW) - The performance of our algorithm on this problem is respectable, scheduling all 100 tasks in a nearly optimal fashion in less than a second. Note that the genetic search required little work, one pseudo-generation, beyond the initial population. The use of a steady-state genetic algorithm helps the search proceed this quickly, since the best children can immediately produce their own offspring. Even more important to the performance is the fact that the Prerequisites and Greedy Criterion formulas were specified so as to tightly pack the schedule.

Overall Conclusions - The experimental results do support the premise that our reconfigurable scheduler can provide reasonable performance on a range of problems. The poor performance on the traveling salesman problem is, in some sense, the exception that proves the rule. Not only did it require many researcher-years to discover the much better solutions (which is a magnitude of effort that cannot be devoted to every scheduling problem), but more importantly it is also a highly "atypical" scheduling problem because of its very simple constraints, the lack of any concept of time, and the existence of exploitable structure in its search space.

5.3 Analysis of Capabilities

Comparing different reconfigurable schedulers with regards to their generality and flexibility to solve a large variety of problems is an inexact and difficult task. There are no sets of benchmark problems where we can say that if a reconfigurable scheduler can solve all problems in set X then it achieves level Y of reconfigurability. There is a wide range of different types of scheduling problem features that a reconfigurable problem should handle, many of which were discussed in Section 4.3 (multitasking, multiresourcing, capacities, etc.). At this point, the best that we can do is list some of the capabilities that help distinguish Vishnu from other reconfigurable schedulers.

These features include

- Auxilliary Tasks - While their only use in the example problems was to represent connecting flights, their most powerful use is allowing the decomposition of a complex task into multiple subtasks with different requirements, yet handling them as a single entity in the scheduling process.
- Capacity Resets - Capacities are a more fluid concept than just adding up the individual contributions and comparing to a threshold, and resets are an important part of this extra complexity.
- Dynamically Computed Constraints - Quantities such as task setup and execution times, capabilities, task availabilities, and capacity contributions all can depend on the schedule and can change as the schedule gets built.
- Algorithmically Specified Constraints - Constraints specified using a real, though simple, programming language can express a wider range of possibilities than the mathematically specified constraints of mathematical programming or just picking among some prespecified types of constraints.
- Scheduler Directives - These allow the user to guide the scheduler to produce better solutions faster, and do so in an easily understandable way and within paradigm of the problem specification framework. The only scheduler controls that are not part of this framework are the the population size and number of evaluations, which the user sets by selecting values rather than formulas.
- Additional Capabilities - Dynamic rescheduling, interactive scheduling, and software composability are all important for building real-world scheduling systems.

Even more important than the particular features already in Vishnu is that the infrastructure allows easy addition of new features and capabilities as required. If we encounter a scheduling problem that requires a capability not currently contained in Vishnu, we can add a new hook and modify the greedy scheduler to handle this new constraint. Since the greedy scheduler does not rely on any idiosyncratic search algorithm, it is generally the case that it can be modified to incorporate the new constraints. We have developed Vishnu this way, starting with very simple functionality and expanding the capabilities as needed to solve new problems requiring new features. We have yet to encounter a problem that we have not been able to handle by adding new functionality to Vishnu, without affecting existing capabilities.

6 Conclusion

We have developed a powerful framework for representing scheduling problems, and we have built a reconfigurable scheduler, Vishnu, that can find an optimized solution for any problem specified in this framework. The approach we have used involves a genetic algorithm feeding task orderings to a greedy schedule builder as its method of finding optimized schedules. Hooks and formulas provide a method for users to define customized scheduling logic. This approach allows easy introduction of new capabilities into the scheduler and problem specification framework, thus allowing us to make a reconfigurable scheduler that is particularly powerful in its ability to handle a wide range of scheduling problems with a wide variety of scheduling logic.

The major benefit of Vishnu is that it makes development of optimized scheduling for a wide range of problems simple and inexpensive. There is a vast array of scheduling problems that are currently solved using manual or non-optimized scheduling. For most of these problems, our reconfigurable scheduler could provide a simple and inexpensive optimized scheduling solution.

References

1. Fourer, R., Gay, D., Kernighan, B.: AMPL: A Modeling Language for Mathematical Programming. Duxbury Press, Belmont, CA (1993)
2. Van Hentenryck, P.: The OPL Optimization Programming Language. MIT Press, Cambridge, MA (1999)
3. Montana, D.: A reconfigurable optimizing scheduler. In: Proceedings of the Genetic and Evolutionary Computation Conference. (2001) 1159–1166
4. Montana, D.: Optimized scheduling for the masses. In: Genetic and Evolutionary Computation Conference Workshop Program. (2001) 132–136
5. Bisschop, J., Meeraus, A.: On the development of a general algebraic modeling system in a strategic planning environment. Mathematical Programming Study **20** (1982) 1–29
6. Bixby, R., Fenelon, M., Gu, Z., Rothberg, E., Wunderling, R.: MIP: Theory and practice - closing the gap. In Powell, M., Scholtes, S., eds.: System Modelling and Optimization: Methods, Theory, and Applications. Kluwer (2000) 19–49
7. Dincbas, M., Van Hentenryck, P., Simonis, H., Aggoun, A., Graf, T., Berthier, F.: The constraint logic programming language CHIP. In: Proceedings of the International Conference on Fifth Generation Computer Systems. (1988) 693–702
8. Colmerauer, A.: An introduction to Prolog III. Communications of the ACM **28**(4) (1990) 412–418
9. Davis, G., Fox, M.: ODO: A constraint-based architecture for representing and reasoning about scheduling problems. In: Proceedings of the 3rd Industrial Engineering Research Conference. (1994)
10. Van Hentenryck, P., Perron, L., Puget, J.F.: Search and strategies in OPL. ACM Transactions on Computational Logic **1**(2) (2000) 285–320
11. McIlhagga, M.: Solving generic scheduling problems with a distributed genetic algorithm. In: Proceedings of the AISB Workshop on Evolutionary Computing. (1997) 85–90
12. Raggl, A., Slany, W.: A reusable iterative optimization library to solve combinatorial problems with approximate reasoning. International Journal of Approximate Reasoning **19**(1-2) (1998) 161–191
13. Smith, S., Becker, M.: An ontology for constructing scheduling systems. In: Working Notes of 1997 AAAI Symposium on Ontological Engineering. (1997)
14. Rajpathak, D., Motta, E., Roy, R.: A generic task ontology for scheduling applications. In: Proceedings of the International Conference on Artificial Intelligence. (2001) 1037–1043
15. Montana, D.: Introduction to the special issue: Evolutionary algorithms for scheduling. Evolutionary Computation **6**(1) (1998) v–ix
16. Cantu-Paz, E.: Efficient and Accurate Parallel Genetic Algorithms. Kluwer (2000)

17. Davis, L.: Job shop scheduling with genetic algorithms. In: Proceedings of the First International Conference on Genetic Algorithms. (1985) 136–140
18. Homberger, J., Gehring, H.: Two evolutionary meta-heuristics for the vehicle routing problem with time windows. INFORMS Journal on Computing **37**(3) (1999) 297–318
19. Syswerda, G.: Schedule optimization using genetic algorithms. In Davis, L., ed.: Handbook of Genetic Algorithms. Van Nostrand Reinhold (1991) 332–349
20. Whitley, D., Starkweather, T., Fuquay, D.: Scheduling problems and traveling salesmen: The genetic edge recombination operator. In: Proceedings of the Third International Conference on Genetic Algorithms. (1989) 133–140
21. Goldberg, D., R. Lingle, J.: Alleles, loci, and the traveling salesman problem. In: Proceedings of the First International Conference on Genetic Algorithms. (1985) 154–159
22. Grefenstette, J., Gopal, R., Rosmaita, B., van Gucht, D.: Genetic algorithms for the traveling salesman problem. In: Proceedings of the First International Conference on Genetic Algorithms. (1985) 160–165
23. Giffler, B., Thompson, G.: Algorithms for solving production-scheduling problems. Operations Research **8**(4) (1960) 487–503
24. Beasley, J.: OR-Library: Distributing test problems by electronic mail. Journal of the Operational Research Society **41**(11) (1990) 1069–1072
25. Muth, J., Thompson, G.: Industrial Scheduling. Prentice Hall, Englewood Cliffs, NJ (1963)
26. Solomon, M.: Algorithms for the vehicle routing and scheduling problem with time window constraints. Operations Research **35** (1987) 254–265
27. Montana, D., Brinn, M., Moore, S., Bidwell, G.: Genetic algorithms for complex, real-time scheduling. In: Proceedings of the IEEE International Conference on Systems, Man, and Cybernetics. (1998) 2213–2218
28. Montana, D., Herrero, J., Vidaver, G., Bidwell, G.: A multiagent society for military transporation scheduling. Journal of Scheduling **3**(4) (2000) 225–246
29. Montana, D.: Vishnu reconfigurable scheduler home page (2001) http://vishnu.bbn.com.
30. Fahle, T., Junker, U., Karisch, S., Kohl, N., Sellmann, M., Vaaben, B.: Constraint programming based column generation for crew assignment. Journal of Heuristics **8**(1) (2002) 59–81
31. Hussain, T., Montana, D., Brinn, M., Cerys, D.: Genetic algorithms for UGV navigation, sniper fire localization and unit of action fuel distribution. In: Military and Security Applications of Evolutionary Computation (MSAEC) Workshop, part of GECCO. (2004)
32. Reinelt, G.: TSPLIB (2001) http://www.iwr.uni-heidelberg.de/groups/comopt/software/TSPLIB95/.
33. Carlier, J., Pinson, E.: Adjustment of heads and tails for the job-shop problem. European Journal of Operations Research **78** (1994) 146–161
34. Applegate, D., Bixby, R., Chvatal, V., Cook, W.: TSP cuts which do not conform to the template paradigm. In Junger, M., Naddef, D., eds.: Computational Combinatorial Optimization. Springer (2001) 261–304
35. Watson, J., Ross, C., Eisele, V., Denton, J., Bins, J., Guerra, C., Whitley, D., Howe, A.: The traveling salesrep problem, edge assembly crossover, and 2-opt. In: Parallel Problem Solving from Nature V. (1998) 823–832

Evolutionary Algorithm for an Inventory Location Problem

Ek Peng Chew, Loo Hay Lee and Kanshukan Rajaratnam

Department of Industrial & Systems Engineering, National University of Singapore, Singapore 119260, Singapore

Summary. This paper deals with minimizing the cost in a joint location-inventory model with a single supplier supplying to multiple capacitated distribution centers. The distribution center faces stochastic demands from multiple retailers. The problem is to determine how to assign retailers to distribution center within the service level constraints. The costs considered include the transportation cost, inventory holding cost and ordering cost. We develop an adaptive real-coded genetic algorithm to solve the problem. We conduct few experiment runs to compare the performance of the proposed method with some existing methods which include the simple genetic algorithm, the column generation method and the greedy method. For the non-capacitated case, the method shows very promising results with respect to both time and quality of the solutions. Similarly for the capacitated case, where the column generation method cannot be applied, the model is also significantly better than all the other methods, especially when the problem size is big.

1. Introduction

This paper studies a distribution problem with joint inventory and location decisions. The system consists of a number of capacitated distribution centers (DCs) which are replenished by a single plant and they are serving their assigned retailers (customers) where the demands are stochastic. The DC and its assigned retailers must satisfy two service requirements. The first requirement is that the delivery leadtime (or the distance between the DC and the retailers) should not exceed a certain threshold value. The second requirement is that the DC should keep

E.P. Chew et al.: *Evolutionary Algorithm for an Inventory Location Problem,* Studies in Computational Intelligence (SCI) **49**, 613–628 (2007)

some level of safety stock to meet a certain fill rate. The problem is to determine which retailer is to be allocated to a particular DC and how much stock to carry in each DC such that the overall costs can be minimized and the service requirements can be satisfied.

When a DC is serving many retailers, potential saving can be achieved through the reduction of the safety stock due to risk pooling. Moreover, consolidating many retailers at a single DC will also reduce the total cycle stock. However, this may potentially lead to higher transportation cost since some of the retailers might be far away from the DC. Therefore we need to balance the transportation cost and the inventory holding cost. In addition, the capacity of DCs also has to be considered.

Formally, the problem is stated as follows: Given a set of DCs each having its own capacity and a set of retailers with stochastic demands, determine which retailer is to be allocated to which DC by not violating the service requirements. Furthermore, determine how often to reorder as well as the level of stock to be maintained at each DC at the lowest system cost. The costs consider include the transportation cost, the inventory holding cost and the ordering cost. The transportation cost includes both the costs of inbound and outbound transportation of the DC. The inventory holding cost includes the holding costs for the average cycle stock and the safety stock. The ordering cost is incurred for each order placed by the DC to the plant. The set up cost for the DC is ignored.

Inventory theory deals with developing and evaluating policies for ordering and fulfillment process at the DC. These are generally evaluated on the service levels, inventory and shortage costs (Hopp and Spearman 2000, Silver et al. 1998). On the other hand, location theory focuses on determining the number of DCs and their locations with respect to various customers or retailers. Daskin and Owen (1999) provide a review of facility location modeling. Chan (2001) also provides an overview of the location theory. However, solving these two problems independently will lead to a suboptimal solution.

Some works have been done on considering the inventory and location problems jointly. Erlebacher and Meller (2000) solve a joint location-inventory model but that model is highly non-linear and takes 117 hours to be solved on a Sun Ultra SPARCstation. Shen (2000), Daskin et al. (2001) and Shen et al. (2003) conduct a study at Chicago blood bank which looks into the setting up of regional DCs at the hospitals distributing the platelets on a daily-need basis. However, the proposed methods can only solve two special cases. The first case assumes the ratio between the variance of the demand and the mean is identical for all retailers while the second case assumes the demand has zero variance. The problem is first formulated as a nonlinear integer programming model and it is later transformed into a set-covering integer programming model. The pricing problem that must be solved as part of the column generation method for the set-covering model involves a non-linear term in the objective function of the location-allocation model. Shu et al. (2002) extend the work to a more general demand case. They propose the use of the primal and dual approach to solve the general model and use the variable fixing technique to solve the sub modular function minimization problem.

In this paper, we consider an integrated approach to model both the inventory and location cost simultaneously. We extend the model by considering the capacity of a DC. We propose an adaptive real-coded genetic algorithm (ARGA) as a solution procedure to solve this problem.

The remainder of the paper is organized as follows. In Section 2, we develop the cost model and discuss the properties associated with the model. The computational procedures and results are presented in Sections 3 and 4 respectively. In Section 5, the conclusion and future direction of work are presented.

2. Model Formulation

For our problem, we need to minimize the cost incurred when allocating DCs to serve all retailers. The cost includes the transportation cost, inventory holding cost and ordering cost. Furthermore, the DC capacity needs to be considered when allocating a DC to retailers.

To model this distribution problem, we define the following variables:

Inputs and Parameters

I set of all retailers

J set of all DCs

J_i set of all DCs that can serve retailer i for each $i \in I$ (not violating the service level requirement by the delivery leadtime)

μ_i Annual mean demand at retailer i for each $i \in I$;

σ_i^2 Variance of the daily mean demand at retailer i for each $i \in I$;

α_j Unit transportation cost between plant/supplier to DC j for each $j \in J$;

β_{ij} Unit transportation cost between DC j to retailer i for each $i \in I$ and $j \in J$;

d_{ij} Unit transportation cost per from plant to DC j and from DC j to retailer i for each $i \in I$ and $j \in J$; i.e. $d_{ij} = \alpha_j + \beta_{ij}$ for each $i \in I$ and $j \in J$;
W_j Capacity of DC j for each $j \in J$;
h_j Holding cost at DC j for each $j \in J$;
L_j Lead time at DC j for each $j \in J$;
A_j Ordering cost at DC j for each $j \in J$;
Q_j Cycle stock held at DC j for retailers i for each $j \in J$;

Decision Variables

$X_{ij} = 1$, if retailer i is served by DC j and 0 otherwise for each $i \in I$ and $j \in J$.

The problem can be formulated as follows.

$$Min \sum_{j \in J} \left(\begin{array}{l} \sum_{i \in I} d_{ij} \mu_i X_{ij} + h_j \left(\frac{Q_j}{2} + z_\alpha \sqrt{\sum_{i \in I} \sigma_i^2 L_j X_{ij}} \right) \\ + A_j \frac{\sum_{i \in I} \mu_i X_{ij}}{Q_j} \end{array} \right) \tag{1}$$

subject to

$$0 <= Q_j <= W_j - (z_\alpha + z_\beta) \sqrt{\sum_{i \in I} \sigma_i^2 L_j X_{ij}} \tag{2}$$

$$\sum_{j \in J_i} X_{ij} = 1 \tag{3}$$

$$X_{ij} \in \{0,1\} \tag{4}$$

The objective function (1) minimizes the system cost. The first term in the objective function represents the total transportation cost within the system and it includes both the transportation costs between the plant to the DCs as well as from the DCs to the retailers. The second term represents the holding costs in the system and the third term accounts for the reorder cost for placing an order from the DC to the plant.

Equation (2) represents the capacity constraint for the DC. z_α is the safety factor to buffer against the demand uncertainty (to satisfy the fill rate α) while z_β is a safety factor to prevent against the inventory overflow due to the possibility of low demand (probability of inventory overflow should be less than β). Constraint (3) ensures that one retailer is allocated to only one DC which satisfies the service requirement based on the delivery leadtime.

The solutions to the model will consider the tradeoff between the risk pooling effect and the transportation cost. For example, it may be better to assign many retailers to a DC although some retailers may be far from the DC due to the potential saving from the risk pooling. When there is no capacity limit on DCs, it is always optimal to have cycle stock equals to Economic Order Quantity (EOQ). However, if there is a capacity limit on the DC, and in the event that the EOQ-determined cycle stock exceeds the allowable capacity of the DC (i.e., violating equation 2), the models might provide three possible solutions according to different scenarios. The first one is that some retailers will be reallocated to other DCs, and the cycle stock of the DC remains at the EOQ level. The second one is that no retailers are reallocated, but the cycle stock is reduced to the level which satisfies the DC capacity level (i.e., equation 2). The third one is the hybrid of the first and second solutions.

As this problem is highly nonlinear and involves integer variables, the traditional mathematical programming approach (MIP or NLP) is not able to provide satisfactory solutions. Hence, we propose to use genetic algorithm as a solution approach and the detail discussion on the approach will be presented in the next section.

3. Proposed methodology

Genetic algorithm (GA) is known to be a good method to tackle combinatorial problems. Song et al. (1996) gives a good overview on the general characteristics of GA. In GA procedure, there are two phases. The first phase involves selecting the set of parameters for GA, i.e., type of GA, mutation probability, crossover probability,... etc. The second phase involves finding the solution to the optimization problem by using the GA with parameters chosen in phase 1. A good choice of parameters in the first phase can lead the search to a good local optimum faster while a bad choice might cause the solutions either to be trapped at a poor local optimal solution or take a longer time to converge. The choice of parameter is usually decided by past experience or trial and error. Some of the parameters need to be varied during the run so as to achieve a better solution, and this approach is known as the adaptive GA (Fogarty 1989, Srinivas and Patnaik 1994, Herrera and Lozano 1996, Lis and Lis 1996, Lee and Fan 2002).

In this paper, we will adopt the adaptive real-coded genetic algorithm (ARGA) approach proposed by Lee and Fan (2002). There are two phases in the ARGA procedure. In the first phase, a screening experiment based on the factorial design on GA parameters is conducted to identify critical and time-sensitive parameters which will influence the searching performance of the GA. Then in the second phase, these critical and time-sensitive parameters will be adaptively tuned according the most recent information on the searching performance when ARGA is used to search for the optimal solution.

The implementation details of the two phases of the ARGA is summarized as follows.

Phase 1 Screening procedure

1. Choose high and low levels for all the parameters of GA that we are interested in investigating.
2. Run all the GAs with different parameter combinations to solve the problem at different number of iterations.
3. Identify the parameter that will influence significantly on the search performance of the GA.
4. For those parameters that are significant, compute the variance of their effect across different number of iterations. Those parameters with high variance on the effect will be identified as time-sensitive parameters and they need to be adaptively tuned during the GA search process.

Phase 2 Searching procedure

1. Choose few levels for those time-sensitive parameters identified in the phase 1 screening procedure, and form different combinations of parameters. Denote m_i as parameter combination i, $i = 1,2,\ldots,M$.
2. Generate initial population P_o with N individuals. Fix the computing budget of one iteration at C (generations). For each m_i, the belief index, $b(m_i) = 1/M$, $i = 1,2,\ldots,M$.
3. If the terminating condition is met, end of procedure, otherwise for $i = 1, 2,\ldots,M$, run the GA with parameter combination m_i for $q(m_i) = C.b(m_i)$ generations. All the M GAs have the same initial population P_o. Denote $P(m_i)$ as the final population obtained by GA with parameter combination, m_i.
4. Form a population $P = \bigcup_{m_i} P(m_i)$.

5. Determine whether the best fitness value of population P has improved. If yes, select the top N individuals from P to form the new population P_{top} and goto Step 6. If no, goto Step 7.
6. Calculate the performance of each parameter combination, $f(m_i)$ which equals to the percentage of P_{top} that is from $P(m_i)$ and update belief index.

$$b'(m_i) = \frac{f(m_i).b(m_i)}{\sum_{i=1}^{M} f(m_i).b(m_i)}, i = 1,2,...,M$$

   Let $b(m_i)=b'(m_i)$ and goto step 3.
7. Reset $b(m_i) = 1/M$, $i = 1,2,\ldots,M$ and goto 3.

In running GAs in step 2 of phase 1 and step 3 of phase 2, we use the following procedure.

GA procedure

1. Initialize a population of chromosomes, which is called the parent population.
2. Evaluate the fitness of each chromosome in the population.
3. Randomly select the chromosomes from the population to form a team with a given team size.
4. Select the best two chromosomes from the team to form parent chromosomes.
5. Create new chromosomes using either crossover operator or mutation operator. The crossover probability will determine the choice of the operators to be used.
6. Repeat steps 2,3,4,5 until a children population is formed.
7. Replace the worst performers of the parent population by the top performers of the children population without duplication. The replacement percentage will determine the number of parents to be replaced. The modified population will be used as the new parent population in the next generation.
8. Repeat steps 2-7 until the termination condition is met.

In implementing the ARGA in solving the inventory and location problem, we use the following chromosome representations and operators.

Chromosome Representation

In our paper, we use a direct representation of solution. The position of each gene in the chromosome represents a retailer and stores an integer value that represents a DC number. An example is seen below in Fig. 1.

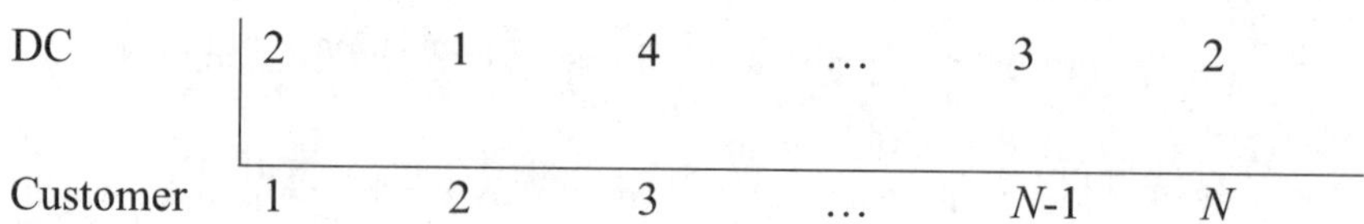

DC	2	1	4	…	3	2
Customer	1	2	3	…	*N*-1	*N*

Fig. 1. Chromosome Representation

The Fig. 1 shows that customer 2 has been allocated to DC 1 and customer *N*-1 has been allocated to DC 3. Note that we have not represented cycle stock quantity in the chromosome. This value can be easily computed in equation (5) once the retailers assignment to DCs is known (i.e., X_{ij}s are all known).

$$Q_j = \min\left(\sqrt{\frac{2\left(\sum_{i\in I}\mu_i X_{ij}\right)\left(A_j\right)}{h_j}},\ \left(W_j - \left(z_\alpha + z_\beta\right)\sqrt{\sum_{i\in I}\sigma^2 L_j X_{ij}}\right)\right) \forall j \in J \quad (5)$$

Crossover Operator

In the crossover operator, we swap certain genes between the parents in the hope that the offspring would have the desirable features of both their parents. One point and two point crossovers have been widely used and presented in the literature. One point crossover is when a single gene is randomly chosen and all genes after the chosen genes are exchanged by the parents to produce the offspring. In a two point crossover, two random genes are chosen and all genes between these two random genes are swapped by the parents to generate the offspring. However, we propose to use the uniform crossover introduced by Syswerda (1989). Uniform crossover uses a mask or template to determine if the genes are to be swapped between the two parents. The mask is of the same length as the chromosomes and is assigned with random numbers between 0 and 1. If the number is less than a predetermined crossover rate, the genes are exchanged and if the number is greater than the crossover rate, the genes are not swapped by the parent chromosomes. Since crossover rates are symmetric at 0.5, we allocate crossover rates between 0 and 0.5. The choice of crossover rate is important as a high crossover rate will lead to wide range of solutions and may overshoot the optimal solution space. A low crossover rate will produce children with little variation from the parents and will take longer time to find the optimal solution.

At each point of crossover, Q_j is calculated using equation (5). If Q_j is negative (this means the capacity constraint is violated), a customer from the DC j is randomly allocated to another DC which does not violate the capacity constraint.

Mutation Operator

Mutation operator brings in fresh solutions to the set of population. We use the mutation process recommended by Thomas Back (1993), where each gene has an equal probability of mutation. For each gene, we generate a random number between 0 and 1 and if this random number is larger than the mutation rate, the gene is mutated and if smaller, the gene is not mutated. The mutation rate is chosen to be a low value (e.g. 0.1) so that the offspring solution will not be too different from the parents. If the gene (retailer) is chosen to be mutated, it will be randomly assigned to a new DC. If capacity constraints are not satisfied, the gene is then allocated to another DC randomly.

Population Initialization

When initializing population, we need to select random assignment of DCs to retailers. However, we must ensure that the DC assignment is within the service level requirements as well as within the capacity constraints of the DCs. To obtain feasible solutions as well as a good representation of the solutions, the following steps are taken to generate the initial population.

1. Start with a gene (retailer) and assign it to a random DC.
2. Check whether the delivery leadtime is within the service level requirement. If no, assign it randomly to another DC and repeat step 2.
3. Use equation (5) to determine if the assignment has violated the capacity constraint of the DC. If yes, select randomly a customer from the current assignment and then assign it randomly to those DCs that are within the service level constraint and repeat step 3.
4. Check if all the genes in the chromosomes have been allocated a DC. If no, allocate a random DC to the gene (retailer) and goto step 2.
5. Check if the predetermined number of initial population has been reached. If no, goto step 1.
6. Stop.

Step 3 ensures that the capacity constraint is always met while allocating retailers to DCs. The above steps will lead to a set of initial feasible solutions which are then used in the ARGA to search for the better solutions.

4. Computational Results

In this section, we would like to evaluate the performance of the ARGA. It will be compared with three different approaches, namely, standard GA (SGA), the column generation method and the greedy method. The brief discussions of these three methods are given below.

SGA

We run the SGA with the same chromosome representation and same operators with ARGA except that the parameters are not adaptive, i.e., it will not be tuned during the runs. However, in order to compare the results fairly, we run all possible parameter combinations for SGA (i.e., the M parameter combinations used in the phase 2 of ARGA procedure), and the best results of these combinations will be selected.

Column Generation Method

Another existing approach for this problem is the column generation method (Shu et al. 2002). In this method, an integer solution will be represented by a column (in this case it is the assignment of retailers to DCs). Then the total-cost of the column can be computed easily by solving the nonlinear problem for the given assignment. If all the possible columns are given, the original problem can be formulated as a set-covering problem and the optimal solution can be found. However, as the number of possible columns can be very large due to the many possible assignments, one efficient way is to start off with a subset of feasible columns, and then add more new columns when it is needed. To find these columns, we need to first solve the relaxed set covering problem given the subset of columns to obtain the dual information. Then we formulate a pricing problem by using the dual information to generate better columns. The approach will be repeated until no better columns can be found. Note that when we solve the relaxed set covering problem, we treat the column variable as a continuous variable and thus the optimal solution cannot be guaranteed. However, past experience shows that it usually gives good solutions in many applications.

For this inventory and location problem, the pricing problem is not trivial as it involves nonlinear terms, and so we use the method proposed by Shu et al. (2002). Due to the complexity of the problem, the method can only solve the non-capacitated case.

Greedy Method

This method uses the marginal cost concept to determine the assignment. It starts with no assignment of DCs to retailers and then assigns a DC to a retailer one at a time. For a given retailer, the DC is selected such that the marginal increase in the total cost is the lowest. For example, retailer 1 is the first retailer to be assigned with a DC. We will pick a DC that gives a lowest cost, and in this case it will be the DC that is closest to the retailer 1. Then retailer 2 will be assigned to a DC that minimizes the marginal increase in the cost function by taking into account of the assignment for retailer 1. In this case, retailer 2 can either choose the same DC as retailer 1 to exploit the benefit of risk pooling or choose a DC that is the closest. The decision depends on which alternative gives the lowest cost increase. This process will be repeated until all the retailers have been assigned with a DC. Note that, the capacity constraint need to be satisfied when determining the assignment.

In the numerical experiment, we first compare the performance of the proposed methodology with the three methods discussed above on the uncapacitated case. As for capacitated case, since the existing column generation method cannot be used, only SGA and the greedy method are used for comparison. The algorithms for the SGA and ARGA are coded in C++ while the column generation method is coded using GAMS and all the experiments are conducted on a PC Pentium 696 MHz.

For the uncapacited case, we consider 5 different data sets as follows. (1) a 30 retailer 6 DC system, (2) a 50 retailer 15 DC system, (3) a 50 retailer system where all the retailer sites can also function as DC, (4) a 100 retailer 30 DC system and (5) a 200 retailer 30 DC system. Table 1 gives the input data used in the experiments.

Table 1. The input data used in the numerical experiment

Parameters	Ranges/values
D_i	[1,100]
σ_i^2	$[1,D_i/3]$
d_{ij}	[1,10]
h_j	1
L_j	1
A_j	100

We set the maximum iteration number for the ARGA and SGA at 20,000, since we observe the algorithm converges around 8,000 iterations at the trial experiments.

For the screening experiment that we conduct during the first phase of ARGA, the high-low levels for all the parameters of GA are given in Table 2.

Table 2. The values for the GA parameters used in the phase 1 ARGA procedure

GA Parameters	Low (-)	High (+)
Population Size	100	200
Team Size	2	4
Crossover Probability	0.5	0.7
Crossover Rate	0.3	0.5
Mutation Rate	0.1	0.3
Replacement Percentage	70%	80%

From the results of the screening experiments, it is found that the population size, the team size, the crossover probability, the crossover rate, the mutation rate, and their interactions have significant impacts on the performance of the search process. Hence, these are critical parameters. On the other hand, among these parameters, the top 5 most time-sensitive parameters or interactions are as follows.

- Interaction between mutation rate and population size
- Interaction between crossover probability and population size
- Population size
- Crossover rate
- Mutation rate

Based on these observations, we choose the population size, the crossover probability, the crossover rate and the mutation rate as the time-sensitive parameters. They will be set at 3 different levels given in Table 3 for forming the parameter combinations. These 81 different parameter combinations will be used in the phase 2 of the ARGA procedure. On the other hand, the replacement percentage is fixed at 70%.

Table 3. The values for the GA parameters used in the phase 2 ARGA procedure

Parameters	Level 1	Level 2	Level 3
Population Size	50	75	100
Crossover Probability	0.3	0.5	0.7
Crossover Rate	0.15	0.3	0.45
Mutation Rate	0.01	0.05	0.1

The final results of the uncapacitated case are shown in Table 4. The first column indicates the problem size. The second, third, fourth and final columns represent the results from the greedy method, the column generation method (CG), SGA and ARGA respectively.

Table 4. Results for the uncapacicated case

Cases	Greedy	CG	SGA	ARGA
30x6	4518	4639	4518	4518
50x15	6500	6734	6353	6375
50x50	4590	4444	4471	4462
100x30	12245	11543	11853	11579
200x30	21604	20229	21822	20304

From Table 4, we notice that the greedy method gives better results when the problem size is small, but becomes worse when the size of the problem is large. This is because the greedy method conducts only the myopic search, and hence usually gives local optimal solution. For a small size problem, the number of alternatives is small and hence the solution given by the greedy method might not be far away from the optimal solution. However, when the number of alternatives becomes very large, the chances of getting a good solution will be low.

The solutions given by the column generation method improves when the problem size gets bigger. Although the solution is not guaranteed to be optimal due to the linear relaxation in the master problem, it does solve the problem holistically and hence the gap between the solution and the optimal solution is expected to be small in most of the cases.

For the two GA algorithms, SGA performs similarly to ARGA for small size problem, but does not perform as well as ARGA for larger size problem. In all the cases that we have conducted, ARGA consistently gives good results. The reason is because the parameters for ARGA are adaptively tuned according to the search performance, and hence it is less dependent on the problem scenarios and initial parameter settings.

In comparing the run times of all the algorithms, the greedy method is always the fastest (less than 1 second). On the other hand, the column generation method is very slow especially when the size of the problem increases. For a 200 retailer 30 DC case, it takes about 7 days to complete the run. For ARGA and one run of SGA, the running time is within 10 minutes. However, we have to run all the 81 possible parameter combinations for SGA in order to obtain the best result given in Table 4. Hence, we can say that the actual running time for SGAs is much larger than ARGA. Based on all these observations, we can conclude that ARGA is very promising in giving consistent results for this problem.

We repeat the experiment for the capacitated case but without the column generation method. We use the same data set as the uncapacitated case except that the capacity of each DC is generated uniformly between 0 and 400. The results of the experiment runs are given in Table 5.

Table 5. Results for the capacitated case

Cases	SGA	ARGA	Greedy
30x6	4552	4552	4552
50x15	6436	6401	6583
50x50	5563	5608	5640
100x30	12327	12195	12858
200x30	22148	20622	22063

The observed trends for all the algorithms are similar to the uncapacitated case. The ARGA again produces better solutions in most of the cases. The greedy method only produces good results for small problems. For the larger problem, ARGA is significantly better than the other methods. Although SGA performs comparably well with ARGA for medium to smaller size problems, the computation time that the SGA takes in fact is much larger than ARGA due to the fact that we have run all the possible 81 parameters combinations in order to get the best answer.

5. Conclusion

In the paper, we propose an ARGA to tackle the transportation inventory problem. Given a set of DCs and retailers, we determine how to assign the retailers to the DCs and how much of cycle and safety stock to be held at each DC to minimize the system cost while satisfying service requirements.

We conduct numerical experiments to compare the performance of ARGA with SGA, the column generation method and the greedy method. ARGA is promising in providing good and consistent results because its parameters are adaptively tuned during the search process, and hence less dependent on the problem scenarios and initial parameter settings. Moreover, the running time of the ARGA is also reasonable compared to the other approaches.

There are two possible extensions to this current work. The first one involves extending the model to include multiple items. The second extension deals with a system with differentiated service requirements for different retailers. This extension is realistic as currently the service providers always give customized services to their customers.

References

Back T (1993) Optimal mutation rates in genetic search. In: Forrest S (Ed), Proceedings of the Fifth International Conference on Genetic Algorithms and their Applications, Morgan-Kaufmann, pp 2-8

Chan Y (2001) Location Theory and Decision Analysis. Cincinnati, Ohio: South-Western College Publishing, a division of Thompson Learning, pp 16-17

Daskin MS and Owen SH (1999) Location models in transportation, R. Hall ed. Handbook of Transportation Science. Kluwer Academic Publishers, Norwell, MA, pp 311-360

Daskin MS, Coullard CR, Shen ZJ (2002) An Inventory Location Model: Formulation, solution, algorithm, and computational results. Annals of Operations Research 110: 83-106

Erlebacher SJ and Meller RD (2000) The interaction of location and inventory in designing distribution systems. IIE Transactions 32:155-166

Fogarty T (1989) Varying the probability of mutation in the genetic algorithm. In: Schaffer JD (Ed.), Proceedings of 3rd International Conference on Genetic Algorithms, San Mateo

Herrera F, Lozano M (1996) Adaptation of Genetic Algorithm Parameters based on Fuzzy Logic Controllers. In: Herrera E, Verdegay JL (Eds), Genetic Algorithms and soft computing, Physical-Verlag, pp 95-125

Hopp W, Spearman ML (2000) Factory Physics: Foundations of Manufacturing Management, Irwin, McGraw-Hill

Lee LH, Fan Y (2002) An Adaptive Real-coded Genetic Algorithm. Applied Artificial Intelligence 16: 457-486

Lis J, Lis M (1996) Self-adapting parallel genetic algorithm with dynamic mutation probability, crossover rate and population size. In: Arabas J (Ed) Proceedings of 1st Polish National Conf. Evolutionary Computation, pp 324-329

Shen ZJ (2000) Efficient algorithms for various supply chain problems. Ph.D. dissertation, Department of Industrial Engineering and Management Sciences, Northwestern University, Evanston, IL

Shen ZJ, Coullard CR, Daskin MS (2003) A joint location-inventory model. Transportation Science 37: 40-55

Shu J, Teo C, Shen ZM (2005) Stochastic Transportation-Inventory Network Design Problem. Operations Research, 53(1): 48-60

Silver E, Pyke DF, Peterson R (1998) Inventory Management and Production Planning and Scheduling. John Wiley & Sons

Song YH, Johns A, Aggarwal R (1996) Computational Intelligence Applications to Power Systems. Science Press, Kluwer Academic Publishers, Chapter 12, pp 128-133

Srinivas M, Patnaik LM (1994) Adaptive Probabilities of Crossover and Mutation in Genetic Algorithms. IEEE Transactions on systems, man and cybernetics 24(4): 656-666

Syswerda G (1989) Uniform crossover in genetic algorithms. In: Proceedings of the Third International Conference on Genetic Algorithms, Fairfax, VA: Morgan Kaufmann, pp 1-9

GP for evolving effective CDRs, both parameters and their combination have been explored.

6 Conclusions

In this paper, a GP-based approach for designing effective composite dispatching rules that minimizes total tardiness in the Flexible Job-Shop model has been presented and analyzed.

CDRs have been studied widely by previous researchers [15-17]. However, all of them were constructed based on the experience of a human scheduler. We employ a GP-framework to generate a CDR based on fundamental terminals that can effectively solve the FJSP (together with a machine assignment rule) by minimizing total tardiness. Two large test samples for training (under our GP framework) and validation were generated. Five evolved rules from each test sample that were most effective were selected to be tested on the validation sets. These rules are based on the combination of parameters such as processing time, release date, due date, current time, number of operations, and average total processing time of each job using basic arithmetical operators for combination. Five other popular rules selected from literature were used as performance benchmarks.

We observed that two composite dispatching rules MDD and SL contain similar parameters (DD and CT), but the performance differential between the results of the two rules were quite large due to use of different algebraic combinations of the parameters. Also, the single dispatching rule EDD contains only one parameter (EDD) but was significantly better than the other rules from literature. This implies that the way to combine the rules can significantly affect the optimality of the schedules; ineffective composite dispatching rules may achieve worse results than the single ones and hence the need for an automated design approach. The experimental results show that our evolved dispatching rules outperforms the most effective human-made rule EDD. In particular, two parameters aTPT and nOps that have received limited study from previous research was found to contribute significantly to the effectiveness of evolved CDRs. We have also proven statistically that our evolved CDRs are sufficiently well-designed through the use of ANOVA (which analyzed the sensitivity to changes in the coefficient values and terminal parameters). Finally, by using a large training data set, we believe that our evolved CDRs can be applied directed in practice without further modifications.

Several possible extensions of this study can be developed. Similar to other applications of GP where the parameters are sensitive, denser terminal sets and more varied ADRs should be investigated to improve the generated rules. The approach of this study can be applied to find the efficient composite dispatching rules for other similar problems, such as a flow shop or the classical job shop. The rules evolved from this GP framework are still quite complex in structure. Therefore, an algebraic simplification tool could be used to make the formula more meaningful. Consideration could even be given to including the number of parameters used as a measure for minimization.

Acknowledgements

This research was funded in part by Nanyang Technological University and CEI Contract Manufacturing Limited Company, Singapore.

References

[1] Hoitomt, D.J., Luh, P.B., Pattipati, K.R.: A Practical Approach to Job-Shop Scheduling Problems. Ieee T Robotic Autom **9** (1993) 1-13

[2] Jain, A.S., Meeran, S.: Deterministic job-shop scheduling: Past, present and future. Eur J Oper Res **113** (1999) 390-434

[3] Carlier, J., Pinson, E.: An Algorithm for Solving the Job-Shop Problem. Manage Sci **35** (1989) 164-176

[4] Kolonko, M.: Some new results on simulated annealing applied to the job shop scheduling problem. Eur J Oper Res **113** (1999) 123-136

[5] Nowicki, E., Smutnicki, C.: A fast taboo search algorithm for the job shop problem. Manage Sci **42** (1996) 797-813

[6] Yamada, T., Nakano, R.: A fusion of crossover and local search. Proceedings of The IEEE International Conference on Industrial Technology (1996) 426-430

[7] Garey, M.R., Johnson, D.S., Sethi, R.: The complexity of flow shop and job-shop scheduling. Mathematics of Operations Research **1** (1976) 117-129

[8] Pinedo, M., Chao, X.: Operations scheduling with applications in manufacturing and services. MCGraw- Hill chapter 3 (1999)

[9] Brandimarte, P.: Routing and scheduling in a flexible job shop by tabu search. Annals of Operations Research (Historical Archive) **41** (1993) 157-183

[10] Mastrolilli, M., Gambardella, L.M.: Effective Neighborhood Functions for the Flexible Job Shop Problem. J Sched **3** (2000) 3-20